Urs Graf

Introduction to Hyperfunctions and Their Integral Transforms

An Applied and Computational Approach

Birkhäuser

Author:

Urs Graf
Rue des Mornets 29
2520 La Neuveville
Switzerland
e-mail: urs.e.graf@bluewin.ch

2000 Mathematics Subject Classification: 26, 30, 33, 35, 44, 45, 97

Library of Congress Control Number: 2010921343

Bibliographic information published by Die Deutsche Bibliothek.
Die Deutsche Bibliothek lists this publication in the Deutsche Nationalbibliografie;
detailed bibliographic data is available in the Internet at http://dnb.ddb.de

ISBN 978-3-0346-0407-9

© 2010 Birkhäuser / Springer Basel AG
P.O. Box 133, CH-4010 Basel, Switzerland
Part of Springer Science+Business Media
Printed on acid-free paper produced from chlorine-free pulp. TCF∞
Printed in Germany

ISBN 978-3-0346-0407-9   e-ISBN 978-3-0346-0408-6

9 8 7 6 5 4 3 2 1   www.birkhauser.ch

Contents

Preface

This textbook is about generalized functions and some of their integral transforms in one variable. It is based on the approach introduced by the Japanese mathematician Mikio Sato. We mention this because the term hyperfunction that Sato has given to his generalization of the concept of function is sometimes used today to denote generalized functions based on other approaches (distributions, Mikusinski's operators etc.). I have written this book because I am delighted by the intuitive idea behind Sato's approach which uses the classical complex function theory to generalize the notion of function of a real variable. In my opinion, Sato's way of introducing the generalized concept of a function is less abstract than the one of Laurent Schwartz who defines his distributions as linear functionals on some space of test functions. On the other hand, I was quickly led to recognize that very few colleagues (mathematicians included) knew anything about Satos's approach. Perhaps Sato and his school is not entirely blameless for this state of affairs. For several decades no elementary textbook addressing a wider audience was available (at least in English). Zealots delighted by the appealing intuitive idea of the approach have probably found their enthusiasm rapidly diminished because of the adopted style of exposition and the highly ambitious abstract mathematical concepts used in the available books and articles. Fortunately, some years ago, I found Isac Imai's Book *Applied Hyperfunction Theory* which explains and applies Sato's hyperfunctions in a concrete, but nontrivial way, and thereby reveals their computational power. Without the help of Imai's book, I would no doubt have been repelled by the sophistication of the available texts, as perhaps many others before me had. So, for the writing of my book I am indebted to Imai, mainly for the first chapter, parts of the second and entirely for the fifth chapter. The objective of my book is to offer an introduction to the theory of hyperfunctions and some of their integral transforms that should be readable by a wider audience (applied mathematicians, physicists, engineers) than to just some few specialists. The prerequisites are some basic notions of complex function theory of one variable and of the classical Laplace and Fourier transformations. Since I am no friend of theories for their own sake, I have inserted throughout the whole book some simple applications mainly to the theory of integral equations.

Chapter 1 is an elementary introduction to generalized functions by the hyperfunction approach of Sato. With a few basic facts on complex function theory, the basic lines of a theory of generalized functions is presented that should be easy to read and easy to be understood. It shows the basic intuitive concept of a hyperfunction of one variable.

Chapter 2 discusses the analytic properties of hyperfunctions. Here, the specific methods of complex analysis come into play. Since a hyperfunction is defined by specifying a defining or generating function, we treat in some detail the question of how to construct a defining function such that the corresponding hyperfunction interprets the given ordinary function (the problem of embedding an ordinary function in the set of hyperfunctions). We shall see that the answer to this problem is not always unique. A hyperfunction which interprets a given ordinary function on a specified interval is said to be a projection of this function to the interval. Given a hyperfunction on an interval (a', b'), the analogous problem of finding another hyperfunction that equals the specified one on a smaller interval $(a, b) \subset (a', b')$ and vanishes outside (a, b) is then treated. This leads to the notion of the so-called standard defining function, a term first defined for hyperfunctions with a compact support, then extended to hyperfunctions defined on an infinite interval. Imai extends the notion of a standard defining function in a different way from Sato's when the hyperfunction is not perfect, i.e. has a non-compact support. Because Imai's extension is found to be useful for the discussion of Hilbert transformation, I shall use the term "strong defining function" to avoid confusion of the two notions. An introduction to periodic hyperfunctions and their Fourier series then follows. The last theoretical part of this chapter discusses convolutions of hyperfunctions. Some informal remarks on integral equations conclude the chapter.

Chapter 3 treats the Laplace transform of hyperfunctions. It is somewhat the main axis of the part about integral transformations. While other texts about generalized functions often treat Fourier transforms in the first place and then use the established Fourier transformation to define the Laplace transformation, I will do it in the converse way. I introduce the Laplace transform of a hyperfunction by using a loop integral of Hankel type over the defining function. Since simplicity of the presentation, together with many concrete examples without digging into finer mathematical points in the arguments, has been our goal, this more elementary approach in the main text is presented rather than Komatsu's theory of Laplace hyperfunctions. But since the Laplace transformation is a central subject of the book, I have presented an outline of Komatsu's approach in Appendix B.

Chapter 4 is about Fourier transforms of hyperfunctions. Fourier transformation is without doubt the greatest beneficiary of the theory of generalized functions. Generally the Fourier transformation is the first to be treated in most approaches to generalized functions. From a mathematico-logical point of view this may be justified, however, from a computational and applied standpoint, Laplace transformation is often more appropriate. I define the Fourier transform of a hyperfunction by using the already established Laplace transform. This approach has the advantage that the available extended tables of Laplace transforms can be used.

Chapter 5 treats Hilbert transforms of hyperfunctions. In this chapter I mainly follow Imai. The concept of a strong defining function plays an important role here because the existence of the Hilbert transform of a hyperfunction is intimately connected to the existence of a strong defining function. A section on analytic signals can also be found there.

For the last two chapters about Mellin and Hankel transformations I had to be entirely self-supporting. While there is an abundant literature about classical Mellin and Hankel transformation of ordinary functions, I could not find anything, at least in English, German or French, on Mellin or Hankel transformation of hyperfunctions. A. H. Zemanian [40] treats the Mellin and Hankel transformation of generalized functions based on Laurent Schwartz's theory of distributions. I have finally succeeded in conveying the Mellin transformation to hyperfunctions by taking advantage of the fact that the Mellin transformation is in some sense a reformulation of the two-sided Laplace transformation. A simple change of variables allows passing from the Laplace to the Mellin transformation, and vice versa. Because the Laplace transformation of hyperfunctions has been firmly established in Chapter 3, it was finally straightforward to establish the Mellin transformation of hyperfunctions by exploiting this connection to the hilt.

It was harder to define the Hankel transform of a hyperfunction. I eventually found a way by working on the line of Mac Robert's proof of the classical Hankel transformation which uses the so-called Lommel integrals of Bessel functions. The established theory about the generalized Hankel transformation then finally works for hyperfunctions of slow growth.

Throughout the book a particular function and its hyperfunction counterpart comes up again and again: the Heaviside function $Y(x)$ and the unit-step hyperfunction $u(x)$. For didactic reasons, and after some hesitation, I have made a distinction between them. This may be pedantic and I agree that it is not absolutely necessary because there is no great danger of confusion by the use of a unified notation for the two. Thus, the reader should feel free to replace everywhere $Y(x)$ by $u(x)$. Systematically, all contours in the complex plane will be positively directed unless noted to the contrary. This will produce minus signs before integrals where some readers will not expect them, perhaps.

The main tool used in this book consists of contour integration in the complex plane. You will find numerous integrals taken on a closed contour or on an infinite loop. We liberally interchange in many places the order of integration in multiple integrals or the integral and an infinite series. In order to keep the flow of the arguments fluid, I do not generally justify these steps in detail. However, for the benefit of readers interested in such technical details, I have collected in Appendix A3 the principal theorems generally used to justify such interchanges of limit operations.

Lastly, a few remarks about what you cannot find in this book and about which I have no pretensions. The theory goes not very deep. No sheafs and other sophisticated concepts are mentioned. The intended message is rather conveyed through many concrete examples. Also, and this is certainly a shortcoming for some, no hyperfunctions of several variables are treated.

La Neuveville, Switzerland, October 2009, U.G.

Chapter 1

Introduction to Hyperfunctions

After a short overview of generalized functions and of the different ways they can be defined, the concept of a hyperfunction is established, followed by an introduction to the most simple and familiar hyperfunctions. Then the elementary operational properties of hyperfunctions are presented. The so-called finite part hyperfunctions are introduced, followed by the important notion of the (definite) integral of a hyperfunction. The chapter closes with a definition of more familiar hyperfunctions frequently used in applications. The books of [19, Imai] and [20, Kaneko] are the principal references for this chapter.

1.1 Generalized Functions

In the mathematical modeling of physical problems, the idealized concepts of a force concentrated at a point $x = a$, or of an impulsive quantity that acts instantaneously at the time $t = a$, were introduced by the physicist Paul Dirac (1902 - 1984) who used such concepts in quantum mechanics. He introduced the so-called delta function $\delta(x - a)$ having the properties

$$\delta(x - a) = 0 \text{ for all } x \neq a,$$

$$\int_\alpha^\beta \delta(x - a)\, dx = \begin{cases} 0, & a \notin (\alpha, \beta), \\ 1, & a \in (\alpha, \beta). \end{cases}$$

Although scientists have widely used this Dirac delta function, classical mathematics has been unable to justify such a function. Indeed, the two properties are contradictory. In every classical theory of integration, a function that is zero everywhere except at one point has an integral which is necessarily zero.

The electrical engineer Oliver Heaviside (1850–1925) may be considered as the father of operational calculus. He introduced and systematically used the unit-step function

$$Y(x - a) = \begin{cases} 0, & x < a \\ 1, & x > a \end{cases}$$

and its "derivatives", irrespective of the fact that this so-called Heaviside function is not at all differentiable at $x = a$. Also, he was operating with series which were obviously not convergent. However, the astonishing fact was that his heuristic operations, erroneous from a mathematical point of view, often yielded correct results, and are much faster and easier to obtain when compared to the classical methods.

The method of classical Fourier transformation has the disadvantage that common and simple functions such as constants or polynomials have no Fourier transform. However, simple limit considerations show that the constant function $f(x) = 1$ should have the Fourier transform $\delta(\omega)$, i.e., the Dirac delta function.

In general, the concept of differentiation is so fundamental and basic in mathematics that it may be annoying to stumble over functions that are not differentiable during a problem solving process. It would be very convenient if all involved functions could always be considered differentiable as many times as you wish.

All these contradictions and shortcomings were finally done away with by several theories generalizing the classical notion of function as well as the classical definition of the derivative. The most famous approach is the theory of distributions of Laurent Schwartz (1915 - 2002). Schwartz defined a distribution as a continuous functional on a space of test functions. His approach is based on functional analysis, especially on the theory of topological vector spaces. It is a somewhat abstract, not very intuitive, but flexible approach to the subject. Readable descriptions of Schwartz's approach to generalized functions useful for scientists and engineers can be found, for example, in [21, Kanwal], [39, Zemanian], [36, Vladimirov], [32, Schwartz]. Numerous books present this approach from a theoretical point of view, for example, [33, Schwartz], [37, Vladimirov], [11, Gel'fand], [13, Gasquet], [40, Zemanian].

Another line of attack on the problem is the sequential approach to generalized functions. Here, a generalized function is defined as a limit of sequences of ordinary functions by means of the concept of equivalence classes. This approach is more intuitive than the one of Schwartz, however, it is seldom used today. Descriptions of this approach can be found, for example, in the books of [3, Antosik–Mikusinski–Sikorski] and [23, Lighthill].

The present book introduces the concept of generalized functions by using a radically different approach. It was invented by the Japanese mathematician Mikio Sato (1928). It is based on complex function theory. The basic idea is to consider a function $f(x)$ of the real variable x as the difference $F(x + i0) - F(x - i0)$ of the boundary values of a function $F(z)$ that is holomorphic in the upper half-plane $\Im z > 0$ and in the lower half-plane $\Im z < 0$. If $F(z)$ is holomorphic at the real point x, i.e., is real analytic at x, then clearly $f(x) = F_+(x + i0) - F(x - i0) = 0$, so in order to produce non-trivial functions $f(x)$, the possible singularities of $F(z)$ on the real axis are important. It turns out that the idea of defining a function $f(x)$ in such a way is general enough that a consistent theory of generalized functions can be built around it. Sato called the generalized functions obtained in this way hyperfunctions. While this book presents an elementary non-technical introduction to the subject, the reader can find more sophisticated and theoretical presentations in the books of [20, Kaneko] and [25, Morimoto].

1.2   The Concept of a Hyperfunction

We denote the upper half-plane by $\mathbb{C}_+ := \{z \in \mathbb{C} \mid \Im z > 0\}$ and the lower half-plane by $\mathbb{C}_- := \{z \in \mathbb{C} \mid \Im z < 0\}$, where $\mathbb{C}$ denotes the Gaussian plane of complex numbers.

Definition 1.1. Let I be a given open interval of the real line, then an open subset $\mathcal{D}(I)$ of $\mathbb{C}$ is called a *complex neighborhood of I*, if I is a closed subset of $\mathcal{D}(I)$.

This means for example that in Figure 1.1 the two endpoints of the interval I do not belong to $\mathcal{D}(I)$ since the subset $\mathcal{D}(I) \setminus I$ is open in $\mathcal{D}(I)$. Because $\mathcal{D}(I)$ is open in $\mathbb{C}$, the subset $\mathcal{D}(I) \setminus I$ is also open in $\mathbb{C}$. By $\mathcal{N}(I)$ we denote the set of all complex neighborhoods of I. The intersection of two or a finite number of complex neighborhoods of I, is again a complex neighborhood of I. For any complex neighborhood of I the two open sets $\mathcal{D}_+(I) := \mathcal{D}(I) \cap \mathbb{C}_+$ and $\mathcal{D}_-(I) := \mathcal{D}(I) \cap \mathbb{C}_-$ are called *upper half-neighborhood* and *lower half-neighborhood* of I, respectively. If it is clear which interval I is under consideration, we may drop its notation and simply write $\mathcal{D}$, $\mathcal{D}_+$, $\mathcal{D}_-$. If $\mathcal{D}$ is a given complex neighborhood of I, let $\overline{\mathcal{D}} := \{z \in \mathbb{C} \,|\, \overline{z} \in \mathcal{D}\}$. Then, $S := \mathcal{D} \cap \overline{\mathcal{D}}$ is a symmetrical complex neighborhood of I. If $\mathcal{D}$ is any open subset of $\mathbb{C}$, we denote by $\mathcal{O}(\mathcal{D})$ the *ring of*

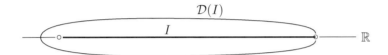

$$\mathcal{D}(I)$$
$$I$$
$$\mathbb{R}$$

Figure 1.1: Complex neighborhood $\mathcal{D}(I)$

all holomorphic functions in $\mathcal{D}$. Thus $\mathcal{O}(\mathcal{D}(I))$ denotes the ring of all holomorphic functions in the complex neighborhood $\mathcal{D}(I)$ of I, while $\mathcal{O}(\mathcal{D}(I) \setminus I)$ is the ring of all holomorphic functions in $\mathcal{D}(I)$ minus I.

For a given interval I, a function $F(z) \in \mathcal{O}(\mathcal{D}(I) \setminus I)$ can be written more explicitly as

$$F(z) := \begin{cases} F_+(z), & z \in \mathcal{D}_+(I) \\ F_-(z), & z \in \mathcal{D}_-(I) \end{cases}$$

with $F_+(z) \in \mathcal{O}(\mathcal{D}_+(I))$ and $F_-(z) \in \mathcal{O}(\mathcal{D}_-(I))$. We call the function $F_+(z)$ the *upper component*, and $F_-(z)$ the *lower component* of $F(z)$.

Note that in general the upper and the lower component of the function $F(z)$ need not be related to each other, i.e., they may be independent holomorphic functions. Nevertheless, the case where the upper and lower components are analytic continuations from each other will often be encountered. In this case $F(z)$ is a global analytic function on $\mathcal{D}(I)$ and we can write $F_+(z) = F_-(z) = F(z)$.

Let us now introduce an equivalence relation on the set of functions

$$\mathcal{F}(I) := \mathcal{O}(\mathcal{D}(I) \setminus I).$$

Definition 1.2. We say that two functions $F(z)$ and $G(z)$ from $\mathcal{F}(I)$ are equivalent, denoted by $F(z) \sim G(z)$, if for $z \in \mathcal{D}_1(I) \cap \mathcal{D}_2(I)$,

$$G(z) = F(z) + \phi(z),$$

with $\phi(z) \in \mathcal{O}(\mathcal{D}(I))$, i.e., $F(z)$ and $G(z)$ differ by a holomorphic function on $\mathcal{D}(I)$. Here $\mathcal{D}_1$ and $\mathcal{D}_2$ are complex neighborhoods of I of $F(z)$ and $G(z)$, respectively.

Let us prove that the relation $\sim$ is an equivalence relation, i.e., that it is reflexive, symmetric and transitive.

Proof. The relation $\sim$ is reflexive, i.e., we have $F(z) \sim F(z)$ since the zero function is a holomorphic function on $\mathcal{D}(I)$. The relation $\sim$ is symmetric, i.e., $F(z) \sim G(z) \Longrightarrow G(z) \sim F(z)$, because if $\phi(z)$ is a holomorphic function on $\mathcal{D}(I)$, then $-\phi(z)$ is also. Now assume that $F(z) \sim G(z)$ and $G(z) \sim H(z)$. This means that $F(z) - G(z) = \phi(z)$, $G(z) - H(z) = \psi(z)$, for two holomorphic functions $\phi(z)$ and $\psi(z)$ on $\mathcal{D}(I)$. Adding the two equations, we obtain $F(z) - H(z) = \phi(z) + \psi(z)$. Since the sum of two holomorphic functions is again holomorphic on $\mathcal{D}(I)$, we have shown that the assumption implies $F(z) \sim H(z)$. This proves that $\sim$ is transitive. $\square$

Therefore, the set $\mathcal{F}(I)$ splits into mutually disjoint equivalence classes. It is just such an equivalence class that defines a hyperfunction. More precisely,

Definition 1.3. We say that an equivalence class of functions $F(z) \in \mathcal{F}(I) = \mathcal{O}(\mathcal{D}(I) \setminus I)$ defines a *hyperfunction $f(x)$ on I*, which is denoted by $f(x) = [F(z)]$. If the upper and the lower component of $F(z)$ should be emphasized, we also use the more explicit notation $f(x) = [F_+(z), F_-(z)]$. The function

$$F(z) := \begin{cases} F_+(z), & z \in \mathcal{D}_+(I) \\ F_-(z), & z \in \mathcal{D}_-(I) \end{cases}$$

is called a *defining or generating function of the hyperfunction $f(x)$*. The set of all hyperfunctions defined on the interval I is denoted by $\mathcal{B}(I)$.

In a more abstract way we can write

$$\mathcal{B}(I) := \mathcal{O}(\mathcal{D}(I) \setminus I)/\mathcal{O}(\mathcal{D}(I)), \tag{1.1}$$

i.e., the quotient space of all functions holomorphic in a complex neighborhood $\mathcal{D}(I)$ minus the interval I over the space of all holomorphic functions in $\mathcal{D}(I)$.

There is no reason to prefer a particular choice of neighborhood $\mathcal{D}(I)$. Indeed, if $\mathcal{D}'(I)$ is another complex neighborhood of I such that $\mathcal{D}(I) \supset \mathcal{D}'(I)$, then $\mathcal{O}(\mathcal{D}'(I) \setminus I)/\mathcal{O}(\mathcal{D}'(I))$ works as well. This shows that what is essential to the definition of hyperfunctions is the behavior of the defining functions in a narrow vincinity of I. Now in (1.1) let $\mathcal{D}(I)$ become narrower and narrower. Intuitively we then write

$$\mathcal{B}(I) := \varinjlim_{\mathcal{D}(I) \supset I} \mathcal{O}(\mathcal{D}(I) \setminus I)/\mathcal{O}(\mathcal{D}(I)), \tag{1.2}$$

and the definition of the space of hyperfunctions has become independent of any particular complex neighborhood of I. The reader should not worry about (1.2) and the symbol used there which denotes the so-called inductive limit. It is beyond the scope of this book to dig into such sophisticated details. An interested reader can find more detailed information about the inductive limit and its ramifications in the book of [20, Kaneko]. For us an intuitive grasp is sufficient. However, we still add that the intuitive idea of rendering the complex neighborhoods narrower and narrower around I leads to another consequence:

A real analytic function $\phi(x)$ on I is defined by the fact that $\phi(x)$ can analytically be continued to a full neighborhood $U \supset I$, i.e., we then have $\phi(z) \in \mathcal{O}(U)$. For any complex neighborhood $\mathcal{D}(I)$ containing U we may then write

$$\mathcal{B}(I) := \mathcal{O}(\mathcal{D}(I) \setminus I)/\mathcal{A}(I), \tag{1.3}$$

where $\mathcal{A}(I))$ now denotes the ring of all real analytic functions on I. Finally, we can say that *a hyperfunction $f(x) \in \mathcal{B}(I)$, denoted by $f(x) = [F(z)]$, is determined by a defining function $F(z)$ which is holomorphic in an adjacent (small) neighborhood above and below the interval I, but is only determind up to a real analytic function on I.*

For a $F(z) \in \mathcal{F}(I)$ it may happen that at $x \in I$ the limit,

$$\lim_{\epsilon \to 0+} \{F_+(x + i\epsilon) - F_-(x - i\epsilon) =: F(x + i0) - F(x - i0),$$

exists. Let I_0 denote the subset of all real $x \in I$ for which this limit exists. Then the given function $F(z)$ defines an ordinary function $x \mapsto f(x)$, $x \in I_0$, where the function-value $f(x)$ is given by $F(x + i0) - F(x - i0)$. Therefore, the so-defined ordinary function is given by the difference of the boundary values of the two holomorphic functions $F_+(z)$ and $F_-(z)$. Note that another function, $G(z)$ from $\mathcal{F}(I)$ which is equivalent to $F(z)$, determines the same ordinary function. In fact, for an $x \in I_0$, and $\epsilon > 0$ sufficiently small, we have

$$G_+(x + i\epsilon) = F_+(x + i\epsilon) + \phi(x + i\epsilon)$$
$$G_-(x - i\epsilon) = F_-(x - i\epsilon) + \phi(x - i\epsilon).$$

Therefore,

$$\begin{aligned}
g(x) &= \lim_{\epsilon \to 0+} \{G_+(x + i\epsilon) - G_-(x - i\epsilon)\} \\
&= \lim_{\epsilon \to 0+} \{F_+(x + i\epsilon) - F_-(x - i\epsilon)\} \\
&\quad + \lim_{\epsilon \to 0+} \{\phi(x + i\epsilon) - \phi((x - i\epsilon)\} \\
&= f(x) + 0 = f(x),
\end{aligned}$$

because $\phi(x) \in \mathcal{A}(I)$.

The set $I \setminus I_0$, consists of all real points where one or both of the limits $F_+(x + i0)$ and $F_-(x - i0)$ do not exist. This set is formed of the isolated singularities of $F(z)$.

Remark. Bearing in mind the case where a given defining function $F(z)$ defines an ordinary function $x \mapsto f(x)$, $x \in I_0$ for some subinterval $I_0 \subset I$, the notation

$$f(x) = F(x + i0) - F(x - i0)$$

is also used for the hyperfunction $f(x)$ defined by the defining function $F(z)$.

Once again, any function equivalent to $F(z)$ defines the same hyperfunction, i.e., the defining function of a hyperfunction is determined only up to a real analytic function on I. A point $x \in I \setminus I_0$ is called a *singular point*, and a point $x \in I_0$ is called a *regular* point of the hyperfunction. At a regular point, a hyperfunction $f(x)$ has a function-value as an ordinary function. At a singular point it does not make sense to speak of a function-value of the hyperfunction.

In the general case upper and lower components of defining functions are not related, i.e., they are independent holomorphic functions in $\mathcal{D}_+(I)$ and $\mathcal{D}_-(I)$, respectively. Also, the specified holomorphic functions for the upper and lower components are often defined in far larger domains than just in a small upper and lower neighborhood of I. Often, $F_+(z)$ is defined in $\mathbb{C}_+$, and $F_-(z)$ in $\mathbb{C}_-$. The existence of a narrow upper and lower neighborhood of I only constitutes a minimal requirement. In other cases the specified defining function may be given by one global analytic function, as for example $F(z) = \log(-z)$. In this case we have $\mathcal{D}(I = \mathbb{R}) = \mathbb{C} \setminus \mathbb{R}_+$.

Since the aim of the theory of hyperfunctions is to generalize the notion of an ordinary function, we shall be confronted with the task of finding a defining function that defines a hyperfunction such that the latter can be considered as a representation in the realm of hyperfunctions of the specified ordinary function. This task is called the *problem of interpretation of an ordinary function by a hyperfunction or the problem of embedding an ordinary function into the space of hyperfunctions*. As we shall see, this problem may not have a unique answer, i.e., there may be more than one hyperfunction representing reasonably a given ordinary function.

Example 1.1. The following three defining functions will be used throughout the whole text. They are denoted as follows:

$$\mathbf{1}_+(z) := \begin{cases} 1, & \text{if } \Im z > 0; \\ 0, & \text{if } \Im z < 0. \end{cases} \tag{1.4}$$

$$\mathbf{1}(z) := \begin{cases} 1/2, & \text{if } \Im z > 0; \\ -1/2, & \text{if } \Im z < 0. \end{cases} \tag{1.5}$$

$$\mathbf{1}_-(z) := \begin{cases} 0, & \text{if } \Im z > 0; \\ -1, & \text{if } \Im z < 0. \end{cases} \tag{1.6}$$

These three defining functions define the same hyperfunction $f(x)$ which interprets the ordinary constant function $x \mapsto 1, x \in I_0 = \mathbb{R}$:

$$f(x) = 1 = [\mathbf{1}_+(z)] = [\mathbf{1}(z)] = [\mathbf{1}_-(z)]$$
$$= [1, 0] = [1/2, -1/2] = [0, -1].$$

The reason is that $\mathbf{1}_+(z) \sim \mathbf{1}(z) \sim \mathbf{1}_-(z)$.

By using these standardized defining functions we can sub-summarize the notation $f(x) = [F_+(z), F_-(z)]$ to $f(x) = [F(z)]$ by defining

$$F(z) := \mathbf{1}_+(z)F_+(z) - \mathbf{1}_-(z)F_-(z). \tag{1.7}$$

Example 1.2. The hyperfunction $f(x) = [z, 0] = [z\,\mathbf{1}_+(z)]$ has for every x the value x, thus interprets the ordinary function $x \mapsto x, I_0 = \mathbb{R}$. The same hyperfunction is obtained by $f(x) = [z/2, -z/2] = [z\,\mathbf{1}(z)]$.

Example 1.3. The hyperfunction $f(x) = [\sin z, 0] = [\sin(z)\,\mathbf{1}_+(z)]$ has for every x the value $\sin x$ and thus interprets the ordinary function $f(x) = \sin x$.

Example 1.4. Generally, we may interpret any given real analytic function $\phi(x) \in \mathcal{A}(\mathbb{R})$ by a hyperfunction again denoted by $\phi(x)$. We have

$$\phi(x) = [\phi(z), 0] = [\phi(z)\,\mathbf{1}_+(z)]$$
$$= [\phi(z)/2, -\phi(z)/2] = [\phi(z)\,\mathbf{1}(z)]$$
$$= [0, -\phi(z)] = [\phi(z)\,\mathbf{1}_-(z)].$$

We may also write

$$\phi(x) = [F(z)], \quad F(z) = \frac{\phi(z)}{2}\{\mathbf{1}_+(z) + \mathbf{1}_-(z)\}. \tag{1.8}$$

Example 1.5. The hyperfunction

$$f(x) = \left[\frac{1}{2i}\frac{1}{(z-i)}, \frac{1}{2i}\frac{1}{(z+i)}\right]$$

has for any real x a value and is an interpretation of the ordinary function

$$
\begin{aligned}
x \mapsto f(x) &:= \lim_{\epsilon \to 0+} \frac{1}{2i}\{\frac{1}{x+i\epsilon-i} - \frac{1}{x-i\epsilon+i}\} \\
&= \frac{1}{2i}\{\frac{1}{x-i} - \frac{1}{x+i}\} \\
&= \frac{1}{1+x^2}, \ x \in I_0 = \mathbb{R}.
\end{aligned}
$$

In the above examples the upper and lower components are different independent holomorphic functions. In fact, the defining functions are specified by giving directly the upper and lower components. Of special interest are hyperfunctions where the defining function is a global analytic function given by one expression. The upper and the lower component is then the restriction of the defining function to the upper and lower complex half-planes, respectively. In this case the upper component is the analytic continuation of the lower component, and vice versa. This always happens if there is a real interval where both components are continuous and have identical values.

Example 1.6. One of the most important hyperfunctions is

$$\delta(x) := \left[-\frac{1}{2\pi i z}\right], \tag{1.9}$$

called the *Dirac impulse at $x = 0$*. Its defining function $F(z) = -1/2\pi i z$ is defined in the entire complex plane and has an isolated singularity at $z = 0$, which is a pole of order 1 (a meromorphic function). For every real $x \neq 0$ the limit $\lim_{\epsilon \to 0+}\{F_+(x + i\epsilon) - F_-(x - i\epsilon)\}$ exists and equals 0. Hence we can write $x \mapsto \delta(x) = 0, x \in I_0 = \mathbb{R} \setminus 0$. For $x = 0$, the limit does not exist, it is a singular point of the hyperfunction, and it has no value there.

Let $f(x) = [F_+(z), F_-(z)]$ be a specified hyperfunction. The sequence of ordinary functions

$$f_n(x) := \lim_{n \to \infty}\{F_+(x + i/n) - F_-(x - i/n)\}$$

is always defined for sufficiently large n. The family of these functions yields, for increasing n, an intuitive picture of the hyperfunction $f(x)$. For the Dirac impulse hyperfunction we obtain

$$x \mapsto \delta_n(x) := \frac{n}{\pi(1 + n^2 x^2)}.$$

Figure 1.2 shows the graphics of some members of the sequence.

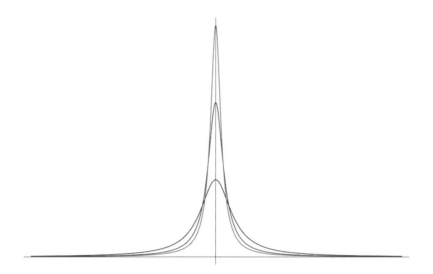

Figure 1.2: Intuitive picture of the Dirac impulse

Example 1.7. Another important hyperfunction is the *unit-step* hyperfunction. It is defined and denoted by

$$u(x) := \left[-\frac{1}{2\pi i}\log(-z)\right], \tag{1.10}$$

and is an interpretation of the *ordinary unit-step function or Heaviside function* $Y(x)$ which is vanishing on the negative part of the real axis, and has the constant value 1 on the positive part. The function $\log z$ will occur frequently in this text. It is a many-valued function with an infinity of branches:

$$\log(z) = \log|z| + i\arg z + 2\pi i k, \quad k \in \mathbb{Z}.$$

We have to single out one branch to obtain a well-defined singled-valued function. This is achieved by cutting the complex plane along an outward drawn ray from 0 to infinity, and by indicating by one way or another what branch is selected. For definiteness let us agree, if nothing else is explicitly stated, that if we write $\log z$ we always mean the function having as its domain the complex plane with a cut along the negative part of the real axis, and with the selected branch being the principal one, i.e., the branch where $\log x \in \mathbb{R}$ for $x > 0$ ($k = 0$ in the above formula). Then $\log(-z)$ takes the value $\log|z| - \pi i$ on the upper side, and $\log|z| + \pi i$ on the lower side of the positive real axis. It is holomorphic or real analytic on the negative real axis. Hence, we obtain, with $F(z) = -1/(2\pi i)\log(-z)$, for every real $x \neq 0$,

$$u(x) = \lim_{\epsilon \to 0+} \{F_+(x + i\epsilon) - F_-(x - i\epsilon)\} = \begin{cases} 0 & \text{if } x < 0 \\ 1 & \text{if } x > 0. \end{cases}$$

For $x = 0$ this hyperfunction has no value, it is a singular point of the hyperfunction. Figure 1.3 shows the graphics of the family of the approaching ordinary function sequence.

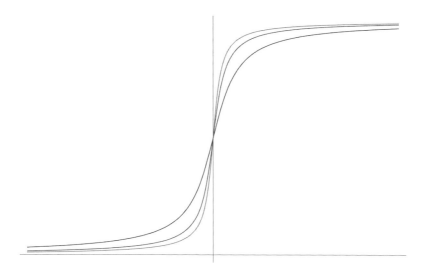

Figure 1.3: Intuitive picture of the unit-step

Example 1.8. Similarly to the unit-step hyperfunction is its mirror with respect to the vertical coordinate axis

$$u(-x) := \left[\frac{1}{2\pi i} \log(z) \right] \tag{1.11}$$

having the value 1 for all negative x and zero for all positive x. It interprets the ordinary function $Y(-x)$. The hyperfunction

$$u(x) + u(-x) := \left[-\frac{1}{2\pi i} \log(-z) \right] + \left[\frac{1}{2\pi i} \log(z) \right]$$

$$= \left[\frac{1}{2\pi i} \left(\log(z) - \log(-z) \right) \right]$$

is an interpretation of the ordinary function $x \mapsto 1$, $x \neq 0$. Observe that the defining function is defined on the disconnected domain $\mathbb{C} \setminus \mathbb{R}$.

Keep in mind that $\log z$ denotes the principle branch of the logarithm defined on the complex plane cut along the negative real axis. Thus, if $\Im z > 0$, i.e., $0 < \arg z < \pi$, we have

$$\log z = \log|z| + i \arg z,$$
$$\log(-z) = \log|z| + i \arg z - i\pi,$$

and, if $\Im z < 0$, i.e., $-\pi < \arg z < 0$, we have

$$\log z = \log|z| + i \arg z,$$
$$\log(-z) = \log|z| + i \arg z + i\pi.$$

Thus, we can write

$$\log(-z) = \log z - i\pi \operatorname{sgn}(\Im z). \tag{1.12}$$

Therefore,

$$u(x) + u(-x) = \left[\frac{1}{2\pi i}\left(\log(z) - \log(-z)\right)\right] = [1/2, -1/2] = 1,$$

i.e., we have found the constant hyperfunction $f(x) = 1$, as it was to be expected. Another important example is the *sign-hyperfunction*, defined and denoted by

$$\mathrm{sgn}(x) := -u(-x) + u(x) = -\frac{1}{2\pi i}\left[\log(z) + \log(-z)\right]. \qquad (1.13)$$

It has the value -1 for negative x, the value 1 for positive x, and no value for $x = 0$ and interprets the ordinary sign-function $x \mapsto -1, x < 0$, and $x \mapsto 1, x > 0$ also denoted by $\mathrm{sgn}(x)$. The hyperfunction defined by (1.13) is an example where the defining function is defined by one functional expression which defines in the upper and lower half-planes distinct holomorphic functions not being an analytic continuation from each other.

Remark. By the same reason you have to be cautious with expressions such as $\log(z) \pm \log(-z)$. Such an expression defines a function $F_+(z)$ holomorphic in the upper half-plane, and another function $F_-(z)$ holomorphic in the lower half-plane. They are not analytic continuation from each other despite the fact that the defining expression is formally the same. Since the domain of $\log z$ is the complex plane with a cut along the negative real axis, we have

$$\arg(-z) := \begin{cases} \arg z - \pi, & \Im z > 0 \\ \arg z + \pi, & \Im z < 0. \end{cases}$$

Therefore we may write $\arg(-z) = \arg z - \mathrm{sgn}(\Im z)\pi$. We then obtain

$$\begin{aligned} \log z + \log(-z) &= \log|z| + i \arg z + \log|-z| + i \arg(-z) \\ &= 2\log|z| + 2i \arg z - i\pi \, \mathrm{sgn}(\Im z) \\ &= \log(z^2) - i\pi \, \mathrm{sgn}(\Im z). \end{aligned}$$

This relation is valid for $\Im z > 0$ and for $\Im z < 0$; it is meaningless for real z.

Now for $x < 0$, we have with $F(z) = \log z + \log(-z)$:

$$F(x + i0) - F(x - i0) = 2\log|x| + 2i\pi - i\pi - \{2\log|x| - 2i\pi + i\pi\} = 2\pi i,$$

whereas for $x > 0$ we obtain

$$F(x + i0) - F(x - i0) = 2\log|x| - i\pi - \{2\log|x| + i\pi\} = -2\pi i.$$

Remark. If we speak just of a *real analytic function* $\phi(x)$ we always mean a function which is holomorphic in a full neighborhood of the entire real axis, i.e., $\phi(x) \in \mathcal{A}(\mathbb{R})$. The function $\exp(x)$ is real analytic, for it can be analytically continued to the entire function $\exp(z)$ holomorphic in the entire complex plane. The same holds for functions such as $\sin(x)$, $\cos(x)$, polynomials and all rational functions having no poles on the real axis.

The hyperfunction $f(x) = [\phi(z)]$, where $\phi(x) \in \mathcal{A}(\mathbb{R})$ is any real analytic function, represents the *zero hyperfunction*. We denote the zero hyperfunction by 0 since it can be identified with the ordinary zero function.

Two hyperfunctions $f(x) \in \mathcal{B}(I)$ and $g(x) \in \mathcal{B}(I)$ may be added. In fact, if $f(x) = [F(z)]$, $g(x) = [G(z)]$, we define

$$f(x) + g(x) := [F(z) + G(z)]. \qquad (1.14)$$

Also any hyperfunction may be multiplied by a complex constant c by putting

$$c\, f(x) = c\,[F(z)] := [c\, F(z)]. \qquad (1.15)$$

For all definitions of this kind we have to verify that they are well defined. In fact, the same hyperfunctions are also given by $f(x) = [F(z) + \phi(z)]$ and $g(x) = [G(z) + \psi(z)]$, where $\phi(x)$ and $\psi(x)$ are arbitrary real analytic functions. By performing the linear combinations on the defining functions, we obtain

$$[c_1\, F(z) + c_2\, G(z) + c_1\, \phi(z) + c_2\, \psi(z)].$$

Since $c_1\, \phi(z) + c_2\, \psi(z)$ is again a real analytic function, the above expression defines the same hyperfunction. This proves that the definition is well defined. Proofs of this kind are generally straightforward and left to the reader in the sequel. The above definition implies an often used formula:

$$[c_1 F_1(z) + c_2 F_2(z)] = c_1\, [F_1(z)] + c_2\, [F_2(z)].$$

It is readily verified that the set $\mathcal{B}(I)$ is a linear space.

Besides the two vector space operations, we have the important operation that hyperfunctions can be multiplied by real analytic functions.

Definition 1.4. If $f(x) = [F(z)] \in \mathcal{B}(I)$ is a hyperfunction and $\phi(x) \in \mathcal{A}(I)$ is a real analytic function on I, the product is again a hyperfunction, i.e., $\phi(x)f(x) \in \mathcal{B}(I)$. It is defined by

$$\phi(x)f(x) := [\phi(z)F(z)].$$

Hence, we can multiply hyperfunctions with polynomials and functions as $\sin x$, $\cos x$, $\exp x$ and so on and the result is again a hyperfunction.

Example 1.9. We have

$$x\, \delta(x) = 0.$$

Because $\delta(x) = \left[-\frac{1}{2\pi i z}\right]$ one has

$$x\, \delta(x) = \left[-\frac{z}{2\pi i z}\right] = [-2\pi i] = 0,$$

because a constant is a real analytic function. Of course, we also have

$$x^n \delta(x) = 0, \quad n \in \mathbb{N} \qquad (1.16)$$

since

$$x^n \delta(x) = \left[-\frac{z^{n-1}}{2\pi i}\right] = 0,$$

and a positive integral power of z is real analytic.

We now prove

Proposition 1.1. *If $\phi(x)$ is a real analytic function, then we have*

$$\phi(x)\delta(x) = \phi(0)\delta(x). \tag{1.17}$$

Proof. We can write $\phi(z) = \phi(0) + \phi'(0)z + \psi(z)z^2$ where $\psi(z)$ is real analytic. Hence

$$\phi(x)\delta(x) = \left[-\frac{\phi(z)}{2\pi i z}\right]$$

$$= \left[-\frac{\phi(0)}{2\pi i z}\right] + \left[-\frac{\phi'(0)}{2\pi i}\right] + \left[-\frac{z\psi(z)}{2\pi i}\right]$$

$$= \phi(0)\left[-\frac{1}{2\pi i z}\right] + [\text{const}] - \frac{1}{2\pi i}\left[z\,\psi(z)\right].$$

The last two terms combine to a real analytic function and can be omitted. Therefore we obtain

$$\phi(x)\delta(x) = \phi(0)\left[-\frac{1}{2\pi i z}\right] = \phi(0)\delta(x) . \qquad \square$$

For example, we have

$$\sin x\,\delta(x) = 0, \quad \cos x\,\delta(x) = \delta(x).$$

The hyperfunction $\exp(-|x|)$ can be defined by

$$\exp(-|x|) := u(-x)\exp(x) + u(x)\exp(-x).$$

Problem 1.1. Show that the hyperfunction

$$f(x) = \left[\frac{1}{2}\frac{1}{(z-i)}, -\frac{1}{2}\frac{1}{(z+i)}\right]$$

is an interpretation of the ordinary function $x \mapsto x/(x^2+1)$.

Problem 1.2. Use the relation

$$\arg(-z) = \arg z - \text{sgn}(\Im z)\pi$$

to establish another important representation for the sign-hyperfunction:

$$\text{sgn}(x) = -\frac{1}{\pi i}\left[\log(-iz), \log(iz)\right]. \tag{1.18}$$

Problem 1.3. Show also that

$$\text{sgn}(x)\,\log|x| = -\frac{1}{2\pi i}\left[\log^2(-iz), \log^2(iz)\right], \tag{1.19}$$

$$2\log^2|x| - \frac{\pi^2}{2} = \left[\log^2(-iz), -\log^2(iz)\right]. \tag{1.20}$$

1.3   Properties of Hyperfunctions

In this section we shall study further basic properties of hyperfunctions. Unless otherwise stated the considered hyperfunctions are assumed to be defined on the entire real axis, i.e., $I = \mathbb{R}$.

1.3.1 Linear Substitution

Remark. Modern mathematics has generally adopted the notation

$$f : \mathcal{D} \longrightarrow \mathbb{R}$$
$$x \longmapsto f(x)$$

for a function or mapping. The function is denoted with the symbol f, whereas the function-value at x is denoted by $f(x)$. The reader has certainly remarked that we do not follow this convention. We maintain the older notation $f(x)$ for a function or hyperfunction bearing in mind that the chosen symbol x is immaterial. In this way we have, for example, by writing $f(ax + b)$ a concise notation for the composition of the two functions $x \mapsto ax + b$ and $x \mapsto f(x)$. Also, we use the same notation $f(x)$ for the function-value $f(x)$ at the point x. There will be no confusion since the meaning will generally be clear from the context.

Let $f(x)$ be a hyperfunction defined by a defining function

$$F(z) := \begin{cases} F_+(z), & z \in \mathcal{D}_+ \\ F_-(z), & z \in \mathcal{D}_- \end{cases}$$

with the upper component $F_+(z)$ and the lower component $F_-(z)$. Let us consider the substitution $z \mapsto az + b$. If $a < 0$, the upper neighborhood $\mathcal{D}_+$ will be transformed into a lower neighborhood, and the lower neighborhood $\mathcal{D}_-$ into an upper neighborhood.

Definition 1.5. Let $f(x) = [F_+(z), F_-(z)]$ be a hyperfunction; then the hyperfunction $f(ax + b)$, with $a, b \in \mathbb{R}$, is defined by

$$f(ax + b) := \begin{cases} [F_+(az + b), F_-(az + b)] & \text{if } a > 0 \\ [-F_-(az + b), -F_+(az + b)] & \text{if } a < 0 \, . \end{cases}$$

For the special case where the upper and lower components of the defining function are function elements of the same global analytic function, i.e., $f(x) = [F(z)] := [F(z), F(z)]$, we can write the above definition as

$$f(ax + b) := [\text{sgn}(a)F(az + b)] \, .$$

Example 1.10. The Dirac impulse hyperfunction at $x = 0$ is defined by $\delta(x) = [-1/(2\pi i z)]$. Hence, for $a \neq 0$,

$$\delta(ax + b) = \left[-\text{sgn}(a)\frac{1}{2\pi i(az + b)} \right]$$
$$= \frac{\text{sgn}(a)}{a} \left[-\frac{1}{2\pi i(z + b/a)} \right]$$
$$= \frac{1}{|a|} \left[-\frac{1}{2\pi i(z + b/a)} \right],$$

i.e., we obtain the formula

$$\delta(ax + b) = \frac{1}{|a|} \delta(x + \frac{b}{a}). \tag{1.21}$$

Note that this implies $\delta(b - x) = \delta(x - b)$, and for $b = 0$, $\delta(-x) = \delta(x)$, i.e., the Dirac impulse is an even hyperfunction.

Example 1.11. The unit-step hyperfunction with step at $x = 0$ is defined by $u(x) = [-1/(2\pi i) \, \log(-z)]$. Hence, for $a > 0$,

$$
\begin{aligned}
u(ax + b) &= \left[-\frac{1}{2\pi i} \log(-az - b) \right] \\
&= \left[-\frac{1}{2\pi i} \log(a(-z - \frac{b}{a})) \right] \\
&= \left[-\frac{1}{2\pi i} \log(-z - \frac{b}{a}) - \frac{\log a}{2\pi i} \right] \\
&= u(x + \frac{b}{a}).
\end{aligned}
$$

The constant $-\log a/(2\pi i)$ is a real analytic function and has been omitted in the last step. For $a < 0$ we write, by taking into account that $\text{sgn}(a) = -1$,

$$
\begin{aligned}
u(ax + b) &= \left[\frac{1}{2\pi i} \log(-az - b) \right] \\
&= \left[\frac{1}{2\pi i} \log((-a)(z + \frac{b}{a})) \right] \\
&= \left[\frac{1}{2\pi i} \log(z + \frac{b}{a}) + \frac{\log(-a)}{2\pi i} \right] \\
&= u(-x - \frac{b}{a})
\end{aligned}
$$

which is the reversed unit-step hyperfunction taking the value 1 for $x < -b/a$ and 0 for $x > -b/a$. We may summarize the two cases as

$$
u(ax + b) = u(\text{sgn}(a) \, x + \frac{b}{|a|}), \quad a \neq 0. \tag{1.22}
$$

Example 1.12. Consider the hyperfunction $\phi(x)\delta(ax - b)$ with $a \neq 0$. We can write

$$
\begin{aligned}
\phi(x)\delta(ax - b) &= -\frac{\text{sgn}\, a}{2\pi i} \left[\frac{\phi(z)}{az - b} \right] = -\frac{\text{sgn}\, a}{2\pi i a} \left[\frac{\phi(z)}{z - \frac{b}{a}} \right] \\
&= -\frac{1}{2\pi i |a|} \left[\frac{\phi(\frac{b}{a} + (z - \frac{b}{a}))}{z - \frac{b}{a}} \right] \\
&= -\frac{1}{2\pi i |a|} \left[\frac{\phi(\frac{b}{a}) + \phi'(\frac{b}{a})(z - \frac{b}{a}) + \cdots}{z - \frac{b}{a}} \right] \\
&= \frac{\phi(\frac{b}{a})}{|a|} \left[-\frac{1}{2\pi i} \frac{1}{z - \frac{b}{a}} + \psi(z) \right].
\end{aligned}
$$

Thus,

$$
\phi(x)\delta(ax - b) = \frac{\phi(\frac{b}{a})}{|a|} \delta(x - \frac{b}{a}), \tag{1.23}
$$

where the real analytic function $\psi(x)$ has been omitted. The special case

$$\phi(x)\,\delta(x-b) = \phi(b)\delta(x-b)$$

is often used.

1.3.2 Hyperfunctions of the Type $f(\phi(x))$

The formula

$$\delta(a(x-b)) = \frac{1}{|a|}\,\delta(x-b), \quad (a \neq 0)$$

has an important generalization. Let us investigate the question whether or not a meaning can be given to $\delta(\phi(x))$. Clearly, we are tempted to write

$$\delta(\phi(x)) = \left[-\frac{1}{2\pi i}\,\frac{1}{\phi(z)}\right].$$

We impose the following requirements about $\phi(x)$:

  (i) $\phi(x)$ must be a real analytic and real-valued function for real x,

  (ii) upper or lower neighborhoods of the real axis must be mapped into such neighborhoods (the two types of neighborhoods may be swapped by the mapping as it is the case with $\phi(z) = -z$), however, the image by ϕ must not be a mixture of upper and lower neighborhoods.

Let us commence with a real analytic function $\phi(x)$ which has only one simple zero at $x = b$, i.e., $\phi(b) = 0$ and $\phi'(b) \neq 0$. Because

$$\phi(z) = (z-b)\phi'(b)\{1 + 1/2\,\phi''(b)/\phi'(b)\,(z-b) + \cdots\}$$

we may write

$$-\frac{1}{2\pi i}\,\frac{1}{\phi(z)} = -\frac{1}{2\pi i}\,\frac{1}{\phi'(b)\,(z-b)} + \psi_1(z)$$

where $\psi_1(x)$ is a real analytic function. Therefore, by Definition 1.5 we obtain

$$\delta(\phi(x)) = \frac{1}{|\phi'(b)|}\delta(x-b).$$

Suppose now that $\phi(x)$ has two different real and simple zeros b_1 and b_2, and no others. Then we can write

$$\phi(z) = (z-b_1)(z-b_2)\psi(z), \quad (\psi(b_1) \neq 0, \psi(b_2) \neq 0)$$

where $\psi(x)$ is real analytic. Taking logarithms followed by differentiation yields

$$\frac{\phi'(z)}{\phi(z)} = \frac{1}{z-b_1} + \frac{1}{z-b_2} + \frac{\psi'(z)}{\psi(z)},$$

$$\frac{1}{\phi(z)} = \frac{1}{\phi'(z)}\frac{1}{z-b_1} + \frac{1}{\phi'(z)}\frac{1}{z-b_2} + \frac{1}{\phi'(z)}\frac{\psi'(z)}{\psi(z)}$$

$$= \frac{1}{\phi'(b_1)}\frac{1}{z-b_1} + \frac{1}{\phi'(b_2)}\frac{1}{z-b_2} + \psi_2(z)$$

where $\psi_2(x)$ is another real analytic function that can be discarded. By applying the above result we obtain

$$\delta(\phi(x)) = \frac{1}{|\phi'(b_1)|}\delta(x - b_1) + \frac{1}{|\phi'(b_2)|}\delta(x - b_2).$$

The generalization to a finite number of simple zeros is now straightforward. We have established

Proposition 1.2. *Assume that the real analytic and real-valued function $\phi(x)$ has n different, simple, real zeros $b_1, \ldots, b_n$ and no others; then*

$$\delta(\phi(x)) = \sum_{k=1}^{n} \frac{\delta(x - b_k)}{|\phi'(b_k)|}. \tag{1.24}$$

The proposition remains true if $\phi(x)$ has an infinite set $b_1, b_2, \ldots$ of real and simple zeros provided the infinite series converges in the sense of hyperfunctions (see Chapter 2), thus

$$\delta(\phi(x)) = \sum_{k=1}^{\infty} \frac{\delta(x - b_k)}{|\phi'(b_k)|}. \tag{1.25}$$

However, the assumption that all zeros must be simple is crucial. For example $\delta(x^2)$ has no meaning! Why? The reader may be convinced that the above condition (ii) is violated if $\phi(x)$ has a real zero of order higher than unity. A map of the type $\phi(z) = (z - b)^2$, for example, maps an upper neighborhood of the real axis to a mixture of upper and lower neighborhoods which is of no use for our purpose.

Example 1.13. Let $a \neq b$, then

$$\delta((x - a)(x - b)) = \frac{1}{|a - b|}\{\delta(x - a) + \delta(x - b)\}.$$

With $a = -b$,

$$\delta(x^2 - a^2) = \frac{1}{2|a|}\{\delta(x + a) + \delta(x - a)\}.$$

Also,

$$\delta(\sin x) = \sum_{k=-\infty}^{\infty} \frac{\delta(x - k\pi)}{|\cos k\pi|} = \sum_{k=-\infty}^{\infty} \delta(x - k\pi). \tag{1.26}$$

Problem 1.4. Show that

$$\delta((x^3 - x)e^x) = \frac{e}{2}\delta(x + 1) + \delta(x) + \frac{1}{2e}\delta(x - 1).$$

Suppose again that the real analytic and real-valued function $\phi(x)$ has only one simple zero at $x = b$. Let us investigate $\delta'(\phi(x))$. We have

$$\delta'(\phi(x)) = \delta'(y)|_{y=\phi(x)} = \frac{1}{\phi'(x)}D\delta(\phi(x))$$

$$= \frac{1}{\phi'(x)}D\frac{\delta(x - b)}{|\phi'(b)|} = \frac{1}{|\phi'(b)|}\frac{1}{\phi'(x)}\delta'(x - b)$$

$$= \frac{1}{|\phi'(b)|}\{\frac{1}{\phi'(b)}\delta'(x - b) + \frac{\phi''(b)}{[\phi'(b)]^2}\delta(x - b)\},$$

where in the last step we have anticipated Proposition 1.3 of the next section. Thus,

$$\delta'(\phi(x)) = \frac{1}{\phi'(b)|\phi'(b)|} \{\delta'(x-b) + \frac{\phi''(b)}{\phi'(b)}\delta(x-b)\}. \tag{1.27}$$

Example 1.14. $\delta'(\sinh x) = \delta'(x)$ because $\phi'(0) = 1$, $\phi''(0) = 0$.

In the case of several simple and distinct zeros, we have the same generalization as for $\delta(\phi(x))$, for example

$$\delta'(\sin x) = \sum_{k=-\infty}^{\infty} \frac{1}{\cos(\pi k)|\cos(\pi k)|} \delta'(x - k\pi)$$

$$= \sum_{k=-\infty}^{\infty} (-1)^k \, \delta'(x - k\pi).$$

This result can also be obtained by differentiating (1.26) and using again Proposition 1.3 of the next section.

The requirement that $\phi(x)$ should be real analytic on $\mathbb{R}$ can be replaced by an assumption of the kind $\phi(x) \in \mathcal{A}(I)$. For example, if $\phi(x)$ is real analytic for $x > 0$ and has a simple zero at $b > 0$, then we have, for $x > 0$,

$$\delta(\phi(x)) = \frac{1}{|\phi'(b)|}\delta(x - b).$$

Example 1.15. (compare [21, Kanwal p. 68]). Show that for $x > 0$ we have

$$\delta(a - \frac{1}{x}) = \begin{cases} (1/a^2)\,\delta(x - 1/a), & a > 0 \\ 0, & a \le 0. \end{cases}$$

Example 1.16. Let us investigate $u(e^{-x} - a)$. If $a \le 0$ we expect that $u(e^{-x} - a) = 1$ for all x since $\phi(x) = e^{-x} - a$ is strictly positive. However, if $a > 0$ the equation $u(e^{-x} - a = 0$ has the unique solution $x_0 = -\log a$. For $x < x_0$ we have $\phi(x) > 0$ thus $u(\phi(x)) = 1$, whereas for $x > x_0$ we have $\phi(x) < 0$ thus $u(\phi(x)) = 0$. Hence we expect $u(e^{-x} - a) = u(x_0 - x) = u(-\log a - x)$. Can we formally establish this? We write

$$u(e^{-x} - a) = \left[(-1)(-1)\frac{1}{2\pi i}\log(a - e^{-z})\right]$$

because $\phi(x)$ is decreasing. On the other hand we have

$$u(-\log a - x) = \left[\frac{1}{2\pi i}\log(z + \log a)\right].$$

If the two hyperfunctions are the same the two defining functions differ by a real analytic function, so let us examine this difference:

$$\psi(z) := \frac{1}{2\pi i}\log(z + \log a) - \frac{1}{2\pi i}\log(a - e^{-z})$$

$$= \frac{1}{2\pi i}\log\frac{z + \log a}{a - e^{-z}}.$$

The only singularity of $\psi(x)$ is at $x = -\log a$. There, the fraction inside the logarithm becomes undetermined of type $0/0$. But the expansion of $a - \exp(-x)$ about $x_0 = -\log a$ shows that the singularity is removable and $\psi(x)$ is in fact real analytic. This proves that for $a > 0$ we have in the sense of hyperfunctions

$$u(e^{-x} - a) = u(-(x + \log a)). \tag{1.28}$$

Similarly, one may establish for $a > 0$,

$$u(a - e^{-x}) = u(x + \log a). \tag{1.29}$$

In Section 6.2 about Mellin transforms of hyperfunctions, we shall take up again the discussion of hyperfunctions of the kind $f(\phi(x))$ in the special case $\phi(x) = \exp(-x)$.

1.3.3   Differentiation

We now come to a cornerstone of every theory of generalized functions. In Sato's hyperfunction approach this is very simple. We use the fact that upper and lower components of the defining function are holomorphic, thus having derivatives of any order.

Definition 1.6. For any given hyperfunction $f(x) = [F_+(z), F_-(z)]$ its *derivative (in the sense of hyperfunctions)* is defined and denoted as

$$Df(x) = f'(x) := \left[\frac{dF_+}{dz}, \frac{dF_-}{dz}\right],$$

and similarly, for higher derivatives of order n,

$$D^n f(x) = f^{(n)}(x) := \left[\frac{d^n F_+}{dz^n}, \frac{d^n F_-}{dz^n}\right].$$

Notation. Henceforth we denote derivatives in the sense of hyperfunctions, for short derivatives, with the operator symbols D, D^n or with the prime notations $'$, $''$ .... Ordinary derivatives of ordinary differentiable functions will generally be denoted by the Leibniz notation d/dx, d^n/dx^n. Because real analytic functions can always be considered as hyperfunctions by means of $\phi(x) = [\phi(z)/2, -\phi(z)/2]$, we might use both types of notation for them.

Derivatives in the sense of hyperfunctions are also called generalized derivatives. We leave it as an exercise to the reader to show that the above definition is well defined, i.e., if we choose another equivalent defining function for $f(x)$, its derivative will be the same. The above definition now implies that *hyperfunctions are always indefinitely differentiable.*

Example 1.17. Let us differentiate the unit-step hyperfunction. The ordinary unit-step function $Y(x)$ is not differentiable at $x = 0$. Now we have

$$u'(x) = \left[\frac{d}{dz}\left(-\frac{1}{2\pi i}\log(-z)\right)\right] = \left[-\frac{1}{2\pi i}\frac{-1}{(-z)}\right]$$

$$= \left[-\frac{1}{2\pi i z}\right] = \delta(x).$$

Thus,

$$u'(x) = \delta(x), \text{ and also } u'(x - a) = \delta(x - a). \tag{1.30}$$

The generalized derivative of the unit-step hyperfunction is the Dirac impulse hyperfunction.

Example 1.18. Let us now consider the derivatives of the Dirac impulse hyperfunction. Applying the definition, we immediately obtain

$$\delta^{(n)}(x - a) = u^{(n+1)}(x - a) = \left[-\frac{(-1)^n n!}{2\pi i (z - a)^{n+1}} \right]. \tag{1.31}$$

In order to get an intuitive feeling of the derivatives of the Dirac impulse hyperfunction, we take a look at the family of ordinary functions

$$x \mapsto -\frac{(-1)^n n!}{2\pi i} \left(\frac{1}{(x + i/m)^{n+1}} - \frac{1}{(x - i/m)^{n+1}} \right),$$

for large values for m. For $n = 1$, we obtain Figure 1.4, and for $n = 2$ Figure 1.5. Consider now $\phi(x)\, \delta^{(n)}(x - a)$ where $\phi(x)$ is a real analytic function.

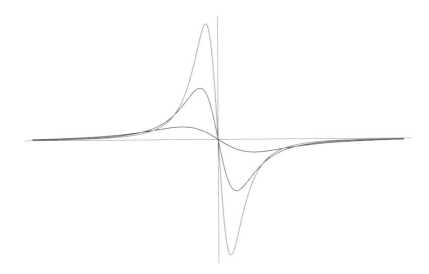

Figure 1.4: Intuitive picture of the first derivative of the Dirac impulse

$$
\begin{aligned}
\phi(x)\, \delta^{(n)}(x - a) &= \left[-\frac{\phi(z)(-1)^n n!}{2\pi i (z - a)^{n+1}} \right] = -\frac{(-1)^n n!}{2\pi i} \left[\frac{\phi(z)}{(z - a)^{n+1}} \right] \\
&= -\frac{(-1)^n n!}{2\pi i} \left[\sum_{k=0}^{\infty} \frac{\phi^{(k)}(a)}{k!} (z - a)^{k-n-1} \right] \\
&= -\frac{(-1)^n}{2\pi i} \left[\sum_{k=0}^{n} \binom{n}{k} (n - k)! \frac{\phi^{(k)}(a)}{(z - a)^{n-k+1}} + \psi(z) \right] \\
&= -\left[\sum_{k=0}^{n} \binom{n}{k} (n - k)! (-1)^k \phi^{(k)}(a) \frac{(-1)^{n-k}}{2\pi i (z - a)^{(n-k)+1}} \right]
\end{aligned}
$$

$$= \sum_{k=0}^{n} (-1)^k \left(\begin{array}{c} n \\ k \end{array} \right) \phi^{(k)}(a) \left[-\frac{(-1)^{n-k}(n-k)!}{2\pi i (z-a)^{(n-k)+1}} \right]$$

$$= \sum_{k=0}^{n} (-1)^k \left(\begin{array}{c} n \\ k \end{array} \right) \phi^{(k)}(a) \, \delta^{(n-k)}(x-a).$$

Here $\psi(x)$ is a real analytic function which has been omitted in the following steps. Replacing k by $n-k$, we obtain

Proposition 1.3. *For any real analytic function $\phi(x)$ we have*

$$\phi(x) \, \delta^{(n)}(x-a) = \sum_{k=0}^{n} (-1)^{n-k} \left(\begin{array}{c} n \\ k \end{array} \right) \phi^{(n-k)}(a) \, \delta^{(k)}(x-a). \qquad (1.32)$$

Special cases are:

$$\phi(x) \, \delta(x-a) = \phi(a) \, \delta(x-a) \, ,$$
$$\phi(x) \, \delta'(x-a) = \phi(a) \, \delta'(x-a) - \phi'(a) \, \delta(x-a) \, ,$$
$$\phi(x) \, \delta''(x-a) = \phi(a) \, \delta''(x-a) - 2\phi'(a) \, \delta'(x-a) + \phi''(a) \, \delta(x-a) \, .$$

Example 1.19. The formula

$$\phi(x) \, \delta^{(n)}(x-a) = \sum_{k=0}^{n} (-1)^k \left(\begin{array}{c} n \\ k \end{array} \right) \phi^{(k)}(a) \, \delta^{(n-k)}(x-a) \qquad (1.33)$$

applied for $\phi(x) = x^n$ and $a = 0$ yields, by taking into account that $\phi^{(k)}(0) = \delta_m^k \, m!$, where δ_m^k is Kronecker's symbol,

$$x^m \, \delta^{(n)}(x) = \sum_{k=0}^{n} (-1)^k \left(\begin{array}{c} n \\ k \end{array} \right) \delta_m^k m! \, \delta^{(n-k)}(x).$$

Finally, we obtain

$$x^m \, \delta^{(n)}(x) = \left\{ \begin{array}{ll} (-1)^m \frac{n!}{(n-m)!} \, \delta^{(n-m)}(x) & \text{for } m \leq n \\ 0 & \text{for } m > n. \end{array} \right. \qquad (1.34)$$

A hyperfunction can be multiplied with a real analytic function and the product is again a hyperfunction. For such products the product rule holds.

Proposition 1.4. *For any hyperfunction $f(x)$ and any real analytic function $\phi(x)$ we have the product rule*

$$D\phi(x)f(x) = \phi'(x) \, f(x) + \phi(x) \, f'(x) \, . \qquad (1.35)$$

Proof. Let $f(x) = [F_+(z), F_-(z)]$ and $\phi(x)$ be any real analytic function. Then,

$$D\phi(x)f(x) = \left[\frac{d}{dz} \{\phi(z)F_+(z)\}, \frac{d}{dz} \{\phi(z)F_-(z)\} \right]$$

$$= \left[\frac{d\phi}{dz} F_+(z) + \phi(z) \frac{dF_+}{dz}, \frac{d\phi}{dz} F_-(z) + \phi(z) \frac{dF_-}{dz} \right]$$

$$= \left[\frac{d\phi}{dz} F_+(z), \frac{d\phi}{dz} F_-(z) \right] + \left[\phi(z) \frac{dF_+}{dz}, \phi(z) \frac{dF_-}{dz} \right]$$

$$= \phi'(x)f(x) + \phi(x)f'(x). \qquad \square$$

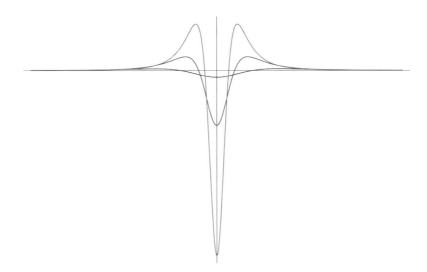

Figure 1.5: Intuitive picture of the second derivative of the Dirac impulse

Example 1.20.

$$(u(x)\cos x)' = -u(x)\sin x + \cos x\,\delta(x) = -u(x)\sin x + \delta(x),$$
$$(u(x)\cos x)'' = (-u(x)\sin x + \delta(x))' = -\delta(x)\sin x - u(x)\cos x + \delta'(x).$$

Repeated application of the above proposition implies Leibniz's rule:

$$D^n(\phi(x)f(x)) = \sum_{k=0}^{n} \binom{n}{k} \frac{d^{n-k}\phi(x)}{dx^{n-k}} f^{(k)}(x) . \tag{1.36}$$

We shall now establish an important formula concerning the generalized derivative of a piece-wise defined hyperfunction. Let

$$f(x) = u(a-x)\phi_1(x) + \{u(x-a) - u(x-b)\}\phi_2(x) + u(x-b)\phi_3(x)$$

be a given hyperfunction with the three real analytic functions $\phi_1(x), \phi_2(x), \phi_3(x)$ and $a < b$. If $f(x)$ is considered as an ordinary function, it is a piece-wise real analytic function with jump discontinuities at a and b, see Figure1.6. By using the product rule we obtain

$$f'(x) = \delta(a-x)(-1)\,\phi_1(x) + u(a-x)\frac{d\phi_1(x)}{dx}$$
$$+ \{\delta(x-a) - \delta(x-b)\}\,\phi_2(x) + \{u(x-a) - u(x-b)\}\frac{d\phi_2(x)}{dx}$$
$$+ \delta(x-b)\,\phi_3(x) + u(x-b)\frac{d\phi_3(x)}{dx}$$
$$= u(a-x)\frac{d\phi_1(x)}{dx} + \{u(x-a) - u(x-b)\}\frac{d\phi_2(x)}{dx} + u(x-b)\frac{d\phi_3(x)}{dx}$$
$$+ \{\phi_2(a) - \phi_1(a)\}\,\delta(x-a) + \{\phi_3(b) - \phi_2(b)\}\,\delta(x-b) .$$

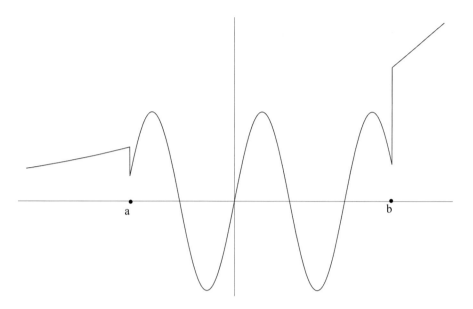

Figure 1.6: Piecewise real analytic function.

By differentiating again, we find

$$f''(x) = u(a - x) \frac{d^2 \phi_1(x)}{dx^2} + \{u(x - a) - u(x - b)\} \frac{d^2 \phi_2(x)}{dx^2} + u(x - b) \frac{d^2 \phi_3(x)}{dx^2}$$
$$+ \{\phi_2'(a) - \phi_1'(a)\} \delta(x - a) + \{\phi_3'(b) - \phi_2'(b) \ \delta(x - b)$$
$$+ \{\phi_2(a) - \phi_1(a)\} \delta'(x - a) + \{\phi_3(b) - \phi_2(b) \ \delta'(x - b).$$

By introducing the *jump magnitudes of $f(x)$ and their derivatives at a point $x = a$* defined by

$$J_0[a] := \phi_2(a) - \phi_1(a) = f(a+) - f(a-),$$
$$J_1[a] := \phi_2'(a) - \phi_1'(a) = f'(a+) - f'(a-),$$
$$J_\ell[a] := \phi_2^{(\ell)}(a) - \phi_1^{(\ell)}(a) = f^{(\ell)}(a+) - f^{(\ell)}(a-),$$

we obtain

Proposition 1.5. *With the above notation for the jump magnitudes, the derivatives of the hyperfunction*

$$f(x) = u(a - x)\phi_1(x) + \{u(x - a) - u(x - b)\}\phi_2(x) + u(x - b)\phi_3(x)$$

are given by

$$f'(x) = u(a - x) \frac{d\phi_1(x)}{dx} + \{u(x - a) - u(x - b)\} \frac{d\phi_2(x)}{dx} + u(x - b) \frac{d\phi_3(x)}{dx}$$
$$+ J_0[a] \, \delta(x - a) + J_0[b] \, \delta(x - b),$$

$$f''(x) = u(a-x)\frac{d^2\phi_1(x)}{dx^2} + \{u(x-a) - u(x-b)\}\frac{d^2\phi_2(x)}{dx^2} + u(x-b)\frac{d^2\phi_3(x)}{dx^2}$$
$$+ J_1[a]\,\delta(x-a) + J_1[b]\,\delta(x-b)$$
$$+ J_0[a]\,\delta'(x-a) + J_0[b]\,\delta'(x-b).$$

If the real axis is partitioned by m points $a_1 < a_2 < \cdots < a_m$, and the piecewise real analytic hyperfunction is specified by

$$f(x) = u(a_1 - x)\phi_1(x)$$
$$+ \sum_{k=2}^{m}\{u(x-a_{k-1}) - u(x-a_k)\}\phi_k(x) + u(x-a_m)\phi_{m+1}(x), \tag{1.37}$$

where $\phi_k(x)$, $k = 1, \ldots m+1$, are given real analytic functions, we have for the nth derivative (in the sense of hyperfunctions)

$$D^n f(x) = u(a_1 - x)\frac{d^n\phi_1(x)}{dx^n}$$
$$+ \sum_{k=2}^{m}\{u(x-a_{k-1}) - u(x-a_k)\}\frac{d^n\phi_k(x)}{dx^n} + u(x-a_m)\frac{d^n\phi_{m+1}(x)}{dx^n}$$
$$+ \sum_{\ell=0}^{n-1}\sum_{k=1}^{m} J_{n-1-\ell}[a_k]\,\delta^{(\ell)}(x-a_k). \tag{1.38}$$

Every jump discontinuity of $f(x)$ produces its own Dirac impulse with its derivatives.

We conclude this section with a remark that shows that the product of two hyperfunctions cannot generally be defined without further restrictions.

Remark. One would be tempted to define the product of two hyperfunctions $f(x) = [F_+(z), F_-(z)]$ and $g(x) = [G_+(z), G_-(z)]$ by putting $f(x)\,g(x) := [F_+(z)G_+(z), F_-(z)G_-(z)]$. But this is not a well-defined definition. It becomes apparent when we differentiate the product. We would have

$$D(f(x)\,g(x)) = \left[\frac{d}{dz}(F_+(z)G_+(z)), \frac{d}{dz}(F_-(z)G_-(z))\right]$$
$$= \left[\frac{dF_+}{dz}G_+(z) + F_+(z)\frac{dG_+}{dz}, \frac{dF_-}{dz}G_-(z) + F_-(z)\frac{dG_-}{dz}\right]$$
$$= \left[\frac{dF_+}{dz}G_+(z), \frac{dF_-}{dz}G_-(z)\right] + \left[F_+(z)\frac{dG_+}{dz}, F_-(z)\frac{dG_-}{dz}\right]$$
$$= Df(x)\,g(x) + f(x)\,Dg(x). \tag{1.39}$$

No problem so far, but $f(x) = [F_+(z) + \phi(z), F_-(z) + \phi(z)]$ is also a valid representation of the same hyperfunction $f(x)$, where $\phi(x)$ is an arbitrary real analytic function. Proceeding as above with this representation, we obtain

$$D(f(x)\,g(x)) = \left[\frac{d}{dz}((F_+(z) + \phi(z))G_+(z)), \frac{d}{dz}((F_-(z) + \phi(z))G_-(z))\right]$$
$$= \left[\frac{dF_+}{dz}G_+(z) + \frac{d\phi}{dz}G_+(z), \frac{dF_-}{dz}G_-(z) + \frac{d\phi}{dz}G_-(z)\right]$$

$$+ \left[F_+(z)\frac{dG_+}{dz} + \phi(z)\frac{dG_+}{dz}, F_-(z)\frac{dG_-}{dz} + \phi(z)\frac{dG_-}{dz} \right]$$

$$= \left[\frac{dF_+}{dz}G_+(z), \frac{dF_-}{dz}G_-(z) \right] + \left[F_+(z)\frac{dG_+}{dz}, F_-(z)\frac{dG_-}{dz} \right]$$

$$+ \left[\frac{d\phi}{dz}G_+(z) + \phi(z)\frac{dG_+}{dz}, \frac{d\phi}{dz}G_-(z) + \phi(z)\frac{dG_-}{dz} \right]. \qquad (1.40)$$

But $\frac{d\phi}{dz}G_+(z) + \phi(z)\frac{dG_+}{dz}$ and $\frac{d\phi}{dz}G_-(z) + \phi(z)\frac{dG_-}{dz}$ are not necessarily the same functions. Even in the case where the upper and lower components of $g(x)$ are given by the same expression, the function $\frac{d\phi}{dz}G(z) + \phi(z)\frac{dG}{dz}$ may still fail to be real analytic since $G(z)$ is not necessarily real analytic. So the right-hand side of (1.40) defines another hyperfunction than the right-hand side of (1.39). This shows that *the product of two hyperfunctions is generally not defined.* Later on in Chapter 2, however, we shall examine the question under what restrictions the product of two hyperfunctions may still be defined.

Problem 1.5. Prove that

$$D \operatorname{sgn}(x) = 2\, \delta(x).$$

Problem 1.6. Specialize the above proposition to establish

$$D^n \{ u(x-a)\phi(x) \} = u(x-a)\frac{d^n \phi}{dx^n} + \sum_{k=0}^{n-1} \phi^{(k)}(a)\, \delta^{(n-1-k)}(x-a), \qquad (1.41)$$

$$D^n \{ u(a-x)\phi(x) \} = u(a-x)\frac{d^n \phi}{dx^n} - \sum_{k=0}^{n-1} \phi^{(k)}(a)\, \delta^{(n-1-k)}(x-a). \qquad (1.42)$$

Problem 1.7. Establish the following formulas, valid for any real analytic function $\phi(x)$:

$$(a_n D^n + a_{n-1} D^{n-1} + \cdots + a_1 D + a_0)\{ u(x-a)\phi(x) \}$$

$$= u(x-a)(a_n \frac{d^n}{dx^n} + a_{n-1}\frac{d^{n-1}}{dx^{n-1}} + \cdots + a_1 \frac{d}{dx} + a_0)\,\phi(x)$$

$$+ \sum_{\nu=0}^{n-1} \phi^{(\nu)}(a) \sum_{\mu=\nu+1}^{n} a_\mu\, \delta^{(\mu-1-\nu)}(x-a),$$

$$(a_n D^n + a_{n-1} D^{n-1} + \cdots + a_1 D + a_0)\{ u(a-x)\phi(x) \}$$

$$= u(a-x)(a_n \frac{d^n}{dx^n} + a_{n-1}\frac{d^{n-1}}{dx^{n-1}} + \cdots + a_1 \frac{d}{dx} + a_0)\,\phi(x)$$

$$- \sum_{\nu=0}^{n-1} \phi^{(\nu)}(a) \sum_{\mu=\nu+1}^{n} a_\mu\, \delta^{(\mu-1-\nu)}(x-a).$$

1.3.4　The Shift Operator as a Differential Operator

Consider the linear differential operator of order n acting on a hyperfunction $f(x)$,

$$(c_0 + c_1 D + c_2 D^2 + \cdots + c_n D^n)f(t) := c_0 f(x) + c_1 f'(x) + c_2 f''(x) + \cdots + c_n f^{(n)}(x).$$

The c_k's are real or complex constants. If $f(x) = [F_+(z), F_-(z)]$, the resulting hyperfunction is given by

$$\left[\sum_{k=0}^{n} c_k \frac{d^k F_+(z)}{dz^k}, \sum_{k=0}^{n} c_k \frac{d^k F_-(z)}{dz^k} \right].$$

Nothing prevents us from defining a differential operator of infinite order acting on a hyperfunction

$$\left(\sum_{k=0}^{\infty} c_k D^k \right) f(x) := \left[\sum_{k=0}^{\infty} c_k \frac{d^k F_+(z)}{dz^k}, \sum_{k=0}^{\infty} c_k \frac{d^k F_-(z)}{dz^k} \right], \qquad (1.43)$$

provided the infinite series converge uniformly in every compact subdomain of the upper and the lower complex half-plane, respectively. Take now $c_k = a^k/k!$, then the above expression becomes

$$\left(\sum_{k=0}^{\infty} \frac{a^k}{k!} D^k \right) f(x) := \left[\sum_{k=0}^{\infty} \frac{a^k}{k!} \frac{d^k F_+(z)}{dz^k}, \sum_{k=0}^{\infty} \frac{a^k}{k!} \frac{d^k F_-(z)}{dz^k} \right]$$
$$= [F_+(z+a), F_-(z+a)] = f(x+a).$$

For the operator on the left-hand side we write formally $\exp(aD)$. Therefore, we obtain

Proposition 1.6. *For any real constant a we have*

$$\exp(aD)f(x) = e^{aD} f(x) = f(x+a). \qquad (1.44)$$

Example 1.21.

$$e^{-aD}\delta(x) = \delta(x-a), \quad e^{-aD}\delta^{(k)}(t) = \delta^{(k)}(x-a). \qquad (1.45)$$

1.3.5 Parity, Complex Conjugate and Realness

As a special case of the linear substitution rule with $a = -1, b = 0$ (see Definition 1.5) we have for a given hyperfunction $f(x) = [F_+(z), F_-(z)]$ that the hyperfunction $f(-x)$ is defined by

$$f(-x) := [-F_-(-z), -F_+(-z)]. \qquad (1.46)$$

It is now easy to define even and odd hyperfunctions.

Definition 1.7. If $f(-x) = f(x)$, the hyperfunction $f(x) = [F_+(z), F_-(z)]$ is said to be an *even hyperfunction*. If $f(-x) = -f(x)$, it is called an *odd hyperfunction*.

In the special case where the upper and lower component of the defining function are restrictions of one global analytic function, i.e., $f(x) = [F(z), F(z)] = [F(z)]$, an odd defining function defines an even hyperfunction and an even defining function defines an odd one. This means that hyperfunction and defining function have opposite parity.

Example 1.22. The Dirac impulse hyperfunction and its derivative of even order are even hyperfunctions, the derivatives of odd order are odd hyperfunctions because

$$\delta^{(n)}(x) = \left[-\frac{(-1)^n n!}{2\pi i z^{n+1}} \right].$$

Example 1.23. The sign-hyperfunction

$$\text{sgn}(x) := \frac{1}{2\pi i} \left[\log(z) + \log(-z) \right]$$

is an odd hyperfunction.

Proposition 1.7. *Any hyperfunction $f(x) = [F_+(z), F_-(z)]$ can be decomposed into an even and an odd hyperfunction, i.e.,*

$$f(x) = f_e(x) + f_o(x),$$

where $f_e(x)$ is even and $f_o(x)$ is odd.

Proof. We simply set

$$f_o(x) := \frac{f(x) - f(-x)}{2}, \; f_e(x) := \frac{f(x) + f(-x)}{2}. \qquad \square$$

Problem 1.8. Prove that $f(-(-x)) = f(x)$.

Problem 1.9. Prove:

(i) If $f(x)$ is an even (odd) hyperfunction, then $f'(x)$ is an odd (even) hyperfunction.

(ii) Let $\phi(x)$ be a real analytic function. If $\phi(x)$ and $f(x)$ have the same (opposite) parity, then the hyperfunction $\phi(x)f(x)$ is even (odd).

Problem 1.10. Show that if $f(x) = [F_+(z), F_-(z)]$ is an even hyperfunction, we can write

$$f(x) = \left[\frac{F_+(z) - F_-(-z)}{2}, \frac{F_-(z) - F_+(-z)}{2} \right],$$

and if $f(x)$ is an odd hyperfunction, we can write

$$f(x) = \left[\frac{F_+(z) + F_+(-z)}{2}, \frac{F_-(z) + F_+(-z)}{2} \right].$$

Our hyperfunctions are supposed to be complex-valued at the points where they have values. Let us now tackle the problem how the complex-conjugate hyperfunction $\overline{f(x)}$ has to be defined if $f(x) = [F_+(z), F_-(z)]$ is given. The upper component $F_+(z)$ is a holomorphic function in $\mathcal{D}_+$, and the lower component is holomorphic in $\mathcal{D}_-$. Let Δ_- be the symmetric image with respect to the real axis of $\mathcal{D}_+$, and Δ_+ the symmetric image of $\mathcal{D}_-$. The function $G_+(z) := \overline{F_-(\bar{z})}$ is a function defined on Δ_+. Similarly, $G_-(z) := \overline{F_+(\bar{z})}$ is a function defined on Δ_-. Are these functions in their respective domains holomorphic in z? This is the case if their partial derivatives with respect to $\bar{z}$ vanish. Let us show this. If we set

$F_-(z) = U(x,y) + iV(x,y)$, then $\overline{F_-(\overline{z})} = U(x,-y) - iV(x,-y)$. Furthermore, we have (see [2, Ahlfors p. 27])

$$\frac{\partial}{\partial z} = \frac{1}{2}\left(\frac{\partial}{\partial x} - i\frac{\partial}{\partial y}\right), \quad \frac{\partial}{\partial \overline{z}} = \frac{1}{2}\left(\frac{\partial}{\partial x} + i\frac{\partial}{\partial y}\right),$$

hence, we can write

$$\frac{\partial G_+(z)}{\partial \overline{z}} = \frac{\partial \overline{F_-(\overline{z})}}{\partial \overline{z}} = \frac{1}{2}\left(\frac{\partial}{\partial x} + i\frac{\partial}{\partial y}\right)(U(x,-y) - iV(x,-y))$$

$$= \frac{1}{2}\left(\frac{\partial U(x,-y)}{\partial x} - i\frac{\partial V(x,-y)}{\partial x} - i\frac{\partial U(x,-y)}{\partial(-y)} - \frac{\partial V(x,-y)}{\partial(-y)}\right).$$

Since $F_-(z) = U(x,y) + iV(x,y)$ is a holomorphic function, the Cauchy-Riemann differential equations

$$\frac{\partial U(x,y)}{\partial x} = \frac{\partial V(x,y)}{\partial y}, \quad \frac{\partial U(x,y)}{\partial y} = -\frac{\partial V(x,y)}{\partial x}$$

hold. Hence,

$$\frac{\partial G_+(z)}{\partial \overline{z}} = \frac{1}{2}\left(\frac{\partial V(x,-y)}{\partial(-y)} + i\frac{\partial U(x,-y)}{\partial(-y)} - i\frac{\partial U(x,-y)}{\partial(-y)} - \frac{\partial V(x,-y)}{\partial(-y)}\right) = 0.$$

This shows that $G_+(z) := \overline{F_-(\overline{z})}$ is a holomorphic function of z. Similarly, one shows that $G_-(z) := \overline{F_+(\overline{z})}$ is a holomorphic function of z too. Therefore, $g(x) := -[G_+(z), G_-(z)] = \left[-\overline{F_-(\overline{z})}, -\overline{F_+(\overline{z})}\right]$ defines a hyperfunction. Let us show that this hyperfunction is the complex-conjugate hyperfunction of $f(x)$. For that we assume that the given hyperfunction has a value at x. Let us compute the value of $g(x)$ at x.

$$g(x) = -\lim_{\epsilon \to 0+}\left\{\overline{F_-(\overline{x+i\epsilon})} - \overline{F_+(\overline{x-i\epsilon})}\right\}$$

$$= -\lim_{\epsilon \to 0+}\left\{\overline{F_-(x-i\epsilon)} - \overline{F_+(x+i\epsilon)}\right\}$$

$$= -\overline{\lim_{\epsilon \to 0+}\left\{F_-(x-i\epsilon) - F_+(x+i\epsilon)\right\}}$$

$$= \overline{\lim_{\epsilon \to 0+}\left\{F_+(x+i\epsilon) - F_-(x-i\epsilon)\right\}} = \overline{f(x)}.$$

We therefore adopt the following definition.

Definition 1.8. If $f(x) = [F_+(z), F_-(z)]$ is a given hyperfunction, the *complex-conjugate hyperfunction* of $f(x)$ is defined and denoted by

$$\overline{f(x)} := \left[-\overline{F_-(\overline{z})}, -\overline{F_+(\overline{z})}\right]. \tag{1.47}$$

We can now define real and pure imaginary hyperfunctions.

Definition 1.9. A hyperfunction $f(x) = [F_+(z), F_-(z)]$ is *real*, if $\overline{f(x)} = f(x)$; the hyperfunction is *pure imaginary* if $\overline{f(x)} = -f(x)$.

Problem 1.11. We define the real part and imaginary part of a hyperfunction by

$$\Re f(x) := \frac{1}{2}\{f(x) + \overline{f(x)}\}, \quad \Im f(x) := \frac{1}{2i}\{f(x) - \overline{f(x)}\}.$$

Hence we have $f(x) = \Re f(x) + i\,\Im f(x)$. What are the defining functions of $\Re f(x)$ and $\Im f(x)$?

Problem 1.12. Show that the hyperfunctions $f(x)$ and $\overline{f(x)}$ have the same parity.

Problem 1.13. Show that if the hyperfunction $f(x)$ is real, then $f'(x)$ is also real.

Problem 1.14. Show that if $f(x)$ is real and if $\phi(x)$ is a real-valued and real analytic function (then it satisfies $\phi(z) = \overline{\phi(\bar z)}$), $\phi(x)f(x)$ is again a real hyperfunction.

1.3.6   The Equation $\phi(x)f(x) = h(x)$

Let us consider the equation $\phi(x)f(x) = h(x)$ where $h(x) = [H(z)]$ is a given hyperfunction and $\phi(x) \in \mathcal{A}(\mathbb{R})$ a given real analytic function. We shall consider the following four cases:

(i) $\phi(x)$ has no real zero;

(ii) $\phi(x)$ has one real zero $x = a$ of order m;

(iii) $\phi(x)$ has n zeros $x = a_k,\ k = 1, 2, \ldots, n$ each of order m_k;

(iv) $\phi(x)$ has infinitely many simple zeros.

    (i) $\phi(x)$ **has no real zero**. In this case $1/\phi(x)$ is a real analytic function, and the equation $\phi(x)f(x) = h(x)$ has the unique solution

$$f(x) = \left[\frac{H(z)}{\phi(z)}\right].$$

For example, the solution of $\exp(x)\,f(x) = \delta(x-c)$ is just $f(x) = \exp(-c)\,\delta(x-c)$.
    (ii) $\phi(x)$ **has one real zero** $x = a$ **of order** m. Take $\phi(x)$ in the form $(x - a)^m\,\psi(x)$, $\psi(x) \neq 0$ and real analytic. Thus, we consider the equation

$$(x - a)^m\,\psi(x)\,f(x) = h(x), \tag{1.48}$$

where we are looking for a hyperfunction solution $f(x)$. It is now convenient to introduce the following hyperfunctions.

Definition 1.10. A linear combination (with arbitrary coefficients) of Dirac impulses and their derivatives at $x = a$,

$$\delta_{(N)}(x - a) := \sum_{j=0}^{N} c_j\,\delta^{(j)}(x - a), \quad N \in \mathbb{N}, \tag{1.49}$$

is called a *generalized delta-hyperfunction of order N at a*.

    If we write

$$\psi(x) = \sum_{k=0}^{\infty} \psi_k\,(x - a)^k, \quad \psi_0 \neq 0,$$

and introduce $\delta_{(m-1)}(x-a)$ for $f(x)$ in (1.48), we obtain

$$x-a)^m \, \psi(x) \, \delta_{(m-1)}(x-a) = \sum_{k=0}^{\infty} \sum_{j=0}^{m-1} \psi_k \, c_j \, (x-a)^{k+m} \, \delta^{(j)}(x-a) = 0.$$

This implies, by using (1.34),

Proposition 1.8. *If the hyperfunction $f_1(x)$ is any particular solution of equation* (1.48), *then the hyperfunction*

$$f(x) = f_1(x) + \sum_{j=0}^{m-1} c_j \, \delta^{(j)}(x-a) = f_1(x) + \delta_{(m-1)}(x-a)$$

is also a solution of (1.48).

For a particular solution we may choose

$$f_1(x) = \left[\frac{H(z)}{\psi(z)\,(z-a)^m} \right].$$

Therefore, solutions of equation (1.48) are only determined up to a generalized delta-hyperfunction of order $m-1$. This will have implications if we are looking for interpretations of certain familiar ordinary functions in terms of hyperfunctions. Take for example the ordinary function $x \mapsto 1/x^m$. Suppose now that we have defined a suitable interpretation $f_1(x)$ of $x \mapsto 1/x^m$. Reasonably it should satisfy the equation $x^m \, f(x) = 1$. But then $f(x) = f_1(x) + \delta_{(m-1)}(x)$ would serve equally as well. In other words, the question of interpretation of given ordinary functions as hyperfunctions does not always have a unique answer, i.e., a choice has to be made. In Section 2.3.4 we shall take up again the discussion concerning this arbitrariness in the choice of hyperfunctions interpreting a given ordinary function. Also, in Chapters 3 and 4 we shall see that the Laplace and Fourier transforms of a generalized delta-hyperfunction are not zero. This implies that the choice in the interpretation of ordinary functions such as $x \mapsto 1/x^m$ in terms of hyperfunctions will affect their Laplace and Fourier transforms. Depending on the approach in a theory of generalized functions, different choices may present themselves as the natural ones. Keep this in mind in the following section when we define hyperfunctions interpreting ordinary functions such as $x \mapsto 1/x^m$.

Example 1.24. Consider the equation $x^m \, f(x) = \delta^{(n)}(x)$. The solution is given by

$$f(x) = \left[-\frac{(-1)^n \, n!}{2\pi i \, z^{n+m+1}} \right] + \sum_{j=0}^{m-1} c_j \, \delta^{(j)}(x).$$

Therefore, we can write

$$\frac{\delta^{(n)}(x)}{x^m} = \frac{(-1)^m n!}{(n+m)!} \, \delta^{(n+m)}(x) + \delta_{(m-1)}(x). \tag{1.50}$$

Particularly for $m = 1, n = 0$, we have for the hyperfunction $\delta(x)/x$,

$$\frac{\delta(x)}{x} := -\delta'(x) + c\,\delta(x), \tag{1.51}$$

where c is an arbitrary constant.

Remark. By doing so we have defined $\delta(x)/x$ as being the hyperfunction $f(x)$ satisfying $x\,f(x) = \delta(x)$. This notation must be considered as a whole entity and not as a product of $1/x$ with $\delta(x)$ that cannot be defined! Different approaches to generalized functions may differ in treating such subtleties. For example, the sequential approach to distributions (see [3, Antosik–Mikusinski–Sikorkski p. 249]) allows us to define the product of the distribution $1/x$ with $\delta(x)$. The authors obtain $(1/x)\cdot\delta(x) = -\delta'(x)/2$. The price they have to pay is that a non-associativity $x \cdot ((1/x)\cdot\delta(x)) \neq (x\cdot 1/x)\cdot\delta(x) = \delta(x)$ has then to be accepted. In Chapter 2 we shall define products of hyperfunctions under certain circumstances, but that hypotheses needed to form the product of $1/x$ and $\delta(x)$ will not be met.

Problem 1.15. Show that the solution of the equation $(x - a)\,f(x) = \delta(x)$, $a \neq 0$ is given by $f(x) = -\delta(x)/a + c\,\delta(x - a)$, where c is an arbitrary constant.

(iii) $\phi(x)$ **has a finite number of different real zeros**. We put $\phi(x)$ into the form

$$\phi(x) = \psi(x) \prod_{k=1}^{n} (x - a_k)^{m_k},\tag{1.52}$$

where the real analytic function $\psi(x)$ has no real zeros. Because $1/\psi(x)$ is real analytic, $h_1(x) := h(x)/\psi(x)$ is a hyperfunction again, and we consider the problem

$$\prod_{k=1}^{n} (x - a_k)^{m_k}\, f(x) = h_1(x).\tag{1.53}$$

As a particular solution of (1.53) we may take the hyperfunction

$$f_1(x) := \left[\frac{H(z)}{\psi(z)\,\prod_{k=1}^{n}(z - a_k)^{m_k}} \right].\tag{1.54}$$

By putting

$$f(x) = f_0(x) + f_1(x)$$

the hyperfunction $f_0(x) = [F_0(z)]$ satisfies the homogeneous equation

$$\prod_{k=1}^{n} (x - a_k)^{m_k}\, f(x) = 0.\tag{1.55}$$

This equation amounts to

$$\prod_{k=1}^{n} (z - a_k)^{m_k}\, F_0(z) = \Phi(z),\tag{1.56}$$

where $\Phi(x) \in \mathcal{A}(\mathbb{R})$ is an arbitrary real analytic function, thus,

$$F_0(z) = \frac{1}{\prod_{k=1}^{n}(z - a_k)^{m_k}}\, \Phi(z).$$

Partial fraction decomposition of the first factor yields

$$\frac{1}{\prod_{k=1}^{n}(z - a_k)^{m_k}} = \sum_{k=1}^{n} \sum_{\ell=1}^{m_k} \frac{c_{k\ell}}{(z - a_k)^{\ell}},$$

and the series expansion at $x = a_k$ of $\Phi(z)$ gives

$$\Phi(z) = \sum_{\mu=0}^{\infty} C_{k\mu} (z - a_k)^{\mu},$$

with arbitrary coefficients $C_{k\mu}$ due to the arbitrariness of $\Phi(z)$. By forming the product of the finite sum and the series and rearranging the terms, we obtain that $F_0(z)$ can be written in the form

$$F_0(z) = \sum_{k=1}^{n} \sum_{\nu=1}^{m_k} \frac{\gamma_{k\nu}}{(z - a_k)^{\nu}} + \Psi(z),$$

where the coefficients $\gamma_{k\nu}$ are arbitrary constants, and $\Psi(x)$ is a real analytic function that will be discarded in the sequel. Remembering the definition of the Dirac impulse and its derivatives, we obtain

$$f_0(x) = \sum_{k=1}^{n} \sum_{\nu=0}^{m_k-1} \alpha_{k\nu} \, \delta^{(\nu)}(x - a_k) = \sum_{k=1}^{n} \delta_{(m_k-1)}(x - a_k). \qquad (1.57)$$

We have obtained

Proposition 1.9. *If the hyperfunction $f_1(x)$ in (1.54) is any particular solution of the equation*

$$\psi(x) \prod_{k=1}^{n} (x - a_k)^{m_k} f(x) = h(x), \qquad (1.58)$$

where $\psi(x)$ is assumed to have no real zero, then the hyperfunction

$$f(x) = f_1(x) + \sum_{k=1}^{n} \delta_{(m_k-1)}(x - a_k)$$

is also a solution.

Therefore, every zero a_k of order m_k of the left-hand side generates a generalized delta-hyperfunction of order $m_k - 1$ at a_k with arbitrary coefficients.

Example 1.25. Solve the equation $(x^2 - a^2) f(x) = 1$, where $a \neq 0$. Because $1 = [\mathbf{1}(z)] = [1/2, -1/2]$ a particular solution is given by

$$f_1(x) = \left[\frac{1}{z^2 - a^2} \mathbf{1}(z) \right] = \frac{1}{2a} \left\{ \left[\frac{1}{z - a} \mathbf{1}(z) \right] - \left[\frac{1}{z + a} \mathbf{1}(z) \right] \right\}$$

$$= \frac{1}{2a} \left\{ \mathrm{fp} \frac{1}{x - a} - \mathrm{fp} \frac{1}{x + a} \right\}.$$

(For finite part hyperfunctions see the next section.) By the above proposition we have $f_0(x) = \alpha_1 \, \delta(x + a) + \alpha_2 \, \delta(x - a)$, where α_1, α_2 are arbitrary constants.

(iv) $\phi(x)$ **has infinitely many zeros.** We illustrate this case by an example. Consider the problem of finding the hyperfunction solutions of the equation

$$\sin(\pi x) f(x) = 0. \qquad (1.59)$$

We need the infinite partial fraction representation

$$\frac{\pi}{\sin(\pi z)} = \frac{1}{z} + \sum_{k=-\infty, k\neq 0}^{\infty} (-1)^k \left(\frac{1}{z-k} + \frac{1}{k} \right) = z \sum_{k=-\infty}^{\infty} \frac{(-1)^k}{z^2 - k^2}$$

which follows from Mittag-Leffler's Theorem (see, for example, [4, Behnke-Sommer p. 246]). Notice that the term $1/k$ in the first of the above infinite series is needed to produce convergence; the series

$$\sum_{k=-\infty, k\neq 0}^{\infty} (-1)^k \frac{1}{z-k}$$

alone would diverge. Let $f(x) = [F(z)]$, then equation (1.59) means

$$\sin(\pi z) F(z) = \psi(z)$$

since $\sin(\pi x)$ is real analytic. $\psi(x)$ is an arbitrary real analytic function. We obtain

$$F(z) = \frac{\psi(z)}{\pi} \frac{\pi}{\sin(\pi z)}$$

$$= \frac{\psi(z)}{\pi} \left\{ \frac{1}{z} + \sum_{k=-\infty, k\neq 0}^{\infty} (-1)^k \left(\frac{1}{z-k} + \frac{1}{k} \right) \right\},$$

and note that

$$\left[\frac{1}{z-k} + \frac{1}{k} \right] = \left[\frac{1}{z-k} \right] = -2\pi i \left[\frac{-1}{2\pi i (z-k)} \right] = -2\pi i\, \delta(x-k)$$

because the constant $1/k$ is real analytic and can be discarded. Then, we get

$$f(x) = -2\pi i \frac{\psi(x)}{\pi} \left\{ \delta(x) + \sum_{k=-\infty, k\neq 0}^{\infty} (-1)^k \delta(x-k) \right\}$$

$$= -2i \sum_{k=-\infty}^{\infty} (-1)^k \psi(x)\, \delta(x-k)$$

$$= -2i \sum_{k=-\infty}^{\infty} (-1)^k \psi(k)\, \delta(x-k).$$

The real analytic function $\psi(x)$, however, is arbitrary, and so are the coefficients $-2i(-1)^k \psi(k) =: c_k$. Thus, the solution of equation (1.59) is given by

$$f(x) = \sum_{k=-\infty}^{\infty} c_k\, \delta(x-k), \tag{1.60}$$

where the c_k's form an arbitrary sequence of complex numbers. Series of hyperfunctions as in (1.60) will be discussed in Section 2.1.

1.4 Finite Part Hyperfunctions

We shall now define the so-called finite part hyperfunctions. Let $p(x)$ be a polynomial having some real zeros. We are again concerned with the question of looking for a hyperfunction which may interpret the ordinary function $x \mapsto 1/p(x)$. One special choice will then be called the finite part of $1/p(x)$. We single out the real zeros of $p(x)$ by writing

$$p(x) = \prod_{k=1}^{n} (x - a_k)^{m_k} \, \psi(x)$$

where the a_k's denote all the real zeros and the polynomial $\psi(x)$ no longer has any real zeros. We consider the equation

$$p(x) \, f(x) = \psi(x) \prod_{k=1}^{n} (x - a_k)^{m_k} \, f(x) = 1. \tag{1.61}$$

Proposition 1.9 tells us that the solution of the above equation is given by the family

$$f(x) = \left[\frac{\mathbf{1}(z)}{\psi(z) \prod_{k=1}^{n} (z - a_k)^{m_k}} \right] + \sum_{k=1}^{n} \delta_{(m_k-1)}(x - a_k)$$

containing the generalized delta-hyperfunctions $\delta_{(m_k-1)}(x - a_k)$. Any member of this family has the right to be considered as a hyperfunction interpretation of the ordinary function $x \to 1/p(x)$. However, in many situations it would be convenient to have a unique one at hand. Therefore, we single out the following member.

Definition 1.11. Let $p(z)$ be a polynomial with the real zeros a_k of order m_k. The special solution of the equation (1.61) containing no generalized delta-hyperfunctions is called the *finite part of* $1/p(x)$ and is denoted by $\mathrm{fp}(1/p(x))$. Thus,

$$\mathrm{fp}\frac{1}{p(x)} := \left[\frac{1}{p(z)} \mathbf{1}(z) \right] = \left[\frac{1}{2\,p(z)}, -\frac{1}{2\,p(z)} \right]. \tag{1.62}$$

We may generalize this concept to any rational function with at least one real pole and set

Definition 1.12. Let $f(x)$ be a rational function with at least one real pole, then the finite part of $f(x)$ is defined by

$$\mathrm{fp}f(x) := [f(z)\,\mathbf{1}(z)] = \left[\frac{1}{2}f(z), -\frac{1}{2}f(z) \right]. \tag{1.63}$$

Note that the concept of the finite part has involved still another choice: the use of $\mathbf{1}(z)$ rather than $\mathbf{1}_+(z)$ or $\mathbf{1}_-(z)$ in the above discussion.

A rational function $f(x)$ with no poles on the real axis is real analytic. In this case (1.63) yields, by dropping the fp,

$$f(x) := \left[\frac{f(z)}{2}, -\frac{f(z)}{2} \right]$$

which is the hyperfunction that interprets the ordinary function $x \mapsto f(x)$, $x \in \mathbb{R}$, as for example,

$$\frac{1}{1+x^2} = \left[\frac{1}{2(1+z^2)}, -\frac{1}{2(1+z^2)} \right].$$

Simple candidates of rational functions giving familiar finite part hyperfunctions are:

$$\frac{1}{z}, \frac{1}{z-a}, \frac{1}{z^m}, \frac{1}{(z-a)^m},$$

where m is a positive integer and a is a real number.

We shall see later on that the finite part hyperfunction $\mathrm{fp}(1/x^m)$ will not always be the most convenient choice in the family of hyperfunctions that interpret the ordinary function $x \mapsto 1/x^m$. For this reason we shall still keep other interpretations at hand too, and define the following notation.

Definition 1.13. We denote any interpretation of the ordinary function $x \mapsto 1/(x-a)^m$ with $m \in \mathbb{N}$ by $(x-a)^{-m}$.

Thus, we can write

$$(x-a)^{-m} = \mathrm{fp}\frac{1}{(x-a)^m} + \delta_{(m-1)}(x-a), \qquad (1.64)$$

where the generalized delta-hyperfunction contains m arbitrary constants.

Example 1.26.

$$\mathrm{fp}\frac{1}{x} := \left[\frac{1}{2z}, -\frac{1}{2z} \right].$$

In order to get an intuitive idea of this hyperfunction, we consider the approaching family of ordinary functions

$$f_n(x) := F_+(x + i/n) - F_-(x - i/n),$$

i.e.,

$$f_n(x) := \frac{1}{2} \{ \frac{1}{x+i/n} + \frac{1}{x-i/n} \} = \frac{x}{x^2 + (1/n)^2}.$$

Figure 1.7 shows three members of the sequence together with the graph of $1/x$. The $f_n(x)$ approach the ordinary function $1/x$ with increasing n coming closer and closer, and each $f_n(x)$ is a smooth function. Note that a hyperfunction $1/x$ does not exist in a straightforward manner, but the hyperfunction $\mathrm{fp}(1/x)$ comes very close in the sense that

$$\lim_{n \to \infty} f_n(x) = \lim_{n \to \infty} \frac{x}{x^2 + (1/n)^2} = \begin{cases} 1/x & \text{if } x \neq 0 \\ \infty & \text{if } x = 0. \end{cases}$$

The convergence, however, is not uniform. The hyperfunction $\mathrm{fp}(1/x)$ is just one possible interpretation of the ordinary function $x \mapsto 1/x$.

Example 1.27.

$$\mathrm{fp}\frac{1}{(x-a)^2} := \left[\frac{1}{2(z-a)^2}, -\frac{1}{2(z-a)^2} \right].$$

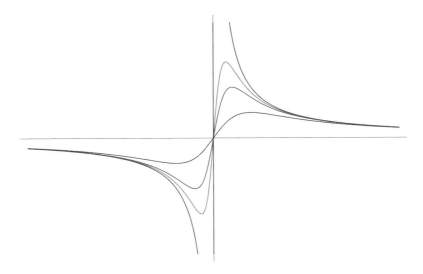

Figure 1.7: Intuitive picture of fp(1/x)

To get an intuitive picture of this hyperfunction, we again consider the family

$$f_n(x) := F_+(x + i/n) - F_-(x - i/n),$$

i.e.,

$$f_n(x) := \frac{1}{2}\{\frac{1}{(x - a + i/n)^2} + \frac{1}{(x - a - i/n)^2}\} = \frac{(x - a)^2 - 1/n^2}{((x - a)^2 + (1/n)^2)^2}.$$

Figure 1.8 shows three members of the family together with the graph of $1/(x-a)^2$.

Concerning differentiation, finite part hyperfunctions behave quite normally, i.e., we have

Proposition 1.10.
$$D\,\mathrm{fp}f(x) = \mathrm{fp}f'(x). \tag{1.65}$$

Proof.

$$D\,\mathrm{fp}f(x) = \left[\frac{d}{dz}\frac{f(z)}{2}, \frac{d}{dz}(-\frac{f(z)}{2})\right] = \left[\frac{1}{2}\frac{df}{dz}, -\frac{1}{2}\frac{df}{dz}\right] = \mathrm{fp}f'(x) \qquad \square$$

For example,

$$D\,\mathrm{fp}\frac{1}{x^m} = -m\,\mathrm{fp}\frac{1}{x^{m+1}}, \quad D^n\,\mathrm{fp}\frac{1}{x} = (-1)^n\,n!\,\mathrm{fp}\frac{1}{x^{n+1}}.$$

Example 1.28. The *Heisenberg hyperfunctions* are defined and denoted by

$$\delta_+(x) = -\frac{1}{2\pi i}\frac{1}{x + i0} := \left[-\frac{1}{2\pi i}\frac{1}{z}, 0\right], \tag{1.66}$$

$$\delta_-(x) = \frac{1}{2\pi i}\frac{1}{x - i0} := \left[0, -\frac{1}{2\pi i}\frac{1}{z}\right]. \tag{1.67}$$

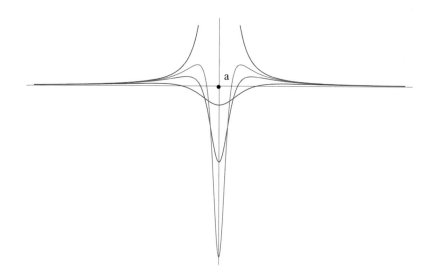

Figure 1.8: Intuitive picture of $\mathrm{fp}(1/(x-a)^2$

If follows immediately that

$$\delta_+(x) + \delta_-(x) = \delta(x), \quad \delta_+(x) - \delta_-(x) = \frac{i}{\pi}\,\mathrm{fp}\frac{1}{x}, \tag{1.68}$$

which implies

$$\delta_+(x) = \frac{1}{2}\,\delta(x) - \frac{1}{2\pi i}\,\mathrm{fp}\frac{1}{x}, \tag{1.69}$$

$$\delta_-(x) = \frac{1}{2}\,\delta(x) + \frac{1}{2\pi i}\,\mathrm{fp}\frac{1}{x}. \tag{1.70}$$

More generally, we introduce

$$\frac{1}{(x+i0)^n} := \left[\frac{1}{z^n}, 0\right],$$

$$\frac{1}{(x-i0)^n} := \left[0, -\frac{1}{z^n}\right].$$

By using

$$\delta^{(n-1)}(x) = \left[-\frac{(-1)^{n-1}(n-1)!}{2\pi i z^n}\right] = -\frac{(-1)^{n-1}(n-1)!}{2\pi i}\left[\frac{1}{z^n}, \frac{1}{z^n}\right]$$

we obtain

$$\left[-\frac{1}{2z^n}, -\frac{1}{2z^n}\right] = \frac{(-1)^{n-1}\pi i}{(n-1)!}\delta^{(n-1)}(x),$$

and finally,

$$\frac{1}{(x+i0)^n} := -\frac{(-1)^{n-1}\pi i}{(n-1)!}\delta^{(n-1)}(x) + \mathrm{fp}\frac{1}{x^n}, \tag{1.71}$$

$$\frac{1}{(x-i0)^n} := \frac{(-1)^{n-1}\pi i}{(n-1)!}\delta^{(n-1)}(x) + \mathrm{fp}\frac{1}{x^n}. \tag{1.72}$$

Remark. In Laurent Schwartz's theory, a distribution is defined as a continuous linear functional on a space of test functions. A finite part distribution is a functional defined by Hadamard's finite part of a divergent integral. Our concept of a finite part hyperfunction is similar in the sense that the same ordinary functions are interpreted, however the theoretical background is different. What we call an interpretation by a hyperfunction of a given ordinary function such as $x \mapsto 1/x$ is treated in Schwartz's theory under the notion of regularisation (see the books of [21, Kanwal] and [10, Kanwal - Estrada]). The reader should be aware that the natural choice involved for a finite part hyperfunction and that of a finite part distribution (pseudo function) may produce a different interpretation (regularisation) of the given ordinary function.

1.5 Integrals

1.5.1 Integrals with respect to the Independent Variable

In Laurent Schwartz's theory of distribution the notation of an integral over a generalized function, as for example

$$\int_{-\infty}^{\infty} \phi(x)\delta(x-a)dx,$$

is generally avoided in favor of the notation $\langle \delta(x-a), \phi(x) \rangle$ which sets forth that a distribution is a functional on a space of test functions. Nevertheless physicists and engineers still continue to use the integral notation that does not quite conform with the framework of Schwartz's theory. In contrast, Sato's theory of hyperfunctions defines the integral over a hyperfunction from the outset.

Later on in Section 2.4.1, we shall thoroughly introduce holomorphic (or analytic) and microanalytic hyperfunctions. A hyperfunction $f(x)$ is called holomorphic at $x = a$, if the lower and upper component of the defining function can analytically be continued to a full (two-dimensional) neighborhood of the real point a, i.e., the upper / lower component can analytically be continued across a into the lower / upper half-plane.

Let the hyperfunctions $f(x)$ be defined on an open interval I containing the compact interval $[a, b]$. Let $\mathcal{D}_+ = \mathcal{D}_+(I)$ and $\mathcal{D}_- = \mathcal{D}_-(I)$ be the upper and lower half-neighborhoods of I as defined in Section 1.2, respectively.

Definition 1.14. Let $f(x) = [F_+(z), F_-(z)]$ be a hyperfunction, holomorphic at both endpoints of the finite interval $[a, b]$, then the *(definite) integral of $f(x)$ over $[a, b]$* is defined and denoted by

$$\int_a^b f(x)dx := \int_{\gamma_{a,b}^+} F_+(z)dz - \int_{\gamma_{a,b}^-} F_-(z)dz := -\oint_{(a,b)} F(z)\,dz, \qquad (1.73)$$

where the contour $\gamma_{a,b}^+$ runs in $\mathcal{D}_+$ from a to b above the real axis, and the contour $\gamma_{a,b}^+$ is in $\mathcal{D}_-$ from a to b below the real axis (see Figure 1.9).

For the above o-integral notation remember that

$$F(z) := \begin{cases} F_+(z), & z \in \mathcal{D}_+ \\ F_-(z), & z \in \mathcal{D}_-. \end{cases}$$

We shall use the indicated notation of integration contours throughout this book: $\gamma_{a,b}^{+}$ will always denote a path running from a to b above the real axis, and similarly, $\gamma_{a,b}^{-}$ a path from a to b running below the real axis. Also the notation (a, b) under the o-integral sign will always denote the positively directed contour $\gamma_{a,b}^{-} - \gamma_{a,b}^{+}$.

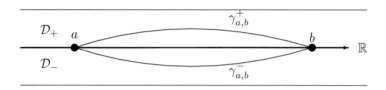

Figure 1.9: Contours $\gamma_{a,b}^{+}$ and $\gamma_{a,b}^{-}$

Cauchy's integral theorem ensures that the line integrals along $\gamma_{a,b}^{+}$ and $\gamma_{a,b}^{-}$ are independent of the chosen integration paths as long as they remain in $\mathcal{D}_{+}$ and $\mathcal{D}_{-}$, respectively. In the general case, where the upper and the lower components of the defining function are independent functions, you have to be careful when using the notation

$$\oint_{(a,b)} F(z)\, dz$$

because the integrand $F(z)$ is not holomorphic inside the closed contour. However, when the defining function $F(z)$ is one meromorphic global analytic function in $\mathcal{D}_{+} \cup \mathcal{D}_{-}$, where the real poles are different from the integration limits, and possibly other non-real poles are outside of the contour, the integral of $f(x) = [F(z)]$ becomes the line integral familiar from complex analysis (which can be evaluated by residues)

$$\int_{a}^{b} f(x)dx = -\oint_{\gamma} F(z)dz = -\oint_{(a,b)} F(z)dz, \tag{1.74}$$

where the closed contour $\gamma = (a, b)$ is directed in the positive sense. See Figure.1.10 where the poles are indicated by small circles.

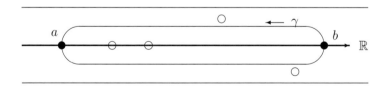

Figure 1.10: Closed contour $\gamma = (a, b)$

Example 1.29. Let us compute the integral of the hyperfunction $\phi(x)\delta(x - a)$ over the interval $[c, d]$ with $c < a < d$. See Figure 1.11. Here $\phi(x)$ is a real analytic function. Using

$$F(z) = -\frac{1}{2\pi i} \frac{\phi(z)}{z - a},$$

we obtain

$$\int_c^d \phi(x)\delta(x-a)dx = \frac{1}{2\pi i} \oint_\gamma \frac{\phi(z)}{z-a} dz = \phi(a) \tag{1.75}$$

by Cauchy's integral formula. The loop γ encircles $[c, d]$ in the positive sense. If the real pole a lies outside the integration interval $[c, d]$ the integrand becomes holomorphic inside γ, and by Cauchy's theorem, its value is zero. Similarly, for the integral of the hyperfunction $\phi(x)\delta^{(n)}(x - a)$ over the interval $[c, d]$. With $c < a < d$, we obtain, by means of

$$F(z) = -\frac{(-1)^n \phi(z) n!}{2\pi i (z - a)^{n+1}},$$

and by Cauchy's integral formula,

$$\int_c^d \phi(x)\delta^{(n)}(x-a)dx := (-1)^n \frac{n!}{2\pi i} \oint_\gamma \frac{\phi(z)}{(z-a)^{n+1}} dz = (-1)^n \phi^{(n)}(a), \tag{1.76}$$

and zero, if $a \notin [c, d]$.

Particularly, we have with $\phi(x) = 1$,

$$\int_{-\infty}^\infty \delta(x-a)\, dx = 1, \tag{1.77}$$

$$\int_{-\infty}^\infty \delta^{(n)}(x-a)\, dx = 0, \quad n \in \mathbb{N}. \tag{1.78}$$

Also, by using (1.34), we find

$$\int_{-\infty}^\infty \frac{x^m}{m!} (-1)^n \delta^{(n)}(x)dx = \delta_n^m, \tag{1.79}$$

where δ_n^m denotes the Kronecker symbol.

The set of all real analytic functions $\mathcal{A}(\mathbb{R})$ and the set of hyperfunctions $\mathcal{B}(\mathbb{R})$ are linear spaces. By defining the scalar product

$$< \cdot, \cdot >: \mathcal{A}(\mathbb{R}) \times \mathcal{B}(\mathbb{R}) \longrightarrow \mathbb{C}$$

$$(\phi(x), f(x)) \longmapsto < \phi(x), f(x) >:= \int_{-\infty}^\infty \phi(x)\, f(x)\, dx,$$

(1.79) shows that the two sequences

$$\frac{x^m}{m!} \quad \text{and} \quad (-1)^n \delta^{(n)}(x) \tag{1.80}$$

form a *biorthogonal system*.

$\delta(x-a)$ and its derivatives are examples of hyperfunctions with meromorphic defining functions. These are examples of meromorphic hyperfunctions (hyperfunctions where the upper component and the lower component of the defining function are both meromorphic).

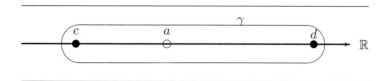

Figure 1.11: Contour for Example 1.29

Proposition 1.11. *Let $f(x) = [F_+(z), F_-(z)]$ be a hyperfunction, holomorphic at the finite real points a and b. Then,*

$$\int_a^b f'(x)dx = f(b) - f(a).$$

Proof. The upper and lower components $F_+(z)$ and $F_-(z)$ are holomorphic at a and b and hence continuous there. Their derivatives are also holomorphic at a and b. We have

$$f'(x) := \left[\frac{dF_+}{dz}(z), \frac{dF_-}{dz}(z)\right]$$

therefore,

$$
\begin{aligned}
\int_a^b f'(x)dx &= \int_{\gamma_{a,b}^+} \frac{dF_+}{dz}dz - \int_{\gamma_{a,b}^-} \frac{dF_-}{dz}dz \\
&= F_+(b) - F_+(a) - (F_-(b) - F_-(a)) \\
&= (F_+(b+i0) - F_-(b-i0)) - (F_+(a+i0) - F_-(a-i0)) \\
&= f(b) - f(a).
\end{aligned}
$$
$\qquad\square$

Example 1.30. Because $u'(x) = \delta(x)$ we have, for $a \neq 0, b \neq 0, a < b$,

$$\int_a^b \delta(x)dx = u(b) - u(a).$$

Sometimes we need integrals over $[a, b]$ where the upper or lower component of the defining function of $f(x)$ fails to be holomorphic at one or both endpoints of the interval. It may then occur that the primitives of $F_+(z)$ in $\mathcal{D}_+$ and of $F_-(z)$ in $\mathcal{D}_-$ are still continuous in a and b. In such a case the integrals

$$\int_{\gamma_{a,b}^+} F_+(z)dz \quad \text{and} \quad -\int_{\gamma_{a,b}^-} F_-(z)dz$$

still may exist, possibly as improper integrals.

Example 1.31. In the next chapter we shall introduce the hyperfunction

$$\frac{u(x)}{\sqrt{x}} = x_+^{-1/2} := \left[\frac{1}{2i}\frac{1}{\sqrt{-z}}\right].$$

Let us consider the integral

$$\int_0^1 \frac{1}{\sqrt{x}}\,dx := \int_{\gamma_{0,1}^+} \frac{1}{2\,i}\frac{1}{\sqrt{-z}}\,dz - \int_{\gamma_{0,1}^-} \frac{1}{2\,i}\frac{1}{\sqrt{-z}}\,dz.$$

The defining function is not holomorphic at $z = 0$, but it is integrable with the primitive $F(z) = -\sqrt{-z}/i = -|z|^{1/2}\exp(i\arg(-z)/2)/i$ which is continuous at $z = 0$. $F(z)$ is defined on the complex plane with a cut along the positive real axis. Evaluation at $z = 1$ and $z = 0$ gives, by passing along the contour $\gamma_{0,1}^+$, the result $-\exp(-i\pi/2)/i - 0 = 1$, and, by passing along $\gamma_{0,1}^-$, we obtain $-\exp(i\pi/2)/i - 0 = -1$. Thus,

$$\int_0^1 \frac{1}{\sqrt{x}}\,dx = 2.$$

Example 1.32. Integrals containing Dirac impulses can also be evaluated using the properties of the Dirac impulse and its derivatives as established in Section 1.3.1 (see also [21, Kanwal p. 49-52]). Evaluate, for example,

$$F(s) = \int_{-\infty}^{\infty} e^{-sx}\delta'(x^3 + x^2 + x)\,dx.$$

The function $\phi(x) = x(x^2 + x + 1)$ has only one simple real zero $x = 0$ with $\phi'(0) = 1$ and $\phi''(0) = 2$. By using (1.27) we obtain

$$\delta'(x^3 + x^2 + x) = \delta'(x) + 2\,\delta(x).$$

Thus,

$$F(s) = \int_{-\infty}^{\infty} e^{-sx}\{\delta'(x) + 2\delta(x)\}\,dx = s + 2.$$

Hence, $s + 2$ is the two-sided Laplace transform of $\delta'(x^3 + x^2 + x)$.

Example 1.33. Another example with an infinite integration interval is

$$F(s) = \int_{-a}^{\infty} u(x)\,e^{-sx}\,x^n dx \quad (a > 0,\ s > 0,\ n \in \mathbb{N}).$$

Of course it equals the one-sided Laplace transform of x^n which is $n!/s^{n+1}$, but the goal here is to illustrate the definition of the integral of a hyperfunction with an infinite integration interval. The hyperfunction integrand is

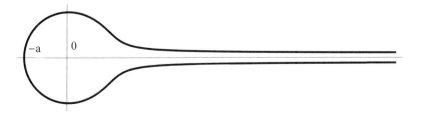

Figure 1.12: Integration loop $(-a, \infty)$

$$u(x)\, x^n\, e^{-sx} := \left[-\frac{1}{2\pi i} z^n\, e^{-sz}\, \log(-z)\right],$$

since $x^n\, e^{-sx}$ is a real analytic function. The defining function $F(z)$ is holomorphic in the complex plane with a cut along the positive real axis. Applying the definition of the integral of a hyperfunction yields

$$-\frac{1}{2\pi i} \int_{\gamma^+_{-a,\infty}} z^n\, e^{-sz}\, \log(-z)\, dz + \frac{1}{2\pi i} \int_{\gamma^-_{-a,\infty}} z^n\, e^{-sz}\, \log(-z)\, dz$$

with $a > 0$. The two integrals on the right-hand side can be collected into one loop integral:

$$\frac{1}{2\pi i} \oint_{\gamma_{(-a,\infty)}} z^n\, e^{-sz}\, \log(-z)\, dz$$

where the integration loop $(-a, \infty)$ starts at $\infty + i0$, runs above the positive real axis towards $-a$, and then, passing through $-a$, runs back below the positive real axis to $\infty - i0$ (Figure 1.12). The classical notation for such a loop integral which we shall also frequently use is

$$\oint_{(-a,\infty)} =: \int_{\infty}^{(0+)}.$$

By the substitution $t = -sz$, $s > 0$, the above integral becomes

$$\frac{(-1)^n}{2\pi i} \int_{-\infty}^{(0+)} (t/s)^n\, e^t\, \log(t/s)\, (-dt/s)$$

where the new integration path now starts at $-\infty - i0$, runs below the negative

Figure 1.13: Integration loop $(-\infty, as)$

real axis towards 0, rounds 0 passing through as, and runs back above the negative real axis to $-\infty + i0$. The integration contour is shown in Figure 1.13. The classical notation for this loop integral is

$$\oint_{(-\infty,a)} := \int_{-\infty}^{(0+)}.$$

Therefore, we get

$$F(s) = \int_{-a}^{\infty} u(x)\, e^{-sx}\, x^n dx = \frac{(-1)^{n+1}}{s^{n+1}} \frac{1}{2\pi i} \int_{-\infty}^{(0+)} t^n\, e^t\, [\log(t) - \log(s)]\, dt$$

$$= \frac{(-1)^{n+1}}{s^{n+1}} \left\{ \frac{1}{2\pi i} \int_{-\infty}^{(0+)} t^n\, e^t\, \log(t)\, dt - \log(s) \frac{1}{2\pi i} \int_{-\infty}^{(0+)} t^n\, e^t\, dt \right\}.$$

We now use the Hankel representation of $1/\Gamma(z)$:

$$\frac{1}{\Gamma(z)} = \frac{1}{2\pi i} \int_{-\infty}^{(0+)} t^{-z} e^t \, dt,$$

valid for all complex z. Since $\Gamma(z)$ has a pole at $z = -n$ the second integral of the right-hand side of the above expression for $F(s)$ is zero. For the first integral we observe

$$\frac{1}{2\pi i} \int_{-\infty}^{(0+)} t^{-z} e^t \log(t) \, dt = -\frac{d}{dz} \frac{1}{2\pi i} \int_{-\infty}^{(0+)} t^{-z} e^t \, dt = -\frac{d}{dz} \frac{1}{\Gamma(z)}.$$

By using the formula

$$\frac{1}{\Gamma(z)} = \frac{\sin(\pi z)}{\pi} \Gamma(1 - z),$$

we readily find

$$\lim_{z \to -n} \frac{d}{dz} \frac{1}{\Gamma(z)} = (-1)^n n!.$$

Finally, we obtain the result

$$F(s) = \int_{-a}^{\infty} u(x) e^{-sx} x^n \, dx = \frac{(-1)^{n+1}}{s^{n+1}} \frac{1}{2\pi i} \int_{-\infty}^{(0+)} t^{-z} e^t \log(t) \, dt$$

$$= \frac{(-1)^{2n+2} n!}{s^{n+1}} = \frac{n!}{s^{n+1}}.$$

We have done this example in some detail not only to illustrate the definition of the integral of a hyperfunction but also to establish

$$F(s) = \int_{-a}^{\infty} u(x) e^{-sx} x^n \, dx = -\int_{\infty}^{(0+)} z^n e^{-sz} \frac{-1}{2\pi i} \log(-z) \, dz. \tag{1.81}$$

In Chapter 3 the Hankel-type integral on the right-hand side will serve to generalize the Laplace transformation to the case where n need no longer be a non-negative integer.

1.5.2 Integrals with respect to a Parameter

We shall also consider hyperfunctions depending on a parameter and their integrals. It is then important to distinguish clearly whether the integration is with respect to the independent variable of the hyperfunction or with respect to the parameter, because the type of the result is quite different. Take the example

$$f(x; t) := \delta(x - t) = \left[-\frac{1}{2\pi i} \frac{1}{z - t} \right],$$

where x is considered to be the variable of the hyperfunction and t a parameter. Let us first integrate with respect to the variable x. By the definition of the integral over a hyperfunction, we obtain

$$\int_a^b \delta(x - t) \, dx = \begin{cases} 1, & a < t < b \\ 0, & t \notin (a, b) \end{cases} = Y(t - a) - Y(t - b).$$

The result must be considered as an ordinary function $t \mapsto Y(t - a) - Y(t - b)$, because when t is fixed a numerical value is assigned to the integral. *If a hyperfunction depending on a parameter is integrated with respect to the independent variable of the hyperfunction, the result is an ordinary function of the parameter.*

Let us now integrate with respect to the parameter. The definition of an integration with respect to a parameter is simply the following

Definition 1.15. Let the hyperfunction $f(x; t) = [F(z; t)]$ depend on the parameter t, then the integral with respect to the parameter is defined by

$$\int_a^b f(x; t)\, dt := \left[G(z) := \int_a^b F(z; t)\, dt \right], \tag{1.82}$$

and the result is a hyperfunction $g(x)$ again.

Thus,

$$\int_a^b \delta(x - t)\, dt = \left[-\frac{1}{2\pi i} \int_a^b \frac{dt}{z - t} \right]$$

$$= \left[\frac{1}{2\pi i} \log(b - z) - \frac{1}{2\pi i} \log(a - z) \right]$$

$$= u(x - a) - u(x - b)$$

and the result is the hyperfunction $u(x - a) - u(x - b)$. For another example of an integral over a hyperfunction with respect to a parameter, see Problem 1.19.

1.6 More Familiar Hyperfunctions

Before going on to more theoretical investigations we shall introduce some more familiar hyperfunctions.

1.6.1 Unit-Step, Delta Impulses, Sign, Characteristic Hyperfunctions

We have already encountered these hyperfunctions before. For the sake of convenience we summarize them here.

The unit-step hyperfunction with step at $x = a$ is defined by

$$u(x - a) := \left[-\frac{1}{2\pi i} \log(-z + a) \right],$$

while the reverse unit-step with step at $x = a$ is

$$u(a - x) := \left[\frac{1}{2\pi i} \log(z - a) \right].$$

The unit-step hyperfunctions $u(x - a)$ and $u(a - x)$ are interpretations of the (ordinary) Heaviside-functions $Y(x - a)$ and $Y(a - x)$, respectively.

The Dirac impulse hyperfunction at $x = a$ and its derivatives are

$$\delta^{(n)}(x - a) = u^{(n+1)}(x) = \left[-\frac{(-1)^n n!}{2\pi i (z - a)^{n+1}} \right].$$

The sign-hyperfunction is defined by

$$\text{sgn}(x) := -u(-x) + u(x) = -\frac{1}{2\pi i} \left[\log(z) + \log(-z) \right]$$

$$= -\frac{1}{\pi i} \left[\log(-iz), \log(iz) \right],$$

and interprets the ordinary sign-function $-Y(-x) + Y(x)$. A generalization of the sign-hyperfunction having the values -1 for $x < a$, 0 on the interval (a, b), and +1 for $x > b$ is denoted by $\text{sgn}(x; a, b)$. It is defined by

$$\text{sgn}(x; a, b) := -\frac{1}{2\pi i} \left[\log((b - z)(z - a)) \right], \quad a < b. \tag{1.83}$$

A hyperfunction that interprets the constant ordinary function $x \mapsto 1$ may be defined by

$$u(-x) + u(x) = \frac{1}{2\pi i} \left[\log(z) - \log(-z) \right],$$

or more directly by

$$1 = [\mathbf{1}(z)] = \left[\frac{1}{2}, -\frac{1}{2} \right] = [\mathbf{1}_+(z)] = [1, 0] = [\mathbf{1}_-(z)] = [0, -1].$$

The *characteristic hyperfunction on the interval* (a, b) is define and denoted by

$$\chi_{a,b}(x) := u(x - a) - u(x - b) = \left[-\frac{1}{2\pi i} \log \frac{a - z}{b - z} \right]. \tag{1.84}$$

This hyperfunction has the value 1 for $x \in (a, b)$ and 0 for $x \notin [a, b]$. As for the hyperfunctions $\text{sgn}(x), \text{sgn}(x; a, b)$ we use the same notation for the corresponding ordinary functions. Also observe that $\text{sgn}(x) = 2u(x) - 1$.

1.6.2 Integral Powers

For $m \in \mathbb{N}$, we have already introduced the finite part integral power hyperfunction

$$\text{fp} \frac{1}{x^m} := \left[z^{-m} \mathbf{1}(z) \right] = \left[\frac{1}{2 z^m}, -\frac{1}{2 z^m} \right],$$

and the more general one (in fact a whole family)

$$x^{-m} = \text{fp} \frac{1}{x^m} + \delta_{(m-1)}(x).$$

A right-sided integral power hyperfunction $f(x) = [F(z)]$ should interpret the ordinary function $x \to Y(x)x^{-m} =: g(x)$, $m \in \mathbb{N}$, i.e., the equation $x^m g(x) = Y(x)$ should be satisfied. The corresponding equation in terms of hyperfunctions is $x^m f(x) = u(x)$, i.e.,

$$z^m F(z) = -\frac{1}{2\pi i} \log(-z) + \phi(z)$$

where $\phi(x)$ is an arbitrary real analytic function. This implies

$$F(z) = -\frac{1}{2\pi i z^m} \log(-z) + \frac{\phi(z)}{z^m}$$

or,

$$f(x) = \left[-\frac{1}{2\pi i z^m} \log(-z) \right] + \delta_{(m-1)}(x)$$

where the arbitrary delta-hyperfunction contains m arbitrary constants.

   We define and denote the *right-sided finite part negative integral power hyperfunction* by

$$\text{fp}\,\frac{u(x)}{x^m} := \left[-\frac{1}{2\pi i\, z^m} \log(-z) \right], \tag{1.85}$$

and the *more general right-sided negative integral power hyperfunction* by

$$x_+^{-m} = u(x)\, x^{-m} := \text{fp}\,\frac{u(x)}{x^m} + \delta_{(m-1)}(x). \tag{1.86}$$

All these hyperfunctions have values for every $x \neq 0$ that equals the value of the ordinary function $x \mapsto Y(x)/x^m$.

Remark. Do not interpret the symbols $u(x)/x^m$, $u(x)x^{-m}$ with $m \in \mathbb{N}$ as a product of two hyperfunctions, because the product of hyperfunctions is generally not defined. The symbol must be read as a whole entity without seeing a product in it. This is the reason for the also used notation x_+^{-m} for $u(x)x^{-m}$.

   Similarly, we define the *left-sided finite part negative integral power hyperfunction* as

$$\text{fp}\,\frac{u(-x)}{x^m} := \left[\frac{1}{2\pi i\, z^m} \log(z) \right], \tag{1.87}$$

and the *more general left-sided negative integral power hyperfunction* by

$$x_-^{-m} = \text{fp}\,\frac{u(-x)}{x^m} := \left[-\frac{1}{2\pi i\, z^m} \log(-z) \right] + \delta_{(m-1)}(x). \tag{1.88}$$

Caution. Look at the following pitfall related with the notation $\text{fp}(u(x)/x^m)$. If you are tempted to consider the expression as a product of $u(x)$ and $1/x^m$ you will be tempted to write as well, for $a > 0$ and $m \in \mathbb{N}$,

$$\text{fp}\,\frac{u(ax)}{(ax)^m} = \frac{1}{a^m}\,\text{fp}\,\frac{u(x)}{x^m} \text{ or } \text{fp}\,\frac{u(ax)}{x^m} = \text{fp}\,\frac{u(x)}{x^m}$$

since $u(ax) = u(x)$. But this is erroneous! We have

$$\text{fp}\frac{u(ax)}{(ax)^m} = \left[-\frac{1}{2\pi i}\frac{\log(-az)}{(az)^m}\right]$$

$$= \frac{1}{a^m}\left[-\frac{1}{2\pi i}\frac{\log(-z)+\log a}{z^m}\right]$$

$$= \frac{1}{a^m}\left\{\text{fp}\frac{u(x)}{x^m} + \frac{(-1)^{m-1}\log a}{(m-1)!}\left[-\frac{1}{2\pi i}\frac{(-1)^{m-1}(m-1)!}{z^m}\right]\right\}$$

$$= \frac{1}{a^m}\left\{\text{fp}\frac{u(x)}{x^m} + \frac{(-1)^{m-1}\log a}{(m-1)!}\delta^{(m-1)}(x)\right\}.$$

Thus, for $a > 0$ we have

$$\text{fp}\frac{u(ax)}{(ax)^m} = \frac{1}{a^m}\text{fp}\frac{u(x)}{x^m} + \frac{(-1)^{m-1}\log a}{a^m(m-1)!}\delta^{(m-1)}(x),$$

or, because there is no harm by simplifying the factor a^{-m} on both sides,

$$\text{fp}\frac{u(ax)}{x^m} = \text{fp}\frac{u(x)}{x^m} + \frac{(-1)^{m-1}\log a}{(m-1)!}\delta^{(m-1)}(x). \tag{1.89}$$

However, the equation

$$u(ax)(ax)^{-m} = \frac{1}{a^m}u(x)x^{-m}$$

is correct in view of (1.86).

We shall show below that the hyperfunction

$$\text{fp}\frac{u(x)}{x^m} + \text{fp}\frac{u(-x)}{x^m} \tag{1.90}$$

is the same as

$$\text{fp}\frac{1}{x^m} := \left[\frac{1}{2z^m}, -\frac{1}{2z^m}\right].$$

Similarly, we define the hyperfunction

$$\text{fp}\frac{\text{sgn}(x)}{x^m} := \text{fp}\frac{u(x)}{x^m} - \text{fp}\frac{u(-x)}{x^m} \tag{1.91}$$

which interprets the ordinary function $x \mapsto \text{sgn}(x)/x^m$.

Let us compute some derivatives, for example,

$$D^n\,\text{fp}\frac{u(x)}{x} = \left[-\frac{1}{2\pi i}\frac{d^n}{dz^n}\left(\frac{\log(-z)}{z}\right)\right].$$

Leibniz's rule gives

$$\frac{d^n}{dz^n}\left(\frac{\log(-z)}{z}\right) = \sum_{k=0}^{n}\binom{n}{k}\frac{d^{n-k}}{dz^{n-k}}(z^{-1})\frac{d^k}{dz^k}\log(-z)$$

$$= \frac{d^n}{dz^n}(z^{-1})\log(-z) + \sum_{k=1}^{n}\binom{n}{k}\frac{d^{n-k}}{dz^{n-k}}(z^{-1})\frac{d^{k-1}}{dz^{k-1}}\frac{(-1)}{(-z)}$$

$$= (-1)^n n!\frac{\log(-z)}{z^{n+1}} + \sum_{k=1}^{n}\frac{n!\,(-1)^{n-k}(n-k)!}{k!\,(n-k)!\,z^{n+1-k}}\frac{(-1)^{k-1}(k-1)!}{z^k}$$

$$= (-1)^n n!\frac{\log(-z)}{z^{n+1}} - \frac{(-1)^n n!}{z^{n+1}}\sum_{k=1}^{n}\frac{1}{k},$$

and finally

$$\left[-\frac{1}{2\pi i}\frac{d^n}{dz^n}\left(\frac{\log(-z)}{z^{n+1}}\right)\right] = (-1)^n n!\left[-\frac{1}{2\pi i}\frac{\log(-z)}{z^{n+1}}\right]$$

$$-\left[-\frac{1}{2\pi i}\frac{(-1)^n n!}{z^{n+1}}\right]\sum_{k=1}^{n}\frac{1}{k},$$

or

$$D^n\,\mathrm{fp}\,\frac{u(x)}{x} = (-1)^n n!\,\mathrm{fp}\,\frac{u(x)}{x^{n+1}} - \delta^{(n)}(x)\sum_{k=1}^{n}\frac{1}{k}. \tag{1.92}$$

Similarly, we may find

$$D^n\,\mathrm{fp}\,\frac{u(-x)}{x} = (-1)^n n!\,\mathrm{fp}\,\frac{u(-x)}{x^{n+1}} + \delta^{(n)}(x)\sum_{k=1}^{n}\frac{1}{k}. \tag{1.93}$$

From these two results and (1.90),

$$D^n\mathrm{fp}\frac{1}{x} = (-1)^n n!\,\mathrm{fp}\frac{1}{x^{n+1}}, \tag{1.94}$$

and

$$D^n\mathrm{fp}\,\frac{\mathrm{sgn}(x)}{x} = (-1)^n n!\,\mathrm{fp}\,\frac{\mathrm{sgn}(x)}{x^{n+1}} - 2\,\delta^{(n)}(x)\sum_{k=1}^{n}\frac{1}{k}. \tag{1.95}$$

We shall now prove that the hyperfunctions

$$\mathrm{fp}\,\frac{u(x)}{x^m} + \mathrm{fp}\,\frac{u(-x)}{x^m} \quad\text{and}\quad \mathrm{fp}\,\frac{1}{x^m} := \left[\frac{1}{2z^m},\,-\frac{1}{2z^m}\right]$$

are the same (see also Proposition 2.39).

Proof. In the complex plane with a cut along the negative real axis the relations

$$\arg(-z) = \arg(z) - \mathrm{sgn}(\Im z)\pi, \quad\text{and}\quad \log(-z) = \log z - \mathrm{sgn}(\Im z)\,i\pi$$

hold for all $z \notin \mathbb{R}_-$. Therefore, we may write

$$\mathrm{fp}\,\frac{u(x)}{x^m} + \mathrm{fp}\,\frac{u(-x)}{x^m} = \left[-\frac{1}{2\pi i z^m}\{\log(-z) - \log(z)\}\right]$$

$$= \left[-\frac{1}{2\pi i z^m}\{\log(z) - \mathrm{sgn}(\Im z)\,i\pi - \log(z)\}\right]$$

$$= \left[\frac{\mathrm{sgn}(\Im z)}{2z^m}\right] = \left[\frac{1}{2z^m},\,-\frac{1}{2z^m}\right] = \mathrm{fp}\,\frac{1}{x^m}. \qquad \square$$

Problem 1.16. Prove that for $m \in \mathbb{N}$,

$$D^n\mathrm{fp}\,\frac{u(x)}{x^m} = (-1)^n (m)_n\,\mathrm{fp}\,\frac{u(x)}{x^{n+m}}$$

$$+ (-1)^m\frac{n!}{(n+m-1)!}\sum_{k=1}^{n}\frac{1}{k}\binom{n+m-k-1}{n-k}\delta^{(n+m-1)}(x),$$

and

$$D^n \text{fp} \frac{u(-x)}{x^m} = (-1)^n (m)_n \, \text{fp} \frac{u(-x)}{x^{n+m}}$$

$$- (-1)^m \frac{n!}{(n+m-1)!} \sum_{k=1}^{n} \frac{1}{k} \left(\begin{array}{c} n+m-k-1 \\ n-k \end{array} \right) \delta^{(n+m-1)}(x)$$

where $(m)_n := m(m+1)(m+2) \cdots (m+n-1)$ is Pochhammer's symbol.

1.6.3 Non-integral Powers

We shall look for hyperfunctions that interpret the ordinary functions $Y(x) x^\alpha$ and $Y(-x) |x|^\alpha$ where, and this is important, $\alpha \in \mathbb{C} \setminus \mathbb{Z}$. Remember that $\log z$ denotes the principal branch of the logarithm, defined on the complex plane with a cut along the negative real axis: $\log z = \log |z| + i \arg z$, $-\pi < \arg z < \pi$. We may write

$$\lim_{\epsilon \to 0+} \log(x \pm i\epsilon) = \log |x| \pm i\pi \, Y(-x)$$

and

$$\lim_{\epsilon \to 0+} \log(-x \pm i\epsilon) = \log |x| \pm i\pi \, Y(x).$$

The principal branch of the power z^α is defined by $\exp(\alpha \log z)$. Therefore we have

$$\lim_{\epsilon \to 0+} (x \pm i\epsilon)^\alpha = \exp(\alpha(\log |x| \pm i\pi \, Y(-x))) = |x|^\alpha \exp(\pm i\pi\alpha \, Y(-x)), \quad (1.96)$$

$$\lim_{\epsilon \to 0+} (-x \pm i\epsilon)^\alpha = \exp(\alpha(\log |x| \pm i\pi \, Y(x))) = |x|^\alpha \exp(\pm i\pi\alpha \, Y(x)). \quad (1.97)$$

This implies

$$\lim_{\epsilon \to 0+} \{(x + i\epsilon)^\alpha - (x - i\epsilon)^\alpha\} = |x|^\alpha \, 2i \, \sin(\alpha\pi \, Y(-x)) = 2i \, Y(-x)|x|^\alpha \, \sin(\alpha\pi),$$

$$\lim_{\epsilon \to 0+} \{(-x + i\epsilon)^\alpha - (-x - i\epsilon)^\alpha\} = |x|^\alpha \, 2i \, \sin(\alpha\pi \, Y(x)) = 2i \, Y(x)|x|^\alpha \, \sin(\alpha\pi).$$

Hence

$$Y(x)x^\alpha = -\frac{1}{2i \sin(\alpha\pi)} \lim_{\epsilon \to 0+} \{(-(x + i\epsilon))^\alpha - (-(x - i\epsilon))^\alpha\},$$

$$Y(-x)|x|^\alpha = \frac{1}{2i \sin(\alpha\pi)} \lim_{\epsilon \to 0+} \{(x + i\epsilon)^\alpha - (x - i\epsilon)^\alpha\}.$$

This leads to the following definition.

Definition 1.16. The hyperfunctions corresponding to the ordinary functions $Y(x) x^\alpha$ and $Y(-x) |x|^\alpha$, where $\alpha \in \mathbb{C} \setminus \mathbb{Z}$, are defined and denoted by

$$x_+^\alpha = u(x) \, x^\alpha := \left[-\frac{1}{2i \sin(\alpha\pi)} (-z)^\alpha \right], \quad (1.98)$$

$$|x|_-^\alpha = u(-x) \, |x|^\alpha := \left[\frac{1}{2i \sin(\alpha\pi)} z^\alpha \right], \quad (1.99)$$

where the powers $(-z)^\alpha$ and z^α, respectively, denote the principal branch of the power function.

With these two hyperfunctions at our disposal, we may define

$$|x|^\alpha = |x|^\alpha_- + x^\alpha_+ = \left[\frac{1}{2i\sin(\alpha\pi)}(z^\alpha - (-z)^\alpha)\right], \qquad (1.100)$$

$$|x|^\alpha \, \mathrm{sgn}(x) := -|x|^\alpha_- + x^\alpha_+ = \left[-\frac{1}{2i\sin(\alpha\pi)}(z^\alpha + (-z)^\alpha)\right]. \qquad (1.101)$$

These hyperfunctions interpret the ordinary functions with the same notation.

Let us differentiate these hyperfunctions. We have

$$D(|x|^\alpha_\pm) = \left[\mp\frac{1}{2i\sin(\alpha\pi)}\frac{d}{dz}((\mp z)^\alpha)\right]$$

$$= \left[\mp\frac{1}{2i\sin(\alpha\pi)}\alpha(\mp z)^{\alpha-1}(\mp 1)\right].$$

Since $\sin(\alpha\pi) = -\sin((\alpha-1)\pi))$, we may write

$$D(|x|^\alpha_\pm) = \pm\alpha\left[\mp\frac{1}{2i\sin((\alpha-1)\pi)}(\mp z)^{\alpha-1}\right].$$

Hence, we have established the rules

$$D(x^\alpha_+) = \alpha x^{\alpha-1}_+, \qquad (1.102)$$

$$D(|x|^\alpha_-) = -\alpha|x|^{\alpha-1}_-. \qquad (1.103)$$

By using these results, we obtain

$$D(|x|^\alpha) = \alpha|x|^{\alpha-1}\,\mathrm{sgn}(x), \qquad (1.104)$$

$$D(|x|^\alpha\,\mathrm{sgn}(x)) = -\alpha|x|^{\alpha-1}. \qquad (1.105)$$

Again, expressions such as $|x|^\alpha\,\mathrm{sgn}(x)$ are to be considered as a whole and not as a product (that cannot be defined).

Problem 1.17. Establish the following alternative representation for a non-integral α

$$|x|^\alpha \, \mathrm{sgn}(x) = -\frac{1}{2i\sin(\frac{\pi}{2}\alpha)}\left[(-iz)^\alpha, (iz)^\alpha\right], \qquad (1.106)$$

and

$$|x|^\alpha = \frac{1}{2\cos(\frac{\pi}{2}\alpha)}\left[(-iz)^\alpha, -(iz)^\alpha\right]. \qquad (1.107)$$

Problem 1.18. If n is a positive integer, establish the following rules:

$$x^n\,x^\alpha_+ = x^{n+\alpha}_+,$$

$$x^n\,|x|^\alpha_- = (-1)^n\,|x|^{n+\alpha}_-,$$

$$x^n\,|x|^\alpha = \begin{cases} |x|^{n+\alpha}, & n \text{ even} \\ |x|^{n+\alpha}\,\mathrm{sgn}(x), & n \text{ odd,} \end{cases}$$

$$x^n\,|x|^\alpha\,\mathrm{sgn}(x) = \begin{cases} |x|^{n+\alpha}\,\mathrm{sgn}(x), & n \text{ even} \\ |x|^{n+\alpha}, & n \text{ odd.} \end{cases}$$

Problem 1.19. The hyperfunctions x_+^α, x_-^α, $|x|^\alpha \operatorname{sgn}(x)$, $|x|^\alpha$ are meromorphic functions with respect to the parameter $\alpha \in \mathbb{C} \setminus \mathbb{Z}$ having simple poles at $\alpha \in \mathbb{Z}$. Show that the following residues hold. For $\ell = 1, 2, \ldots$ we have

$$\operatorname*{Res}_{\alpha=-\ell} \{x_+^\alpha\} = \frac{(-1)^{\ell-1}}{(\ell-1)!} \delta^{(\ell-1)}(x),$$

$$\operatorname*{Res}_{\alpha=-\ell} \{|x|_-^\alpha\} = \frac{1}{(\ell-1)!} \delta^{(\ell-1)}(x),$$

$$\operatorname*{Res}_{\alpha=-2\ell-1} \{|x|^\alpha\} = 2\frac{1}{(2\ell)!} \delta^{(2\ell)}(x),$$

$$\operatorname*{Res}_{\alpha=-2\ell} \{|x|^\alpha \operatorname{sgn}(x)\} = -2\frac{1}{(2\ell-1)!} \delta^{(2\ell-1)}(x).$$

For $\ell = 0, -1, -2$ the residues are the zero hyperfunction. We shall do the first one. We have to show

$$\frac{1}{2\pi i} \oint_{C(-\ell)} x_+^\alpha \, d\alpha = \frac{(-1)^{\ell-1}}{(\ell-1)!} \delta^{(\ell-1)}(x),$$

where the contour $C(-\ell)$ encircles the point $-\ell$ once in the positive sense. By the way, this illustrates that the integral of a hyperfunction with respect to a parameter results in a hyperfunction. So, let us investigate the contour integral

$$\frac{1}{2\pi i} \oint_{C(k)} x_+^\alpha \, d\alpha, \tag{1.108}$$

where $k \in \mathbb{Z}$. By Definition 1.15 we have to compute

$$-\frac{1}{2\pi i} \oint_{C(k)} \frac{(-z)^\alpha}{2i \sin(\pi\alpha)} \, d\alpha = -\operatorname*{Res}_{\alpha=k} \left\{ \frac{(-z)^\alpha}{2i \sin(\pi\alpha)} \right\} = -\frac{1}{2\pi i} z^k$$

in order to obtain the defining function of (1.108). Observe that $(-z)^\alpha$ is for a fixed $z, \Im z \neq 0$, an entire function of α, and the computation of the residue is straightforward. Finally, we obtain for $k = -\ell$,

$$\left[\frac{1}{2\pi i} \oint_{C(-\ell)} \frac{(-z)^\alpha}{2i \sin(\pi\alpha)} \, d\alpha \right] = \left[-\frac{1}{2\pi i} \frac{1}{z^\ell} \right] = \frac{(-1)^{\ell-1}}{(\ell-1)!} \delta^{(\ell-1)}(x).$$

For a non-negative integer k, the residue becomes the zero hyperfunction. The Gamma function $\Gamma(\alpha+1)$ has also simple poles at $\alpha = -\ell = -1, -2, \ldots$ with the residues $(-1)^{\ell-1}/(\ell-1)!$. This allows normalization of the hyperfunction x_+^α by $x_+^\alpha/\Gamma(\alpha+1)$, and similar for the three other hyperfunctions. This will be accomplished in detail in Section 1.6.6.

1.6.4 Logarithms

We wish to define hyperfunctions that interpret the ordinary functions $x \mapsto \log|x|, Y(x)\log(x), \operatorname{sgn}(x)\log|x|$. Inspired by the idea of the definition of the finite part hyperfunctions, we may define

$$\operatorname{fp} \log|x| := \left[\frac{\log(z)}{2}, -\frac{\log(z)}{2} \right]. \tag{1.109}$$

Indeed, the reader may verify by computing $F_+(x+i0) - F_-(x-i0)$ that for all $x \neq 0$ this hyperfunction has a value equal to $\log|x|$. Furthermore $\log z$ is holomorphic in the complex plane with a cut along the negative part of the real axis.

Next we start again with the relations

$$\lim_{\epsilon \to 0+} \log(x \pm i\epsilon) = \log|x| \pm i\pi\, Y(-x),$$

and

$$\lim_{\epsilon \to 0+} \log(-x \pm i\epsilon) = \log|x| \pm i\pi\, Y(x).$$

By forming the square of the latter equation, we obtain

$$\lim_{\epsilon \to 0+} \log^2(-x \pm i\epsilon) = \log^2|x| \pm 2\pi i\, Y(x)\log|x| - \pi^2 Y^2(x)$$

$$= \log^2|x| \pm 2\pi i\, Y(x)\log(x) - \pi^2 Y(x).$$

Taking the difference gives

$$\lim_{\epsilon \to 0+} \{\log^2(-x+i\epsilon) - \log^2(-x-i\epsilon)\} = 4\pi i\, Y(x)\log(x).$$

Hence,

$$Y(x)\log(x) = -\frac{1}{4\pi i} \lim_{\epsilon \to 0+} \{\log^2(-(x+i\epsilon)) - \log^2(-(x-i\epsilon))\}.$$

Definition 1.17. The hyperfunction that interprets the ordinary function $x \mapsto Y(x)\log(x)$ is defined and denoted by

$$u(x)\log(x) := \left[-\frac{1}{4\pi i}\log^2(-z)\right]. \tag{1.110}$$

Similarly, we obtain the hyperfunction corresponding to the ordinary function $x \mapsto Y(-x)\log|x|$ by

$$u(-x)\log|x| := \left[\frac{1}{4\pi i}\log^2(z)\right]. \tag{1.111}$$

From these two hyperfunctions we may again define

$$\log|x| := u(-x)\log|x| + u(x)\log|x|$$
$$= \left[-\frac{1}{4\pi i}\{\log^2(-z) - \log^2(z)\}\right]. \tag{1.112}$$

We shall show below that (1.112) and (1.109) define the same hyperfunctions. Also, we get

$$\mathrm{sgn}(x)\log|x| := -u(-x)\log|x| + u(x)\log|x|$$
$$= \left[-\frac{1}{4\pi i}\{\log^2(-z) + \log^2(z)\}\right]. \tag{1.113}$$

Let us differentiate the hyperfunction $\log |x|$,

$$
\begin{aligned}
D(\log|x|) &= \left[-\frac{1}{4\pi i}\frac{d}{dz}\{\log^2(-z) - \log^2(z)\} \right] \\
&= \left[-\frac{1}{4\pi i}\{2\log(-z)\frac{(-1)}{(-z)} - 2\log(z)\frac{1}{z}\} \right] \\
&= \left[-\frac{1}{2\pi i}\frac{\log(-z)}{z} \right] + \left[\frac{1}{2\pi i}\frac{\log(z)}{z} \right] \\
&= \text{fp}\frac{u(x)}{x} + \text{fp}\frac{u(-x)}{x} = \text{fp}\frac{1}{x}.
\end{aligned}
$$

Therefore the hyperfunction $\log|x|$ behaves with respect to differentiation like the ordinary function $x \mapsto \log|x|$. Similarly, we have

$$
D(\text{sgn}(x)\log|x|) = \text{sgn}(x)x^{-1}.
$$

Again, by using

$$
\log(-z) = \log z - \text{sgn}(\Im z)\,i\pi,
$$

we may write

$$
\begin{aligned}
\log^2(-z) - \log^2(z) &= \{\log(z) - \text{sgn}(\Im z)\,i\pi\}^2 - \log^2(z) \\
&= -2\pi i\,\text{sgn}(\Im z)\,\log(z) - \text{sgn}^2(\Im z)\pi^2 \\
&= -2\pi i\,\text{sgn}(\Im z)\,\log(z) - \pi^2
\end{aligned}
$$

hence, by dropping the constant $-\pi^2$, i.e., by passing to an equivalent defining function, we have

$$
\begin{aligned}
\log|x| &= \left[-\frac{1}{4\pi i}\{\log^2(-z) - \log^2(z)\} \right] = \left[\frac{1}{2}\text{sgn}(\Im z)\,\log(z) \right] \\
&= \left[\frac{\log(z)}{2}, -\frac{\log(z)}{2} \right] = \text{fp}\log|x|.
\end{aligned}
$$

This shows that the two hyperfunctions defined by (1.112) and (1.109) are indeed the same.

Problem 1.20. Establish the rules

$$
D(u(x)\log(x)) = \text{fp}\frac{u(x)}{x},
$$

$$
D(u(-x)\log|x|) = \text{fp}\frac{u(-x)}{x}.
$$

Problem 1.21. Show that for $n \in \mathbb{N}$ we have

$$
D(x^n\log|x|) = nx^{n-1}\log|x| + x^{n-1},
$$
$$
D(x^n\text{sgn}(x)\log|x|) = nx^{n-1}\text{sgn}(x)\log|x| + x^{n-1}\text{sgn}(x),
$$
$$
D(x^n u(x)\log(x)) = nx^{n-1}u(x)\log(x) + x^{n-1}u(x),
$$
$$
D(x^n\log|x|) = nx^{n-1}u(-x)\log|x| + x^{n-1}u(-x).
$$

Again, notation such as $u(x) \log(x)$ should be taken as a whole entity and not as a product of the hyperfunctions $u(x)$ and $\log x$. The same remark applies for the following definitions.

For $m \in \mathbb{N}$, we now look for hyperfunctions that interpret ordinary functions as $Y(x)x^{-m}\log(x)$. We shall see again that this task will not have a unique answer. See also the discussion in Sections 1.3.6 and 1.4. Suppose that $f(x) = [F(z)]$ is a hyperfunction that interprets $Y(x)x^{-m}\log(x)$. Then, $x^m f(x) = [z^m F(z)]$ should be an interpretation of the hyperfunction $u(x) \log x$ having the defining function $-1/(4\pi i)\log^2(-z) + \phi(z)$, where $\phi(x)$ is an arbitrary real analytic function. Thus, we have the equation

$$z^m F(z) = -\frac{1}{4\pi i)} \log^2(-z) + \phi(z),$$

or

$$f(x) = \left[-\frac{1}{4\pi i} \frac{\log^2(-z)}{z^m} + \frac{\phi(z)}{z^m} \right],$$

giving eventually

$$f(x) = \left[-\frac{1}{4\pi i} \frac{\log^2(-z)}{z^m} \right] + \delta_{(m-1)}(x),$$

with the generalized Dirac delta-hyperfunction having m arbitrary coefficients. We shall define and denote this whole family of hyperfunctions by

$$u(x)\, x^{-m}\, \log(x) := \left[-\frac{1}{4\pi i} \frac{\log^2(-z)}{z^m} \right] + \delta_{(m-1)}(x), \quad m \in \mathbb{N}. \tag{1.114}$$

However, we single out the member containing no Dirac impulses by the notation

$$\mathrm{fp}\frac{u(x)\, \log(x)}{x^m} := \left[-\frac{1}{4\pi i} \frac{\log^2(-z)}{z^m} \right]. \tag{1.115}$$

Similarly, we introduce

$$\mathrm{fp}\frac{u(-x)\, \log|x|}{x^m} := \left[\frac{1}{4\pi i} \frac{\log^2(z)}{z^m} \right], \tag{1.116}$$

$$\mathrm{fp}\frac{\log|x|}{x^m} := \mathrm{fp}\frac{u(-x)\, \log|x|}{x^m} + \mathrm{fp}\frac{u(x)\, \log(x)}{x^m}$$

$$= \left[\frac{1}{4\pi i} \{ \frac{\log^2(z)}{z^m} - \frac{\log^2(-z)}{z^m} \} \right], \tag{1.117}$$

$$\mathrm{fp}\frac{\mathrm{sgn}(x)\, \log|x|}{x^m} := -\mathrm{fp}\frac{u(-x)\, \log|x|}{x^m} + \mathrm{fp}\frac{u(x)\, \log(x)}{x^m}$$

$$= \left[-\frac{1}{4\pi i} \{ \frac{\log^2(z)}{z^m} + \frac{\log^2(-z)}{z^m} \} \right]. \tag{1.118}$$

An intuitive picture of the latter hyperfunction for $m = 1$ is given by the sequence of ordinary functions given in Figure 1.14.

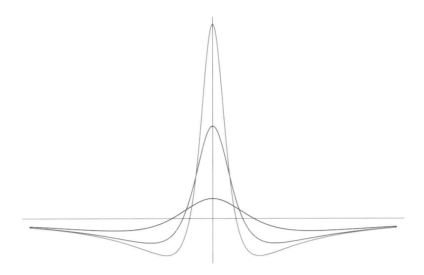

Figure 1.14: Intuitive picture of $\mathrm{fp}(\mathrm{sgn}(x)\log|x|/x)$

Problem 1.22. If α is not an integral number, establish

$$u(x)x^\alpha \log x = \left[-\frac{(-z)^\alpha}{2i\sin(\pi\alpha)} \{\log(-z) - \pi\cot(\pi\alpha)\} \right],$$

$$u(x)x^\alpha \log^2(x) = \left[-\frac{(-z)^\alpha}{2i\sin(\pi\alpha)} \left\{ \left(\frac{\pi}{\sin(\pi\alpha)}\right)^2 + (\log(-z) - \pi\cot(\pi\alpha))^2 \right\} \right].$$

Problem 1.23. Find a recursion formula for $u(x)\log^m(x)$ and use it to establish

$$u(x)\log^2(x) = \left[-\frac{1}{6\pi i}\log^3(-z) + i\frac{\pi}{6}\log(-z) \right],$$

$$u(x)\log^3 x) = \left[-\frac{1}{8\pi i}\log^4(-z) + i\frac{\pi}{4}\log^2(-z) \right],$$

$$u(x)\log^4 x) = \left[-\frac{1}{10\pi i}\log^5(-z) + i\frac{\pi}{3}\log^3(-z) + i\frac{7\pi^3}{30}\log^3(-z) \right].$$

Problem 1.24. Show that the indicated defining function defines the given hyperfunctions ($\alpha \in \mathbb{C} \setminus \mathbb{Z}$),

$$[(-iz)^\alpha \log(-iz), (iz)^\alpha \log(iz)]$$
$$= -i|x|^\alpha \, \mathrm{sgn}(x) \left\{ \pi\cos(\frac{\pi}{2}\alpha) + 2\sin(\frac{\pi}{2}\alpha)\log|x| \right\}, \tag{1.119}$$

$$[(-iz)^\alpha \log(-iz), -(iz)^\alpha \log(iz)]$$
$$= |x|^\alpha \left\{ i\pi\sin(\frac{\pi}{2}\alpha) + 2\cos(\frac{\pi}{2}\alpha)\log|x| \right\}. \tag{1.120}$$

Problem 1.25. Show that for $n = 0, 1, 2, \ldots$

$$\left[\frac{\log(-iz)}{z^n}, 0\right] = \frac{\log|x|}{x^n} - i\frac{\pi}{2}\frac{\operatorname{sgn} x}{x^n}, \tag{1.121}$$

$$\left[0, \frac{\log(iz)}{z^n}\right] = -\frac{\log|x|}{x^n} - i\frac{\pi}{2}\frac{\operatorname{sgn} x}{x^n}, \tag{1.122}$$

thus

$$\left[\frac{\log(-iz)}{z^n}, \frac{\log(iz)}{z^n}\right] = -i\pi\frac{\operatorname{sgn} x}{x^n}, \tag{1.123}$$

and

$$\left[\frac{\log(-iz)}{z^n}, -\frac{\log(iz)}{z^n}\right] = 2\frac{\log|x|}{x^n}. \tag{1.124}$$

1.6.5   Upper and Lower Hyperfunctions

At the end of Section 1.4 we introduced the Heisenberg hyperfunctions $\delta_+(x)$ and $\delta_-(x)$. These are examples of upper and lower hyperfunctions. Let us investigate this type of hyperfunctions in more detail.

Again let I be a real interval, and $\mathcal{D} = \mathcal{D}(I)$ a complex neighborhood of I. Let $\mathcal{D}_+ = \mathcal{D} \cap \mathbb{C}_+$ be the upper half-neighborhood and $\mathcal{D}_- = \mathcal{D} \cap \mathbb{C}_-$ be the lower half-neighborhood of I.

Definition 1.18. Let $F(z)$ be a given holomorphic function in $\mathcal{D}_+$, i.e., $F(z) \in \mathcal{O}(\mathcal{D}_+)$. Then, a hyperfunction of the type

$$F(x + i0) := [F(z), 0] = [F(z)\mathbf{1}_+(z)] \tag{1.125}$$

is called an *upper hyperfunction on* I. Similarly, for a holomorphic function $F(z) \in \mathcal{O}(\mathcal{D}_-)$, a hyperfunction of the type

$$F(x - i0) := [0, -F(z)] = [F(z)\mathbf{1}_-(z)] \tag{1.126}$$

is called a *lower hyperfunction on* I.

We denote by $\mathcal{B}_+(I)$ and $\mathcal{B}_-(I)$ the subspace of all upper and lower hyperfunctions on the interval I, respectively.

Observe that any hyperfunction $f(x) = [F_+(z), F_-(z)]$ can be decomposed into an upper and a lower hyperfunction.

$$f(x) = [F_+(z), F_-(z)] = [F_+(z), 0] - [0, -F_-(z)]$$
$$= F_+(x + i0) - F_-(x - i0).$$

Proposition 1.12. *Let* $f(x) = [F(z)]$ *be a meromorphic hyperfunction on* $I = \mathbb{R}$ *with some poles on the real axis. Then, we have*

$$F(x + i0) = \operatorname{fp} f(x) + \frac{1}{2}f(x), \tag{1.127}$$

$$F(x - i0) = \operatorname{fp} f(x) - \frac{1}{2}f(x). \tag{1.128}$$

Proof. We may write

$$F(x+i0) := [F(z), 0]$$

$$= \left[\frac{1}{2}F(z), -\frac{1}{2}F(z)\right] + \left[\frac{1}{2}F(z), \frac{1}{2}F(z)\right]$$

$$= [F(z)\mathbf{1}(z)] + \frac{1}{2}[F(z)]$$

$$= \mathrm{fp}f(x) + \frac{1}{2}f(x).$$

Similarly for $F(x-i0)$. □

Example 1.34. Let $F(z) = 1/z$. Then,

$$\frac{1}{x+i0} = [\frac{1}{z}, 0] = \mathrm{fp}\frac{1}{x} + \frac{1}{2}[\frac{1}{z}]$$

$$= \mathrm{fp}\frac{1}{x} - \pi i \left[-\frac{1}{2\pi i}\frac{1}{z}\right]$$

$$= \mathrm{fp}\frac{1}{x} - \pi i\,\delta(x).$$

The following examples show that the idea of the above decomposition can be generalized to other defining functions which are not necessarily meromorphic.

Example 1.35. Let $F(z) = \log(-z)$. Then

$$\log(-(x+i0)) = [\log(-z), 0]$$

$$= \left[\frac{1}{2}\log(-z), -\frac{1}{2}\log(-z)\right] + \left[\frac{1}{2}\log(-z)\right]$$

$$= \left[\frac{1}{2}\log(-z), -\frac{1}{2}\log(-z)\right] - \pi i \left[-\frac{1}{2\pi i}\log(-z)\right]$$

$$= \log|x| - \pi i u(x).$$

Similarly, we find

$$\log(-(x-i0)) = \log|x| + \pi i u(x).$$

Problem 1.26. Establish

$$\log(x+i0) = \log|x| + \pi i u(-x), \quad \log(x-i0) = \log|x| - \pi i u(-x).$$

Example 1.36. By using the relations (1.96) and (1.97) we have immediately for a non-integral α,

$$(x+i0)^\alpha = |x|^\alpha\, e^{i\alpha\,\pi Y(-x)}, \tag{1.129}$$

$$(x-i0)^\alpha = |x|^\alpha\, e^{-i\alpha\,\pi Y(-x)}, \tag{1.130}$$

$$(-(x+i0))^\alpha = |x|^\alpha\, e^{-i\alpha\,\pi Y(x)}, \tag{1.131}$$

$$(-(x-i0))^\alpha = |x|^\alpha\, e^{i\alpha\,\pi Y(x)}. \tag{1.132}$$

For example, we have for $\alpha = 1/2, -1/2$,

$$\sqrt{-(x\pm i0)} = \sqrt{|x|}\,\{Y(-x) \mp iY(x)\}, \tag{1.133}$$

$$\frac{1}{\sqrt{-(x\pm i0)}} = \frac{1}{\sqrt{|x|}}\,\{Y(-x) \pm iY(x)\}. \tag{1.134}$$

Example 1.37. By using the power hyperfunctions

$$x_+^\alpha = u(x)\, x^\alpha = \left[-\frac{1}{2i\sin(\alpha\pi)}(-z)^\alpha \right],$$

$$|x|_-^\alpha = u(-x)\, |x|^\alpha = \left[\frac{1}{2i\sin(\alpha\pi)} z^\alpha \right]$$

and the formula

$$\log(-z) = \log z - \operatorname{sgn}(\Im z)\, i\pi,$$

we have

$$(-z)^\alpha = z^\alpha \, \exp(-i\alpha\pi \, \operatorname{sgn}(\Im z)).$$

Let us compute

$$-\exp(-i\alpha\frac{\pi}{2})\,(-z)^\alpha + \exp(i\alpha\frac{\pi}{2})\,z^\alpha$$

$$= \exp(i\alpha\frac{\pi}{2})\,z^\alpha\{1 - \exp(-i\alpha\pi - i\alpha\pi \operatorname{sgn}(\Im z)\}$$

$$= \exp(i\alpha\frac{\pi}{2})\,z^\alpha\,(1 - \exp(-2i\alpha\pi)Y(\Im z)) = 2i\sin(\alpha\pi)\,\exp(-i\alpha\frac{\pi}{2})\,z^\alpha\,\mathbf{1}_+(z)$$

$$= 2i\sin(\alpha\pi)\,(-iz)^\alpha\,\mathbf{1}_+(z).$$

Similarly, we find

$$-\exp(i\alpha\frac{\pi}{2})\,(-z)^\alpha + \exp(-i\alpha\frac{\pi}{2})\,z^\alpha = 2i\sin(\alpha\pi)(iz)^\alpha\,\mathbf{1}_-(z).$$

Therefore, we have established the following upper and lower power hyperfunctions:

$$h(x) := u(x)x^\alpha\, e^{-i\alpha\pi/2} + u(-x)|x|^\alpha\, e^{i\alpha\pi/2} = [(-iz)^\alpha\,\mathbf{1}_+(z)], \qquad (1.135)$$

$$g(x) := u(x)x^\alpha\, e^{i\alpha\pi/2} + u(-x)|x|^\alpha\, e^{-i\alpha\pi/2} = [(iz)^\alpha\,\mathbf{1}_-(z)]. \qquad (1.136)$$

From these upper and lower power hyperfunctions we may reconstitute

$$u(x)x^\alpha = -\frac{1}{2i\sin(\alpha\pi)}\{e^{-i\alpha\pi/2}h(x) - e^{i\alpha\pi/2}g(x)\}, \qquad (1.137)$$

$$u(-x)|x|^\alpha = \frac{1}{2i\sin(\alpha\pi)}\{e^{i\alpha\pi/2}h(x) - e^{-i\alpha\pi/2}g(x)\},. \qquad (1.138)$$

which is left to the reader.

1.6.6   The Normalized Power $x_+^\alpha/\Gamma(\alpha+1)$

So far all introduced hyperfunctions have been specified by an explicit algebraic expression for the defining function. At present the defining function will be defined by an integral. We consider the family of functions defined by

$$F(z,\alpha) := (-z)^\alpha \int_{-\infty}^z e^t\, (-t)^{-\alpha-1} dt$$

depending on a complex parameter α. For a fixed α the function is holomorphic with respect to z in the complex plane with a cut along the positive real axis. For a fixed z, $\Im z \neq 0$, we have an entire function with respect to α. Now,

$$(-z)^\alpha := e^{\alpha \log(-z)}, \quad (-t)^{-\alpha-1} = e^{-(\alpha+1)\log(-t)}.$$

First, we assume $\Re\alpha < 0$ since the integrand is then integrable at $t = 0$. We decompose the integration path into a contour comprising the negative part of the real axis $(-\infty, 0)$ and the line segment $(0, z)$ and obtain

$$F(z, \alpha) := (-z)^\alpha \int_{-\infty}^{0} e^t(-t)^{-\alpha-1} dt + (-z)^\alpha \int_{0}^{z} e^t(-t)^{-\alpha-1} dt.$$

The first integral on the right-hand side can be written as

$$\int_{-\infty}^{0} e^t(-t)^{-\alpha-1} dt = \int_{0}^{\infty} e^{-t} t^{-\alpha-1} dt = \Gamma(-\alpha)$$

according to the definition of the gamma function. In the second integral we expand the exponential function and integrate term by term.

$$\Phi(z, \alpha) := (-z)^\alpha \int_{0}^{z} e^t(-t)^{-\alpha-1} dt = \sum_{k=0}^{\infty} \frac{z^k}{k!\,(\alpha - k)}.$$

The function $\Phi(z, \alpha)$ defined by the series can analytically be continued to $\mathbb{C} \times (\mathbb{C} \setminus \mathbb{N}_0)$. We now have

$$F(z, \alpha) = (-z)^\alpha \Gamma(-\alpha) + \Phi(z, \alpha),$$

or

$$\Gamma(\alpha + 1)F(z, \alpha) = (-z)^\alpha \Gamma(\alpha + 1)\Gamma(-\alpha) + \Gamma(\alpha + 1)\Phi(z, \alpha).$$

If α is not an integer, we can use

$$\Gamma(\alpha + 1)\Gamma(-\alpha) = \frac{\pi}{\sin(-\pi\alpha)} = -\frac{\pi}{\sin(\pi\alpha)},$$

and obtain

$$\Gamma(\alpha + 1)F(z, \alpha) = -(-z)^\alpha \frac{\pi}{\sin(\pi\alpha)} + \Gamma(\alpha + 1)\Phi(z, \alpha).$$

Finally, we obtain

$$\frac{F(z, \alpha)}{2\pi i} = -\frac{1}{\Gamma(\alpha + 1)} \frac{(-z)^\alpha}{2i\sin(\pi\alpha)} + \frac{1}{2\pi i}\Phi(z, \alpha).$$

If $\alpha \notin \mathbb{N}_0$ the second term on the right-hand side is a real analytic function with respect to z, and, when considered as a defining function, may be discarded. Recalling the definition of the hyperfunction x_+^α, we have for $\Re\alpha < 0$ and $\alpha \notin \mathbb{Z}$:

$$\frac{1}{\Gamma(\alpha + 1)} x_+^\alpha = \left[\frac{(-z)^\alpha}{2\pi i} \int_{-\infty}^{z} e^t(-t)^{-\alpha-1} dt \right]. \tag{1.139}$$

Equation (1.139) is established for a complex α located in the left half-plane and different from an integer. Let us still show that the relation also holds when $\alpha = n \in \mathbb{N}$. We have then

$$F(z,n) = (-z)^n \int_{-\infty}^{z} e^t(-t)^{-n-1} dt = -z^n \int_{-\infty}^{z} \frac{e^t}{t^{n+1}} dt$$

$$= -z^n \int_{-\infty}^{-1} \frac{e^t}{t^{n+1}} dt - z^n \int_{-1}^{z} \frac{e^t}{t^{n+1}} dt.$$

The first integral on the right-hand side is a constant, say $-K_1$. In the second integral, after replacing t by $-t$, the exponential term is expanded, hence

$$F(z,n) = K_1 z^n + z^n \int_{1}^{-z} \sum_{k=0}^{\infty} \frac{(-t)^{k-n-1}}{k!} dt$$

$$= K_1 z^n + z^n \int_{1}^{-z} \{\sum_{k=0}^{n-1} \frac{(-t)^{k-n-1}}{k!} + \frac{1}{n!(-t)} + \sum_{k=n+1}^{\infty} \frac{(-t)^{k-n-1}}{k!}\} dt$$

$$= K_1 z^n + z^n \{-\sum_{k=0}^{n-1} \frac{z^{k-n}}{k!(k-n)} + K_2 - \frac{\log(-z)}{n!} + K_3$$

$$- \sum_{k=n+1}^{\infty} \frac{z^{k-n}}{k!(k-n)} + K_4\}$$

$$= (K_1 + K_2 + K_3 + K_4)z^n - \sum_{k=0}^{n-1} \frac{z^k}{k!(k-n)} - \frac{z^n \log(-z)}{n!}$$

$$- \sum_{k=n+1}^{\infty} \frac{z^k}{k!(n-k)} = -\frac{z^n \log(-z)}{n!} + \psi(z),$$

where $\psi(z)$ is a real analytic function. Therefore, we have

$$\left[\frac{1}{2\pi i} F(z,n)\right] = \left[-\frac{1}{2\pi i} \frac{z^n \log(-z)}{n!}\right] = \frac{1}{n!} x_+^n = \frac{1}{\Gamma(n+1)} x_+^n.$$

If α is a negative integer, $\Gamma(\alpha+1)$ has a pole, hence $1/\Gamma(\alpha+1)$ can be considered as zero, and therefore $x_+^\alpha/\Gamma(\alpha+1)$ has the value zero for $x \neq 0$. For $x = 0$, however, we expect an undetermined situation. In fact we have the following result:

Proposition 1.13. *The hyperfunction*

$$f(x,\alpha) := \frac{1}{\Gamma(\alpha+1)} x_+^\alpha = \frac{u(x)x^\alpha}{\Gamma(\alpha+1)}$$

depending on the complex parameter α is defined for $\Re\alpha < 0$, $\alpha \notin \{0, -1, -2, \ldots\}$ or $\alpha \in \mathbb{N}$ by the defining function

$$F(z,\alpha) = \frac{(-z)^\alpha}{2\pi i} \int_{-\infty}^{z} e^t(-t)^{-\alpha-1} dt.$$

However, if α tends to $-n-1, n \in \mathbb{N}_0$, then $f(x)$ tends to $\delta^{(n)}(x)$, i.e.,

$$\lim_{\alpha \to -n-1} \frac{x_+^\alpha}{\Gamma(\alpha+1)} = \delta^{(n)}(x). \tag{1.140}$$

Proof.

$$\frac{F(z,-n-1)}{2\pi i} = \frac{(-z)^{-n-1}}{2\pi i} \int_{-\infty}^{z} e^t(-t)^n dt = -\frac{1}{2\pi i} \frac{1}{z^{n+1}} \int_{-\infty}^{z} e^t t^n dt$$

$$= -\frac{1}{2\pi i} \frac{1}{z^{n+1}} \int_{-\infty}^{0} e^t t^n dt - \frac{1}{2\pi i} \frac{1}{z^{n+1}} \int_{0}^{z} e^t t^n dt$$

$$= -\frac{1}{2\pi i} \frac{(-1)^n n!}{z^{n+1}} - \frac{1}{2\pi i z^{n+1}} \int_{0}^{z} \sum_{k=0}^{\infty} \frac{t^{k+n}}{k!} dt$$

$$= -\frac{1}{2\pi i} \frac{(-1)^n n!}{z^{n+1}} - \frac{1}{2\pi i} \sum_{k=0}^{\infty} \frac{z^k}{k!(k+n+1)}.$$

The series in the right-hand side of the last term is a real analytic function, hence it can be omitted, and we obtain

$$\left[\frac{F(z,-n-1)}{2\pi i} \right] = \left[-\frac{1}{2\pi i} \frac{(-1)^n n!}{z^{n+1}} \right] = \delta^{(n)}(x). \qquad \Box$$

The reason for (1.140) is that the gamma function $\Gamma(\alpha+1)$ and the hyperfunction x_+^α have the same singularities at $\alpha = -n-1$ (see also Problem 1.19.

Problem 1.27. Use the other residues computed in Problem 1.19 to establish the following results, similarly to the above ones

$$\lim_{\alpha \to -n-1} \frac{|x|_-^\alpha}{\Gamma(\alpha+1)} = (-1)^n \, \delta^{(n)}(x),$$

$$\lim_{\alpha \to -2n-1} \frac{|x|^\alpha}{\Gamma((\alpha+1)/2)} = (-1)^n \frac{n!}{(2n)!} \delta^{(2n)}(x),$$

$$\lim_{\alpha \to -2n} \frac{|x|^\alpha \operatorname{sgn}(x)}{\Gamma((\alpha+2)/2)} = (-1)^n \frac{(n-1)!}{(2n-1)!} \delta^{(2n-1)}(x).$$

1.6.7 Hyperfunctions Concentrated at One Point

We conclude this chapter with some remarks on hyperfunctions concentrated at one point $x = a$, i.e., which vanish for $x < a$ and $x > a$. Such a hyperfunction must be specified by a defining function having an upper and a lower component with one isolated singularity at $z = a$, thus, by Laurent's theorem

$$F_+(z) = \sum_{k=-\infty}^{\infty} c_k^+ (z-a)^k, \quad F_-(z) = \sum_{k=-\infty}^{\infty} c_k^- (z-a)^k.$$

Because for any $\epsilon \neq 0$ and $x \neq a$ we must have $F_+(x+\epsilon) - F_-(x+\epsilon) = 0$ we immediately obtain

$$0 = \sum_{k=-\infty}^{\infty} (c_k^+ - c_k^-) \epsilon^k$$

which implies $c_k^+ - c_k^- = 0$ for all k. This means that the upper and lower components must be the same function. Therefore, after having discarded the real analytic part of $F(z)$, we have

$$F(z) = \sum_{k=-\infty}^{-1} c_k (z-a)^k, \quad c_k = \frac{1}{2\pi i} \oint_C \frac{F(z)}{(z-a)^{k+1}} \, dz,$$

where C is a small a-centered circle of radius $r > 0$. Then, we have

$$F(z) = \sum_{k=1}^{\infty} \frac{c_{-k}}{(z-a)^k} = \sum_{k=0}^{\infty} \frac{c_{-k-1}}{(z-a)^{k+1}}$$

$$= \sum_{k=0}^{\infty} \frac{(-1)^k \, (-2\pi i \, c_{-k-1})}{k!} \left\{ -\frac{1}{2\pi i} \frac{(-1)^k k!}{(z-a)^{k+1}} \right\}$$

with

$$-2\pi i \, c_{-k-1} = -\oint_C F(z) \, (z-a)^k \, dz = \int_{a-r}^{a+r} f(x) \, (x-a)^k \, dx.$$

The last integral is the integral over the hyperfunction $f(x)(x-a)^k = [F(z)(z-a)^k]$. We have established

Proposition 1.14. *A hyperfunction concentrated at one point $x = a$ has the general form*

$$f(x) = \sum_{k=0}^{\infty} \frac{(-1)^k \, A_k}{k!} \delta^{(k)}(x-a), \quad A_k = \int_{a-r}^{a+r} f(x) \, (x-a)^k \, dx \qquad (1.141)$$

where r is any positive number.

Example 1.38. Consider the hyperfunction $f(x) = [-\exp(1/z)/(2\pi i)]$. The point $z = 0$ is an essential singularity of the defining function. Because

$$e^{1/z} = \sum_{k=0}^{\infty} \frac{1}{k!} \frac{1}{z^k} = 1 + \sum_{k=0}^{\infty} \frac{1}{(k+1)!} \frac{1}{z^{k+1}}$$

$$= 1 + (-2\pi i) \sum_{k=0}^{\infty} \frac{(-1)^k}{k! \, (k+1)!} \frac{(-1) \, k! \, (-1)^k}{2\pi i} \frac{1}{z^{k+1}}$$

we have directly

$$\left[-\frac{1}{2\pi i} e^{1/z} \right] = \sum_{k=0}^{\infty} \frac{(-1)^k}{k! \, (k+1)!} \delta^{(k)}(x).$$

Chapter 2

Analytic Properties

Having presented the elementary properties of hyperfunctions in the first chapter, we shall now enter into more subtle topics. First, sequences and series of hyperfunctions are investigated, then Cauchy-type integrals that play an important part in the theory of hyperfunctions are discussed. The basic question of any theory of generalized functions, namely, how ordinary functions can be embedded in their realm, is investigated (projection of an ordinary function). The book of [12, Gakhov] has been very helpful for the treatment of this question. The subject of the projection or restriction of a hyperfunction to a smaller interval is then exposed. The important notions of holomorphic hyperfunction, analytic and micro-analytic hyperfunctions are discussed, and the more technical concepts such as support, singular support and singular spectrum are introduced. The product of two generalized functions is always a difficult point in any theory about generalized functions. Generally, the product of two generalized function cannot be defined. We shall discuss under what circumstances the product of two hyperfunctions makes sense. The sections on periodic hyperfunctions and their Fourier series and the important subject of convolution of hyperfunctions form the ending material of this chapter. Also, the track of applications to integral and differential equations starts here.

2.1 Sequences, Series, Limits

Let us consider hyperfunctions depending on a continuous parameter α or an integral parameter k. We assume that the continuous parameter α varies in some open region Λ of the complex plane, and that α_0 is a limit point of Λ. In the case of an integral parameter k, it may vary in $\mathbb{N}$ or $\mathbb{Z}$. Hence,

$$f(x, \alpha) = [F(z, \alpha)] \quad \alpha \in \Lambda, \quad f_k(x) = [F_k(z)] \quad k \in \mathbb{N} \text{ or } k \in \mathbb{Z}.$$

We say that a family of holomorphic functions $F(z, \alpha)$, or a sequence of holomorphic functions $F_k(z)$, defined on a common domain $\mathcal{D} \subset \mathbb{C}$ *converges uniformly in the interior of* $\mathcal{D}$ to $F(z)$ as $\alpha \to \alpha_0$, or $k \to \infty$, respectively, if $F(z, \alpha)$, or $F_k(z)$, converges uniformly to $F(z)$ in every compact sub-domain of $\mathcal{D}$. This uniform convergence in the interior of $\mathcal{D}$ is also called *compact convergence in* $\mathcal{D}$. From complex variables theory, we know that the limit function $F(z)$ is then again holomorphic in $\mathcal{D}$. In the setting that follows, the role of the mentioned domain $\mathcal{D}$ is taken by the upper and lower half-neighborhood of the complex neighborhood $\mathcal{D}(I)$, respectively, when the hyperfunctions under consideration are defined on I.

Definition 2.1. Let $f(x) = [F_+(z), F_-(z)]$ be a hyperfunction and $f(x, \alpha) = [F(z, \alpha)]$ a family of hyperfunctions depending on the parameter α. Assume that for every α an equivalent defining function $G(z, \alpha)$ of $F(z, \alpha)$ exists, such that $G_+(z, \alpha)$ and $G_-(z, \alpha)$ converge uniformly in the interior of $\mathcal{D}_+(I)$ and $\mathcal{D}_-(I)$ to $F_+(z)$ and $F_-(z)$, respectively. Then, we write

$$f(x) = \lim_{\alpha \to \alpha_0} f(x, \alpha),$$

and say that the family of hyperfunctions $f(x, \alpha)$ *converges in the sense of hyperfunctions* to $f(x)$.

Definition 2.2. Let $f(x) = [F_+(z), F_-(z)]$ be defined on I such that, for every k, equivalent defining functions $G_k(z)$ of $F_k(z)$ exist, such that $G_{+k}(z)$ and $G_{-k}(z)$ are uniformly convergent in the interior of $\mathcal{D}_+(I)$ and $\mathcal{D}_-(I)$ to $F_+(z)$ and $F_-(z)$, respectively. Then, we write

$$f(x) = \lim_{k \to \infty} f_k(x),$$

and say that the sequence of hyperfunctions $f_k(x)$ *converges in the sense of hyperfunctions* to $f(x)$.

For series, we set

Definition 2.3. We write

$$f(x) = \sum_{k=0}^{\infty} f_k(x)$$

if the sequence of partial sums converges in the sense of hyperfunctions to $f(x)$.

Remark. For the compact sub-domains of $\mathcal{D}_+(I)$ and $\mathcal{D}_-(I)$ think of the compact rectangles $\mathcal{D}_+(\epsilon, \delta; a, b) := \{z \in \mathbb{C} \,|\, \delta \geq \Im z \geq \epsilon > 0, \, a \leq \Re z \leq b\}$, $[a, b] \subset I$ and $\mathcal{D}_-(\epsilon, \delta; a, b) := \{z \in \mathbb{C} \,|\, -\delta \leq \Im z \leq -\epsilon < 0, \, a \leq \Re z \leq b, [a, b] \subset I\}$, respectively.
The reader may check that if a limit in the sense of hyperfunctions exists, it is unique.

Example 2.1. The ordinary limits $\lim_{\alpha \to \infty} \sin(\alpha x)$ and $\lim_{\alpha \to \infty} \cos(\alpha x)$ do not exist. Now, we may represent the real analytic function $e^{i\alpha x}$ in three distinct but equivalent ways:

$$e^{i\alpha x} = \left[e^{i\alpha z}\, \mathbf{1}(z)\right] = \left[e^{i\alpha z}/2, -e^{i\alpha z}/2\right] = \left[e^{i\alpha z}\, \mathbf{1}_-(z)\right] = \left[0, -e^{i\alpha z}\right]$$
$$= \left[e^{i\alpha z}\, \mathbf{1}_+(z)\right] = \left[e^{i\alpha z}, 0\right] := [G(z, \alpha)].$$

Similarly,

$$e^{-i\alpha x} = \left[e^{-i\alpha z}\, \mathbf{1}(z)\right] = \left[e^{-i\alpha z}/2, -e^{-i\alpha z}/2\right] = \left[e^{-i\alpha z}\, \mathbf{1}_+(z)\right] = \left[e^{-i\alpha z}, 0\right]$$
$$= \left[e^{-i\alpha z}\, \mathbf{1}_-(z)\right] = \left[0, -e^{-i\alpha z}\right] =: [G(z, \alpha)].$$

In the above formulas only the representation in the second line produces convergence as $\alpha \to \infty$, because for $z \in \mathcal{D}_+$ and $z \in \mathcal{D}_-$, respectively, we have for some $\epsilon > 0$,

$$e^{i\alpha z} = e^{i\alpha(x+i\epsilon)} = e^{-\alpha\epsilon}e^{i\alpha x}, \quad e^{-i\alpha z} = e^{-i\alpha(x-i\epsilon)} = e^{-\alpha\epsilon}e^{-i\alpha x}.$$

Therefore we choose the defining functions in the second lines and obtain, for $z \in \mathcal{D}_+$, $\lim_{\alpha\to\infty} e^{i\alpha z} = 0$, and for $z \in \mathcal{D}_-$, $\lim_{\alpha\to\infty} e^{-i\alpha z} = 0$ (uniformly convergent in the interior of $\mathcal{D}_\pm$). Hence, we have in the sense of hyperfunctions $\lim_{\alpha\to\infty} e^{i\alpha x} = [0,0] = 0$, $\lim_{\alpha\to\infty} e^{-i\alpha x} = [0,0] = 0$. Also, by Euler's formula, we conclude that

$$\lim_{\alpha\to\infty} \sin(\alpha x) = 0, \quad \lim_{\alpha\to\infty} \cos(\alpha x) = 0. \tag{2.1}$$

Example 2.2. The ordinary limit $\sin(\alpha x)/(\pi x), \alpha \to \infty$ does not exist. Let us compute the limit in the sense of hyperfunctions. Since the expression represents a real analytic function we may write

$$\frac{\sin(\alpha x)}{\pi x} = \left[\frac{\sin(\alpha z)}{\pi z}, 0\right] = \left[\frac{\sin(\alpha z)}{\pi z}\mathbf{1}_+(z)\right]$$

or also

$$\frac{\sin(\alpha x)}{\pi x} = \left[\frac{\sin(\alpha z)}{2\pi z}, -\frac{\sin(\alpha z)}{2\pi z}\right] = \left[\frac{\sin(\alpha z)}{\pi z}\mathbf{1}(z)\right].$$

For both choices of the defining functions the limits of the right-hand side do not exist. However, there is another equivalent defining function that produces a limit. We may write

$$\frac{\sin(\alpha x)}{\pi x} = \frac{1}{\pi}\left[\frac{\sin(\alpha z)}{z}, 0\right] = \frac{1}{2\pi i}\left[\frac{e^{i\alpha z} - e^{-i\alpha z}}{z}, 0\right]$$

$$= \frac{1}{2\pi i}\left[\frac{e^{i\alpha z} - 1}{z} - \frac{e^{-i\alpha z} - 1}{z}, 0\right]$$

$$= \frac{1}{2\pi i}\left[\frac{e^{i\alpha z} - 1}{z}, \frac{e^{-i\alpha z} - 1}{z}\right] =: [G(z,\alpha)],$$

since $(e^{-i\alpha z} - 1)/z$ is a real analytic function which can be added to the upper and lower component of the defining function. As in the previous example, we have

$$e^{i\alpha z} = e^{i\alpha(x+i\epsilon)} = e^{-\alpha\epsilon}e^{i\alpha x}, \quad e^{-i\alpha z} = e^{-i\alpha(x-i\epsilon)} = e^{-\alpha\epsilon}e^{-i\alpha x}.$$

This implies

$$\lim_{\alpha\to\infty} \frac{\sin(\alpha x)}{\pi x} = \frac{1}{2\pi i}\left[\lim_{\alpha\to\infty}\frac{e^{i\alpha z} - 1}{z}, \lim_{\alpha\to\infty}\frac{e^{-i\alpha z} - 1}{z}\right]$$

$$= \left[-\frac{1}{2\pi i}\frac{1}{z}, -\frac{1}{2\pi i}\frac{1}{z}\right] = \left[-\frac{1}{2\pi i}\frac{1}{z}\right].$$

Finally, we have in the sense of hyperfunctions,

$$\lim_{\alpha\to\infty} \frac{\sin(\alpha x)}{\pi x} = \delta(x). \tag{2.2}$$

Problem 2.1. By using the same technique, show that

$$\lim_{\alpha \to \infty} \frac{1 - \cos(\alpha x)}{x} = \mathrm{fp}\frac{1}{x}. \tag{2.3}$$

Problem 2.2. By using

$$f(x; \alpha) := \int_0^\alpha 2\cos(2\pi tx)\, dt = \frac{\sin(2\pi \alpha x)}{\pi x}$$

show that we have

$$\lim_{\alpha \to \infty} f(x; \alpha) = \int_0^\infty \cos(2\pi tx)\, dt = \frac{\delta(x)}{2}.$$

Example 2.3. The function

$$f(z, \alpha) = \frac{2}{\pi} \frac{\alpha}{1 + \alpha^2 z^2}$$

is holomorphic for $|\Im z| < 1/\alpha$. Therefore, we can write

$$f(x, \alpha) = \left[\frac{2}{\pi} \frac{\alpha}{1 + \alpha^2 z^2}, 0\right] = \frac{1}{\pi}\left[\frac{\alpha}{1 + i\alpha z} + \frac{\alpha}{1 - i\alpha z}, 0\right].$$

Since $-\alpha/(1 + i\alpha x)$ is a real analytic function for $\alpha > 0$, we can add this function to the upper and lower component to obtain the equivalent defining function:

$$\frac{1}{\pi}\left[\frac{\alpha}{1 - i\alpha z}, -\frac{\alpha}{1 + i\alpha z}\right] = -\frac{i}{\pi}\left[\frac{i\alpha}{1 - i\alpha z}, -\frac{i\alpha}{1 + i\alpha z}\right]$$

$$= -\frac{i}{\pi}\left[\frac{1}{1/(i\alpha) - z}, -\frac{1}{1/(i\alpha) + z}\right] =: [G(z, \alpha)].$$

For $\alpha \to \infty$ the right-hand side has the limit

$$\frac{i}{\pi}[1/z, 1/z]/ = \frac{i}{\pi}\left[\frac{1}{z}\right] = 2\left[-\frac{1}{2\pi i}\frac{1}{z}\right],$$

thus we have in the sense of hyperfunctions,

$$\lim_{\alpha \to \infty} \frac{1}{\pi}\frac{\alpha}{1 + \alpha^2 x^2} = \delta(x). \tag{2.4}$$

Example 2.4. Let us now show that the sequence

$$\delta_n(x) := \sqrt{\frac{n}{\pi}}\, e^{-nx^2}$$

which represents a Gaussian probability density of mean 0 and variance $1/2n$ tends to $\delta(x)$ as $n \to \infty$ in the sense of hyperfunctions. We shall show that, for $\Im z > 0$ and for $\Im z < 0$,

$$F_n(z) := \sqrt{\frac{n}{\pi}}\, e^{-nz^2} \to -\frac{1}{2\pi i\, z}$$

as $n \to \infty$. To show this, we anticipate Proposition 2.14 which says that the defining function of our sequence can be represented as

$$F_n(z) = \sqrt{\frac{n}{\pi}}\, e^{-nz^2} = \sqrt{\frac{n}{\pi}}\, \frac{1}{2\pi i} \int_{-\infty}^{\infty} \frac{e^{-nt^2}}{t-z}\, dt.$$

The change of variables $u = \sqrt{n}\,t$ yields

$$F_n(z) = \sqrt{\frac{n}{\pi}}\, \frac{1}{2\pi i} \int_{-\infty}^{\infty} \frac{e^{-u^2}}{u-\sqrt{n}\,z}\, du.$$

We now use a result from [1, Abramowitz-Stegun Chapter 7, formulas 7.1.3, 7.4.13, 7.4.14]

$$\frac{1}{i\pi} \int_{-\infty}^{\infty} \frac{e^{-u^2}}{u-z}\, du = e^{-z^2} \operatorname{erfc}(-iz). \tag{2.5}$$

Therefore, we may write

$$F_n(z) = \frac{1}{2} \sqrt{\frac{n}{\pi}}\, e^{-nz^2} \operatorname{erfc}(-i\sqrt{n}\,z),$$

and all that we need is the asymptotic expansion of the error function (see [1, Abramowitz-Stegun formula 7.1.23])

$$\sqrt{\pi}\, z e^{z^2} \operatorname{erfc} z \sim 1, \quad (z \to \infty, |\arg z| < 3\pi/2). \tag{2.6}$$

Thus,

$$F_n(z) \to \frac{1}{2} \sqrt{\frac{n}{\pi}}\, e^{-nz^2} \frac{1}{\sqrt{\pi}(-i)\sqrt{n}\,z} e^{-(-i\sqrt{n}z)^2} = -\frac{1}{2\pi i\, z}$$

as $n \to \infty$. This shows that the Gaussian sequence tends in the sense of hyperfunctions to the Dirac impulse hyperfunction:

$$\lim_{n\to\infty} \sqrt{\frac{n}{\pi}}\, e^{-nx^2} = \delta(x). \tag{2.7}$$

Example 2.5. Consider the series of hyperfunctions

$$\sum_{n=0}^{\infty} \{u(x-n) - u(x-n-1)\}$$

with its sequence of partial sums

$$g_n(x) = \sum_{k=0}^{n-1} \{u(x-k) - u(x-k-1)\} = [G_n(z)].$$

Of course,

$$(-2\pi i)\, G_n(z) = \sum_{k=0}^{n-1} \log \frac{k-z}{k+1-z}$$

$$= \log \frac{-z}{1-z} + \log \frac{1-z}{2-z} + \log \frac{2-z}{3-z} + \cdots + \log \frac{n-1-z}{n-z}$$

$$= \log \frac{-z}{n-z}.$$

Hence,

$$g_n(x) = \left[-\frac{1}{2\pi i} \{\log(-z) - \log(n-z)\} \right] = u(x) - u(x-n)$$

as expected. The partial sum sequence

$$g_n(x) = \left[-\frac{1}{2\pi i} \log \frac{-z}{n-z} \right]$$

does not converge in the ordinary sense as $n \to \infty$. However, we can write

$$\log \frac{-z}{n-z} = \log(-z) - \log(n-z) = \log(-z) - \log[n(1-z/n)]$$

$$= \log(-z) - \log(1 - z/n) - \log n \sim \log(-z) - \log[(1 - z/n)],$$

where the new sequence of equivalent defining functions is now convergent,

$$g_n(x) = \left[-\frac{1}{2\pi i} \log \frac{-z}{1 - z/n} \to -\frac{1}{2\pi i} \log(-z) \right]$$

as $n \to \infty$. Therefore we have, in the sense of hyperfunctions,

$$\lim_{n\to\infty} g_n(x) = \sum_{k=0}^{\infty} \{u(x-k) - u(x-k-1)\} = u(x). \tag{2.8}$$

Example 2.6 (Dirac comb). Let us consider the series

$$\sum_{k=-\infty}^{\infty} \delta(x - kT)$$

for $T > 0$. We have $\delta(x - kT) = [-1/(2\pi i\,(z - kT))]$, but the series

$$-\frac{1}{2\pi i} \sum_{k=-\infty}^{\infty} \frac{1}{z - kT} = \sum_{k=-\infty}^{\infty} F_k(z)$$

does not converge. However, the following functions $G_k(z)$ are equivalent to the $F_k(z)$:

$$G_k(z) = -\frac{1}{2\pi i} \{ \frac{1}{z - kT} + \frac{1}{kT} \} = -\frac{1}{2\pi i} \frac{z}{(z - kT)kT}$$

$$= \frac{1}{2\pi i} \frac{1}{(kT)^2} \frac{z}{1 - \frac{z}{kT}},$$

and the resulting series converges in every compact sub-domain which does not contain an integer number. Furthermore, it is known that

$$\pi \cot(\pi z) = \frac{1}{z} + \sum_{k=-\infty, k\neq 0}^{\infty} \{ \frac{1}{z - k} + \frac{1}{k} \}$$

(see, for example, [4, Behnke-Sommer p. 245]). Therefore, we have in the sense of hyperfunctions

$$\sum_{k=-\infty}^{\infty} \delta(x - kT) = \left[\frac{i}{2T} \cot(\frac{\pi}{T} z) \right]. \tag{2.9}$$

Problem 2.3. By using the following partial fraction expansions,

$$\pi \tan(\pi z) = -\sum_{k=-\infty}^{\infty} \{\frac{1}{z - k + 1/2} + \frac{1}{k - 1/2}\},$$

$$\frac{\pi}{\sin(\pi z)} = \frac{1}{z} + \sum_{k=-\infty,k\neq 0}^{\infty} (-1)^k \{\frac{1}{z - k} + \frac{1}{k}\},$$

$$\frac{\pi}{\cos(\pi z)} = \pi + \sum_{k=-\infty,k\neq 0}^{\infty} (-1)^k \{\frac{1}{z - k + 1/2} + \frac{1}{k - 1/2}\},$$

$$\frac{\pi^2}{\sin^2(\pi z)} = \sum_{k=-\infty}^{\infty} \frac{1}{(z - k)^2},$$

$$\frac{\pi^2}{\cos^2(\pi z)} = \sum_{k=-\infty}^{\infty} \frac{1}{(z - k + 1/2)^2}$$

establish

$$\sum_{k=-\infty}^{\infty} \delta(x - (k - 1/2)T) = \left[-\frac{i}{2T} \tan(\frac{\pi}{T} z)\right], \qquad (2.10)$$

$$\sum_{k=-\infty}^{\infty} (-1)^k \delta(x - kT) = \left[\frac{i}{2T} \frac{1}{\sin(\frac{\pi}{T} z)}\right], \qquad (2.11)$$

$$\sum_{k=-\infty}^{\infty} (-1)^k \delta(x - (k - 1/2)T) = \left[\frac{i}{2T} \frac{1}{\cos(\frac{\pi}{T} z)}\right], \qquad (2.12)$$

$$\sum_{k=-\infty}^{\infty} \delta'(x - kT) = \left[-\frac{i\pi}{2T^2} \frac{1}{\sin^2(\frac{\pi}{T} z)}\right], \qquad (2.13)$$

$$\sum_{k=-\infty}^{\infty} \delta'(x - (k - 1/2)T) = \left[-\frac{i\pi}{2T^2} \frac{1}{\cos^2(\frac{\pi}{T} z)}\right]. \qquad (2.14)$$

Example 2.7. For the defining function at the end of the previous example we may write

$$\frac{i}{2T} \cot(\frac{\pi}{T} z) = -\frac{1}{2T} + \frac{1}{T} \frac{1}{1 - e^{2\pi i z/T}},$$

and also

$$\frac{i}{2T} \cot(\frac{\pi}{T} z) = -\frac{1}{2T} - \frac{1}{T} \frac{e^{-2\pi i z/T}}{1 - e^{-2\pi i z/T}}.$$

The first expression becomes

$$\frac{i}{2T} \cot(\frac{\pi}{T} z) = -\frac{1}{2T} + \frac{1}{T} \sum_{k=0}^{\infty} e^{i2\pi k z/T} = \frac{1}{2T} + \frac{1}{T} \sum_{k=1}^{\infty} e^{i2\pi k z/T},$$

and the series converges uniformly in the interior of the upper complex half-plane due to exponential damped terms when $\Im z > 0$. Similarly, we obtain for the second expression the compact convergent series

$$\frac{i}{2T} \cot(\frac{\pi}{T} z) = -\frac{1}{2T} - \frac{1}{T} \sum_{k=1}^{\infty} e^{-i2\pi k z/T},$$

in the lower complex half-plane, due to exponential damped terms when $\Im z < 0$. Thus, we may write

$$\left[\frac{i}{2T}\cot(\frac{\pi}{T}z)\right] = \left[\frac{1}{2T}, -\frac{1}{2T}\right] + \left[\frac{1}{T}\sum_{k=1}^{\infty}e^{i2\pi kz/T}, -\frac{1}{T}\sum_{k=1}^{\infty}e^{-i2\pi kz/T}\right]$$

$$= \frac{1}{T}\mathbf{1}(z) + \frac{1}{T}\sum_{k=1}^{\infty}e^{i2\pi kz/T}\,\mathbf{1}_{+}(z) + \frac{1}{T}\sum_{k=1}^{\infty}e^{-i2\pi kz/T}\mathbf{1}_{-}(z).$$

In Section 2.6 we shall show that the last expression equals the Fourier series

$$\sum_{k=-\infty}^{\infty}e^{i\frac{2\pi}{T}kx}.$$

Finally, we obtain in the sense of hyperfunctions

$$\sum_{k=-\infty}^{\infty}\delta(x-kT) = \frac{1}{T}\sum_{k=-\infty}^{\infty}e^{i\frac{2\pi}{T}kx}. \qquad (2.15)$$

Let us now investigate the question of convergence in the sense of hyperfunctions of a general infinite Dirac comb of the form

$$\sum_{k=0}^{\infty}\sum_{j=0}^{m_k-1}c_{kj}\,\delta^{(j)}(x-\alpha_k),$$

where the c_{kj} form a given sequence of complex numbers, and α_k a sequence of real numbers. We assume that $|\alpha_k| \to \infty$ as $k \to \infty$. The corresponding series of defining functions

$$-\frac{1}{2\pi i}\sum_{k=0}^{\infty}\sum_{j=0}^{m_k-1}c_{kj}\frac{(-1)^j j!}{(z-\alpha_k)^{j+1}}$$

does generally not converge. However, the theorem of Mittag-Leffler from complex variable theory indicates that there exists a meromorphic function

$$F(z) := \sum_{k=0}^{\infty}\{g_k(z) - p_k(z)\} + F_0(z)$$

having its poles at the points α_k with the corresponding singular parts

$$g_k(z) = \sum_{j=0}^{m_k-1}c_{kj}\frac{(-1)^j j!/(-2\pi i)}{(z-\alpha_k)^{j+1}},$$

where the series converges uniformly in any compact domain, and where $F_0(z)$ is an entire function. Here, the $p_k(z)$ are polynomials introduced for ensuring convergence of the series. For a proof of Mittag-Leffler's Theorem see for example [4, Behnke-Sommer p. 236], or [2, Ahlfors p. 185]. Because the $g_k(z) - p_k(z)$ are equivalent defining functions of

$$\sum_{j=0}^{m_k-1}c_{kj}\,\delta^{(j)}(x-\alpha_k),$$

and $F_0(z)$ is real analytic, we have convergence in the sense of hyperfunctions, i.e., we may always write

$$\sum_{k=0}^{\infty} \sum_{j=0}^{m_k-1} c_{kj}\, \delta^{(j)}(x - \alpha_k) = \left[\sum_{k=0}^{\infty} \{g_k(z) - p_k(z)\}\right] \tag{2.16}$$

with appropriate polynomials $p_k(z)$ in order to ensure convergence.

Problem 2.4. It may be believed that $\delta(x) = \infty$ at $x = 0$ must always hold. This is wrong as it may be shown by the following example:

$$\delta_n(x) := \begin{cases} -n, & |x| < 1/2n \\ 2n, & 1/2n < |x| < 1/n \\ 0, & \text{otherwise.} \end{cases}$$

Show that $\delta_n(x) \to \delta(x)$ as $n \to \infty$ in spite of the fact that $\delta_n(0) \to -\infty$. Hint: Show that the sequence of the defining function of $\delta_n(x)$ has the form

$$\frac{1}{2\pi i} \log\left\{\left(1 + \frac{(-1/z)}{n}(1 + O(1/n))\right)^n\right\}.$$

2.2   Cauchy-type Integrals

By a Cauchy integral is generally meant a line integral of the type

$$\frac{1}{2\pi i} \int_C \frac{f(t)}{t - z}\, dt \tag{2.17}$$

with a closed positively directed contour C. In the theory of hyperfunctions various integrals of this type, but with contours being no more necessarily a closed curve, are of great importance. We call an integral of the above type with a contour (open or closed) to be specified a *Cauchy-type integral*. The function $f(t)$ is called its *density*. C is generally assumed to be a smooth and simple contour (a closed or open line without points of intersection with itself, and continuously varying tangent). Furthermore, if the contour is closed, we adopt once and for all the convention that the contour C is always taken in the positive sense. This means that the interior of the domain enclosed by C is to the left when C is traversed in the positive sense. If the curve C is not closed we speak of an *interior point of C*, if it is different from the two endpoints.

The Cauchy-type integral is a special case of a more general curvilinear integral for which we cite the following

Proposition 2.1. *Let C be a closed or open contour situated wholly in the finite part of the complex plane, $f(t, z)$ a function continuous with respect to $t \in C$ and holomorphic in z in some domain $\mathcal{D}$ for all values of $t \in C$. Then, the function of the complex variable z defined by the curvilinear integral*

$$F(z) := \int_C f(t, z)\, dt$$

is holomorphic in the domain $\mathcal{D}$.

A proof can be found in any textbook of complex variable theory. For a contour of infinite length see Appendix A.3.

For the Cauchy-type integral with continuous density $f(t)$, the only points where the integrand fails to be holomorphic with respect to z are points on the contour C. The contour C thus constitutes a singular line for the function $F(z)$.

Proposition 2.2. *Let C be a simple curve of finite length from $a \in \mathbb{C}$ to $b \in \mathbb{C}$. Assume that $f(t)$ is holomorphic on C and let*

$$F(z) := \frac{1}{2\pi i} \int_C \frac{f(t)}{t - z} \, dt.$$

Then, $F(z)$ has the following properties:

(i) *$F(z)$ is holomorphic for all $z \notin C$.*

(ii) *$F(z)$ can analytically be continued from the left to an interior point $z \in C, z \neq a, z \neq b$, thereby defining a function $F_+(z)$ being also holomorphic at interior points of C. Similarly, $F(z)$ can analytically be continued from the right to an interior point $z \in C$, and defines a function $F_-(z)$, holomorphic at interior points of C.*

(iii) *For points $t \in C$ we have $f(t) = F_+(t) - F_-(t)$.*

Proof. The first statement (i) follows from the previous proposition. We shall prove the second and third property. Let $\mathcal{D}_f$ be the (open) domain containing C where $f(z)$ is holomorphic. We join b with a by a curve C' running in $\mathcal{D}_f$ and having no intersection with C. C' is chosen such that the domain $\mathcal{D}_1$ bounded by $C + C'$ is on the left-hand side of $C + C'$. For a point $z \in \mathcal{D}_1$, we have by Cauchy's integral theorem,

$$f(z) = \frac{1}{2\pi i} \int_C \frac{f(t)}{t - z} \, dt + \frac{1}{2\pi i} \int_{C'} \frac{f(t)}{t - z} \, dt,$$

thus

$$F(z) = f(z) - \frac{1}{2\pi i} \int_{C'} \frac{f(t)}{t - z} \, dt. \tag{2.18}$$

For a point z not in $\mathcal{D}_1$,

$$0 = \frac{1}{2\pi i} \int_C \frac{f(t)}{t - z} \, dt + \frac{1}{2\pi i} \int_{C'} \frac{f(t)}{t - z} \, dt.$$

The latter equation gives for $z \notin \mathcal{D}_1$,

$$F(z) = -\frac{1}{2\pi i} \int_{C'} \frac{f(t)}{t - z} \, dt. \tag{2.19}$$

The function

$$\frac{1}{2\pi i} \int_{C'} \frac{f(t)}{t - z} \, dt$$

is holomorphic for all $z \in C, z \neq a, z \neq b$. Equation (2.18) shows that $F(z)$ can analytically be continued from the left across the points lying in the interior of the curve C, thereby defining the the function $F_+(z)$. Similarly, by (2.19) we see that $F(z)$ can be analytically continued from the right across the points lying in the interior of the curve C, defining the the function $F_-(z)$. The difference of (2.18) and (2.19) just yields $f(z)$. $\qquad\square$

Applying this proposition to the theory of hyperfunction where the curve C now is a real interval, yields

Corollary 2.3. *Let $[a, b]$ be a compact real interval. Let $f(x) \in \mathcal{A}(I)$ be a real analytic function on the interval I containing $[a, b]$. Then,*

$$G(z) := \frac{1}{2\pi i} \int_a^b \frac{f(t)}{t - z} \, dt$$

is a defining function of a hyperfunction $g(x) = [G(z)]$ which equals $f(x)$ for $a < x < b$ and which is zero for $x < a$ and $x > b$.

This hyperfunction $g(x)$ is a hyperfunction with compact support and is called the projection of $f(x)$ on the interval $[a, b]$. In the next section we shall extend this notion of projection to a wider class of functions not necessarily real analytic. Next, we shall prove a general lemma about Cauchy-type integrals that

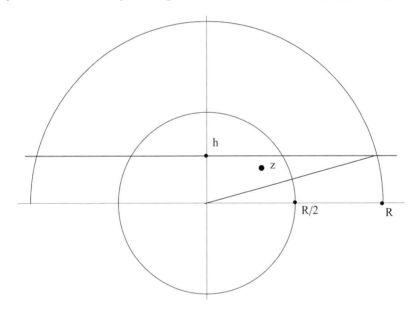

Figure 2.1: For the proof of Lemma 2.4

we shall need in the sequel.

Lemma 2.4. *Let the function $F(z)$, $z = x + iy$, be continuous in the upper half-plane $\mathbb{C}_+$ satisfying the two conditions*

(i) *$F(x + iy)$ tends to zero as $y \to \infty$ uniformly in x of any compact interval, i.e., $(\forall \epsilon > 0)(\forall [a, b] \subset \mathbb{R})(\exists h > 0)(y > h \Longrightarrow |F(x + iy)| < \epsilon, \forall x \in [a, b])$,*

(ii) *$F(x + iy)$ is uniformly bounded with respect to x of any compact interval and where y varies in a horizontal strip of finite width, i.e., $(\forall h > 0)(\forall [a, b] \subset \mathbb{R})(\exists M > 0)(0 < y < h \Longrightarrow |F(x + iy)| < M, \forall x \in [a, b])$.*

Then, we have

$$\lim_{R \to \infty} \frac{1}{2\pi i} \int_{C_R} \frac{F(t)}{t - z} \, dt = 0,$$

where C_R is the semicircle $t = R \exp(i\tau)$, $0 \le \tau \le \pi$ with radius R in the upper half-plane.

Proof. See Figure 2.1. Let $z \in \mathbb{C}$ be fixed, then choose R so large that $|z| < R/2$. For any $\epsilon > 0$ there is an $h = h(\epsilon)$ such that for all $y > h > 0$ $|F(x+iy)| < \epsilon$ holds for all $x \in [-R, R]$. Let γ_R be the part of the contour C_R above the horizontal line $\Im z = h$. The following estimates then hold:

$$\left| \int_{\gamma_R} \frac{F(t)}{t-z} \, dt \right| \le \int_{\gamma_R} \frac{|F(t)|}{|t-z|} |dt| \le \epsilon \int_{\gamma_R} \frac{1}{|t-z|} |dt|$$

$$\le \epsilon \int_{\gamma_R} \frac{1}{|t|-|z|} |dt| = \epsilon \int_{\gamma_R} \frac{1}{R-|z|} |dt| \le \epsilon \int_{\gamma_R} \frac{1}{R-R/2} |dt|$$

$$= \frac{2\epsilon}{R} \int_{\gamma_R} |dt| \le \frac{2\epsilon}{R} R\pi = 2\epsilon\pi.$$

Therefore, we obtain

$$\left| \frac{1}{2\pi i} \int_{\gamma_R} \frac{F(t)}{t-z} \, dt \right| < \epsilon.$$

For the selected $h = h(\epsilon)$ there is an $M = M(h(\epsilon))$ such that $|F(x+iy)| < M$ for all $0 < y < M$ and all $x \in [-R, R]$. Let γ_h be one of the two parts of the contour C_R between the real axis and the horizontal line $\Im z = h$. The following estimates then hold:

$$\left| \int_{\gamma_h} \frac{F(t)}{t-z} \, dt \right| \le M \int_{\gamma_h} \frac{1}{|t-z|} |dt| \le M \int_{\gamma_h} \frac{1}{\sqrt{R^2 - h^2} - R/2} |dt|$$

$$\le \frac{M}{R} \int_{\gamma_h} \frac{1}{\sqrt{1-(h/R)^2} - 1/2} |dt|.$$

For sufficiently large R we then have the estimate

$$\left| \frac{1}{2\pi i} \int_{\gamma_h} \frac{F(t)}{t-z} \, dt \right| < \frac{M}{2\pi R} \int_{\gamma_h} \frac{1}{3/4 - 1/2} |dt|$$

$$= \frac{2M}{\pi R} \int_{\gamma_h} |dt| \le \frac{2M}{\pi R} 2h = \frac{4Mh}{\pi R} \to 0$$

as $R \to \infty$. This completes the proof. □

Clearly, an analogous lemma holds for the symmetric situation with respect to the real axis.

It is worth singling out the following subclass of holomorphic functions.

Definition 2.4. We call a function $F_+(z)$ *un upper half-plane function*, if

(i) $F_+(z)$ is holomorphic in the upper half-plane $0 < \Im z < \infty$,

(ii) for any fixed z,

$$\int_{C_R} \frac{F_+(t)}{t-z} \, dt \to 0 \ , \text{ as } R \to \infty,$$

where C_R is the semicircle $t = R \exp(i\tau)$, $0 \le \tau \le \pi$ with radius R in the upper half-plane. The set of all upper half-plane functions is denoted by $\mathcal{O}_0(\mathbb{C}_+)$.

The last term on the right-hand side is integrated and simplifies (for all $\epsilon > 0$) to

$$f(x) \log \frac{b-x}{x-a}.$$

The first two terms on the right-hand side are estimated as follows:

$$\left| \int_a^{x-\epsilon} \frac{f(t)-f(x)}{t-x} \, dt \right| \leq \int_a^{x-\epsilon} \frac{|f(t)-f(x)|}{|t-x|} \, dt$$

$$\leq C \int_a^{x-\epsilon} \frac{1}{(x-t)^{1-\alpha}} \, dt.$$

Similarly, we have

$$\left| \int_{x+\epsilon}^b \frac{f(t)-f(x)}{t-x} \, dt \right| \leq C \int_{x+\epsilon}^b \frac{1}{(t-x)^{1-\alpha}} \, dt.$$

This shows that the integrals are convergent. By letting $\epsilon \to 0+$ we obtain the assertion of the proposition. □

We know that for a holomorphic function defined by a Cauchy-type integral the integration contour constitutes a singular line where the function ceases to be holomorphic. Let us now investigate in detail the behavior of a function defined by a Cauchy-type integral on the integration contour. First, we establish the following basic lemma.

Lemma 2.8. *Let $a < x < b$ and assume that the density $f(t) \in \mathcal{H}_\alpha(a,b)$, i.e., satisfies the Hölder condition, then the function of the complex variable z,*

$$h(z) := \int_a^b \frac{f(t)-f(x)}{t-z} \, dt, \qquad (2.25)$$

behaves as a continuous function at the point $t \in (a,b)$ as z approaches x from above or from below along any path, i.e.,

$$\lim_{z \to x} h(z) = \int_a^b \frac{f(t)-f(x)}{t-x} \, dt. \qquad (2.26)$$

We present the proof of the lemma for a non-tangential approach. For the extension to a tangential approach of z to x the reader may consult [12, Gakhov p. 22].

Proof. Consider the difference

$$h(z) - h(x) = \int_a^b (z-x) \, \frac{f(t)-f(x)}{(t-z)(t-x)} \, dt =: I_1 + I_2$$

where I_1 is the integral from $x-\delta$ to $x+\delta$, and I_2 the sum of the two integrals from a to $x-\delta$ and from $x+\delta$ to b; $\delta > 0$ is chosen sufficiently small. First we estimate I_1 assuming that z approaches x in a non-tangential way to the real axis. Consider

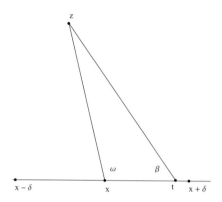

Figure 2.2: To Lemma 2.8

Figure 2.2. According to the non-tangential approach we have $0 < \omega_0 \leq \omega \leq \pi - \omega_0$ which implies, by the sine theorem, the estimate A:

$$\frac{|z - x|}{|t - z|} = \frac{\sin \beta}{\sin \omega} \leq \frac{1}{\sin \omega_0} =: K > 0.$$

The Hölder condition implies the estimate B:

$$\left| \frac{f(t) - f(x)}{t - x} \right| < C|t - x|^{\alpha - 1} = C r^{\alpha - 1}$$

where $r = |t - x|$. By using both estimates A and B we can write

$$|I_1| \leq \int_{x-\delta}^{x+\delta} \left| \frac{z - x}{t - z} \right| \left| \frac{f(t) - f(x)}{t - x} \right| dt < KC \int_{x-\delta}^{x+\delta} r^{\alpha - 1} dt$$

$$= 2KC \int_0^\delta r^{\alpha - 1} dr = 2KC \frac{\delta^\alpha}{\alpha}.$$

This shows that given $\epsilon > 0$, we can choose a corresponding $\delta > 0$ such that $|I_1| < \epsilon/2$. Next we estimate I_2. We observe that the two integrals in the sum of I_2 are continuous functions of z since the denominator $(t - z)(t - x)$ of the integrand remains bounded away from zero. Thus, for the already chosen $\delta > 0$ we can choose $|z - x|$ sufficiently small such that $|I_2| < \epsilon/2$. This establishes the assertion of the lemma.   □

It should be mentioned that the Hölder condition was only used locally in a neighborhood of the point x. The estimation for I_2 remains valid for an only piecewise continuous function on $(a, x - \delta) \cup (x + \delta, b)$. Consider now

$$\int_a^b \frac{dt}{t - z} dt$$

as a function of the complex variable z. If z is real and $a < z < b$, then Proposition 2.7 for $f(t) = 1$ yields

$$\text{pv} \int_a^b \frac{dt}{t - z}\, dt = \log \frac{b - z}{z - a}.$$

If z is not real, i.e., $\Im z \neq 0$, then

$$\int_a^b \frac{dt}{t - z}\, dt = \log \frac{b - z}{a - z}$$

which now involves the logarithm of a complex quantity. We select the principal branch of log and make an incision along the negative part of the real axis which implies the relation

$$\log w = \log(-w) + i \operatorname{sgn}(\Im w)\, \pi.$$

Thus, in this case we can write with $w = (b - z)/(a - z)$,

$$\int_a^b \frac{dt}{t - z}\, dt = \log \frac{b - z}{a - z} = \log w = \log(-w) + i \operatorname{sgn}(\Im w)\, \pi$$

$$= \log \frac{b - z}{z - a} + i \operatorname{sgn}(\Im w)\, \pi = \log \frac{b - z}{z - a} + i\pi \operatorname{sgn}(\Im z),$$

and we have established the following lemma.

Lemma 2.9. *If $z \in \mathbb{C}$ we have*

$$(\text{pv}) \int_a^b \frac{dt}{t - z}\, dt = \begin{cases} \log \frac{b-z}{z-a} + i\pi, & \Im z > 0 \\ \log \frac{b-z}{z-a}, & \Im z = 0 \\ \log \frac{b-z}{z-a} - i\pi, & \Im z < 0. \end{cases} \tag{2.27}$$

We have written (pv) because the pv is only necessary for the middle line in the above statement. Consider now the Cauchy-type integral

$$w(z) := \frac{1}{2\pi i} \int_a^b \frac{f(t)}{t - z}\, dt$$

$$= \frac{1}{2\pi i} \int_a^b \frac{f(t) - f(x)}{t - z}\, dt + \frac{f(x)}{2\pi i} (\text{pv}) \int_a^b \frac{dt}{t - z}\, dt.$$

This eventually implies the relation

$$\lim_{z \to x} \frac{1}{2\pi i} \int_a^b \frac{f(t) - f(x)}{t - z}\, dt + \frac{f(x)}{2\pi i} \log \frac{b - z}{z - a} = \begin{cases} w(z) - \frac{f(x)}{2}, & \Im z > 0 \\ w(z), & \Im z = 0 \\ w(z) + \frac{f(x)}{2}, & \Im z < 0. \end{cases}$$

By Lemma 2.8, the left-hand side remains in all three cases the same regardless of whether z tends from above, from the left or right, or from below to x. Therefore,

$$w(x + i0) - \frac{f(x)}{2} = \frac{1}{2\pi i} \text{pv} \int_a^b \frac{f(t)}{t - x}\, dt = w(x - i0) + \frac{f(x)}{2}, \tag{2.28}$$

and we have established the *Sokhotski formulas*.

Proposition 2.10. *Let $f(t) \in \mathcal{H}_\alpha(a, b)$ satisfy the Hölder condition and $[a, b]$ be a compact interval. Then, the Cauchy-type integral*

$$w(z) = \frac{1}{2\pi i} \int_a^b \frac{f(t)}{t - z} \, dt \tag{2.29}$$

has limiting values $w(x+i0$ and $w(x-i0)$ at all points of the open interval (a, b) on approaching the real line from above and from below along an arbitrary path. Furthermore, between the density $f(x)$, these limiting values, and the Cauchy principal value integral (2.23) the Sokhotski formulas hold:

$$w(x + i0) - w(x - i0) = f(x), \tag{2.30}$$

$$w(x + i0) + w(x - i0) = \frac{1}{\pi i} \, \mathrm{pv} \int_a^b \frac{f(t)}{t - x} \, dt. \tag{2.31}$$

Because $w(z)$ is holomorphic in $\mathbb{C} \setminus [a, b]$, thus $w(x)$ is real analytic to the left and to the right of $[a, b]$, the function $w(z)$ defined by the Cauchy-type integral with density $f(x)$ qualifies for the defining function of a hyperfunction whose values on (a, b) equals those of the given ordinary function $f(x)$ and vanishes to the left and to the right of $[a, b]$. Thus, we have established the proposition.

Proposition 2.11. *If $[a, b]$ is a compact interval, and $f(x) \in \mathcal{H}_\alpha(a, b)$, then a hyperfunction*

$$\chi_{(a,b)} f(x) := \left[\frac{1}{2\pi i} \int_a^b \frac{f(x)}{x - z} \, dx \right] \tag{2.32}$$

is defined called the projection of $f(x)$ on the interval (a, b). This projection equals $f(x)$ on (a, b) and vanishes outside of $[a, b]$.

Example 2.8. Take the constant function $x \mapsto 1$. We shall compute its projection on the finite interval (a, b). Because

$$\frac{1}{2\pi i} \int_a^b \frac{dx}{x - z} = \frac{1}{2\pi i} \left\{ \log(b - z) - \log(a - z) \right\} = -\frac{1}{2\pi i} \log \frac{a - z}{b - z}$$

we obtain the hyperfunction $\chi_{(a,b)} 1 =: \chi_{(a,b)} = u(x - a) - u(x - b)$.

Corollary 2.3 and Proposition 2.11 give sufficient conditions that the hyperfunction

$$\left[\frac{1}{2\pi i} \int_a^b \frac{f(x)}{x - z} \, dx \right]$$

yields an interpretation of the given function $f(x)$ on (a, b). Let us now investigate the case of a semi-infinite interval $[a, \infty)$. In this event we require that

(i) the function $f(x)$ obeys the Hölder condition in every finite sub-interval of $[a, \infty)$,

(ii) for large x the inequality

$$|f(x)| < \frac{A}{|x|^\mu}, \quad \mu > 0, \; A > 0 \tag{2.33}$$

holds, which implies $f(\infty) = 0$.

The Cauchy-type integral

$$w(z) := \frac{1}{2\pi i} \int_a^\infty \frac{f(t)}{t - z}\, dt \qquad (2.34)$$

then converges and is a holomorphic function in $\mathbb{C} \setminus [a, \infty)$ due to the uniform convergence at its upper limit. Suppose now that $z = x$ is real and $a < x < \infty$. Then

$$\mathrm{pv} \int_a^\infty \frac{f(t)}{t - x}\, dt := \lim_{\epsilon \to 0+} \left\{ \int_a^{x-\epsilon} \frac{f(t)}{t - x}\, dt + \int_{x+\epsilon}^\infty \frac{f(t)}{t - x}\, dt \right\}, \qquad (2.35)$$

and, by writing for any $A > x + \epsilon$,

$$\int_a^{x-\epsilon} + \int_{x+\epsilon}^\infty = \int_A^\infty + \int_a^{x-\epsilon} + \int_{x+\epsilon}^A,$$

where the first integral is independent of ϵ, and the limit of the sum of the latter two integrals exists, as was shown above. We see that the principal-value integral

$$\mathrm{pv} \int_a^\infty \frac{f(t)}{t - x}\, dt, \quad x \in (a, \infty]$$

exists. Repeating the reasoning preceding Proposition 2.10 we have the Sokhotski formulas

$$w(x + i0) - w(x - i0) = f(x), \qquad (2.36)$$

$$w(x + i0) + w(x - i0) = \frac{1}{\pi i} \mathrm{pv} \int_a^\infty \frac{f(t)}{t - x}\, dt. \qquad (2.37)$$

We have proved

Proposition 2.12. *Let $f(x)$ satisfy the Hölder condition in every finite subinterval of $[a, \infty)$. For large $|x|$ we require that the inequality*

$$|f(x)| < \frac{A}{|x|^\mu}, \quad \mu > 0, \; A > 0$$

holds. Then a hyperfunction

$$\chi_{(a,\infty)} f(x) := \left[\frac{1}{2\pi i} \int_a^\infty \frac{f(x)}{x - z}\, dx \right] \qquad (2.38)$$

exists which is again called the projection of $f(x)$ on (a, ∞). This projection equals $f(x)$ on (a, ∞) and vanishes for $x < a$.

A similar result holds in the event of a semi-infinite interval $(-\infty, b]$. In the sequel we also use the notation $u(x - a) f(x)$ for $\chi_{(a,\infty)} f(x)$, and $u(b - x) f(x)$ for $\chi_{(-\infty,b)} f(x)$.

Example 2.9. Let us compute the projection on $(0, \infty)$ of the ordinary function $f(x) = x^\alpha$. For $-1 < \alpha < 0$ the improper integral

$$F(z) = \frac{1}{2\pi i} \int_0^\infty \frac{x^\alpha}{x - z}\, dx$$

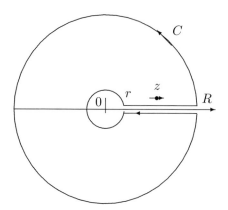

Figure 2.3: Contour in Example 2.9

is convergent. In order to compute this integral, we consider the contour integral

$$\frac{1}{2\pi i} \oint_C \frac{(-t)^\alpha}{t - z} \, dt,$$

where the contour is shown in Figure 2.3. Because $(-t)^\alpha$ is holomorphic inside C, the residue at $t = z$ is $(-z)^\alpha$ which is the value of the contour integral. Because $-1 < \alpha < 0$, the contribution to the contour integral of the small circle with radius r, and of the large circle with radius R tends to zero as $r \to 0+, R \to \infty$, respectively. Therefore, it remains

$$(-z)^\alpha = \frac{1}{2\pi i} \Big\{ \int_0^\infty \frac{(-(x + i0))^\alpha}{x - z} \, dx - \int_0^\infty \frac{(-(x - i0))^\alpha}{x - z} \, dx \Big\}$$

$$= \frac{e^{-\alpha\pi i} - e^{\alpha\pi i}}{2\pi i} \int_0^\infty \frac{x^\alpha}{x - z} \, dx = -\frac{\sin(\alpha\pi)}{\pi} \int_0^\infty \frac{x^\alpha}{x - z} \, dx.$$

Thus,

$$F(z) = \frac{1}{2\pi i} \int_0^\infty \frac{x^\alpha}{x - z} \, dx = -\frac{1}{2i \sin(\pi\alpha)} (-z)^\alpha.$$

This is just the defining function of the familiar hyperfunction $x_+^\alpha = u(x)x^\alpha$. The temporary restriction $-1 < \alpha < 0$ can now be removed, because the expression

$$-\frac{1}{2i \sin(\pi\alpha)} (-z)^\alpha$$

as a function of α with z fixed can analytically be continued to $\mathbb{C} \setminus \mathbb{Z}$. Therefore, for $\alpha \in \mathbb{C} \setminus \mathbb{Z}$ we obtain

$$\chi_{(0,\infty)} x^\alpha = u(x) \, x^\alpha.$$

Problem 2.5. Show that, for $\alpha \in \mathbb{C} \setminus \mathbb{Z}$,

$$\chi_{(-\infty,0)} |x|^\alpha = u(-x)|x|^\alpha.$$

Next we consider the case where the interval (a, b) becomes $\mathbb{R}$. We follow the exposition in [12, Gakhov]. In this case the behavior at infinity of $f(x)$ is somewhat less restrictive. We assume that

(i) the function $f(x)$ obeys the Hölder condition in every finite subinterval, and tends to a definite limit f_∞ as $x \to \pm\infty$, i.e., $f(-\infty) = f(\infty) = f_\infty$.

(ii) for large $|x|$ the following inequality holds:

$$|f(x) - f_\infty| < \frac{A}{|x|^\mu}, \quad \mu > 0, \ A > 0. \tag{2.39}$$

If $f_\infty \neq 0$, the Cauchy-type integral

$$w(z) := \frac{1}{2\pi i} \int_{-\infty}^{\infty} \frac{f(t)}{t - z} dt \tag{2.40}$$

is divergent. However, we shall establish that for $\Im z \neq 0$ it will exist as a principal-value integral.

Lemma 2.13. *For a non-real $z \in \mathbb{C}$ we have*

$$\frac{1}{2\pi i} \mathrm{pv} \int_{-\infty}^{\infty} \frac{f(t)\,dt}{t - z} = \frac{1}{2\pi i} \int_{-\infty}^{\infty} \frac{f(t) - f_\infty}{t - z} dt + \mathrm{sgn}(\Im z) \frac{f_\infty}{2}. \tag{2.41}$$

The principal-value integral on the left-hand side means here

$$\mathrm{pv} \int_{-\infty}^{\infty} \frac{f(t)}{t - z} dt := \lim_{R \to \infty} \int_{-R}^{R} \frac{f(t)}{t - z} dt,$$

whereas the integral on the right-hand side is convergent. If $f_\infty = 0$ the integral (2.40) is convergent in the ordinary sense and the pv is not necessary.

Proof. Consider

$$\int_{-R}^{R} \frac{f(t)}{t - z} dt = \int_{-R}^{R} \frac{f(t) - f_\infty}{t - z} dt + f_\infty \int_{-R}^{R} \frac{dt}{t - z},$$

where in the first integral on the right-hand side the integrand for large $|t|$ is of order $|t|^{-1-\mu}$ which implies convergence of the integral with infinite limits. The second integral is readily computed and we obtain

$$\int_{-R}^{R} \frac{dt}{t - z} = \log \frac{R - z}{-R - z} = \log \left| \frac{R - z}{-R - z} \right| + i\,\mathrm{sgn}(\Im z)\,|\beta|$$

where $\beta = \arg(R - z) - \arg(-R - z)$ is the angle between the straight lines connecting the point z with the points R and $-R$ on the real axis. If $R \to \infty$ the log term on the right-hand side tends to zero whereas $|\beta| \to \pi$. This proves what is asserted. $\square$

If now $z = x$ becomes real, the Cauchy-type integral has to be taken as a principal-value integral in a twofold sense, with respect to x and with respect to the infinite integration limits:

$$\mathrm{pv} \int_{-\infty}^{\infty} \frac{f(t)}{t - x} dt := \lim_{R \to \infty} \lim_{\epsilon \to 0+} \left\{ \int_{-R}^{x-\epsilon} \frac{f(t)}{t - x} dt + \int_{x+\epsilon}^{R} \frac{f(t)}{t - x} dt \right\}. \tag{2.42}$$

In particular,

$$\text{pv} \int_{-\infty}^{\infty} \frac{1}{t-x} \, dt = \lim_{R\to\infty} \log\{\frac{(-\epsilon)}{-R-x} \frac{R-x}{\epsilon}\} = 0. \tag{2.43}$$

This last relation enables us to write

$$\text{pv} \int_{-\infty}^{\infty} \frac{f(t)}{t-x} \, dt = \lim_{R\to\infty} \lim_{\epsilon\to 0+} \left\{ \int_{-R}^{x-\epsilon} \frac{f(t)-f_\infty}{t-x} \, dt + \int_{x+\epsilon}^{R} \frac{f(t)-f_\infty}{t-x} \, dt \right\}$$

$$= \lim_{\epsilon\to 0+} \left\{ \int_{-\infty}^{x-\epsilon} \frac{f(t)-f_\infty}{t-x} \, dt + \int_{x+\epsilon}^{\infty} \frac{f(t)-f_\infty}{t-x} \, dt \right\}$$

where the last two integrals converge in the ordinary sense. As in the case of a semi-infinite interval concerning the limit $\epsilon \to 0+$, the fact that the integration contour is infinite is immaterial, since we can return to the case of a finite interval by grouping the integrals as follows:

$$\int_{-\infty}^{x-\epsilon} + \int_{x+\epsilon}^{\infty} = \left\{ \int_{-\infty}^{x-A} + \int_{x+A}^{\infty} \right\} + \int_{x-A}^{x-\epsilon} + \int_{x+\epsilon}^{x+A} \,,$$

where the limit in the curly brackets is independent of ϵ, and the limit of the sum of the two latter integrals exists. The remaining steps are the same as in the event of a finite interval (a, b) above, and we may establish again the Sokhotski formulas for the case of the infinite interval,

$$w(x + i0) - w(x - i0) = f(x), \tag{2.44}$$

$$w(x + i0) + w(x - i0) = \frac{1}{\pi i} \text{pv} \int_{-\infty}^{\infty} \frac{f(t)}{t-x} \, dt. \tag{2.45}$$

We finally have obtained

Proposition 2.14. *Let $f(x)$ satisfy the Hölder condition in every finite subinterval of $\mathbb{R}$ and assume that $f(x)$ tends to a finite limit f_∞ as $x \to \pm\infty$. For large $|x|$ we require that the inequality*

$$|f(x) - f_\infty| < \frac{A}{|x|^\mu}, \quad \mu > 0, \ A > 0$$

holds. Then a hyperfunction

$$\chi_{(-\infty,\infty)} f(x) := \left[\frac{1}{2\pi i} \text{pv} \int_{-\infty}^{\infty} \frac{f(x)}{x-z} \, dx \right] \tag{2.46}$$

is defined having the same values as the given ordinary function $f(x)$.

Observe that in this case the defining function might have independent upper and lower components:

$$F_+(z) := \frac{1}{2\pi i} \text{pv} \int_{-\infty}^{\infty} \frac{f(x)}{x-z} \, dx, \quad \Im z > 0,$$

$$F_-(z) := \frac{1}{2\pi i} \text{pv} \int_{-\infty}^{\infty} \frac{f(x)}{x-z} \, dx, \quad \Im z < 0. \tag{2.47}$$

The pv is only necessary if the common limit value $f_\infty \neq 0$.

The Hilbert transform of an ordinary function $f(x)$ is defined and denoted by

$$\mathcal{H}[f](x) := \frac{1}{\pi} \, \text{pv} \int_{-\infty}^{\infty} \frac{f(t)}{t - x} \, dx \tag{2.48}$$

(see Chapter 5). Thus, the Sokhotski formulas may also be written as

$$w(x + i0) - w(x - i0) = f(x), \tag{2.49}$$
$$i \{w(x + i0) + w(x - i0)\} = \mathcal{H}[f](x). \tag{2.50}$$

Example 2.10. Let $f(x) = 1$ for all x, then $f_\infty = 1$. Assuming $\Im z > 0$ we obtain

$$\frac{1}{2\pi i} \int_{-R}^{R} \frac{1}{x - z} \, dx = \frac{1}{2\pi i} \log \frac{R - z}{-R - z}$$

$$:= \frac{1}{2\pi i} \{\log \frac{|R - z|}{|-R - z|} + i \arg \frac{|R - z|}{|-R - z|}\} \to 1/2$$

as $R \to \infty$. It $\Im z < 0$ the same limit tends to $-1/2$ (the reader is advised to draw a figure). Therefore, we obtain

$$\chi_{(-\infty,\infty)} 1 := [1/2, -1/2] = [\mathbf{1}(z)].$$

Remark. If $f(x)$ is a real-valued function, then the upper and lower component of the defining function defined by (2.47) satisfy, for $\Im z > 0$ and $\Im z < 0$,

$$F_-(\bar{z}) = \overline{F_+(z)}, \text{ and } F_+(\bar{z}) = \overline{F_-(z)},$$

respectively. In this event we say the *one function is the reflection of the other*, or that they are *conjugate holomorphic functions*. Thus, if $f(x)$ is real-valued, the upper and lower components of the defining function of the corresponding hyperfunction are conjugate holomorphic functions.

2.3.3 Convergence Factors

We again consider the case of an infinite interval (a, b). For definiteness we take a finite and $b = \infty$. It may happen that for a given function $f(x)$ the defining integral for the projection

$$\chi_{(a,\infty)} f(x) = \left[\frac{1}{2\pi i} \int_{a}^{\infty} \frac{f(x)}{x - z} \, dx\right]$$

does not converge as an improper integral. In this case the introduction of a convergence factor in the integrand may be helpful. A *convergence factor of the first kind is a real analytic and non-vanishing function* $\phi(x)$ on $[a, \infty)$ such that the improper integral is rendered convergent upon the introduction of $\phi(x)$ in the denominator of the integrand. Candidates for convergence factors of the first kind are functions such as

$$(1 + x^2)^n, \quad (x - c)^n \ (\Im c \neq 0), \quad \exp(x^2).$$

Instead of

$$F(z) = \frac{1}{2\pi i} \int_a^\infty \frac{f(x)}{x - z}\, dx,$$

we then consider the improper integral

$$G(z) = \frac{\phi(z)}{2\pi i} \int_a^\infty \frac{f(x)}{\phi(x)\,(x - z)}\, dx$$

which may now converge uniformly at ∞ with respect to z in any compact sub domain of $\mathbb{C} \setminus [a, \infty)$. Let us show that $G(z)$ is indeed a defining function. By (2.36) we have, for $a < x < \infty$,

$$\lim_{y \to 0+} \frac{G_+(x + iy) - G_-(x - iy)}{\phi(x)} = \frac{f(x)}{\phi(x)}$$

which immediately yields

$$\lim_{y \to 0+} \{G_+(x + iy) - G_-(x - iy)\} = f(x).$$

Next, we show that the choice of the convergence factor does not matter. Let $\psi(x)$ be another convergence factor of the first kind that renders the integral convergent, giving another defining function

$$H(z) = \frac{\psi(z)}{2\pi i} \int_a^\infty \frac{f(x)}{\psi(x)\,(x - z)}\, dx$$

for $f(x)$. Then,

$$H(z) - G(z) = \frac{1}{2\pi i} \int_a^\infty \{\frac{\psi(z)}{\psi(x)} - \frac{\phi(z)}{\phi(x)}\} \frac{f(x)}{(x - z)}\, dx.$$

Expanding $\phi(x)$ and $\psi(x)$ at $x = z$ yields

$$\frac{\psi(z)}{\psi(x)} - \frac{\phi(z)}{\phi(x)} = \frac{\psi(z)\phi(x) - \psi(x)\phi(z)}{\psi(x)\phi(x)}$$

$$= (x - z)\{\psi(z)[\phi'(z) + (\phi''(z)/2)\,(x - z) + \cdots]$$
$$- \phi(z)[\psi'(z) + (\psi''(z)/2)\,(x - z) + \cdots]\}/(\phi(x)\psi(x))$$
$$=: (x - z)g(x, z)/(\phi(x)\psi(x)),$$

where $g(x, z)$ is holomorphic with respect to z (and to x). Therefore,

$$H(z) - G(z) = \frac{1}{2\pi i} \int_a^\infty \frac{g(x, z)f(x)}{\psi(x)\phi(x)}\, dx$$

converges uniformly and represents a real analytic function on $\mathbb{R}$. This proves that $H(z) \sim G(z)$. Noting that the Hölder condition of $f(x)/\phi(x)$ differs at most by a constant factor of those of $f(x)$, we have established the following

Proposition 2.15. *If a convergence factor of the first kind $\phi(x)$ exists such that $f(x)/\phi(x)$ meets the hypotheses of Proposition 2.12, and such that the integral in the right-hand side of*

$$G(z) := \frac{\phi(z)}{2\pi i} \int_a^\infty \frac{f(x)}{\phi(x)\,(x - z)}\, dx$$

is rendered convergent, then $G(z)$ yields a defining function of the hyperfunction $\chi_{(a,\infty)}f(x)$. Moreover, the defined hyperfunction is irrespective of the choice of the convergence factor $\phi(x)$.

Example 2.11. Let $f(x) = x$ be the ordinary given function. We want to compute the projection $\chi_{(0,\infty)}x$. The function $\phi(x) = (x+i)^2$ is a suitable convergence factor. We compute

$$
\begin{aligned}
F(z) &= \frac{(z+i)^2}{2\pi i} \int_0^\infty \frac{t}{(t+i)^2\,(t-z)}\,dt \\
&= \frac{(z+i)^2}{2\pi i} \int_0^\infty \frac{z}{(z+i)^2}\{\frac{1}{t-z} - \frac{1}{t+i}\} + \frac{i}{(z+i)(t+i)^2}\,dt \\
&= \frac{(z+i)^2}{2\pi i}\{\frac{z}{(z+i)^2}\log(\frac{t-z}{t+i}) - \frac{i}{z+i}\frac{1}{t+i}\}\,|_{t=0}^{t\to\infty} \\
&= -\frac{z}{2\pi i}\log(-z) - \frac{z\log(i) + z + i}{2\pi i}.
\end{aligned}
$$

In the first step we have used partial fraction decomposition of the integrand. The finally resulting expression is an equivalent defining function to

$$
-\frac{z}{2\pi i}\log(-z)
$$

which is the defining function of the hyperfunction $u(x)x$. Therefore, we have found that $\chi_{(0,\infty)}x = u(x)x$.

Problem 2.6. Show that

$$
\chi_{(-\infty,0)}x^2 = u(-x)x^2.
$$

2.3.4 Homologous and Standard Hyperfunctions

In Section 1.3.6 we pointed out that the choice of a hyperfunction which interprets a given ordinary function may not be unique. We are confronted with this fact when we want to define the projection of an ordinary function of the type

$$
x \mapsto f(x) = \frac{g(x)}{(x-c)^m}, \quad x \neq c, \quad (m \in \mathbb{N})
$$

on (a,b). We suppose that $g(x) \in \mathcal{H}_\alpha(a,b)$ and $c \in [a,b]$. Note that we cannot simply define the projection of $f(x)$ on (a,b) by

$$
\frac{1}{2\pi i} \int_a^b \frac{g(x)}{(x-c)^m\,(x-z)}\,dx
$$

owing to lack of convergence of the integral. Assume that $\chi_{(a,b)}g(x) = [G(z)]$ is the projection of the ordinary function $g(x)$ on (a,b). For $x \in (a,b) \setminus \{c\}$ the equation

$$
f(x) = \frac{g(x)}{(x-c)^m}
$$

should be equivalent with the equation

$$
(x-c)^m\,f(x) = \chi_{(a,b)}g(x)
$$

which we have treated in Section 1.3.6. We know that if

$$f_1(x) = \left[\frac{G(z)}{(z-c)^m} \right]$$

is a solution, then

$$f(x) = f_1(x) + \sum_{k=0}^{m-1} \gamma_k \, \delta^{(k)}(x-c) = f_1(x) + \delta_{(m-1)}(x-c)$$

is a solution too, where the added term which contains m arbitrary constants γ_k is a generalized delta-hyperfunction of order $m-1$ at $x = c$. The hyperfunction $f(x)$ is vanishing outside $[a, b]$ and equals the ordinary function $x \mapsto f(x)$ on $[a, b] \setminus c$, thus would qualify for the projection we are looking for. However, we want the projection operator to give us a unique answer. For this reason we define the concept of homologous hyperfunctions.

Definition 2.6. Let $f(x)$ and $f_1(x)$ be two hyperfunctions on (a, b) interpreting in the range $(a, b) \setminus \{c\}$ the ordinary functions $x \mapsto f(x)$. The two hyperfunctions are called *homologous*, if $f(x) - f_1(x)$ is a generalized delta-hyperfunction $\delta_{(.)}(x - c)$ of some order.

Thus, different interpretations of a given ordinary function are homologous. Among these homologous hyperfunctions we now pick out the hyperfunction that contains no generalized delta-hyperfunction. It is called the *standard hyperfunction* (do not confuse it with the notion of the standard defining function) and will serve to define the projection of the given ordinary function $x \mapsto f(x) = g(x)/(x-c)^m$.

Definition 2.7. The projection $\chi_{(a,b)} f(x)$ of the ordinary function $x \mapsto f(x) = g(x)/(x-c)^m$ on (a, b), is defined to be the *standard hyperfunction*.

Example 2.12. Let us compute $\chi_{(0,\infty)}(1/x^2)$. Here $a = c = 0$, $b = \infty$, $g(x) = 1$. We have $\chi_{(0,\infty)} g(x) = u(x) = [-1/(2\pi i) \log(-z)]$, thus

$$\chi_{(0,\infty)} \frac{1}{x^2} = \left[-\frac{1}{2\pi i} \frac{\log(-z)}{z^2} \right] = \mathrm{fp}\, \frac{u(x)}{x^2}. \tag{2.51}$$

The above situation may also be covered in another way. For definiteness, we seek the projection of the ordinary function $x \mapsto (x - a)^\alpha (b - x)^\beta$ on (a, b). If $\alpha > -1$ and $\beta > -1$ the answer is the hyperfunction

$$f(x) = \left[\frac{1}{2\pi i} \int_a^b \frac{(x - a)^\alpha (b - x)^\beta}{x - z} \, dx \right].$$

However, if $\alpha \le -1$ or $\beta \le -1$, the integral would diverge and the method fails. A remedy is to introduce a convergence factor as we have done in the event of an infinite interval. Here the convergence factor is inserted into the numerator of the integrand, and called a *convergence factor of the second kind*. We choose $N, M \in \mathbb{N}$ such that $\alpha + N > -1$, $\beta + M > -1$ and insert the factor $((x-a)^N (b-x)^M$ thereby rendering the integral

$$H(z) = \frac{1}{2\pi i} \int_a^b \frac{x - a)^{\alpha+N} (b - x)^{\beta+M}}{x - z} \, dx$$

convergent. $H(z)$ is the defining function of a hyperfunction $h(x)$ that takes the values $(x - a)^{\alpha+N}(b - x)^{\beta+M}$ for $x \in (a, b)$ and vanishes outside (a, b). The same holds for $H(z) + \phi(z)$, where $\phi(z)$ is an arbitrary real analytic function. Thus, the hyperfunction

$$g(x) = \left[\frac{H(z) + \phi(z)}{((z - a)^N (b - z)^M} \right]$$

$$= \left[\frac{1}{2\pi i\, (z - a)^N (b - z)^M} \int_a^b \frac{x - a)^{\alpha+N} (b - x)^{\beta+M}}{x - z} dx \right]$$

$$+ \left[\frac{\phi(z)}{(z - a)^N (b - z)^M} \right]$$

has the values $(x - a)^{\alpha}(b - x)^{\beta}$ for $x \in (a, b)$ and is vanishing outside (a, b). The latter term produces two generalized delta-hyperfunctions containing $N + M$ arbitrary coefficients:

$$\left[\frac{\phi(z)}{(z - a)^N (b - z)^M} \right] = \delta_{(N-1)}(x - a) + \delta_{(M-1)}(x - b).$$

All members of the family of hyperfunctions

$$g(x) = \left[\frac{1}{2\pi i\, (z - a)^N (b - z)^M} \int_a^b \frac{x - a)^{\alpha+N} (b - x)^{\beta+M}}{x - z} dx \right.$$

$$\left. + \delta_{(N-1)}(x - a) + \delta_{(M-1)}(x - b) \right.$$

are homologous and are interpretations on (a, b) of the given ordinary function. This remains true for different N and M. However, another choice of N or M will produce another number of arbitrary coefficients in the generalized delta-hyperfunctions. We shall denote a generalized delta hyperfunction at $x = a$ of an arbitrary or unknown order and arbitrary coefficients by $\delta_{(\cdot)}(x - a)$. We now assume that all possibly real analytic components in the above integral term have been removed, then

$$\left[\frac{1}{2\pi i\, (z - a)^N (b - z)^M} \times \int_a^b \frac{x - a)^{\alpha+N} (b - x)^{\beta+M}}{x - z} dx \right] \qquad (2.52)$$

defines the *projection of $x \mapsto (x - a)^{\alpha}(b - x)^{\beta}$ on the interval (a, b)*. The second component

$$\delta_{(N-1)}(x - a) + \delta_{(M-1)}(x - b)$$

is sometimes called the projection of $x \mapsto (x - a)^{\alpha}(b - x)^{\beta}$ to the points a, b.

Example 2.13. Let $x \mapsto 1/x^m$, and $a < 0 < b$. By choosing the converging factor of the second kind as x^m, we obtain

$$\chi_{(a,b)} \frac{1}{x^m} = \left[\frac{1}{2\pi i z^m} \int_a^b \frac{x^m}{x^m (x - z)} dx \right]$$

$$= \left[\frac{1}{2\pi i} \frac{1}{z^m} \log \frac{b - z}{a - z} \right],$$

hence

$$\chi_{(a,b)}\frac{1}{x^m} = \left[\frac{1}{2\pi i}\frac{1}{z^m}\log\frac{b-z}{a-z}\right] = \mathrm{fp}\frac{u(x-a)}{x^m} - \mathrm{fp}\frac{u(x-b)}{x^m}.$$

All members of the family of hyperfunctions

$$\mathrm{fp}\frac{u(x-a)}{x^m} - \mathrm{fp}\frac{u(x-b)}{x^m} + \delta_{(.)}(x)$$

are equally good interpretations of $x \mapsto 1/x^m$ on (a, b).

2.4 Projections of Hyperfunctions

2.4.1 Holomorphic and Meromorphic Hyperfunctions

The reader may have observed that many familiar hyperfunctions defined on an interval I have defining functions that are holomorphic in much larger domains than just in a narrow upper and lower half-neighborhood $\mathcal{D}_+(I)$ and $\mathcal{D}_-(I)$. Of special interest is now when the upper and lower component can analytically be extended across the real axis into portions of the lower and upper half-plane, respectively.

For the following discussion remember that a *full (two-dimensional) neighborhood* of a real point $a \in I$ is a subset of the form $\{z \in \mathbb{C} \mid |z - a| < \epsilon\}$ for some positive ϵ, whereas a *real neighborhood* of a is a subset of the form $\{x \in \mathbb{R} \mid |x - a| < \epsilon\}$ for some positive ϵ.

Definition 2.8. Let the hyperfunction $f(x) = [F(z)]$ be specified on the interval I. This means that there is a complex neighborhood $\mathcal{D}$ containing I such that $F(z) \in \mathcal{O}(\mathcal{D} \setminus I)$. We say that the hyperfunction $f(x) = [F(z)]$ is *holomorphic at* $x = a \in I$, if the upper *and* the lower component $F_+(z)$ and $F_-(z)$ can analytically be continued across the real axis to a full neighborhood of a. A hyperfunction is called *holomorphic or analytic in an open interval* $J = (a, b) \subset I$, denoted by $f(x) \in \mathcal{B}_\mathcal{O}(J)$, if it is holomorphic at all $x \in (a, b)$.

This definition implies that if $f(x)$ is a holomorphic hyperfunction at $x = a$, there exists a real neighborhood of a where $f(x)$ is holomorphic.

Definition 2.9. A hyperfunction $f(x) = [F(z)]$ is *entire* if the upper and lower component of the defining function $F(z)$ are both entire functions.

Definition 2.10. A hyperfunction $f(x) = [F(z)]$ is called *meromorphic* if the upper and lower component of the defining function $F(z)$ are both meromorphic functions (having poles on the real axis).

Remember the representation

$$f(x) = [F_1(z), F_2(z)] = [F_1(z)\,\mathbf{1}_+(z) - F_2(z)\,\mathbf{1}_-(z)]. \tag{2.53}$$

The special case where $F_1(z) = F(z)$, $F_2(z) = -F(z)$ is called the finite part hyperfunction of $f(x) = F(x)$. A meromorphic hyperfunction is holomorphic at all x, except possibly at the poles on the real axis of $F_1(z)$ and $F_2(z)$.

Example 2.14. The hyperfunction

$$\phi(x) := [\phi(z)\,\mathbf{1}_+(z)] = [\phi(z)\,\mathbf{1}_-(z)] = [\phi(z)\,\mathbf{1}(z)],$$

where $\phi(x)$ is a real analytic function on $\mathbb{R}$, is a holomorphic hyperfunction on the entire real axis. Polynomials, $\sin x, \cos x, \exp x$ are entire hyperfunctions.

Example 2.15. The Dirac impulse hyperfunction $\delta^{(k)}(x-a)$ is holomorphic in the intervals $(-\infty, a)$ and (a, ∞) and constitutes an example of a meromorphic hyperfunction.

Example 2.16. As mentioned before, the finite part hyperfunction $\mathrm{fp}(1/x^m) = 1/z^m\,[1/z)]$ is a meromorphic hyperfunction which is holomorphic in $(-\infty, 0)$ and $(0, \infty)$.

In Section 1.2 we have said that a hyperfunction $f(x) = [F_+(z), F_-(z)]$ has a value at x, if the limit

$$\lim_{\epsilon \to 0+} \{F_+(x+i\epsilon) - F_-(x-i\epsilon)\}$$

exists (this limit is then again denoted by $f(x)$). Assume now that $f(x)$ is holomorphic in the interval (a, b). In this event the hyperfunction has a value for all $x \in (a, b)$, and

$$f(z) := F_+(z) - F_-(z)$$

is an ordinary holomorphic function defined on a full neighborhood of (a, b). The ordinary function $x \mapsto F_+(x) - F_-(x)$ constitutes then a real analytic function on (a, b). This simple observation will be the clue for defining the product of two hyperfunctions in a later section.

Example 2.17. Consider the unit-step hyperfunction defined by

$$u(x) = \left[-\frac{1}{2\pi i}\,\log(-z)]\right].$$

Here log denotes the principal part of the logarithm defined in the complex plane with a cut along the negative part of the real axis. On this part $\log(-x)$ is real analytic and $u(x)$ has the value zero and is a holomorphic hyperfunction on $(-\infty, 0)$. For $x > 0$, the function $F_+(z) = \log(-z)$ can be analytically continued across x to the following sheet of the Riemann surface of $\log(-z)$ (which denotes now the complete multifunction). Similarly, $F_-(z) = \log(-z)$ can be analytically continued across x into the adjoining sheet of the Riemann surface. Hence, the hyperfunction $u(x)$ is also holomorphic on $(0, \infty)$.

Example 2.18. The projection $\chi_{(a,b)} f(x)$ of the ordinary function $f(x)$ on the interval (a, b) is a holomorphic hyperfunction on $\mathbb{R} \setminus [a, b]$.

Next, let us quote two well-known theorems of complex variable theory. A proof for these theorems can be found in any textbook of the subject, see for example [4, Behnke - Sommer p. 138, 178].

Proposition 2.16 (Identity Theorem). *Let $F_1(z)$ and $F_2(z)$ be holomorphic functions in a domain $\mathcal{D}$ (an open and connected subset of $\mathbb{C}$). If $F_1(z) = F_2(z)$ on a segment γ of a curve in $\mathcal{D}$, then $F_1(z) = F_2(z)$ for all $z \in \mathcal{D}$.*

Proposition 2.17. *Let $\mathcal{D}_1$ and $\mathcal{D}_2$ be two overlapping domains, and let $F_1(z)$ and $F_2(z)$ be holomorphic functions in $\mathcal{D}_1$ and $\mathcal{D}_2$, respectively. If $F_1(z) = F_2(z)$ on a segment γ of a curve in $\mathcal{D}_1 \cap \mathcal{D}_2 \neq \emptyset$, then*

$$F(z) := \left\{ \begin{array}{ll} F_1(z), & z \in \mathcal{D}_1 \\ F_2(z), & z \in \mathcal{D}_2 \end{array} \right.$$

constitutes one holomorphic function on $\mathcal{D}_1 \cup \mathcal{D}_2$.

In the event of this proposition, $F_1(z)$ and $F_2(z)$ are said to be the *analytic continuation of each other*. The globally defined function $F(z)$ on $\mathcal{D}_1 \cup \mathcal{D}_2$ is called the analytic continuation of $F_1(z)$ and $F_2(z)$.

Now let $f(x) = [F_+(z), F_-(z)]$ be a holomorphic hyperfunction defined on the interval I. Then the upper half-neighborhood $\mathcal{D}_+(I)$ where the upper component $F_+(z)$ is holomorphic can be extended across I into the lower half-plane to a full neighborhood $\mathcal{U}_1$, and similarly the lower half-neighborhood $\mathcal{D}_-(I)$ where $F_-(z)$ is holomorphic, can be extended into the upper half-plane to a full neighborhood $\mathcal{U}_2$. Both functions $F_+(z)$ and $F_-(z)$ and thus $F_+(z) - F_-(z)$ then are holomorphic in the intersection $\mathcal{U} = \mathcal{U}_1 \cap \mathcal{U}_2$. On an interval $(a,b) \subset \mathcal{U}$, we suppose now that $f(x) = 0$. Then $F_+(z) = F_-(z) + \phi(z)$ for all $z \in (a,b)$, where $\phi(x)$ is a real analytic function on (a,b). Thus, by the identity theorem, we must have $F_+(z) = F_-(z) + \phi(z)$ for all $z \in \mathcal{U}$. This proves

Proposition 2.18. *Let $f(x) = [F_+(z), F_-(z)]$ be a holomorphic hyperfunction on the interval I. If there is an open subinterval $(a,b) \subset I$ of positive length where $f(x) = 0$, then $f(x) = 0$ for all $x \in I$.*

Another important theorem of complex variable theory is Schwarz's Reflection Principle.

Proposition 2.19. *Let $\mathcal{D}$ be a domain in the upper complex half-plane such that its boundary contains the real interval (a,b). Let $X_+(z)$ be holomorphic in $\mathcal{D}$, continuous and real-valued on (a,b). Then, $X_+(z)$ can analytically be continued across (a,b) into the domain $\overline{\mathcal{D}}$ of the lower complex half-plane where the domain $\overline{\mathcal{D}}$ is symmetric to $\mathcal{D}$ with respect to the real axis. The analytic continuation $X_-(z)$ of $X_+(z)$ in $\overline{\mathcal{D}}$ is given by the conjugate analytic function $X_-(z) = \overline{X_+(\overline{z})}$.*

A similar statement holds for a holomorphic function $X_-(z)$ in the lower half-plane. For a proof, see for example [4, Behnke-Sommer p.188]. Now a lot of familiar defining functions are imaginary-valued on the real axis. If $Y_+(z)$ is such a function, then $X_+(z) = iY_+(z)$ satisfies Proposition 2.19 and we obtain

Proposition 2.20. *Let $\mathcal{D}$ be a domain in the upper complex half-plane such that its boundary contains the real interval (a,b). Let $Y_+(z)$ be holomorphic in $\mathcal{D}$, continuous and imaginary-valued on (a,b). Then, $Y_+(z)$ can analytically be continued across (a,b) into the domain $\overline{\mathcal{D}}$ of the lower complex half-plane where the domain $\overline{\mathcal{D}}$ is symmetric to $\mathcal{D}$ with respect to the real axis. The analytic continuation $Y_-(z)$ of $Y_+(z)$ in $\overline{\mathcal{D}}$ is given $Y_-(z) = -\overline{Y_+(\overline{z})}$.*

Applying these propositions to the upper and lower component of a hyperfunction gives

Proposition 2.21. *Let* $f(x) = [F_+(z), F_-(z)]$ *be a given hyperfunction on the open interval* I. *If both the upper and lower component* $F_+(z)$ *and* $F_-(z)$ *are either continuous and real-valued or continuous and imaginary-valued on* I, *they can be analytically continued across* I *into the lower and upper complex half-plane, respectively, and* $f(x)$ *is a holomorphic hyperfunction on* I. *If in addition* $F_+(x) = F_-(x)$ *for all* $x \in (a, b) \subset I$ *holds, the defining function*

$$F(z) := \begin{cases} F_+(z), & z \in \mathcal{D} \cup I \\ F_-(z), & z \in \overline{\mathcal{D}} \cup I \end{cases}$$

constitutes one holomorphic function, i.e., a global analytic function in $\mathcal{D} \cup \overline{\mathcal{D}} \cup I$.

2.4.2 Standard Defining Functions

The notion of a projection of an ordinary function will now be carried over to hyperfunctions. First, let us start with a hyperfunction $f(x) = [F(z)]$ defined on an interval I containing the compact interval $[a, b]$.

We shall build another hyperfunction, again denoted by $\chi_{(a,b)} f(x)$ and also called the *projection of* $f(x)$ *on the interval* (a, b). It will be a hyperfunction that equals the given $f(x)$ on the open interval (a, b), and vanishes on $(-\infty, a) \cup (b, \infty)$. So this projection of $f(x)$ on (a, b) may also be regarded as the *restriction of* $f(x)$ *to* (a, b), or we may say that $f(x)$ *has been localized to* (a, b).

The defining function for the projection will formally be similar to the definition of the projection of an ordinary function. If $\Im z \neq 0$, then $1/(x - z)$ is a real analytic function of x, thus the hyperfunction $f(x)/(x - z)$ is well defined, and its integral over $[a, b]$ makes sense.

Proposition 2.22. *Let the hyperfunction* $f(x) = [F(z)] = [F_+(z), F_-(z)]$ *be defined on an interval* I *containing the compact interval* $[a, b]$. *Assume that* $f(x)$ *is holomorphic at* a *and* b. *Then, the projection of* $f(x)$ *on* (a, b) *is given by*

$$\chi_{(a,b)} f(x) := \left[\frac{1}{2\pi i} \int_a^b \frac{f(x)}{x - z} \, dx \right]. \tag{2.54}$$

Proof. Let us show that the so-defined hyperfunction has the announced properties. Let (see Figure 2.4)

$$G(z) = \frac{1}{2\pi i} \int_a^b \frac{f(x)}{x - z} \, dx = -\frac{1}{2\pi i} \oint_{(a,b)} \frac{F(t)}{t - z} \, dt$$

$$= \frac{1}{2\pi i} \int_{\gamma_{a,b}^+} \frac{F_+(t)}{t - z} \, dt - \frac{1}{2\pi i} \int_{\gamma_{a,b}^-} \frac{F_-(t)}{t - z} \, dt. \tag{2.55}$$

The contours $\gamma_{a,b}^+$ and $\gamma_{a,b}^-$ are arbitrary near the real axis such that the point z is between the two paths $\gamma_{a,b}^+$ and γ_1^+, if $\Im z > 0$, or between $\gamma_{a,b}^-$ and γ_1^-, if $\Im z < 0$. By Cauchy's integral formula we have in the event of $\Im z > 0$,

$$F_+(z) = \frac{1}{2\pi i} \left\{ \int_{\gamma_{a,b}^+} \frac{F_+(t)}{t - z} \, dt - \int_{\gamma_1^+} \frac{F_+(t)}{t - z} \, dt \right\}$$

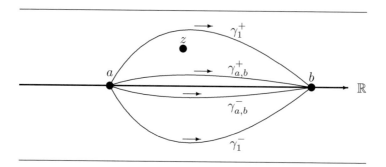

Figure 2.4: Contours for Definition 2.22

and also,

$$\frac{1}{2\pi i} \int_{\gamma_1^-} \frac{F_-(t)}{t-z}\, dt = \frac{1}{2\pi i} \int_{\gamma_{a,b}^-} \frac{F_-(t)}{t-z}\, dt.$$

Using these relations, (2.55) gives

$$G(z) := F_+(z) + \frac{1}{2\pi i} \int_{\gamma_1^+} \frac{F_+(t)}{t-z}\, dt - \frac{1}{2\pi i} \int_{\gamma_1^-} \frac{F_-(t)}{t-z}\, dt,$$

i.e.,

$$G(z) = F_+(z) + \psi(z), \tag{2.56}$$

where, see Proposition 2.2,

$$\psi(z) := \frac{1}{2\pi i}\Big\{ \int_{\gamma_1^+} \frac{F_+(t)}{t-z}\, dt + \int_{\gamma_1^-} \frac{F_-(t)}{t-z}\, dt \Big\} \tag{2.57}$$

is a holomorphic, hence real analytic function on $\mathbb{R} \setminus \{a, b\}$. Similarly, it is shown that

$$G(z) = F_-(z) + \psi(z) \tag{2.58}$$

holds, if $\Im z < 0$. By (2.56) or (2.58) we notice that for $a < \Re z < b$ the defining functions $G(z)$ and

$$F(z) := \begin{cases} F_+(z), & z \in D_+ \\ F_-(z), & z \in D_- \end{cases}$$

are equivalent, thus define the same hyperfunction on the open interval (a, b). By (2.55) it is seen that $G(z)$ is holomorphic on the real axis outside the interval $[a, b]$, thus $g(x) = [G(z)]$ vanishes outside $[a, b]$ (see Proposition 2.2). □

Example 2.19. Let us compute the projection of the constant hyperfunction $1 = [1, 0] = [\mathbf{1}_+(z)]$ on the interval $[a, b]$.

$$G(z) = -\frac{1}{2\pi i} \int_{\gamma_{a,b}^+} \frac{1}{t-z}\, dt = -\frac{1}{2\pi i} \log(t - z)|_b^a$$

$$= -\frac{1}{2\pi i}\{\log(a - z) - \log(b - z)\}.$$

Hence, the projection of the hyperfunction $f(x) = [1, 0]$ on (a, b) is $u(x - a) - u(x - b)$. Thus, $\chi_{(a,b)}1 = u(x - a) - u(x - b)$.

Proposition 2.23. *For any $a < b$ we have*

$$\chi_{(a,b)}\delta^{(m)}(x) = \begin{cases} \delta^{(m)}(x), & 0 \in (a, b) \\ 0, & 0 \notin [a, b], \end{cases} \tag{2.59}$$

also, we have for any generalized delta-hyperfunction,

$$\chi_{(a,b)}\delta_{(M)}(x - c) = \begin{cases} \delta_{(M)}(x - c), & c \in (a, b) \\ 0, & c \notin [a, b]. \end{cases} \tag{2.60}$$

Proof. By using (2.55), and

$$\delta^{(m)}(x) = \left[-\frac{1}{2\pi i} \frac{(-1)^m \, m!}{z^{m+1}} \right]$$

we have

$$\chi_{(a,b)}\delta^{(m)}(x) = -\frac{1}{2\pi i}\left(-\frac{(-1)^m \, m!}{2\pi i}\right) \oint_{(a,b)} \frac{dt}{t^{m+1}(t - z)}.$$

If the closed contour encircles 0 we have

$$\frac{1}{2\pi i} \oint_{(a,b)} \frac{dt}{t^{m+1}(t - z)} = \operatorname{Res}_{t=0}\left\{ \frac{1}{t^{m+1}(t - z)} \right\} = -\frac{1}{z^{m+1}},$$

and, if 0 is exterior to the closed contour, we have

$$\frac{1}{2\pi i} \oint_{(a,b)} \frac{dt}{t^{m+1}(t - z)} = 0.$$

This proves the proposition. □

Example 2.20. Compute the projection of $\mathrm{fp}(1/x)$ on (a, b) with $a < 0 < b$. The defining function of the projection is

$$G(z) = \frac{1}{2\pi i} \int_{\gamma_{a,b}^+} \frac{1}{2t} \frac{1}{t - z} \, dt - \frac{1}{2\pi i} \int_{\gamma_{a,b}^-} \left(-\frac{1}{2t}\right)\frac{1}{t - z} \, dt$$

$$= \frac{1}{4\pi i} \left\{ \int_{\gamma_{a,b}^+} \frac{1}{t(t - z)} \, dt + \int_{\gamma_{a,b}^-} \frac{1}{t(t - z)} \, dt \right\}.$$

Partial fraction decomposition of the integrand yields

$$\int_{\gamma_{a,b}^+} \frac{1}{t(t - z)} \, dt = \frac{1}{z}\{ -\log t + \log(t - z)\}\Big|_a^b = \frac{1}{z} \log \frac{t - z}{t}\Big|_a^b.$$

The function $\log((t - z)/t)$ defines a singled-valued, holomorphic branch in the t-plane with an incision along the segment from 0 to z. Consider now the closed contour $\gamma_a^+ + \gamma_0 + \gamma_b^+ - \gamma_{a,b}^-$, where γ_0 is the contour that rounds in the positive sense 0 from one side of the cut to the other side (see Figure 2.5). Inside this closed contour the integrand is holomorphic, so, by Cauchy's theorem, we have

$$\int_{\gamma_a^+} + \int_{\gamma_0} + \int_{\gamma_b^+} - \int_{\gamma_{a,b}^-} = 0.$$

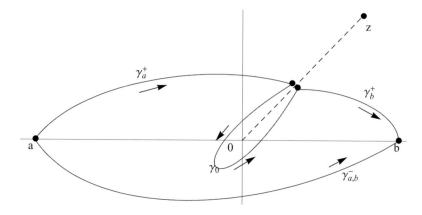

Figure 2.5: Contours of Example 2.20

Since $\gamma_{a,b}^+ = \gamma_a^+ + \gamma_b^+$ we have

$$\int_{\gamma_{a,b}^+} = \int_{\gamma_a^+} + \int_{\gamma_b^+} = \int_{\gamma_{a,b}^-} - \int_{\gamma_0}.$$

Because

$$\int_{\gamma_{a,b}^-} = \frac{1}{z}\{\log \frac{b-z}{b} - \log \frac{a-z}{a}\}, \qquad \int_{\gamma_0} = -2\pi i,$$

we finally obtain for $\Im z > 0$,

$$G(z) = \frac{1}{4\pi i z}\{2 \log \frac{(b-z)\,a}{(a-z)\,b} + 2\pi i\}.$$

A similar line of thought gives, for $\Im z < 0$,

$$G(z) = \frac{1}{4\pi i z}\{2 \log \frac{(b-z)\,a}{(a-z)\,b} - 2\pi i\}.$$

Therefore, we may write for all $\Im z \neq 0$,

$$G(z) = \frac{1}{2\pi i z} \log \frac{(b-z)a}{(a-z)b} + \mathrm{sgn}(\Im z)\,\frac{1}{2\,z}$$

$$= -\frac{1}{2\pi i z} \log \frac{z-a}{b-z} - \frac{1}{2\pi i\,z} \log \frac{b}{|a|} + \mathrm{sgn}(\Im z)\,\frac{1}{2\,z}.$$

This gives

$$\chi_{(a,b)}\mathrm{fp}\frac{1}{x} = \log \frac{b}{|a|}\,\delta(x) + \mathrm{fp}\frac{1}{x} + \left[-\frac{1}{2\pi i z} \log \frac{z-a}{b-z}\right] \tag{2.61}$$

which can also be written as (see Section 1.4)

$$\chi_{(a,b)}\mathrm{fp}\frac{1}{x} = x^{-1} - \mathrm{fp}\frac{u(a-x)}{x} - \mathrm{fp}\frac{u(x-b)}{x}.$$

Remark. Compare this result with Example 2.13 where we have computed the projection of the ordinary function $x \mapsto 1/x$. There we had found

$$\chi_{(a,b)} \frac{1}{x} = \left[-\frac{1}{2\pi i} \frac{1}{z} \log \frac{z-a}{z-b} \right] + \delta_{(\cdot)}(x)$$

$$= \mathrm{fp} \frac{u(x-a)}{x} - \mathrm{fp} \frac{u(x-b)}{x} + \delta_{(\cdot)}(x).$$

We have the following simple decomposition property.

Proposition 2.24. *Let $f(x)$ be a hyperfunction specified on an interval I. Let $[a,b] \subset I$ be a compact interval, and let $a = c_0 < c_1 < c_2 < \cdots c_{n-1} < b = c_n$ be a finite number of points where $f(x)$ is assumed to be holomorphic. We then have*

$$\chi_{(a,b)} f(x) = \chi_{(a,c_1)} f(x) + \chi_{(c_1,c_2)} f(x) + \cdots + \chi_{(c_{n-1},b)} f(x). \tag{2.62}$$

Proof. It suffices to prove the proposition for one point c. Thus, let us show that

$$\chi_{(a,b)} f(x) = \chi_{(a,c)} f(x) + \chi_{(c,b)} f(x). \tag{2.63}$$

This follows from the equation

$$\frac{1}{2\pi i} \int_a^b \frac{f(x)}{x-z} \, dx = \{ \frac{1}{2\pi i} \int_a^c \frac{f(x)}{x-z} \, dx + \frac{1}{2\pi i} \int_c^b \frac{f(x)}{x-z} \, dx \}. \qquad \square$$

Be aware that the above decomposition is not unique in the following sense: let $g_k(x)$ be any family of hyperfunctions where each member is concentrated at the point c_k, $k = 0, 1, \ldots, n$ and $g_0(x) = g_n(x) = 0$. Then,

$$\chi_{(a,b)} f(x) = \sum_{k=1}^{n} \{ -g_{k-1}(x) + \chi_{(c_{k-1},c_k)} f(x) + g_k(x) \} \tag{2.64}$$

also holds. The reason is simply

$$\sum_{k=1}^{n} \{ -g_{k-1}(x) + \chi_{(c_{k-1},c_k)} f(x) + g_k(x) \} = \sum_{k=1}^{n} \chi_{(c_{k-1},c_k)} f(x)$$

$$+ (-0 + g_1(x)) + (-g_1(x) + g_2(x)) + \cdots$$
$$+ (-g_{n-2}(x) + g_{n-1}(x)) + (-g_{n-1}(x) + 0) = \chi_{(a,b)} f(x).$$

Therefore, *the decomposition of a hyperfunction is only determined modulo any sequence of hyperfunctions concentrated at the partition points.*

Sato's Decomposition Theorems

Proposition 2.24 has several important generalizations. For example, our technical hypothesis that $f(x)$ has to be holomorphic at all points c_i can be dropped (Sato's decomposition theorem). Proofs of this and the following general theorem are beyond the scope of this book. It needs sophisticated developments in order to construct all the involved defining functions. For an application oriented use of hyperfunctions these theorems are not so essential as they are from a mathematico-logical point of view. The proofs can be found in [31, Sato p. 160 - 167]. Sato's decomposition theorem can also be generalized to an open interval that may even be $\mathbb{R}$, $\mathbb{R}_+$ or $\mathbb{R}_-$.

Proposition 2.25. *Let $f(x)$ be a hyperfunction defined on I. Let $(a,b) \subset I$ be an open interval in I, and let c_k, $k = 0, \pm 1, \pm 2, \dots$ be any sequence in (a,b) such that $a \leftarrow c_k$ as $k \to -\infty$, and $c_k \to b$ as $k \to +\infty$. Then, we have*

$$\chi_{(a,b)}f(x) = \sum_{k=-\infty}^{\infty} \chi_{(c_{k-1},c_k)}f(x) \tag{2.65}$$

in the following sense: for any finite a' and b' such that $a < a' < b' < b$, we have

$$\chi_{(a',b')}f(x) = \sum_{k=-N}^{N} \chi_{(c_{k-1},c_k)}f(x) \tag{2.66}$$

for sufficiently large N. Conversely, if we have a sequence of hyperfunctions $f_k(x)$ defined on intervals I_k such that $(c_{k-1}, c_k) \subset I_k$ for all k, then there exists a unique defined hyperfunction $f(x)$ such that (2.65) holds.

An illustration of this proposition is given in Example 2.5.

Remark. We have seen that for functions or hyperfunctions distinct from Dirac impulses or generalized delta-hyperfunctions, we have

$$\chi_{(0,\infty)}f(x) = u(x)f(x), \qquad\qquad \chi_{(a,\infty)}f(x) = u(x-a)f(x),$$
$$\chi_{(-\infty,0)}f(x) = u(-x)f(x), \qquad\qquad \chi_{(-\infty,a)}f(x) = u(a-x)f(x),$$
$$\chi_{(a,b)}f(x) = u(x-a)f(x) - u(x-b)f(x).$$

In later chapters when dealing with integral transforms, we shall often identify a hyperfunction $f(x)$ defined on the real axis with $\chi_{(-\infty,\infty)}f(x)$ and then proceed to the decomposition

$$f(x) = \chi_{(-\infty,\infty)}f(x) = \chi_{(-\infty,0)}f(x) + \chi_{(0,\infty)}f(x) := f_1(x) + f_2(x),$$

but keeping in mind that it is only determind modulo a hyperfunction concentrated at $x = 0$.

Perfect Hyperfunctions and Their Standard Defining Functions

Consider now that the hyperfunction $f(x)$ is specified from the outset as having a compact support $[a,b]$. This is the case if its defining function satisfies $F(z) \in \mathcal{O}(\mathcal{D} \setminus [a,b])$. Then, its upper and lower components are analytic continuations from each other, i.e., we have $F_+(z) = F_-(z) = F(z)$. Such a hyperfunction is called *a perfect hyperfunction*. Let now $a' < a < b < b'$, then $f(x)$ is holomorphic in a' and b', and

$$-\frac{1}{2\pi i} \oint_{(a',b')} \frac{F(t)}{t-z}\, dt$$

defines a defining function for $\chi_{(a',b')}f(x)$. Let now $a' \to a - 0$ and $b' \to b + 0$, then

Definition 2.11.

$$\tilde{F}(z) := -\frac{1}{2\pi i} \oint_{(a,b)} \frac{F(t)}{t-z}\, dt \tag{2.67}$$

is called the *standard defining function of $f(x)$*.

Note that while $F(z)$ needs only be defined on a complex neighborhood $\mathcal{D}$ containing (a', b'), the standard defining function is holomorphic in $\mathbb{C} \setminus [a, b]$. More precisely,

Proposition 2.26. *The standard defining function of the perfect hyperfunction $f(x)$ is holomorphic at ∞ and has there the expansion*

$$\tilde{F}(z) = \sum_{n=1}^{\infty} \frac{c_n}{z^n}, \quad \text{with} \quad c_n = -\frac{1}{2\pi i} \int_a^b f(x) x^{n-1} \, dx, \tag{2.68}$$

valid for sufficiently large $|z|$. Especially we have,

$$\tilde{F}(z) = \frac{c_1}{z} + O\left(\frac{1}{z^2}\right) \quad \text{as } z \to \infty. \tag{2.69}$$

The above integral is an integral over the hyperfunction $f(x)$ as defined in Section 1.4.1. In fact the above proposition is subsumed in the following

Proposition 2.27. *Let $[a, b]$ be a compact interval. The defining function $G(z)$ of $\chi_{(a,b)} f(x)$, is holomorphic at ∞ and has there the expansion*

$$G(z) = \sum_{n=1}^{\infty} \frac{c_n}{z^n}, \quad \text{with} \quad c_n = -\frac{1}{2\pi i} \int_a^b f(x) x^{n-1} \, dx, \tag{2.70}$$

valid for sufficiently large $|z|$. Especially we have,

$$G(z) = \frac{c_1}{z} + O\left(\frac{1}{z^2}\right) \quad \text{as } z \to \infty. \tag{2.71}$$

Proof. Suppose that $|z| > \max\{|a|, |b|\}$, then

$$G((z) = -\frac{1}{2\pi i} \left\{ -\int_{\gamma_{a,b}^+} \frac{F_+(t)}{t - z} \, dt + \int_{\gamma_{a,b}^-} \frac{F_-(t)}{t - z} \, dt \right\}$$

$$= -\frac{1}{2\pi i} \left\{ -\int_{\gamma_{a,b}^+} \frac{F_+(t)}{z(t/z - 1)} \, dt + \int_{\gamma_{a,b}^-} \frac{F_-(t)}{z(t/z - 1)} \, dt \right\}$$

$$= \frac{1}{2\pi i} \left\{ -\int_{\gamma_{a,b}^+} \frac{F_+(t)}{z} \frac{1}{1 - t/z} \, dt + \int_{\gamma_{a,b}^-} \frac{F_-(t)}{z} \frac{1}{1 - t/z} \, dt \right\}$$

$$= \frac{1}{2\pi i} \left\{ -\int_{\gamma_{a,b}^+} F_+(t) \sum_{n=0}^{\infty} \frac{t^n}{z^{n+1}} \, dt + \int_{\gamma_{a,b}^-} F_-(t) \sum_{n=0}^{\infty} \frac{t^n}{z^{n+1}} \, dt \right\}.$$

Integration and summation can be interchanged due to the uniform convergence of the series, and we obtain

$$G(z) = \sum_{n=0}^{\infty} \frac{1}{z^{n+1}} \times \frac{1}{2\pi i} \left\{ -\int_{\gamma_{a,b}^+} F_+(t) \, t^n \, dt + \int_{\gamma_{a,b}^-} F_-(t) \, t^n \, dt \right\}$$

$$= \sum_{n=0}^{\infty} \frac{1}{z^{n+1}} \times \frac{-1}{2\pi i} \int_a^b f(x) x^n \, dx = \sum_{n=1}^{\infty} \frac{1}{z^n} \times \frac{-1}{2\pi i} \int_a^b f(x) x^{n-1} \, dx. \qquad \square$$

For integrals over a projection we have

Proposition 2.28. *Let the hyperfunction* $f(x) = [F(z)]$ *be defined on the entire real axis, and* $-\infty < c < a < b < d < \infty$, *then*

$$\int_{-\infty}^{\infty} \chi_{(a,b)} f(x)\, dx = \int_c^d \chi_{(a,b)} f(x)\, dx = \int_a^b f(x)\, dx. \qquad (2.72)$$

Proof. By the definition of the projection we have

$$\chi_{(a,b)} f(x) = \left[\frac{1}{2\pi i}\{\int_{\gamma_{a,b}^+} \frac{F_+(t)}{t-z}\, dt + \int_{\gamma_{a,b}^-} \frac{F_-(t)}{t-z}\, dt\}\right].$$

Therefore

$$\int_c^d \chi_{(a,b)} f(x)\, dx = \int_{\gamma_{c,d}^+} \frac{1}{2\pi i}\{\int_{\gamma_{a,b}^+} \frac{F_+(t)}{t-z}\, dt + \int_{\gamma_{a,b}^-} \frac{F_-(t)}{t-z}\, dt\}\, dz$$

$$- \int_{\gamma_{c,d}^-} \{\frac{1}{2\pi i} \int_{\gamma_{a,b}^+} \frac{F_+(t)}{t-z}\, dt + \int_{\gamma_{a,b}^-} \frac{F_-(t)}{t-z}\, dt\}\, dz$$

where the contour $\gamma_{c,d}^+$ is above $\gamma_{a,b}^+$ and $\gamma_{c,d}^-$ is below $\gamma_{a,b}^-$ (see Figure 2.6). Exchanging the order of integration yields

$$\int_c^d \chi_{(a,b)} f(x)\, dx$$

$$= \int_{\gamma_{a,b}^+} \frac{F_+(t)}{2\pi i} \oint_{-\gamma_{c,d}^+ + \gamma_{c,d}^-} \frac{dz}{z-t}\, dt - \int_{\gamma_{a,b}^-} \frac{F_-(t)}{2\pi i} \oint_{-\gamma_{c,d}^+ + \gamma_{c,d}^-} \frac{dz}{z-t}\, dt$$

$$= \int_{\gamma_{a,b}^+} F_+(t)\, dt - \int_{\gamma_{a,b}^-} F_-(t)\, dt$$

$$= \int_a^b f(x)\, dx. \qquad \square$$

Proposition 2.29. *Let* $\tilde{F}(z)$ *be the standard defining function of the perfect hyperfunction* $f(x)$, *then*

$$\int_a^b f(x)\, dx = -2\pi i c_1, \quad c_1 = \lim_{z\to\infty} z\,\tilde{F}(z). \qquad (2.73)$$

Proof. We have from the above proposition

$$\int_a^b f(x)\, dx = \int_c^d \chi_{(a,b)} f(x)\, dx = -\oint_{(c,d)} \tilde{F}(z)\, dz$$

$$= -\oint_{(c,d)} \{\frac{c_1}{z} + \frac{c_2}{z^2} + \frac{c_3}{z^3} + \dots\}\, dz = -2\pi i c_1.$$

From Proposition 2.27 we have $c_1 = \lim_{z\to\infty} z\,\tilde{F}(z)$. $\qquad \square$

Problem 2.7. Show that the defining function of $\chi_{(a,b)} f(x)$, where $f(x) = 1$ is the constant hyperfunction, has for sufficiently large $|z|$ the expansion

$$G(z) = -\frac{1}{2\pi i}\{\frac{b-a}{z} + \frac{b^2 - a^2}{z^2} + \frac{b^3 - a^3}{z^3} + \dots\}.$$

Example 2.22. Consider the hyperfunction $f(x) = \exp(-x^2)\,[\mathbf{1}_+(z)]$ which is holomorphic on the entire real axis and specified as an upper hyperfunction. Let us determine its strong defining function. We have for $\Im z > 0$,

$$
\tilde{F}_+(z) = \frac{1}{2\pi i} \int_{\gamma_{-\infty,\infty}^+} \frac{e^{-t^2}}{t-z}\, dt = \frac{1}{2\pi i} \int_{-\infty}^{\infty} \frac{e^{-x^2}}{x-z}\, dx
$$

$$
= \frac{1}{2}\, e^{-z^2} \operatorname{erfc}(-iz).
$$

Similarly, if $\Im z < 0$, we have

$$
\tilde{F}_-(z) = \frac{1}{2\pi i} \int_{\gamma_{-\infty,\infty}^+} \frac{e^{-t^2}}{t-z}\, dt = \frac{1}{2\pi i} \int_{-\infty}^{\infty} \frac{e^{-x^2}}{x-z}\, dx
$$

$$
= -\frac{1}{2}\, e^{-z^2} \operatorname{erfc}(iz).
$$

(See Example 2.4 and [1, Abramowitz–Stegun p.84].) Thus the given hyperfunction has the strong defining function

$$
e^{-x^2} = \left[\frac{1}{2}\, e^{-z^2} \operatorname{erfc}(-iz), \; -\frac{1}{2}\, e^{-z^2} \operatorname{erfc}(iz) \right]. \tag{2.81}
$$

Proposition 2.31. *If the hyperfunctions $f_1(x)$ and $f_2(x)$ have standard defining functions $\tilde{F}_1(z)$ and $\tilde{F}_2(z)$, respectively, then $c_1 f_1(x) + c_2 f_2(x)$ has the strong defining function $c_1 \tilde{F}_1(z) + c_2 \tilde{F}_2(z)$.*

The proof follows immediately from the above definition.

Proposition 2.32. *If a strong defining function of a hyperfunction exists, it is unique.*

Proof. We must show that $f(x) = [F_+(z) + \psi(z), F_-(z) + \psi(z)]$, where $\psi(x)$ is a real analytic function, i.e., a function that is holomorphic in a full neighborhood of the real axis, produces the same standard defining function as $f(x) = [F_+(z), F_-(z)]$. By using (2.79), we see that for all n a term of the form

$$
-\frac{1}{2\pi i} \oint_{(-R_n, R_n)} \frac{\psi(t)}{t-z}\, dt,
$$

is added to $\tilde{F}(z)$. Since z is outside the closed integration contour $(-R, R)$, it follows by Cauchy's theorem that this term is zero. $\qquad\square$

Let us compute more examples of strong defining functions.

Example 2.23. Let $f(x) = [\mathbf{1}_+(z)] = [1, 0]$ be the constant unit hyperfunction, then for $R_n \to \infty$,

$$
\tilde{F}(z) = \frac{1}{2\pi i} \mathrm{pv} \int_{-\infty}^{\infty} \frac{1}{x-z} dx = \lim_{n\to\infty} \frac{1}{2\pi i} \int_{-R_n}^{R_n} \frac{1}{x-z} dx
$$

$$
= \lim_{n\to\infty} \frac{1}{2\pi i} \log\left(\frac{R_n - z}{-R_n - z}\right) = \lim_{n\to\infty} \frac{1}{2\pi i} \log\left(\frac{1 - z/R_n}{-1 - z/R_n}\right)
$$

$$
= \lim_{n\to\infty} \frac{1}{2\pi i} \{\log\left|\frac{1 - z/R_n}{-1 - z/R_n}\right| + i\arg\left(\frac{1 - z/R_n}{-1 - z/R_n}\right)\}
$$

$$
= \lim_{n\to\infty} \frac{1}{2\pi} \arg\left(\frac{z/R_n - 1}{z/R_n + 1}\right)
$$

$$
= \lim_{n\to\infty} \frac{1}{2\pi} \{\arg(z/R_n - 1) - \arg(z/R_n + 1)\}.
$$

If $\Im z > 0$ the expression of the last line equals π and if $\Im(z) < 0$ it equals $-\pi$. Therefore, we obtain for the strong defining function of $f(x) = 1$

$$
\tilde{F}(z) = \mathbf{1}(z) = \begin{cases} 1/2, & \Im z > 0 \\ -1/2, & \Im z < 0. \end{cases}
$$

Example 2.24. Consider the real analytic function $f(x) = \exp(i\omega x)$ where ω is a real constant. We may write the corresponding hyperfunction as

$$
f(x) = \left[e^{i\omega z}, 0\right] = \left[0, -e^{i\omega z}\right].
$$

If $\omega > 0$, we choose the first one of the two equivalent defining functions,

$$
\tilde{F}(z) = \frac{1}{2\pi i} \mathrm{pv} \int_{-\infty}^{\infty} \frac{e^{i\omega x}}{x-z} dx = \lim_{R\to\infty} \frac{1}{2\pi i} \int_{\gamma_{-R,R}^{+}} \frac{e^{i\omega x}}{x-z} dx.
$$

We close the integration path from $\gamma_{-R,R}^{+}$ by a semicircle in the upper half-plane. The contribution to the integral on this upper semicircle tends to zero by Jordan's lemma as $R \to \infty$. Therefore, if $\Im z > 0$, we have by Cauchy's theorem

$$
\tilde{F}_{+}(z) = \mathrm{Res}_{x=z}\{\frac{e^{i\omega x}}{x-z}\} = e^{i\omega z},
$$

and $\tilde{F}_{-}(z) = 0$, if $\Im z < 0$. If $\omega < 0$ we choose the second defining function, and close the integration path $\gamma_{-R,R}^{-}$ by a semicircle in the lower half-plane. We obtain $\tilde{F}_{+}(z) = 0$ for $\Im z > 0$, and

$$
\tilde{F}_{-}(z) = -\mathrm{Res}_{x=z}\{\frac{e^{i\omega x}}{x-z}\} = -e^{i\omega z}
$$

for $\Im z < 0$. Thus, the strong defining function is given by

$$
\tilde{F}(z) := \begin{cases} e^{i\omega z}\,\mathbf{1}_{+}(z), & \omega > 0 \\ e^{i\omega z}\,\mathbf{1}_{-}(z), & \omega < 0. \end{cases}
$$

Problem 2.8. If $\Im c \neq 0$,

$$
f(x) = \frac{1}{(x-c)^m}, \quad m \in \mathbb{N}
$$

is a real analytic function, thus can be interpreted as a hyperfunction. Show that its strong defining function is given by

$$\tilde{F}(z) := \begin{cases} \frac{1}{(z-c)^m} \mathbf{1}_+(z), & \Im c < 0 \\ \frac{1}{(z-c)^m} \mathbf{1}_-(z), & \Im c > 0. \end{cases}$$

The following proposition gives a sufficient condition in order that the given defining function $F(z)$ and the strong defining function will be the same. It uses the notion of upper and lower half-plane functions (see Definition 2.4).

Proposition 2.33. *If upper and lower components of a given defining function are upper and lower half-plane functions, respectively, i.e., $F_+(z) \in \mathcal{O}_0(\mathbb{C}_+)$ and $F_-(z) \in \mathcal{O}_0(\mathbb{C}_-)$ (see Definition 2.4), then the given defining function and the strong defining function of $f(x)$ are the same.*

Proof. By Cauchy's integral formula, we have for $\Im z > 0$,

$$F_+(z) = \frac{1}{2\pi i}\{\int_{\gamma_{-R,R}^+} \frac{F_+(t)}{t-z}\,dt - \int_{C_R^+} \frac{F_+(t)}{t-z}\,dt\},$$

$$0 = \frac{1}{2\pi i}\{-\int_{\gamma_{-R,R}^-} \frac{F_-(t)}{t-z}\,dt + \int_{C_R^-} \frac{F_-(t)}{t-z}\,dt\},$$

where C_R^+ and C_R^- are semicircles in the upper and lower half-plane directed in the positive sense. It follows that

$$\tilde{F}_R(z) := \frac{1}{2\pi i}\{\int_{\gamma_{-R,R}^+} \frac{F_+(z)}{t-z}\,dt - \int_{\gamma_{-R,R}^-} \frac{F_-(t)}{t-z}\,dt\}$$

$$= \frac{1}{2\pi i}\{\int_{\gamma_{-R,R}^+} \frac{F_+(t)}{t-z}\,dt - \int_{C_R^-} \frac{F_-(t)}{t-z}\,dt\}$$

$$= \frac{1}{2\pi i}\{\int_{\gamma_{-R,R}^+} \frac{F_+(t)}{t-z}\,dt - \int_{C_R^+} \frac{F_+(t)}{t-z}\,dt$$

$$+ \int_{C_R^+} \frac{F_+(t)}{t-z}\,dt - \int_{C_R^-} \frac{F_-(t)}{t-z}\,dt\}$$

$$= F_+(z) + \frac{1}{2\pi i}\{\int_{C_R^+} \frac{F_+(t)}{t-z}\,dt - \int_{C_R^-} \frac{F_-(t)}{t-z}\,dt\}.$$

If $R \to \infty$, the integrals in the last line tends to 0 because $F_+(z)$ and $F_-(z)$ are half-plane functions, while $\tilde{F}_R(z)$ tends to the strong defining function $\tilde{F}(z)$ of $f(x)$. Similarly, we obtain $\tilde{F}_R(z) = F_-(z)$ for $\Im z < 0$. $\square$

Hyperfunctions satisfying the condition in the hypothesis of Proposition 2.33 are worth to be singled out.

Definition 2.14. The subspace of all hyperfunctions $f(x) = [F_-(z), F_+(z)]$, which are defined on the entire real line, and such that $F_+(z) \in \mathcal{O}_0(\mathbb{C}_+)$ and $F_-(z) \in \mathcal{O}_0(\mathbb{C}_-)$, is denoted by $\mathcal{B}_0(\mathbb{R})$.

Thus, a hyperfunction $f(x) = [F(z)]$ being in $\mathcal{B}_0(\mathbb{R})$ has the property that a strong defining function exists, and equals the specified defining function.

Corollary 2.34. *If a function*

$$F(z) := \begin{cases} F_+(z), & \Im z > 0 \\ F_-(z), & \Im z < 0 \end{cases}$$

satisfies $F_+(z) \in \mathcal{O}(\mathbb{C}_+)$, $F_-(z) \in \mathcal{O}(\mathbb{C}_-)$, *and* $F(z) = O(1/z)$ *as* $z \to \infty$, *then the given function* $F(z)$ *qualifies as the strong defining function of the hyperfunction defined by* $F(z)$.

Also, if $f(x)$ is an ordinary function defined on a domain containing the finite interval $[a, b]$, we have

Corollary 2.35. *If* $[a, b]$ *is a compact interval, and* $f(x) \in \mathcal{H}_\alpha(a, b)$, *then the defining function*

$$F(z) = \frac{1}{2\pi i} \int_a^b \frac{f(x)}{x - z} \, dx \tag{2.82}$$

of the projection $\chi(a, b)f(x)$ *is the strong defining function of the hyperfunction* $\chi(a, b)f(x)$.

We shall see later that hyperfunctions from $\mathcal{B}_0(\mathbb{R})$, where the given defining function equals the strong defining function, form an import subclass in the theory of Hilbert transforms.

Proposition 2.36. *Let the hyperfunction* $f(x) = [F_+(z), F_-(z)]$ *satisfy*

(i) $F_+(z) \in \mathcal{O}(\mathbb{C}_+)$, $F_-(z) \in \mathcal{O}(\mathbb{C}_-)$.

(ii) $\lim_{y \to \infty} F_+(x + iy) = c_+$, *and* $\lim_{y \to \infty} F_-(x - iy) = c_-$ *converges uniformly with respect to* $x = \Re z$ *of any compact interval, respectively.*

(iii) $F_+(z)$ *and* $F_-(z)$ *are in every horizontal strip of finite width* $0 < \Im z < h$ *and* $0 > \Im z > -h$, *respectively, uniformly bounded with respect to* $x = \Re z$.

Then the strong defining function of $f(x)$ *is given by*

$$\tilde{F}(z) = F(z) - \frac{c_+ + c_-}{2}.$$

Proof. Consider

$$\tilde{F}_1(z) := F(z) - c_+ \mathbf{1}_+(z) + c_- \mathbf{1}_-(z)$$

$$= \begin{cases} F_+(z) - c_+, & \Im z > 0 \\ F_-(z) - c_-, & \Im z < 0, \end{cases}$$

then $\tilde{F}_1(z)$ satisfies the conditions of Proposition 2.33 by Lemma 2.4, and $\tilde{F}_1(z)$ is the strong defining function of $f_1(x) := f(x) - c_+ + c_-$. Because $\mathbf{1}_+(z) = \mathbf{1}(z) + 1/2$ and $\mathbf{1}_-(z) = \mathbf{1}(z) - 1/2$ we obtain

$$\tilde{F}_1(z) = F(z) - c_+ \left(\mathbf{1}(z) + \frac{1}{2}\right) + c_- \left(\mathbf{1}(z) - \frac{1}{2}\right)$$

$$= F(z) - \frac{c_+ + c_-}{2} - (c_+ - c_-)\mathbf{1}(z),$$

or

$$\tilde{F}(z) := F(z) - \frac{c_+ + c_-}{2} = \tilde{F}_1(z) + (c_+ - c_-)\mathbf{1}(z)$$

is the sum of the strong defining function of $f(x) - c_+ + c_-$ and of the strong defining function of the constant $c_+ - c_-$, hence of $f(x)$ (see Example 2.23). This proves the theorem. □

2.4.3 Micro-analytic Hyperfunctions

We called a hyperfunction $f(x) = [F_+(z), F_-(z)]$ holomorphic or analytic on I, if its upper component $F_+(z)$ can analytically be continued across I into the lower half-plane, and if its lower component $F_-(z)$ can analytically be continued across I into the upper half-plane. If we only require either of these two properties we obtain the notion of micro-analyticity. We say that the hyperfunction $f(x) = [F_+(z), F_-(z)]$ is micro-analytic from above at $x = a \in I$, if the upper component $F_+(z)$ can analytically be continued across the real axis to a full neighborhood of a. Similarly, $f(x)$ is micro-analytic from below at $x = a \in I$, if the lower component $F_-(z)$ can analytically be continued across the real axis to a full neighborhood of a. This notion contains the mention of a direction. We would like attach the whole stipulation to the point a itself. The direction "from above" may be indicated by the vector $-i$ and the direction from below by the vector i. This leads us to a slightly more abstract definition by labeling a point in $\mathbb{R}$ by $-i$ or i.

Definition 2.15. Let the hyperfunction $f(x) = [F_+(z), F_-(z)]$ be defined on the open interval I, and $a \in I$. We say that the hyperfunction $f(x)$ is *micro-analytic at $(a, -i)$*, if the upper component $F_+(z)$ can analytically be continued across the real axis to a full neighborhood of a. Similarly, $f(x)$ is *micro-analytic at (a, i)*, if the lower component $F_-(z)$ can analytically be continued across the real axis to a full neighborhood of a. A hyperfunction is micro-analytic on $(I, -i)$ or (I, i) if for all $x \in I$ it is micro-analytic at $(x, -i)$ or (x, i), respectively.

Thus, a hyperfunction is holomorphic on I if it is micro-analytic at $(x, -i)$ and (x, i) for all $x \in I$. Referring to Proposition 2.21, we see that if the upper component of the defining function is continuous on an interval I and real or imaginary valued there, then it is micro-analytic on $(I, -i)$. Similarly, if the lower component is continuous on I, then it is micro-analytic on (I, i).

Example 2.25. The hyperfunction $f(x) = [-1/(2\pi i) \log(-z)]$ is a holomorphic hyperfunction on $(-\infty, 0) \cup (0, \infty)$. Neither $(0, -i)$ nor $(0, i)$ are points of micro-analyticity. We say that the point 0 is in the singular spectrum of $f(x)$.

2.4.4 Support, Singular Support and Singular Spectrum

In this section we assume that the mentioned hyperfunctions are defined on the entire real axis. We already know many examples of hyperfunctions vanishing on portions of the real axis. Let us now introduce three notions that will help us to be more precise on this matter.

Definition 2.16. Let Σ_0 be the largest open subset of the real line where the hyperfunction $f(x) = [F(z)]$ is vanishing, i.e., has the values zero. Its complement $K_0 := \mathbb{R} \setminus \Sigma_0$ is said to be the *support of the hyperfunction $f(x)$ denoted by* $\mathrm{supp} f(x)$.

Let Σ_1 be the largest open subset of the real line where the hyperfunction $f(x) = [F(z)]$ is holomorphic. Its complement $K_1 := \mathbb{R} \setminus \Sigma_1$ is said to be the *singular support of the hyperfunction $f(x)$ denoted by* $\mathrm{sing\ supp} f(x)$.

Let Σ_2 be the largest open subset of the real line where the hyperfunction $f(x) = [F(z)]$ is micro-analytic (from above or from below). Its complement $K_2 := \mathbb{R} \setminus \Sigma_2$ is said to be *the singular spectrum of the hyperfunction $f(x)$ denoted by* $\mathrm{sing\ spec} f(x)$.

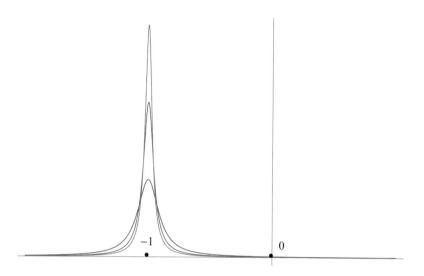

Figure 2.7: Intuitive picture of the hyperfunction of Example 2.26

Support, singular support and singular spectrum are closed subsets. It is clear that $\Sigma_1 \subset \Sigma_2$. Let us show that $\Sigma_0 \subset \Sigma_1$. In fact, for any $x \in \Sigma_0$, $F(x)$ is real analytic at x, thus $F(x)$ can analytically be continued to a function $F(z)$ holomorphic in a full neighborhood of x which means that $x \in \Sigma_1$. Therefore, we have proved

Proposition 2.37. *For a hyperfunction $f(x)$ we have*

$$\operatorname{sing\,spec} f(x) \subset \operatorname{sing\,supp} f(x) \subset \operatorname{supp} f(x).$$

Example 2.26. Consider the hyperfunction $f(x) = [1/(\sqrt{z}-i)]$. This hyperfunction is micro-analytic at $(-1, i)$ but not at $(-1, -i)$, because it can be analytically continued across -1 from below, however not from above. Moreover, $\Sigma_0 = (0, \infty)$, thus, $\operatorname{supp} f(x) = (-\infty, 0]$; $\Sigma_1 = \mathbb{R} \setminus \{-1, 0\}$, thus, sing supp $= \{-1, 0\}$; $\Sigma_2 = \mathbb{R} \setminus \{0\}$, thus, sing spec $= \{0\}$. An intuitive picture of this hyperfunction is shown in Figure 2.7.

Example 2.27. For $n \in \mathbb{N}$ and $\alpha \in \mathbb{R} \setminus \mathbb{Z}$ we have, for example, the following values.

$f(x)$	$\operatorname{sing\,spec} f(x)$	$\operatorname{sing\,supp} f(x)$	$\operatorname{supp} f(x)$
$\delta(x-a)$	$\{a\}$	$\{a\}$	$\{a\}$
$u(x-a)$	$\{a\}$	$\{a\}$	$[a, \infty)$
1	$\emptyset$	$\emptyset$	$\mathbb{R}$
$x^n \delta(x)$	$\emptyset$	$\emptyset$	$\emptyset$
$u(x)x^\alpha$	$\{0\}$	$\{0\}$	$[0, \infty)$
$u(-x)|x|^\alpha$	$\{0\}$	$\{0\}$	$(-\infty, 0]$
$|x|^\alpha$	$\{0\}$	$\{0\}$	$\mathbb{R}$
$\operatorname{fp}\frac{1}{x}$	$\{0\}$	$\{0\}$	$\mathbb{R}$
$u(x) \log x$	$\{0\}$	$\{0\}$	$[0, \infty)$
$u(-x) \log |x|$	$\{0\}$	$\{0\}$	$(-\infty, 0]$
$\log |x|$	$\{0\}$	$\{0\}$	$\mathbb{R}$
$1/(\sqrt{z} - i)$	$\{0\}$	$\{-1, 0\}$	$(-\infty, 0]$

If there is an open interval (a, b) containing the singular support of a hyperfunction $f_1(x)$ and the singular support of a hyperfunction $f_2(x)$, then both hyperfunctions are holomorphic outside of (a, b). If on (a, b) they are equal, we apply Proposition 2.18 to the difference $f_1(x) - f_2(x)$ and obtain

Proposition 2.38. *Let the hyperfunctions $f_1(x)$ and $f_2(x)$ be defined on the interval I. If both $K_1 = $ sing supp$f_1(x)$ and $K_2 = $ sing supp$f_2(x)$ are contained in $(a, b) \subset I$, and if $f_1(x) = f_2(x), x \in (a, b) \setminus (K_1 \cup K_2)$ holds, then $f_1(x) = f_2(x)$ on I.*

This shows, for example, that $f_1(x) = u(-x) + u(x)$ and $f_2(x) = u(-x+1) + u(x-1)$ are the same hyperfunctions.

If sing supp$f_1(x)$ equals sing supp$f_2(x)$, i.e., the two hyperfunctions have the same singularities, we may reformulate the above proposition and get:

Proposition 2.39 (Theorem of identity for hyperfunctions). *If two hyperfunctions $f_1(x)$ and $f_2(x)$ defined on an open interval I have the same singularities, and if there is an open subinterval $(a, b) \subset I$ where $f_1(x) = f_2(x)$, then $f_1(x) = f_2(x)$ holds on I.*

This shows, for example, that the two hyperfunctions

$$\text{fp}\frac{u(x)}{x^m} + \text{fp}\frac{u(-x)}{x^m}, \quad \text{fp}\frac{1}{x^m}$$

are the same.

We now wish to extend the theorem of analytic continuation to hyperfunctions. First, two hyperfunctions holomorphic and equal on a common open interval cannot necessarily be considered as an analytic continuation of one another. Consider the following line of thought: take two distinct hyperfunctions $f_1(x)$ and $f_2(x)$ defined on an interval I containing the subinterval (a, b). Consider then the projections $\chi_{(a,b)} f_1(x)$ and $\chi_{(a,b)} f_2(x)$. Take another open interval (c, d) disjoint to (a, b) and lying inside I. On (c, d) both $\chi_{(a,b)} f_1(x)$ and $\chi_{(a,b)} f_2(x)$ are holomorphic hyperfunctions which are equal there because they are zero. But on (a, b) they are generally different! However, we can prove

Proposition 2.40 (Theorem of analytic continuation of hyperfunctions). *Let $f_1(x)$ and $f_2(x)$ be hyperfunctions defined on (a_1, b_1) and (a_2, b_2) with a non-void overlap $O = (a_1, b_1) \cap (a_2, b_2) \neq \emptyset$. If for any $\delta > 0$:*

(i) *$f_1(x)$ is holomorphic in $(a_1 + \delta, b_1 - \delta)$, and $f_2(x)$ is holomorphic in $(a_2 + \delta, b_2 - \delta)$,*

(ii) *they are equal in the overlap O,*

then there exists a unique hyperfunction $f(x)$ such that

$$f(x) := \begin{cases} f_1(x), & x \in (a_1, b_1) \\ f_2(z), & x \in (a_2, b_2) \end{cases} \tag{2.83}$$

which is called the analytic continuation of $f_1(x)$ and $f_2(x)$. The given $f_1(x)$ and $f_2(x)$ are said to be the analytic continuation of each other.

Proof. For definiteness we take $a_1 < a_2 < b_1 < b_2$ such that $O = [a_2, b_1]$. By the definition of holomorphicity of hyperfunctions there is a positive ϵ such that

$f_1(x)$ is a holomorphic hyperfunction at $a_1 + \epsilon, b_1 - \epsilon$, and $f_2(x)$ is a holomorphic hyperfunction at $a_2 + \epsilon, b_2 - \epsilon$. Since $f_1(x) = f_2(x)$ on the overlap (a_2, b_1), we have that $f_2(x)$ is a holomorphic hyperfunction at $b_1 - \epsilon$, and $f_1(x)$ is a holomorphic hyperfunction at $a_2 + \epsilon$. Then, we can write

$$\chi_{(a_1+\epsilon,b_1-\epsilon)} f_1(x) = \chi_{(a_1+\epsilon,a_2+\epsilon)} f_1(x) + \chi_{(a_2+\epsilon,b_1-\epsilon)} f_1(x),$$
$$\chi_{(a_2+\epsilon,b_2-\epsilon)} f_2(x) = \chi_{(a_2+\epsilon,b_1-\epsilon)} f_2(x) + \chi_{(b_1-\epsilon,b_2-\epsilon)} f_2(x).$$

Because on the overlap (a_2, b_1) we have $g(x) := f_1(x) = f_2(x)$ we may write

$$f(x) := \chi_{(a_1+\epsilon,a_2+\epsilon)} f_1(x) + \chi_{(a_2+\epsilon,b_1-\epsilon)} g(x) + \chi_{(b_1-\epsilon,b_2-\epsilon)} f_2(x). \qquad (2.84)$$

Obviously, for all sufficiently small $\epsilon > 0$ we have

$$f(x) := \begin{cases} f_1(x), & x \in (a_1 + \epsilon, b_1 - \epsilon) \\ f_2(z), & x \in (a_2 + \epsilon, b_2 - \epsilon). \end{cases}$$

By letting $\epsilon \to 0$ we obtain (2.83). $\qquad \qquad \qquad \qquad \qquad \qquad \qquad \qquad \square$

For example, for the two hyperfunctions $f_1(x) = u(x+1) - u(x-2)$, and $f_2(x) = u(x-1) - u(x-4)$ there is one hyperfunction $f(x) = u(x+1) - x(x-4)$ which is the analytic continuation of each other.

From Example 2.18 follows the important

Corollary 2.41. *Let $a_1 < a_2 < b_1 < b_2$ and $f(x)$ be a hyperfunction defined on an interval I containing (a_1, b_2), then the projection $\chi_{(a_1,b_2)} f(x)$ is the analytic continuation of the projections $\chi_{(a_1,b_1)} f(x)$ and $\chi_{(a_2,b_2)} f(x)$.*

2.5   Product of Hyperfunctions

The product of two hyperfunctions cannot generally be defined. However, there are special subclasses of hyperfunctions or special circumstances where the definition of the product of two hyperfunctions is feasible. We already know that the product of a real analytic function and an arbitrary hyperfunction is always defined. Another subclass for which the product can be defined is discussed in the next section. The general investigation of the required circumstances for defining the product of two hyperfunctions then follows.

2.5.1   Product of Upper or Lower Hyperfunctions

It is quite natural to define the product of two upper or two lower hyperfunctions. Since with $f(x) = F(x + i0)$, $g(x) = G(x + i0)$, $x = \Re z$ we have

$$f(x) \cdot g(x) = F(x + i0) \cdot G(x + i0) = \left(\lim_{z \to x+i0} F(z) \right) \left(\lim_{z \to x+i0} G(z) \right)$$
$$= \lim_{z \to x+i0} F(z) G(z),$$

giving the natural

Definition 2.17. Let $f(x) = [F(z)\mathbf{1}_+(z)] = [F(z), 0] \in \mathcal{B}_+(I)$ and $g(x) = [G(z)\mathbf{1}_+(z)]$ $= [G(z), 0] \in \mathcal{B}_+(I)$ be two upper hyperfunctions; then their product is defined and denoted as

$$f(x) \cdot g(x) := [F(z)G(z)\mathbf{1}_+(z)] = [F(z)G(z), 0] \in \mathcal{B}_+(I). \qquad (2.85)$$

Similarly, if $f(x) = [F(z)\mathbf{1}_-(z)] = [0, -F(z)] \in \mathcal{B}_-(I)$ and $[G(z)\mathbf{1}_-(z)] = [0, -G(z)] \in \mathcal{B}_-(I)$ are two lower hyperfunctions, their product is defined by

$$f(x) \cdot g(x) := [F(z)G(z)\mathbf{1}_-(z)] = [0, -F_-(z)G_-(z)] \in \mathcal{B}_-(I). \qquad (2.86)$$

Clearly, the product of two upper or two lower hyperfunctions is commutative, associative and Leibniz's product rule holds, i.e.,

$$D(f(x) \cdot g(x)) = Df(x) \cdot g(x) + f(x) \cdot Dg(x).$$

Let us now indicate why the product of two arbitrary hyperfunctions $f(x) = [F_+(z), F_-(z)]$ and $g(x) = [G_+(z), G_-(z)]$ cannot be defined. Both hyperfunctions can be decomposed into an upper and a lower hyperfunction:

$$f(x) = [F_+(z), 0] - [0, -F_-(z)], \quad g(x) = [G_+(z), 0] - [0, -G_-(z)].$$

By attempting to straightforwardly form the product we would have

$$\begin{aligned} f(x) \cdot g(x) &= \{[F_+(z), 0] - [0, -F_-(z)]\} \cdot \{[G_+(z), 0] - [0, -G_-(z)]\} \\ &= [F_+(z), 0] \cdot [G_+(z), 0] - [F_+(z), 0] \cdot [0, -G_-(z)] \\ &\quad - [0, -F_-(z)] \cdot [G_+(z), 0] + [0, -F_-(z)] \cdot [0, -G_-(z)]. \end{aligned}$$

While the first and the fourth term make sense the second and the third ones are meaningless because $F_+(z)$ and $G_-(z)$ and $F_-(z)$ and $G_+(z)$ cannot be multiplied together since the variable z lies in different half-planes. However, if it is assumed that $f(x) = [F_+(z), F_-(z)]$ is micro-analytic at (x, i) for all $x \in I$, then $F_-(z)$ is in fact real analytic in I, thus the first factor has the equivalent representation $f(x) = [F_+(z) - F_-(z), 0]$ on I. Similarly, by assuming that $g(x) = [G_+(z), G_-(z)]$ is also micro-analytic at (x, i) for all $x \in I$, $G_-(z)$ becomes real analytic on I, and $g(x) = [G_+(z) - G_-(z), 0]]$ is an equivalent representation. Now the product can be formed and we obtain

$$\begin{aligned} f(x) \cdot g(x) &= [F_+(z) - F_-(z), 0] \cdot [G_+(z) - G_-(z), 0] \\ &= [\{F_+(z) - F_-(z)\}\{G_+(z) - G_-(z)\}, 0] \\ &= [F_+(z)G_+(z) - F_-(z)G_+(z) - F_+(z)G_-(z) + F_-(z)G_-(z), 0] \end{aligned}$$

where all products are well defined for z lying in an upper half-neighborhood $\mathcal{D}_+(I)$ due to the analytic continuations of $F_-(z)$ and $G_-(z)$ into the upper half-neighborhood. Similarly, if both $f(x)$ and $g(x)$ are micro-analytic at $(x, -i)$ for all $x \in I$, their product is well defined and we have

$$f(x) \cdot g(x) = [0, -\{F_+(z)G_+(z) - F_-(z)G_+(z) - F_+(z)G_-(z) + F_-(z)G_-(z)\}]$$

where z is now in some lower half-neighborhood $\mathcal{D}_-(I)$. The reader may readily convince himself that if one factor is micro-analytic from above and the other from below in I, or vice-versa, the product cannot be defined. Thus, only if $f(x)$ and $g(x)$ are both micro-analytic at $(x, -i)$ or both at (x, i) for all $x \in I$ their product is defined on I.

There is a third possibility for defining the product of two hyperfunctions, namely if one factor is a holomorphic hyperfunction on the interval I, without any restriction for the other factor. In practice, this is the most important case because most familiar hyperfunctions are holomorphic except at a finite number of singularities. This case is studied in detail in the next section.

2.5.2   Products in the Case of Disjoint Singular Supports

We assume that the two hyperfunctions $f(x)$ and $g(x)$ satisfy the condition

$$\text{sing supp} f(x) \cap \text{sing supp} g(x) = \emptyset. \tag{2.87}$$

Thus, for any $x \in \mathbb{R}$ at least one of the two hyperfunctions is holomorphic at x.

For definiteness, let I be a real interval where $f(x) = [F_+(z), F_-(z)]$ is holomorphic, i.e., $f(x) \in \mathcal{B}_{\mathcal{O}}(I)$. This means that the upper component $F_+(z)$ and the lower component $F_-(z)$ can analytically be continued across I into the lower and into the upper half-plane, respectively. Then, $\phi(x) := F_+(x) - F_-(x)$ becomes a real analytic function on I. This enables us to define the product of $f(x)$ with any other hyperfunction $g(x) = [G(z)] = [G_+(z), G_-(z)]$, also defined on I as

$$\begin{aligned} f(x) \cdot g(x) &:= \phi(z) [G(z)] = \{F_+(z) - F_-(z)\} [G(z)] \\ &= [\{F_+(z) - F_-(z)\} G_+(z), \{F_+(z) - F_-(z)\} G_-(z)], \end{aligned} \tag{2.88}$$

where the order of the factors is important owing to lack of symmetry of the situation. In the above formula the left factor is the holomorphic hyperfunction. In the event that the right factor is the assumed holomorphic hyperfunction, we have

$$\begin{aligned} f(x) \cdot g(x) &:= \{G_+(z) - G_-(z)\} [F(z)] \\ &= [\{G_+(z) - G_-(z)\} F_+(z), \{G_+(z) - G_-(z)\} F_-(z)]. \end{aligned} \tag{2.89}$$

Example 2.28. The hyperfunction $1 = [1(z)] = [1/2, -1/2]$ is a holomorphic hyperfunction on $I = \mathbb{R}$, therefore we may write

$$1 \cdot g(x) := [\{1/2 - (-1/2)\} G(z)] = [G(z)] = g(x).$$

Example 2.29. Let $\phi(x)$ be a real analytic function which is a defining function of the zero hyperfunction, i.e., $0 = [\phi(x)]$. Let $g(x) = [G(z)]$ be an arbitrary hyperfunction in the interval I, then

$$\begin{aligned} 0 \cdot g(x) &= [\{\phi_+(z) - \phi_-(z)\} G(z)] \\ &= [\{\phi(z) - \phi(z)\} G(z)] = [0\, G(z)] \\ &= 0. \end{aligned}$$

Proposition 2.42. *If $f(x), g(x) \in \mathcal{B}_{\mathcal{O}}(I)$, i.e., both factors $f(x)$ and $g(x)$ are holomorphic hyperfunctions on I, then the defined product becomes commutative, i.e., $f(x) \cdot g(x) = g(x) \cdot f(x)$.*

Proof. We have that $f(x) \cdot g(x) - g(x) \cdot f(x)$ equals

$$[\{F_+(z) - F_-(z)\} G(z)] - [\{G_+(z) - G_-(z)\} F(z)]$$
$$= \{F_+(z) - F_-(z)\} [G_+(z), G_-(z)] - \{G_+(z) - G_-(z)\} [F_+(z), F_-(z)]$$
$$= [F_+(z)G_-(z) - F_-(z)G_+(z), F_+(z)G_-(z) - F_-(z)G_+(z)]$$
$$= [\psi(z)]$$

where $\psi(z) = F_+(z)G_-(z) - F_-(z)G_+(z)$ is a holomorphic function in a full neighborhood of I, hence real analytic on I. Therefore,

$$f(x) \cdot g(x) - g(x) \cdot f(x) = [\psi(z)] = 0, \quad x \in I. \qquad \square$$

Example 2.30. If $f(x), g(x) \in \mathcal{B}_{\mathcal{O}}(I)$, then

$$f(x) \cdot g(x) = [\{F_+(z) - F_-(z)\}G_+(z), \{F_+(z) - F_-(z)\}G_-(z)].$$

Since $g(x)$ is a holomorphic hyperfunction too, the hyperfunction $f(x) \cdot g(x)$ has values for all $x \in I$, i.e., defines an ordinary real analytic function on I (again denoted by $f(x) \cdot g(x)$)

$$x \in I \mapsto f(x) \cdot g(x) := \{F_+(x) - F_-(x)\}\{G_+(x) - G_-(x)\}.$$

Therefore, the product of two holomorphic hyperfunctions on I defines an ordinary real analytic function on I.

Example 2.31. The unit-step hyperfunction $u(x) = [-\log(-z)/(2\pi i)]$ is a holomorphic hyperfunction in $(0, \infty)$ and in $(-\infty, 0)$: $u(x) \in \mathcal{B}_{\mathcal{O}}(0, \infty)$, $u(x) \in \mathcal{B}_{\mathcal{O}}(-\infty, 0)$. Therefore, for any other hyperfunction $f(x) = [F(z)]$ defined on the real axis we have, on $I = (-\infty, 0)$,

$$u(x) \cdot f(x) = [0 \cdot F(z)] = [0] = 0,$$

and on $I = (0, \infty)$,

$$u(x) \cdot f(x) = \left[\frac{-\pi i - \pi i}{-2\pi i} F(z)\right] = [F(z)] = f(x).$$

This equals the projection of $f(x)$ on $(0, \infty)$, i.e., $\chi_{(0,\infty)}f(x)$. Therefore, from now on, we may write

$$u(x) \cdot f(x) = \chi_{(0,\infty)}f(x) = u(x)f(x). \qquad (2.90)$$

Next let $f(x), g(x) \in \mathcal{B}(\mathbb{R})$ with sing supp$f(x) = \{\alpha_1, \alpha_2, \ldots, \alpha_m\}$ and sing supp$g(x) = \{\beta_1, \ldots, \beta_p\} \cup \{\beta_{p+1}, \ldots, \beta_n\}$, such that

$$\beta_1 < \cdots, < \beta_p < \alpha_1 < \alpha_2 < \cdots < \alpha_m < \beta_{p+1} < \cdots, \beta_n,$$

i.e., the singular support of $f(x)$ lies wholly between two consecutive singularities of $g(x)$, also sing supp$f(x) \cap$ sing supp$g(x) = \emptyset$. Choose two real separators d_1 and d_2 such that $\beta_p < d_1 < \alpha_1$ and $\alpha_m < d_2 < \beta_{p+1}$. Then $f(x)$ is holomorphic on $(-\infty, \alpha_1) \cup (\alpha_m, \infty)$ and hence on $(-\infty, d_1)$ and (d_2, ∞). The hyperfunction $g(x)$ is holomorphic on $(-\infty, \beta_1) \cup (\beta_p, \beta_{p+1}) \cup (\beta_n, \infty)$, thus on (d_1, d_2). In the open intervals $(-\infty, \beta_1)$, (β_p, α_1), (α_m, β_{p+1}), (β_n, ∞), both hyperfunctions $f(x)$

and $g(x)$ are holomorphic. On the interval $(-\infty, d_1)$ and (d_2, ∞) the product $f(x) \cdot g(x)$ is defined, and on the interval (d_1, d_2) the product $g(x) \cdot f(x)$. By using the concept of projection, we may define the product of $f(x)$ and $g(x)$ by

$$f(x) \cdot g(x) := \chi_{(-\infty, d_1)}\{f(x) \cdot g(x)\} + \chi_{(d_2, \infty)}\{f(x) \cdot g(x)\} \\ + \chi_{(d_1, d_2)}\{g(x) \cdot f(x)\}. \tag{2.91}$$

This product is a hyperfunction defined on the entire real line. If we denote the defining function of the product hyperfunction by $H(z)$ we have, by taking into account the definition of the projection of a hyperfunction:

$$H(z) = \frac{\phi_1(z)}{2\pi i} \int_{-\infty}^{d_1} \frac{f(x) \cdot g(x)}{\phi_1(x)(x-z)} dx + \frac{\phi_2(z)}{2\pi i} \int_{d_2}^{\infty} \frac{f(x) \cdot g(x)}{\phi_2(x)(x-z)} dx \\ + \frac{1}{2\pi i} \int_{d_1}^{d_2} \frac{g(x) \cdot f(x)}{x-z} dx$$

where $\phi_{1,2}(x)$ are possibly suitable converging factors, if necessary. More explicitly, $H(z)$ amounts to

$$\frac{\phi_1(z)}{2\pi i} \{ \int_{\gamma_{-\infty, d_1}^+} \frac{(F_+(\zeta) - F_-(\zeta))G_+(\zeta)}{\phi_1(\zeta)(\zeta - z)} d\zeta - \int_{\gamma_{-\infty, d_1}^-} \frac{(F_+(\zeta) - F_-(\zeta))G_-(\zeta)}{\phi_1(\zeta)(\zeta - z)} d\zeta \}$$

$$+ \frac{\phi_2(z)}{2\pi i} \{ \int_{\gamma_{d_2, \infty}^+} \frac{(F_+(\zeta) - F_-(\zeta))G_+(\zeta)}{\phi_2(\zeta)(\zeta - z)} d\zeta - \int_{\gamma_{d_2, \infty}^-} \frac{(F_+(\zeta) - F_-(\zeta))G_-(\zeta)}{\phi_2(\zeta)(\zeta - z)} d\zeta \}$$

$$+ \frac{1}{2\pi i} \{ \int_{\gamma_{d_1, d_2}^+} \frac{(G_+(\zeta) - G_-(\zeta))F_+(\zeta)}{\zeta - z} d\zeta - \int_{\gamma_{d_1, d_2}^-} \frac{(G_+(\zeta) - G_-(\zeta))F_-(\zeta)}{\zeta - z} d\zeta \}.$$

If we further assume that $f(x)$ is a meromorphic hyperfunction whose defining function has only the real poles $\alpha_1, \alpha_2, \ldots, \alpha_m$, $H(z)$ is simplified and becomes

$$\frac{1}{2\pi i} \{ \int_{\gamma_{d_1, d_2}^+} \frac{(G_+(\zeta) - G_-(\zeta))F(\zeta)}{\zeta - z} d\zeta - \int_{\gamma_{d_1, d_2}^-} \frac{(G_+(\zeta) - G_-(\zeta))F(\zeta)}{\zeta - z} d\zeta \}$$

$$= -\frac{1}{2\pi i} \oint_{(d_1, d_2)} \frac{(G_+(\zeta) - G_-(\zeta))F(\zeta)}{\zeta - z} d\zeta$$

$$= -\text{Res}_{\alpha_1, \alpha_2, \ldots, \alpha_m} \{ \frac{(G_+(\zeta) - G_-(\zeta))F(\zeta)}{\zeta - z} \}$$

$$= -\sum_{k=1}^{m} \text{Res}_{\zeta = \alpha_k} \{ \frac{g(\zeta) F(\zeta)}{\zeta - z} \}.$$

Proposition 2.43. *Let the defining function $F(z)$ of the hyperfunction $f(x)$ be a meromorphic function with the sole real poles $\alpha_1 < \alpha_2 < \cdots < \alpha_m$. Let the hyperfunction $g(x) = [G(z)]$ be holomorphic in the interval (a', b') with $a' < \alpha_1$ and $\alpha_m < b'$. Then, the product $f(x) \cdot g(x)$ is defined and given by*

$$f(x) \cdot g(x) = \left[\sum_{k=1}^{m} \text{Res}_{\zeta = \alpha_k} \{ \frac{g(\zeta) F(\zeta)}{z - \zeta} \} \right]. \tag{2.92}$$

Example 2.32. Take $f(x) = \delta(x - a)$ and suppose that $g(x)$ is any hyperfunction holomorphic at $x = a$. Then $F(\zeta) = -1/(2\pi i(\zeta - a))$ is meromorphic with the residuum at $\zeta = a$ equal to $-1/(2\pi i)$. The above formula becomes

$$H(z) = g(a)\frac{-1}{2\pi i}\frac{1}{z - a}.$$

Thus,

$$\delta(x - a) \cdot g(x) = g(a)\delta(x - a) \tag{2.93}$$

a well-known formula, but until the present state of affairs it was only valid for a real analytic function $g(x)$. If, for example $g(x) = \mathrm{fp}\frac{1}{x-a}$ with $a \neq 0$, we obtain

$$\delta(x) \cdot \mathrm{fp}\frac{1}{x - a} = -\frac{1}{a}\delta(x).$$

We also have

$$\delta(x - a) \cdot \delta(x - b) = 0 \quad (a \neq b). \tag{2.94}$$

If $F(z)$ has just one pole at $x = a$ and $g(x)$ is holomorphic there, we can proceed by a direct calculation. We illustrate this in the next example.

Example 2.33. Consider $x^\alpha \delta^{(m)}(x - a)$, $a \neq 0$. The function $g(x) = x^\alpha$ then is holomorphic at $x = a$.

$$x^\alpha = (a + (x - a))^\alpha = a^\alpha\left(1 + \frac{x - a}{a}\right)^\alpha = \sum_{k=0}^\infty \binom{\alpha}{k} a^{\alpha-k}(x - a)^k.$$

Now,

$$\delta^{(m)}(x - a) \cdot x^\alpha := \sum_{k=0}^\infty \binom{\alpha}{k} a^{\alpha-k}\left[(z - a)^k \frac{-(-1)^m m!}{2\pi i (z - a)^{m+1}}\right]$$

$$= \sum_{k=0}^m \binom{\alpha}{k} a^{\alpha-k}\left[\frac{-(-1)^m m!}{2\pi i (z - a)^{m+1-k}} + \phi(z)\right]$$

where $\phi(x)$ is a real analytic function that can be discarded giving an equivalent defining function. Thus,

$$\delta^{(m)}(x - a) \cdot x^\alpha = \sum_{k=0}^m \binom{\alpha}{k} (-1)^k \frac{m!}{(m - k)!} a^{\alpha-k}\left[\frac{-(-1)^{m-k}(m - k)!}{2\pi i (z - a)^{m+1-k}}\right]$$

$$= \sum_{k=0}^m \binom{\alpha}{k} (-1)^k \frac{m!}{(m - k)!} a^{\alpha-k}\delta^{(m-k)}(x - a).$$

Example 2.34. Let us show that for $a < b$,

$$u(x - a) \cdot u(b - x) = u(x - a) - u(x - b) = \chi_{(a,b)}(x). \tag{2.95}$$

Proof. For definiteness, we assume $a < b$. Then we have

$$u(x - a) = [F(z)] = [-\frac{1}{2\pi i} \log(a - z)], \quad u(b - x) = [G(z)] = [\frac{1}{2\pi i} \log(z - b)],$$

thus, $u(x - a$ is a holomorphic hyperfunction in $(-\infty, a) \cup (a, \infty)$, and $F_+(x) - F_-(x) = 1$ for $x > a$, similarly $G(z)$ is real analytic for $(-\infty, b) \cup (b, \infty)$, and $G_+(x) - G_-(x) = 1$ for $x < b$. Therefore we may write

$$h(x) := u(x - a) \cdot u(b - x) = \chi_{(a,b)}\{u(x - a) \cdot u(b - x)\} = [H(z)].$$

Therefore, if c is a point between a and b, we have

$$
\begin{aligned}
&H(z)\\
&= \frac{1}{2\pi i}\Big\{\int_a^c \frac{[G_+(x) - G_-(x)]F_+(x + i0)}{x - z} - \frac{[G_+(x) - G_-(x)]F_-(x + i0)}{x - z}\, dx\\
&\quad + \int_c^b \frac{[F_+(x) - F_-(x)]G_+(x + i0)}{x - z} - \frac{[F_+(x) - F_-(x)]G_-(x + i0)}{x - z}\, dx\Big\}\\
&= \frac{1}{2\pi i}\Big\{\int_a^c \frac{[F_+(x + i0) - F_-(x - i0)]}{x - z}\, dx + \int_c^b \frac{[G_+(x + i0) - G_-(x - i0)]}{x - z}\, dx\Big\}\\
&= \frac{1}{2\pi i}\Big\{\int_a^c \frac{dx}{x - z} + \int_c^b \frac{dx}{x - z}\Big\}\\
&= -\frac{1}{2\pi i}\log(a - z) + \frac{1}{2\pi i}\log(b - z).
\end{aligned}
$$

This proves that $h(x) = u(x - a) - u(x - b)$.   □

Problem 2.9. Let $f(x) = \delta^{(m)}(x - a)$, and let $g(x) = [G(z)]$ be holomorphic in the interval (a', b') with $a' < a < b'$ Then, the product is given by

$$\delta^{(m)}(x - a) \cdot g(x) = \sum_{k=0}^m (-1)^k g^{(k)}(a) \binom{m}{k} \delta^{(m-k)}(x - a).$$

2.5.3 The Integral of a Product

Often, products of hyperfunctions appear under an integral sign. We shall show how the integral over a product of hyperfunctions can be computed without knowing the defining function of the product, i.e., only by using the defining functions of the factors. Let $f(x) = [F(z)]$ and $g(x) = [G(z)]$ be given hyperfunctions such that sing supp$f(x) \cap$ sing supp$g(x) = \emptyset$.

Let sing supp$f(x) = \{\alpha_1, \alpha_2, \ldots\}$ and sing supp$g(x) = \{\beta_1, \beta_2, \ldots\}$. We place the real separator points $d_0, d_1, \ldots$ such that (d_0, d_1) only encloses sing supp$g(x)$, (d_1, d_2) only sing supp$f(x)$, (d_2, d_3) only sing supp$g(x)$ and so on. Thus,

$$f(x) \in \mathcal{B}_O(d_{2k}, d_{2k+1}), \quad g(x) \in \mathcal{B}_O(d_{2k+1}, d_{2k+2}), \quad (k = 0, 1, \ldots).$$

For definiteness we assume a situation as pictured in Figures 2.8 and 2.9. We define

$$f(x) \cdot g(x) := \sum_{k=0}^{\cdots} \chi_{(d_{2k}, d_{2k+1})}\{f(x) \cdot g(x)\} + \chi_{(d_{2k+1}, d_{2k+2})}\{g(x) \cdot f(x)\}.$$

Let the interval $[a, b]$ enclose all separator points and thereby sing supp$f(x)$ and sing supp$g(x)$. By Proposition 2.28 we have

$$\int_a^b f(x) \cdot g(x)\, dx = \sum_{k=0}^{\cdots} \Big\{\int_{d_{2k}}^{d_{2k+1}} f(x) \cdot g(x)\, dx + \int_{d_{2k+1}}^{d_{2k+2}} g(x) \cdot f(x)\, dx\Big\}$$

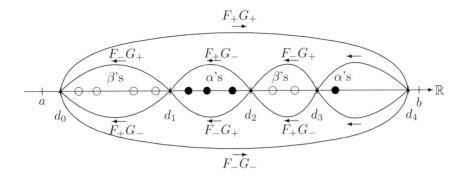

Figure 2.8: Circles: singularities of $g(x)$, filled disks: singularities of $f(x)$

with

$$\int_{d_{2k}}^{d_{2k+1}} f(x) \cdot g(x)\, dx$$

$$= \int_{\gamma_{d_{2k},d_{2k+1}}^{+}} (F_+(\zeta) - F_-(\zeta))\, G_+(\zeta)\, d\zeta - \int_{\gamma_{d_{2k},d_{2k+1}}^{-}} (F_+(\zeta) - F_-(\zeta))\, G_-(\zeta)\, d\zeta,$$

$$\int_{d_{2k+1}}^{d_{2k+2}} g(x) \cdot f(x)\, dx$$

$$= \int_{\gamma_{d_{2k+1},d_{2k+2}}^{+}} (G_+(\zeta) - G_-(\zeta))\, F_+(\zeta)\, d\zeta$$

$$- \int_{\gamma_{d_{2k+1},d_{2k+2}}^{-}} (G_+(\zeta) - G_-(\zeta))\, F_-(\zeta)\, d\zeta.$$

Therefore,

$$\int_{a}^{b} f(x) \cdot g(x)\, dx$$

$$= \sum_{k=0}^{\ldots} \{ \int_{\gamma_{d_{2k},d_{2k+1}}^{+}} F_+(\zeta)\, G_+(\zeta)\, d\zeta + \int_{\gamma_{d_{2k+1},d_{2k+2}}^{+}} F_+(\zeta)\, G_+(\zeta)\, d\zeta$$

$$+ \int_{\gamma_{d_{2k},d_{2k+1}}^{-}} F_-(\zeta)\, G_-(\zeta)\, d\zeta + \int_{\gamma_{d_{2k+1},d_{2k+2}}^{-}} F_-(\zeta)\, G_-(\zeta)\, d\zeta$$

$$+ \int_{-\gamma_{d_{2k},d_{2k+1}}^{+}} F_-(\zeta)\, G_+(\zeta)\, d\zeta + \int_{-\gamma_{d_{2k+1},d_{2k+2}}^{+}} F_+(\zeta)\, G_-(\zeta)\, d\zeta$$

$$+ \int_{-\gamma_{d_{2k},d_{2k+1}}^{-}} F_+(\zeta)\, G_-(\zeta)\, d\zeta + \int_{-\gamma_{d_{2k+1},d_{2k+2}}^{-}} F_-(\zeta)\, G_+(\zeta)\, d\zeta \}.$$

If d_N denotes the most right located separator point, the integrals in the first and second line can be joined together,

$$\sum_{k=0}^{\cdots}\{\int_{\gamma^+_{d_{2k},d_{2k+1}}} F_+(\zeta)\,G_+(\zeta)\,d\zeta + \int_{\gamma^+_{d_{2k+1},d_{2k+2}}} F_+(\zeta)\,G_+(\zeta)\,d\zeta\}$$

$$= \int_{\gamma^+_{d_0,d_N}} F_+(\zeta)\,G_+(\zeta)\,d\zeta,$$

$$\sum_{k=0}^{\cdots}\{\int_{\gamma^-_{d_{2k},d_{2k+1}}} F_-(\zeta)\,G_-(\zeta)\,d\zeta + \int_{\gamma^-_{d_{2k+1},d_{2k+2}}} F_-(\zeta)\,G_-(\zeta)\,d\zeta$$

$$= \int_{\gamma^-_{d_0,d_N}} F_-(\zeta)\,G_-(\zeta)\,d\zeta.$$

We introduce the contours

$$\gamma_1 := \gamma^+_{d_0,d_N}, \; \gamma_2 := -\gamma^+_{d_0,d_1} - \gamma^-_{d_1,d_2} - \gamma^+_{d_2,d_3}\cdots$$

$$\gamma_3 := \gamma^-_{d_0,d_N}, \; \gamma_4 := -\gamma^-_{d_0,d_1} - \gamma^+_{d_1,d_2} - \gamma^-_{d_2,d_3}\cdots$$

which can be joined together giving one closed contour $\gamma = \gamma_1 + \gamma_2 + \gamma_3 + \gamma_4$. Finally, we obtain (see Figure 2.9).

Proposition 2.44. *Let $f(x) = [F(z)]$ and $g(x) = [G(z)]$ be two given hyperfunctions with* sing supp$f(x) \cap$ sing supp$g(x) = \emptyset$, *where* sing supp$f(x) = \{\alpha_1, \alpha_2, \ldots\}$ *and* sing supp$g(x) = \{\beta_1, \beta_2, \ldots\}$. *Place the real separators $d_0, d_1, \ldots$ such that* sing supp$g(x) \subset (d_0, d_1) \cup (d_2, d_3) \cup (d_4, d_5) \cup \cdots$ *and* sing supp$f(x) \subset (d_1, d_2) \cup (d_3, d_4) \cup (d_5, d_6) \cup \cdots$ *which means that $f(x)$ is holomorphic in $(d_0, d_1) \cup (d_2, d_3) \cup (d_4, d_5) \cup \cdots$), and $g(x)$ in $(d_1, d_2) \cup (d_3, d_4) \cup (d_5, d_6) \cup \cdots$). Then, with $a \le d_0$ and $b \ge d_N$, and the closed contour $\gamma := \gamma_1 + \gamma_2 + \gamma_3 + \gamma_4$ we have*

$$\int_a^b f(x) \cdot g(x)\,dx = \int_\gamma P(\zeta)\,d\zeta$$

with

$$P(\zeta) := \begin{cases} F_+(\zeta)G_+(\zeta), & \zeta \in \gamma_1 \\ F_-(\zeta)G_+(\zeta), & \zeta \in \gamma_2 \\ F_-(\zeta)G_-(\zeta), & \zeta \in \gamma_3 \\ F_+(\zeta)G_-(\zeta), & \zeta \in \gamma_4. \end{cases}$$

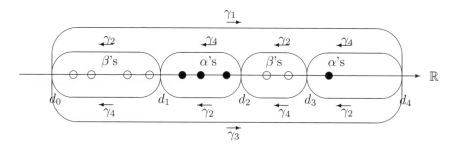

Figure 2.9: Closed contour γ in Proposition 2.44

Notice that for the closed contour γ, the parts γ_1 lying in the upper half-plane and γ_3 in the lower half-plane are always directed from left to right, while the contours γ_2 and γ_4 lying alternatively in the lower and upper half-planes are directed from right to left. Figure 2.10 shows the contours when the number of connected sets of singularities is not the same.

Example 2.35. Let for $a < b$, and non-integral numbers α and β,

$$f(x) = u(x-a)(x-a)^\alpha = \left[-\frac{1}{2i \sin(\alpha\pi)} (a-z)^\alpha \right],$$

$$g(x) = u(b-x)|x-b|^\beta = \left[\frac{1}{2i \sin(\beta\pi)} (z-b)^\beta \right].$$

We want to compute the integral

$$\int_{-\infty}^\infty f(x) \cdot g(x)\, dx.$$

Because $u(x-a) \cdot u(b-x) = u(x-a) - u(x-b) = \chi_{(a,b)}(x)$ we can write, by using Proposition 2.28, and for $a' < a < b < b'$,

$$\int_{-\infty}^\infty f(x) \cdot g(x)\, dx = \int_{-\infty}^\infty \chi_{(a,b)}\{(x-a)^\alpha \cdot (b-x)^\beta\}\, dx$$

$$= \int_{a'}^{b'} \chi_{(a,b)}\{(x-a)^\alpha \cdot (b-x)^\beta\}\, dx.$$

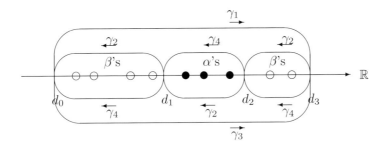

Figure 2.10: Closed contour γ for three groups of singularities

If $\Re\alpha > -1$, $\Re\beta > -1$, the integral

$$I(\alpha, \beta) := \int_a^b (x-a)^\alpha (b-x)^\beta\, dx$$

exists in the ordinary sense and can be computed in terms of known functions. The change of variables $x = a + (b-a)t$ yields

$$I(\alpha, \beta) = (b-a)^{\alpha+\beta+1} \int_0^1 t^\alpha (1-t)^\beta\, dt$$

$$= (b-a)^{\alpha+\beta+1} \int_0^1 t^{(\alpha+1)-1} (1-t)^{(\beta+1)-1}\, dt$$

$$= (b-a)^{\alpha+\beta+1} B(\alpha+1, \beta+1)$$

where $B(\alpha, \beta) = \Gamma(\alpha)\Gamma(\beta)/\Gamma(\alpha + \beta)$ is Euler's beta function. If $\Re\alpha$ or $\Re\beta$ is $\leq$ -1, the integral only exists as a *Hadamard Finite Part Integral,* denoted by FP $\int$. The theory of hyperfunctions can be used to compute the Hadamard Finite Part of a divergent integral (see the next section). We then have

$$\text{FP} \int_a^b (x-a)^\alpha \, (b-x)^\beta \, dx := \int_{-\infty}^\infty \chi_{(a,b)}\{(x-a)^\alpha \cdot (b-x)^\beta\} \, dx. \qquad (2.96)$$

We place the separators d_0, d_1, d_2 such that $d_0 < a < d_1 < b < d_2$. According to Proposition 2.44 we have

$$\int_{d_0}^{d_2} \chi_{(a,b)}\{(x-a)^\alpha \cdot (b-x)^\beta\} \, dx$$

$$= \int_{\gamma_1} F_+(\zeta)\, G_+(\zeta)\, d\zeta + \int_{\gamma_2} F_+(\zeta)\, G_-(\zeta)\, d\zeta$$

$$+ \int_{\gamma_3} F_-(\zeta)\, G_-(\zeta)\, d\zeta \int_{\gamma_4} F_-(\zeta)\, G_+(\zeta)\, d\zeta =: I_c,$$

with

$$F_\pm(z) = \left[-\frac{1}{2i\sin(\alpha\pi)}(a-z)^\alpha \right],$$

$$G_\pm(z) = \left[\frac{1}{2i\sin(\beta\pi)}(z-b)^\beta \right],$$

and where the integration contours are shown in Figure 2.11. In spite of the fact that the upper and lower components of the defining functions have the same functional expressions, they represent different branches of a multi-valued function. In the points d_0 and d_2 the branches have to be selected such that the integrand

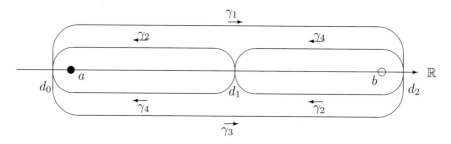

Figure 2.11: Closed contour in Example 2.35

is continuous on the closed contour $\gamma := \gamma_1 + \gamma_2 + \gamma_3 + \gamma_4$. The contour γ can then be deformed into one comprising four parts (the reader is advised to draw a picture):

(i) part 1 starts at $a - r$, encircles a on a small semicircle with radius r and center a in the negative sense to $a + r + i0$, then runs on the upper side of the real axis to $b - r + i0$,

(ii) part 2 encircles b in a small circle with radius r and center b in the negative sense to $b - r - i0$, and goes back thereby crossing the real axis to $a + r + i0$,

(iii) part 3 encircles a in the positive sense on a small circle of radius r and center a to $a + r - i0$, and goes back on the lower side of the real axis to $b - r - i0$,

(iv) part 4 encircles b in the positive sense to $b - r + i0$, goes back thereby crossing the real axis to $a + r - i0$, and finally encircles a on a small semicircle to $a - r$.

We denote this contour by $\gamma(a-, b-, a+, b+)$. Suppose that at the points $a - r$ and $b + r$ the expressions $(a - z)^\alpha$ and $(z - b)^\beta$ have their principal values, respectively, i.e.,

$$(a - z)^\alpha = e^{\alpha \, [\log |a - z| + i \, arg(a - z)]} =: f_1(z),$$
$$(z - b)^\beta = e^{\beta \, [\log |z - b| + i \, arg(z - b)]} =: g_1(z).$$

On the straight line segments of part 1 we have

$$(a - z)^\alpha \, (z - b)^\beta = f_1(z) g_1(z) e^{\pi i (-\alpha + \beta)},$$

of part 2,

$$(a - z)^\alpha \, (z - b)^\beta = f_1(z) g_1(z) e^{\pi i (-\alpha - \beta)},$$

of part 3,

$$(a - z)^\alpha \, (z - b)^\beta = f_1(z) g_1(z) e^{\pi i (\alpha - \beta)},$$

of part 4,

$$(a - z)^\alpha \, (z - b)^\beta = f_1(z) g_1(z) e^{\pi i (\alpha + \beta)}.$$

The total contributions of the segment $(a + r, b - r)$ to the contour integral I_c becomes

$$\frac{e^{-\pi i (\alpha - \beta)} - e^{-\pi i (\alpha + \beta)} + e^{\pi i (\alpha - \beta)} - e^{\pi i (\alpha + \beta)}}{4 \sin(\alpha \pi) \sin(\beta \pi)} \int_{a+r}^{b-r} (x - a)^\alpha \, (b - x)^\beta \, dx.$$

The expression on the left of the integral simplifies to 1. If we temporarily assume that $\Re \alpha > -1$, $\Re \beta > -1$, then the four contributions of the circular arcs to the contour integral I_c tends to zero as $r \to 0$ and the contour integral becomes

$$I_c = I(\alpha, \beta) = \int_a^b (x - a)^\alpha \, (b - x)^\beta \, dx$$
$$= (b - a)^{\alpha + \beta + 1} \, B(\alpha + 1, \beta + 1).$$

But the contour integral I_c is a holomorphic function in α and β, both different from negative integer values. Thus, by analytic continuation, we have achieved that

$$\text{FP} \int_a^b (x - a)^\alpha \, (b - x)^\beta \, dx = (b - a)^{\alpha + \beta + 1} \, B(\alpha + 1, \beta + 1)$$

for all α and β different from any negative integer values.

2.5.4   Hadamard's Finite Part of an Integral

In the above example we have associated to the integrand $(x - a)^\alpha (b - x)^\beta$ the hyperfunction $\chi_{(a,b)}(x - a)^\alpha (b - x)^\beta$, and then we have defined

$$\text{FP} \int_a^b (x - a)^\alpha (b - x)^\beta \, dx := \int_{-\infty}^\infty \chi_{(a,b)}(x - a)^\alpha (b - x)^\beta \, dx$$

$$= \int_a^b \chi_{(a,b)}(x - a)^\alpha (b - x)^\beta \, dx.$$

Let us now consider the concept of Hadamard's finite part of the following integral in some more detail. Consider

$$I(b, \alpha, k) := \int_0^b x^\alpha \log^k(x) \, dx \qquad (2.97)$$

where $\alpha \in \mathbb{R}$, $k \in \mathbb{N}_0$, $b > 0$. If $\alpha < -1$ the integral is divergent. Let us associate the hyperfunction $u(x)x^\alpha \log^k(x)$ to the integrand $x^\alpha \log x$. We then have for an integral $\alpha = n \in \mathbb{Z}$,

$$u(x)x^n = \left[-\frac{z^n}{2\pi i} \log(-z) \right],$$

$$u(x)x^n \log x = \left[-\frac{z^n}{4\pi i} \log^2(-z) \right],$$

$$u(x)x^n \log^2(x) = \left[-\frac{z^n}{6\pi i} \log^3(-z) - \frac{\pi}{6i} \log(-z) \right],$$

and, for a non-integral α,

$$u(x)x^\alpha = \left[-\frac{1}{2i \sin(\alpha\pi)} (-z)^\alpha \right],$$

$$u(x)x^\alpha \log x = \left[-\frac{(-z)^\alpha}{2i \sin(\pi\alpha)} \{\log(-z) - \pi \cot(\pi\alpha)\} \right].$$

Definition 2.18. *Hadamard's finite part of the integral* (2.97) *is defined by*

$$\text{FP} \int_0^b x^\alpha \log^k(x) \, dx := \int_{-\infty}^b u(x)x^\alpha \log^k(x) \, dx = -\oint_{C_b} F(z) \, dz \qquad (2.98)$$

where $F(z)$ is a defining function of $u(x)x^\alpha \log^k(x)$, where α here may be integral or non-integral.

The contour is shown in Figure 2.12. By the given defining functions above, it is seen that we need to compute contour integrals of the type

$$\frac{1}{2i} \oint_{C_b} (-z)^\alpha \log^k(-z) \, dz. \qquad (2.99)$$

This is a straightforward but somewhat lengthy calculation whose result is given by

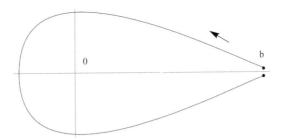

Figure 2.12: Contour C_b

Lemma 2.45. *For $k = 0$ and $\alpha \neq -1$ the contour integral (2.99) is $b^{\alpha+1} \sin(\alpha\pi)/(\alpha + 1)$. For $k = 1, 2, 3, \ldots$ and $\alpha \neq -1$, its value is given by*

$$\frac{b^{\alpha+1}}{\alpha+1} \sum_{\nu=0}^{k} \sum_{\mu=0}^{k-\nu} (-1)^\nu \frac{k!}{\mu!(k-\nu-\mu)!} \frac{\pi^\mu (\log b)^{k-\nu-\mu}}{(\alpha+1)^\nu} \tag{2.100}$$

$$\times \{\epsilon(\mu) \sin(\alpha\pi) + \epsilon(\mu-1) \cos(\alpha\pi)\},$$

where $\epsilon(\mu)$ is zero for μ odd, and equals $(-1)^{\mu/2}$ for μ even.
If $\alpha = -1$, we have,

$$\frac{1}{2i} \oint_{C_b} \frac{\log^k(-z)}{z} dz = \frac{\pi}{k+1} \sum_{\mu=0}^{\left[\frac{k}{2}\right]} (-1)^\mu \pi^{2\mu} \binom{k+1}{2\mu+1} (\log b)^{k-2\mu}. \tag{2.101}$$

Special cases are for $k = 1$, and integral $\alpha = n$:

$$\frac{1}{2\pi i} \int_{C_b} z^n \log(-z)\, dz = \begin{cases} \log b, \ n = -1 \\ \frac{b^{n+1}}{n+1}, \ n \neq -1. \end{cases} \tag{2.102}$$

For for $k = 2$, and integral $\alpha = n$:

$$\frac{1}{2\pi i} \int_{C_b} z^n \log^2(-z)\, dz = \begin{cases} \log^2 b - \pi^2/3, \ n = -1 \\ 2\frac{b^{n+1}}{(n+1)^2}(-1 + \log b + n \log b), \ n \neq -1. \end{cases} \tag{2.103}$$

For $k = 1$ and a non-integral α:

$$\frac{1}{2i} \oint_{C_b} (-z)^\alpha \log(-z)\, dz = \frac{b^{1+\alpha}}{1+\alpha} \{\pi \cos(\alpha\pi) + (\log b - \frac{1}{1+\alpha}) \sin(\alpha\pi)\}, \tag{2.104}$$

and for $k = 2$ and a non-integral α:

$$\frac{1}{2i} \oint_{C_b} (-z)^\alpha \log^2(-z)\, dz$$
$$= -\frac{b^{1+\alpha}}{(1+\alpha)^3} \{2\pi((1+\alpha) - (1+\alpha)^2) \cos(\alpha\pi) \tag{2.105}$$
$$+ (-2 + 2(1+\alpha) \log b + (\pi^2 + \log^2 b)(1+\alpha)^2) \sin(\pi\alpha)\}.$$

By using these results, we have the following Hadamard finite parts for non-integral α:

$$\mathrm{FP} \int_0^b x^\alpha \, dx = \frac{b^{\alpha+1}}{\alpha+1}, \tag{2.106}$$

$$\mathrm{FP} \int_0^b x^\alpha \log x \, dx = \frac{b^{1+\alpha}}{1+\alpha}\{\log b - \frac{1}{1+\alpha}\}. \tag{2.107}$$

For an integral and negative $\alpha = -n$, there will be some ambiguity when we use the above definition in a straightforward way. The reason is that we have to make a choice when associating a hyperfunction to an ordinary function as $Y(x)\log(x)/x^n$. In fact here is a whole family at disposition (see Section 1.6.4). With the above finite part hyperfunctions we shall get

$$\mathrm{FP} \int_0^b \frac{dx}{x} = \int_{-\epsilon}^b \mathrm{fp}\frac{u(x)}{x}\, dx = -\int_{C_b} -\frac{1}{2\pi i}\frac{\log(-z)}{z}\, dz = \log b, \tag{2.108}$$

$$\mathrm{FP} \int_0^b \frac{dx}{x^n} = -\frac{1}{n-1}\frac{1}{b^{n-1}}, \quad (n \neq 1), \tag{2.109}$$

$$\mathrm{FP} \int_0^b \frac{\log x}{x} := \int_{-\infty}^b x^{-1}\log x\, dx = \int_{-\infty}^b \mathrm{fp}\frac{u(x)\log x}{x} + c\,\delta(x)\, dx$$

$$= -\int_{C_b} -\frac{1}{4\pi i}\frac{\log^2(-z)}{z}\, dz + c = \frac{1}{2}\log^2(b) - \frac{\pi^2}{6} + c.$$

In order to get rid of the additive constant we choose $c = \frac{\pi^2}{6}$ and define

$$\mathrm{FP} \int_0^b \frac{\log x}{x} := \frac{1}{2}\log^2(b). \tag{2.110}$$

Similarly,

$$\mathrm{FP} \int_0^b \frac{\log x}{x^n} = \frac{b^{-n+1}}{-n+1}\{(1-n)\log b - 1\}, \quad (n \neq 1). \tag{2.111}$$

Notice that in all cases Hadamard's finite part yields the ordinary integral, if the integral converges, and, if it does not, we simply evaluate the primitive of the integrand at the upper limit b, while the contribution of the lower limit (which would be infinite) is ignored.

2.6   Periodic Hyperfunctions and Their Fourier Series

Definition 2.19. A hyperfunction $f(x) \in \mathcal{B}(\mathbb{R})$ satisfying

$$f(x+T) = f(x),$$

with $T > 0$, is said to be a *periodic hyperfunction with period T*.

Proposition 2.46. Let $f(x) = [F(z)]$ be a hyperfunction with a periodic defining function, i.e., $F(z+T) = F(z)$, $\forall z \in \mathcal{D}(\mathbb{R})$, then $f(x)$ is a periodic hyperfunction with period T.

Remember that $\mathcal{D}(\mathbb{R})$ denotes a complex neighborhood of the real axis. The proof follows at once from Definition 1.5 of Section 1.3.1. Conversely, we now show that a periodic hyperfunction has a periodic strong defining function.

Proposition 2.47. *If $f(x)$ is a periodic hyperfunction with period T, then it has a periodic strong defining function given by*

$$\tilde{F}(z) = \mathrm{pv} \sum_{k=-\infty}^{\infty} F_0(z - kT) := F_0(z) + \sum_{k=1}^{\infty} \{F_0(z - kT) + F_0(z + kT)\}, \quad (2.112)$$

where $F_0(z)$ is the defining function of the projection $f_0(x) := \chi_{(0,T)} f(x)$ on the interval $(0, T)$. Furthermore, this strong defining function $\tilde{F}(z)$ is holomorphic at $z = \infty$.

Proof. For the proof we must assume that the hyperfunction $f(x)$ is holomorphic at $x = 0$ and thus at $x = T$, i.e., that $0, T \notin \mathrm{sing\ supp} f(x)$. This is not a severe restriction, because we can replace the interval $[0, T]$ by any interval $[x_0, x_0 + T]$.

Remark. We shall keep in mind in all subsequent relations concerning periodic hyperfunctions with period T that we may replace the interval $[0, T]$ by any interval $[x_0, x_0 + T]$ without explicitly mentioning it every time.

We have now

$$f_0(x) := \chi_{(0,T)} f(x) = [F_0(z)] = \left[\frac{1}{2\pi i} \int_0^T \frac{f(x)}{x - z} \, dx \right].$$

Observe that $F_0(z)$ is holomorphic in $\mathbb{C} \setminus [0, T]$. By a change of variables we obtain

$$F_0(z \mp kT) = \frac{1}{2\pi i} \int_{\pm kT}^{\pm kT + T} \frac{f(x)}{x - z} \, dx.$$

Thus,

$$\chi_{(\pm kT, \pm kT + T)} f(x) = [F_0(z \mp kT)],$$

and it follows that

$$F_0(z) + \sum_{k=1}^{n-1} \{F_0(z - kT) + F_0(z + kT)\} = \frac{1}{2\pi i} \int_{-nT}^{nT} \frac{f(x)}{x - z} \, dx.$$

The term on the right-hand side is the defining function of $\chi_{(-nT, nT)} f(x)$. By letting $n \to \infty$ we obtain

$$\tilde{F}(z) = \frac{1}{2\pi i} \, \mathrm{pv} \int_{-\infty}^{\infty} \frac{f(x)}{x - z} \, dx$$

$$= F_0(z) + \sum_{k=1}^{\infty} \{F_0(z - kT) + F_0(z + kT)\}$$

which is the strong defining function of $f(x)$, provided the series on the right-hand side is convergent and represents a function whose upper and lower components are holomorphic. Let us prove this. By Proposition 2.26 we have

$$F_0(z) = \frac{c_1}{z} + O\left(\frac{1}{z^2}\right) \quad \text{as } |z| \to \infty.$$

This implies

$$F_0(z - kT) + F_0(z + kT) = \frac{2c_1 z}{z^2 - k^2 T^2} + O(\frac{1}{k^2})$$

$$= -\frac{2c_1}{T^2 k^2} \frac{z}{1 - (z/(kT))^2} + O(\frac{1}{z^2})$$

as $|k| \to \infty$. This shows that the series converges uniformly in every compact subset of the upper and lower half-plane, the point at infinity included. Thus, the series defines a defining function whose upper and lower components are holomorphic in their respective half-planes. We have still to show that $\tilde{F}(z)$ is periodic. Let

$$\tilde{F}_n(z) := \sum_{k=-n}^{n} F_0(z - kT),$$

then

$$\tilde{F}_n(z + T) := \sum_{k=-n}^{n} F_0(z - (k-1)T) = \sum_{k=-n-1}^{n-1} F_0(z - kT)$$

$$= F_0(z) + F_0(z - nT) + F_0(z - (n+1)T)$$

$$+ \sum_{k=1}^{n-1} \{F_0(z - kT) + F_0(z + kT)\}.$$

By letting $n \to \infty$ and and taking into account that

$$F_0(z - nT) \to 0, \quad F_0(z - (n+1)T) \to 0$$

we arrive at

$$\tilde{F}(z + T) = \lim_{n \to \infty} \tilde{F}_n(z + T) = \tilde{F}(z)$$

which proves the periodicity of $\tilde{F}(z)$.   □

Example 2.36. Consider the Dirac impulse comb

$$f(x) = \mathrm{pv} \sum_{k=-\infty}^{\infty} \delta(x - kT) \tag{2.113}$$

which is a periodic hyperfunction with period T. Its projection on the interval $[-T/2, T/2]$ is

$$f_0(x) = \chi_{(-T/2, T/2)} f(x) = \delta(x),$$

hence $F_0(z) = -1/(2\pi i z)$. Therefore, the strong defining function of the Dirac impulse comb becomes

$$\tilde{F}(z) = -\frac{1}{2\pi i} \{\frac{1}{z} + \sum_{k=1}^{\infty} \{\frac{1}{z - kT} + \frac{1}{z + kT}\}\}$$

$$= -\frac{1}{2\pi i} \{\frac{1}{z} + 2z \sum_{k=1}^{\infty} \frac{1}{z^2 - k^2 T^2}\}$$

$$= -\frac{1}{2\pi i} \frac{\pi}{T} \cot(\frac{\pi z}{T}) = \frac{i}{2T} \cot(\frac{\pi z}{T}).$$

Here we have used the partial fraction decomposition of the cotangent function

$$\frac{\pi}{T}\cot(\frac{\pi z}{T}) = \frac{1}{z} + 2z\sum_{k=1}^{\infty}\frac{1}{z^2 - k^2\,T^2}. \tag{2.114}$$

This shows that the defining function of the Dirac impulse comb established in Example 2.6 coincide with the strong defining function.

We now use a theorem about Fourier series of holomorphic functions from complex variable theory.

Proposition 2.48. *Let $F(z)$ be a holomorphic and periodic function on the horizontal straight line $l: z = z_0 + t\,T$, $-\infty < t < \infty$, with period $T > 0$, i.e., for all z on l,*

$$F(z + T) = F(z).$$

Then,

(i) *$F(z)$ is periodic with period T everywhere $F(z)$ is holomorphic.*

(ii) *$F(z)$ can be expanded in a Fourier series*

$$F(z) = \sum_{k=-\infty}^{\infty} c_k\,e^{ik\omega z}, \quad (\omega = \frac{2\pi}{T})$$

valid for all z in the largest horizontal strip

$$-\infty \le \alpha_0 < \Im z < \beta_0 \le +\infty$$

containing the line l, and contained in the domain where $F(z)$ is holomorphic.

(iii) *The Fourier series converges uniformly in any strip $\alpha_0 < \alpha \le \Im z \le \beta < \beta_0$.*

(iv) *The coefficients c_k are given by*

$$c_k = \frac{1}{T}\int_{z_0}^{z_0+T} F(z)e^{-ik\omega z}dz.$$

A proof of this proposition can be found in [4, Behnke-Sommer p.264]. Presently, we apply this proposition to the upper and lower component of the strong defining function $\check{F}(z)$ of the periodic hyperfunction $f(x)$. For the upper and lower component we have $\alpha_0 = 0$ and $\beta_0 = +\infty$ and $\alpha_0 = -\infty$ and $\beta_0 = 0$, respectively, because the standard defining function is holomorphic in the upper and in the lower complex plane (including the point at infinity). In this setting we have

Proposition 2.49. *For the upper and the lower component of the strong defining function $\check{F}(z)$,*

$$\check{F}_+(z) = \sum_{k=0}^{\infty} c_k^+\,e^{ik\omega z}, \quad i.e., \; c_k = 0 \; for \; k < 0, \tag{2.115}$$

$$\check{F}_-(z) = \sum_{k=-\infty}^{0} c_k^-\,e^{ik\omega z}, \quad i.e., \; c_k = 0 \; for \; k > 0. \tag{2.116}$$

Furthermore, the formulas for the coefficients can be replaced by

$$c_k^+ = \frac{1}{T} \int_{\gamma_{x_0,x_0+T}^+} \tilde{F}_+(z)e^{-ik\omega z}dz, \tag{2.117}$$

$$c_k^- = \frac{1}{T} \int_{\gamma_{x_0,x_0+T}^-} \tilde{F}_-(z)e^{-ik\omega z}dz \tag{2.118}$$

with $x_0 = \Re z_0$.

Proof. We prove the statement for the upper component, the proof for the lower component is similar. Let $\mathcal{D}$ be the domain bounded by the contour comprising γ_{x_0,x_0+T}^+ below, the vertical segment from x_0+T to z_0+T at right, the horizontal segment from z_0+T to x_0+T above, and the vertical segment from x_0+T to x_0 at left. In $\mathcal{D}$ the integrand $\tilde{F}_+(z)e^{-ik\omega z}$ is holomorphic and periodic with period T. The periodicity implies that the contributions to the contour integral along the boundary of $\mathcal{D}$ on the two vertical segments cancel. Therefore, by Cauchy's integral theorem, we have

$$c_k^+ = \frac{1}{T} \int_{z_0}^{z_0+T} \tilde{F}_+(z)e^{-ik\omega z}dz = \frac{1}{T} \int_{\gamma_{x_0,x_0+T}^+} \tilde{F}_+(z)e^{-ik\omega z}dz.$$

The first integral on the right-hand side can be written as

$$\int_{z_0}^{z_0+T} \tilde{F}_+(z)e^{-ik\omega z}dz = e^{k\omega y_0}e^{ik\omega x_0}\int_0^T \tilde{F}_+(z_0+t)e^{-ik\omega t}dt.$$

If $k < 0$, and letting $y_0 \to +\infty$, we see that the first factor on the right-hand side tends to zero. This proves that $c_k^+ = 0$ for $k < 0$. Similarly, by considering the symmetric domain with respect to the real axis of the above domain $\mathcal{D}$, we shall have, for $k > 0$ and letting $y_0 \to -\infty$, that $c_k^- = 0$ for $k > 0$. $\square$

The strong defining function can now be written as

$$\tilde{F}(z) = c_0^+\mathbf{1}_+(z) - c_0^-\mathbf{1}_-(z) + \sum_{k=1}^{\infty} c_k^+ e^{ik\omega z}\mathbf{1}_+(z) - \sum_{k=-\infty}^{-1} c_k^- e^{ik\omega z}\mathbf{1}_-(z).$$

If we put $c_k := c_k^+ - c_k^-$ and recall the definition of an integral of a hyperfunction, we obtain the following result.

Proposition 2.50. *The strong defining function for the periodic hyperfunction $f(x) = f(x+T)$ is also given by*

$$\tilde{F}(z) = c_0\mathbf{1}(z) + \sum_{k=1}^{\infty} c_k\, e^{ik\omega z}\mathbf{1}_+(z) + \sum_{k=1}^{\infty} c_{-k}\, e^{-ik\omega z}\mathbf{1}_-(z). \tag{2.119}$$

The coefficients are computed by

$$c_k := \frac{1}{T} \int_{x_0}^{x_0+T} f(x)e^{-ik\omega x}dx, \tag{2.120}$$

provided $f(x)$ is holomorphic at x_0.

By taking into account the results of Examples 2.23 and 2.24, we now introduce the Fourier series of a periodic hyperfunction.

Definition 2.20. The *Fourier series of the periodic hyperfunction* $f(x) = [\tilde{F}(z)]$ with period $T > 0$ is

$$f(x) = c_0 + \sum_{k=1}^{\infty} c_k\, e^{i\,k\,\omega\,x} + \sum_{k=1}^{\infty} c_{-k}\, e^{-i\,k\,\omega\,x} \tag{2.121}$$

$$= \text{pv} \sum_{k=-\infty}^{\infty} c_k\, e^{i\,k\,\omega\,x}. \tag{2.122}$$

Example 2.37. In Example 2.36 the strong defining function of the Dirac impulse comb was computed. Let us now determine its Fourier series. We have

$$\tilde{F}(z) = -\frac{1}{2Ti} \cot\left(\frac{\pi z}{T}\right) = -\frac{1}{2T}\, \frac{e^{i\pi z/T} + e^{-i\pi z/T}}{e^{i\pi z/T} - e^{-i\pi z/T}}$$

$$= \frac{1}{2T}\, \frac{1 + e^{i2\pi z/T}}{1 - e^{i2\pi z/T}}\, \mathbf{1}_+(z) + \frac{1}{2T}\, \frac{1 + e^{-i2\pi z/T}}{1 - e^{-i2\pi z/T}}\, \mathbf{1}_-(z)$$

$$= \frac{1}{2T}(1 + e^{i2\pi z/T}) \sum_{k=0}^{\infty} e^{2\pi i k z/T}\, \mathbf{1}_+(z)$$

$$\qquad + \frac{1}{2T}(1 + e^{-i2\pi z/T}) \sum_{k=0}^{\infty} e^{-2\pi i k z/T}\, \mathbf{1}_-(z)$$

$$= \frac{1}{2T}(1 + 2\sum_{k=1}^{\infty} e^{2\pi i k z/T})\, \mathbf{1}_+(z) + \frac{1}{2T}(1 + 2\sum_{k=1}^{\infty} e^{-2\pi i k z/T})\, \mathbf{1}_-(z)$$

$$= \frac{1}{T}\{\mathbf{1}(z) + \sum_{k=1}^{\infty} e^{2\pi i k z/T}\, \mathbf{1}_+(z) + \sum_{k=1}^{\infty} e^{-2\pi i k z/T}\, \mathbf{1}_-(z)\}$$

$$= \frac{1}{T}\mathbf{1}(z) + \sum_{k=1}^{\infty} \frac{1}{T} e^{2\pi i k z/T}\, \mathbf{1}_+(z) + \sum_{k=1}^{\infty} \frac{1}{T} e^{-2\pi i k z/T}\, \mathbf{1}_-(z).$$

Therefore, $c_k = 1/T$, $c_{-k} = 1/T$ and the Fourier series of the Dirac impulse comb hyperfunction becomes

$$\frac{1}{T} \sum_{k=-\infty}^{\infty} e^{2\pi i k x/T}.$$

We also obtained the remarkable formula

$$\text{pv} \sum_{k=-\infty}^{\infty} \delta(x - kT) = \frac{1}{T}\, \text{pv} \sum_{k=-\infty}^{\infty} e^{2\pi i k x/T}. \tag{2.123}$$

Of course, by using (2.120) we also immediately obtain

$$c_k = \frac{1}{T} \int_{-T/2}^{T/2} \delta(x) e^{2\pi i k x/T}\, dx = \frac{1}{T}.$$

Remark. The significant point in Proposition 2.50 is that the series

$$\sum_{k=1}^{\infty} c_k \, e^{ik\omega z} \, \mathbf{1}_+(z) + \sum_{k=1}^{\infty} c_{-k} \, e^{-ik\omega z} \, \mathbf{1}_-(z)$$

converges as long as $|c_{|k|}| = O(|k|^m)$ for any positive integer m. Thereby, the standard defining function $\tilde{F}(z)$ can be differentiated term by term as many times as you want. The reason for this is that, for $\Im z = y > 0$,

$$\left| \sum_{k=1}^{\infty} c_k \, e^{ik\omega z} \right| \le \sum_{k=1}^{\infty} |c_k| \, |e^{ik\omega (x+iy)}|$$

$$\le \sum_{k=1}^{\infty} A \, k^m \, e^{-k\omega y} < \infty,$$

for any $m \in \mathbb{N}$. A similar estimate can be done for the second term where $\Im z = y < 0$. We have obtained

Proposition 2.51. *The Fourier series of a periodic hyperfunction converges (in the sense of hyperfunctions), if the Fourier coefficients satisfy*

$$|c_k| = O(k^m), \quad k \to \pm\infty$$

for some positive integer m. Furthermore, the Fourier series can be differentiated term by term, i.e., if

$$f(x) = \text{pv} \sum_{k=-\infty}^{\infty} c_k \, e^{ik\omega x},$$

then

$$f^{(n)}(x) = \text{pv} \sum_{k=-\infty}^{\infty} c_k (ik\omega)^n \, e^{ik\omega x}.$$

Problem 2.10. Show that if $f(x)$ is an even periodic hyperfunction of period T, its Fourier series can be written as a Fourier cosine series

$$f(x) = \frac{A_0}{2} + \sum_{k=1}^{\infty} A_k \, \cos(k\omega x).$$

Similarly, if $f(x)$ is an odd periodic hyperfunction of period T, its Fourier series can be written as a Fourier sine series

$$f(x) = \sum_{k=1}^{\infty} B_k \, \sin(k\omega x).$$

The real coefficients are given by

$$A_k = c_k + c_{-k} = \frac{2}{T} \int_{x_0}^{x_0+T} f(x) \cos(k\omega x) dx,$$

$$B_k = i\,(c_k - c_{-k}) = \frac{2}{T} \int_{x_0}^{x_0+T} f(x) \sin(k\omega x) dx,$$

$$(k = 1, 2, 3, \dots), \quad \omega = 2\pi/T$$

with $B_0 = 0$ and $f(x)$ is assumed to be holomorphic at x_0.

Since the Dirac impulse comb is an even periodic hyperfunction, we have

$$\sum_{k=-\infty}^{\infty} \delta(x - kT) = \frac{1}{T} + \frac{2}{T} \sum_{k=1}^{\infty} \cos(2\pi kx/T). \tag{2.124}$$

Setting $T = 2\pi$ and replacing x by $x - \pi$, we obtain

$$\sum_{k=-\infty}^{\infty} \delta(x - (2k+1)\pi) = \frac{1}{2\pi} + \frac{1}{\pi} \sum_{k=1}^{\infty} (-1)^k \cos(kx). \tag{2.125}$$

Let us illustrate Proposition 2.51 by an example.

Example 2.38. The Fourier series of the periodic function $f(x)$ of period 2π, obtained from $f_0(x) = x$, $0 \le x \le 2\pi$ by periodic repetition to the left and to the right, is

$$\pi - 2 \sum_{k=1}^{\infty} \frac{\sin(kx)}{k}.$$

As a hyperfunction $f(x)$ is written in the form

$$f(x) = \sum_{k=-\infty}^{\infty} (x - 2k\pi)\{u(x - 2k\pi) - u(x - 2(k+1)\pi)\}.$$

Term by term differentiation of the Fourier series yields

$$-2 \sum_{k=1}^{\infty} \cos(kx),$$

and the derivative in the sense of hyperfunctions of $f(x)$ yields

$$f'(x) = \sum_{k=-\infty}^{\infty} \{u(x - 2k\pi) - u(x - 2(k+1)\pi)\}$$

$$+ \sum_{k=-\infty}^{\infty} (x - 2k\pi)\{\delta(x - 2k\pi) - \delta(x - 2(k+1)\pi)\}$$

$$= 1 - 2\pi \sum_{k=-\infty}^{\infty} \delta(x - 2(k+1)\pi) = 1 - 2\pi \sum_{k=-\infty}^{\infty} \delta(x - 2k\pi).$$

Finally,

$$\sum_{k=-\infty}^{\infty} \delta(x - 2k\pi) = \frac{1}{2\pi} + \frac{1}{\pi} \sum_{k=1}^{\infty} \cos(kx)$$

as we already know. Going on with differentiating, we obtain

$$\sum_{k=-\infty}^{\infty} \delta'(x - 2k\pi) = -\frac{1}{\pi} \sum_{k=1}^{\infty} k \sin(kx), \tag{2.126}$$

$$\sum_{k=-\infty}^{\infty} \delta''(x - 2k\pi) = -\frac{1}{\pi} \sum_{k=1}^{\infty} k^2 \cos(kx). \tag{2.127}$$

Problem 2.11. From the known cosine series for $0 < x < \pi$,

$$\log\left|\sin\frac{x}{2}\right| = -\log 2 - \sum_{k=1}^{\infty}\frac{\cos kx}{k},$$

establish

$$\sum_{k=1}^{\infty}\sin kx = \frac{1}{2}\cot\frac{x}{2}, \quad \sum_{k=1}^{\infty}k\cos kx = -\frac{1}{4}\frac{1}{\sin^2\frac{x}{2}}. \tag{2.128}$$

Example 2.39. Consider the boundary value problem

$$\frac{d^2y}{dx^2} - a^2y(x) = -\pi f(x), \quad y'(-\pi) = y'(\pi) = 0$$

where a is a complex constant. It is known that its solution is given by

$$y(x) = \int_{-\pi}^{\pi} G(x-t)\,f(t)\,dt$$

where $G(x)$ is Green's function satisfying the associated boundary value problem

$$\frac{d^2y}{dx^2} - a^2y(x) = -\pi\delta(x), \quad y'(-\pi) = y'(\pi) = 0.$$

Equation (2.124) with $T = 2\pi$, and x restricted to the interval $(-\pi, \pi)$ becomes

$$\pi\,\delta(x) = \frac{1}{2} + \sum_{k=1}^{\infty}\cos(kx). \tag{2.129}$$

By twice differentiating the function

$$y(x) = \frac{1}{2a^2} + \sum_{k=1}^{\infty}\frac{\cos(kx)}{a^2 + k^2}$$

it can be checked that $y(x)$ solves the above associated boundary value problem, i.e., $y(x)$ is Green's function. On the other hand $y(x)$ satisfies for $x \neq 0$ the homogeneous differential equation $y''(x) - a^2y(x) = 0$ whose general solution is $A\cosh(ax) + B\sinh(ax)$. Integrating the differential equation from $-\epsilon$ to ϵ, shows that Green's function must satisfy the jump condition $y'(0+) - y'(0-) = -\pi$ and the continuity condition $y(0-) = y(0+)$ at $x = 0$. If we write

$$y(x) = A_-\cosh(ax) + B_-\sinh(ax), \quad (-\pi < x < 0),$$
$$y(x) = A_+\cosh(ax) + B_+\sinh(ax), \quad (0 < x < \pi),$$

the jump condition and the continuity condition yields the system of equations

$$-aA_-\sinh(a\pi) + aB_-\cosh(a\pi) = 0,$$
$$aA_+\sinh(a\pi) + aB_+\cosh(a\pi) = 0,$$
$$aB_+ - aB_- = -\pi,$$
$$A_+ = A_-,$$

whose solution is $A_+ = A_- = \pi \cosh(a\pi)/(2a\sinh(a\pi))$, and $B_+ = -B_- = -\pi/(2a)$. This gives another closed expression for Green's function,

$$y(x) = G(x) = \begin{cases} \frac{\pi \cosh a(x+\pi)}{2a \sinh a\pi}, & -\pi < x < 0 \\ \frac{\pi \cosh a(x-\pi)}{2a \sinh a\pi}, & 0 < x < \pi, \end{cases} \tag{2.130}$$

and the result

$$\frac{1}{2a^2} + \sum_{k=1}^{\infty} \frac{\cos(kx)}{a^2 + k^2} = \frac{\pi \cosh a(x - \mathrm{sgn}(x)\pi)}{2a \sinh a\pi} \tag{2.131}$$

for $-\pi < x < \pi$. Replacing a by ia yields

$$\frac{1}{2a^2} + \sum_{k=1}^{\infty} \frac{\cos(kx)}{a^2 - k^2} = \frac{\pi \cos a(x - \mathrm{sgn}(x)\pi)}{2a \sin a\pi} \tag{2.132}$$

which is Green's function of the boundary value problem

$$\frac{d^2 y}{dx^2} + a^2 y(x) = -\pi f(x), \quad y'(-\pi) = y'(\pi) = 0.$$

In (2.130) we may put $x = 0$ and obtain the the partial fraction expansion

$$\frac{1}{2a^2} + \sum_{k=1}^{\infty} \frac{1}{a^2 + k^2} = \frac{\pi}{2a} \coth a\pi, \tag{2.133}$$

and similarly,

$$\frac{1}{2a^2} + \sum_{k=1}^{\infty} \frac{1}{a^2 - k^2} = \frac{\pi}{2a} \cot a\pi. \tag{2.134}$$

2.7 Convolutions of Hyperfunctions

2.7.1 Definition and Existence of the Convolution

Let $f(x) = [F_+(z), F_-(z)]$ and $g(x) = [G_+(z), G_-(z)]$ be two hyperfunctions defined on the entire real axis. If the complex variable z is confined to the upper half-plane $\Im z > 0$, the function $\phi_+(x) := F_+(z-x)$ becomes a real analytic function in x, depending on the parameter z. Likewise, if $\Im z < 0$, $\phi_-(x) := F_-(z-x)$ is a real analytic function of x with parameter z. Therefore, $\phi_\pm(x)g(x) = F_\pm(z-x)g(x)$ are hyperfunctions, and

$$H_\pm(z; a, b) := \int_a^b F_\pm(z-x)g(x)\, dx$$

are well-defined integrals of hyperfunctions depending holomorphically on the parameter z, for any finite a and b. Explicitly written, these integrals of hyperfunctions are

$$H_+(z; a, b) := \int_{\gamma_{a,b}^+} F_+(z-\zeta)G_+(\zeta)\, d\zeta - \int_{\gamma_{a,b}^-} F_+(z-\zeta)G_-(\zeta)\, d\zeta,$$

$$H_-(z; a, b) := \int_{\gamma_{a,b}^+} F_-(z-\zeta)G_+(\zeta)\, d\zeta - \int_{\gamma_{a,b}^-} F_-(z-\zeta)G_-(\zeta)\, d\zeta.$$

The variable z is outside the closed contour $\gamma_{a,b}^- - \gamma_{a,b}^+$ (Figure 2.13). Let now $a \to -\infty$ and $b \to \infty$ and assume that the improper integrals $H_\pm(z; a, b)$ converge uniformly inside $\mathbb{C}_+$ and $\mathbb{C}_-$ with respect to z, respectively. Then, they define two holomorphic functions of z in the upper and lower half-plane, respectively (see Appendix A.3). For $\Im z > 0$, the limit function becomes explicitly

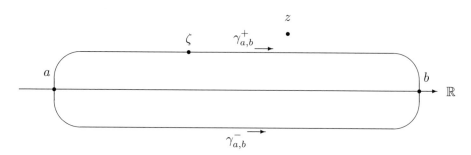

Figure 2.13: Contours in the ζ-plane

$$H_+(z) = \int_{\gamma_{-\infty,\infty}^+} F_+(z - \zeta)G_+(\zeta)\,d\zeta - \int_{\gamma_{-\infty,\infty}^-} F_+(z - \zeta)G_-(\zeta)\,d\zeta,$$

and for $\Im z < 0$,

$$H_-(z) = \int_{\gamma_{-\infty,\infty}^+} F_-(z - \zeta)G_+(\zeta)\,d\zeta - \int_{\gamma_{-\infty,\infty}^-} F_-(z - \zeta)G_-(\zeta)\,d\zeta.$$

Notice that we may write more concisely

$$H_+(z) = \int_{-\infty}^{\infty} F_+(z - x)g(x)\,dx, \quad \Im z > 0 \tag{2.135}$$

$$H_-(z) = \int_{-\infty}^{\infty} F_-(z - x)g(x)\,dx, \quad \Im z < 0, \tag{2.136}$$

according to the definition of an integral over a hyperfunction.

Definition 2.21. Let $f(x) = [F_+(\zeta), F_-(\zeta)]$ and $g(x) = [G_+(\zeta), G_-(\zeta)]$ be hyperfunctions defined on the entire real line. If the two improper integrals (2.135), (2.136) converge uniformly with respect to z inside $\mathbb{C}_+$ and $\mathbb{C}_-$, respectively, the thereby defined hyperfunction

$$f(x) \star g(x) := h(x) := [H_+(z), H_-(z)] \tag{2.137}$$

is said to be the *convolution of $f(x)$ and $g(x)$.*

Note that the definition implies that if $f(x)$ is an upper or lower hyperfunction, then the convolution $f(x) \star g(x)$ is also an upper or lower hyperfunction, respectively.

In general, convolution of hyperfunctions with carriers encompassing the entire real axis need not exist. However, as we shall see later on, the existence of the convolution of two right-sided or two left-sided hyperfunctions will always exist.

Furthermore, if the second factor $g(x)$ is an ordinary function, the two integrals (2.135), (2.136) can be read as ordinary integrals over ordinary functions (and still defining a hyperfunction).

Problem 2.12. Show that $1 \star \mathrm{fp}(1/x)$ and $\mathrm{fp}(1/x) \star 1$ do not exist.

Example 2.40. Let us compute the convolution $\mathrm{fp}(1/x^m) \star \mathrm{fp}(1/x^n)$, $m, n \in \mathbb{N}$. Because $\mathrm{fp}(1/x^m) = [1/(2z^m), -1/(2z^m)]]$, we have to compute, for $\Im z > 0$,

$$H_+(z) = \int_{\gamma_{-\infty,\infty}^+} F_+(z-\zeta)G_+(\zeta)\,d\zeta - \int_{\gamma_{-\infty,\infty}^-} F_+(z-\zeta)G_-(\zeta)\,d\zeta$$

$$= \frac{1}{4}\int_{\gamma_{-\infty,\infty}^+} \frac{d\zeta}{(z-\zeta)^m\,\zeta^n} + \frac{1}{4}\int_{\gamma_{-\infty,\infty}^-} \frac{d\zeta}{(z-\zeta)^m\,\zeta^n}.$$

The first integral can be computed by closing the contour by a semicircle over $[-R, R]$ in the upper half-plane, and by using the theorem of residue and finally letting $R \to \infty$. Similarly, the second integral is computed by closing the contour over $[-R, R]$ by a semicircle in the lower half-plane. Because $\Im z > 0$, it yields no contribution. Therefore, we obtain

$$H_+(z) = \frac{\pi i}{2}\, \mathrm{Res}_{\zeta=z}\left\{\frac{1}{(z-\zeta)^m\,\zeta^n}\right\}.$$

Similarly, for $\Im z < 0$ we have,

$$H_-(z) = \int_{\gamma_{-\infty,\infty}^+} F_-(z-\zeta)G_+(\zeta)\,d\zeta - \int_{\gamma_{-\infty,\infty}^-} F_-(z-\zeta)G_-(\zeta)\,d\zeta$$

$$= -\frac{1}{4}\int_{\gamma_{-\infty,\infty}^+} \frac{d\zeta}{(z-\zeta)^m\,\zeta^n} - \frac{1}{4}\int_{\gamma_{-\infty,\infty}^-} \frac{d\zeta}{(z-\zeta)^m\,\zeta^n}.$$

Since $\Im z < 0$, the first integral, now gives zero contribution, while the second integral gives the same contribution as the first integral in the case $\Im z > 0$. Thus,

$$H(z) = H_+(z) = H_-(z) = \frac{\pi i}{2}\, \mathrm{Res}_{\zeta=z}\left\{\frac{1}{(z-\zeta)^m\,\zeta^n}\right\}$$

$$= -\frac{\pi i}{2}\,\frac{(m+n-2)!}{(m-1)!\,(n-1)!}\,\frac{1}{z^{m+n-1}}$$

$$= -\pi^2\,\frac{(-1)^{m+n-2}}{(m-1)!\,(n-1)!}\left(-\frac{1}{2\pi i}\right)\frac{(-1)^{m+n-2}\,(m+n-2)!}{z^{m+n-1}}.$$

Finally, for $m, n \in \mathbb{N}$,

$$\mathrm{fp}\frac{1}{x^m} \star \mathrm{fp}\frac{1}{x^n} = -\pi^2\,\frac{(-1)^{m+n}}{(m-1)!\,(n-1)!}\,\delta^{(m+n-2)}(x), \tag{2.138}$$

especially,

$$\mathrm{fp}\frac{1}{x} \star \mathrm{fp}\frac{1}{x} = -\pi^2\,\delta(x). \tag{2.139}$$

The above definition of the convolution is not symmetric with respect to $f(x)$ and $g(x)$, hence it is not clear whether or not the convolution is commutative. However, we can prove the following

Proposition 2.52. *Let* $f(x) = [F_+(z), F_-(z)]$ *and* $g(x) = [G_+(z), G_-(z)]$ *be given hyperfunctions. If both convolutions* $f(x) \star g(x)$ *and* $g(x) \star f(x)$ *exist, then the convolution product is commutative, i.e.,*

$$f(x) \star g(x) = g(x) \star f(x) \tag{2.140}$$

holds.

Proof. Let $a < b$ and assume $\Im z > 0$. Then

$$\int_a^b F_+(z - t)g(t)\, dt = - \oint_{(a,b)} F_+(z - \zeta)G(\zeta)\, d\zeta$$

with the closed contour $(a, b) := \gamma_{a,b}^- - \gamma_{a,b}^+$, the point z lying outside (a, b) (see Figure 2.13). A change of variables $\theta := z - \zeta$ yields (Figure 2.14)

$$- \oint_{(a,b)} F_+(z - \zeta)G(\zeta)\, d\zeta = \oint_{\Gamma_z} G(z - \theta)F_+(\theta)\, d\theta$$

$$= \int_{\Gamma_{x-b,x-a}^+} G_+(z - \theta)F_+(\theta)\, d\theta - \int_{\Gamma_{x-b,x-a}^-} G(z - \theta)F_+(\theta)\, d\theta.$$

Now consider

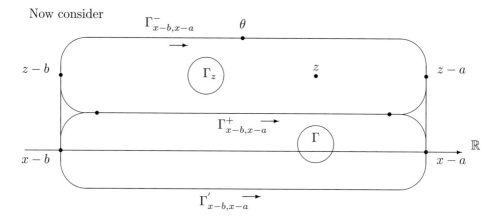

Figure 2.14: Contours in the θ-plane

$$\oint_\Gamma G(z - \theta)F(\theta)\, d\theta = \oint_\Gamma G_+(z - \theta)F(\theta)\, d\theta$$

$$= \int_{\Gamma_{x-b,x-a}'} G_+(z - \theta)F_-(\theta)\, d\theta - \int_{\Gamma_{x-b,z-a}^+} G_+(x - \theta)F_+(\theta)\, d\theta.$$

Eliminating

$$\int_{\Gamma_{x-b,x-a}^+} G_+(z - \theta)F_+(\theta)\, d\theta$$

from the last two formulas gives

$$-\oint_{(a,b)} F_+(z-\zeta)G(\zeta)\,d\zeta = \int_{\Gamma'_{x-b,x-a}} G_+(z-\theta)F_-(\theta)\,d\theta$$

$$-\oint_{\Gamma} G_+(z-\theta)F(\theta)\,d\theta - \int_{\Gamma^-_{x-b,x-a}} G(z-\theta)F_+(\theta)\,d\theta$$

$$= -\oint_{\Gamma} G_+(z-\theta)F(\theta)\,d\theta + \phi(z;a,b)$$

with

$$\phi(z;a,b) = \int_{\Gamma'_{x-b,x-a}} G_+(z-\theta)F_-(\theta)\,d\theta - \int_{\Gamma^-_{x-b,x-a}} G(z-\theta)F_+(\theta)\,d\theta.$$

The integral defining $\phi(z;a,b)$ has the closed contour $\Gamma'_{x-b,x-a} - \Gamma^-_{x-b,x-a}$ containing z in its interior and defines a real analytic function $\phi(t;a,b)$ for $x-b < t < x-a$. Thus, we have established the formula

$$\int_a^b F_+(z-t)g(t)\,dt = \int_{x-b}^{x-a} G_+(z-t)f(t)\,dt + \phi(z;a,b), \qquad (2.141)$$

where $x = \Re z$ and $\phi(t;a,b)$ is real analytic for $x-b < t < x-a$. Now let $a \to -\infty, b \to \infty$. Under the assumption that $g(x) \star f(x)$ exists, we may write

$$-\oint_{(-\infty,\infty)} F_+(z-\zeta)G(\zeta)\,d\zeta = -\oint_{(-\infty,\infty)} G_+(z-\theta)F(\theta)\,d\theta + \phi(z)$$

with

$$\phi(z) = \int_{\Gamma'_{-\infty,\infty}} G_+(z-\theta)F_-(\theta)\,d\theta - \int_{\Gamma^-_{-\infty,\infty}} G(z-\theta)F_+(\theta)\,d\theta,$$

where $\phi(x)$ is a real analytic function. Likewise, details are left to the reader, we have for $\Im z < 0$,

$$\int_a^b F_-(z-t)g(t)\,dt = \int_{x-b}^{x-a} G_-(z-t)f(t)\,dt + \phi_1(z;a,b), \qquad (2.142)$$

where $x = \Re z$ and $\phi'(t;a,b)$ is real analytic for $x-b < t < x-a$. However, in the limit case $a = -\infty$, $b = \infty$ it can be seen that $\phi(x)$ and $\phi_1(x)$ can be identified, thus finally proving $f(x) \star g(x) = g(x) \star f(x)$ $\square$

2.7.2 Sufficient Conditions for the Existence of Convolutions

Proposition 2.53. *The convolution of a hyperfunction $f(x)$ with compact support $[\alpha, \beta]$ and an arbitrary hyperfunction $g(x)$ always exists, and $g(x) \star f(x) = f(x) \star g(x)$.*

Proof. The hypothesis implies that the defining function $F(z)$ of $f(x)$ is real analytic for $\Re z < \alpha$ and for $\Re z > \beta$. This implies, for a fixed ζ, that $F(z - \zeta)$ is a holomorphic function of z, as long as $\Re z < \alpha + \Re \zeta$ and $\Re z > \beta + \Re \zeta$. Likewise, for a fixed z, $F(z - \zeta)$ is a holomorphic function of ζ, as long as $\Re \zeta < \Re z - \beta$ and

$\Re\zeta > \Re z - \alpha$. Choose a point $c_1 < \Re z - \beta$ and a point $c_2 > \Re z - \alpha$. For $a < c_1$ and $b > c_2$ break the integral of the defining function of $f(x) \star g(x)$ into three parts (notice that we have $F_+(z) = F_-(z) = F(z)$ due to the compact support of $f(x)$):

$$\int_a^b F(z-x)g(x)\,dx = \int_a^{c_1} F(z-x)g(x)\,dx$$

$$+ \int_{c_1}^{c_2} F(z-x)g(x)\,dx + \int_{c_2}^b F(z-x)g(x)\,dx.$$

The first and the last integral (as integrals over hyperfunctions) defines two real analytic functions of $\Re z$ and can be discarded. The remaining middle integral then is an equivalent defining function, and we can write, for any $a < c_1$ and $b > c_2$,

$$h(x; a, b) := \left[\int_{c_1}^{c_2} F(z-x)g(x)\,dx \right].$$

By passing to the limit $a \to -\infty, b \to \infty$, we obtain

$$f(x) \star g(x) = \left[\int_{c_1}^{c_2} F(z-x)g(x)\,dx \right] \tag{2.143}$$

where, for a given z, we must choose c_1, c_2 such that $c_1 < \Re z - \beta$ and $c_2 > \Re z - \alpha$.
On the other hand, because

$$\int_{-\infty}^{\infty} G_\pm(z-x)f(x)\,dx = \int_{\alpha'}^{\beta'} G_\pm(z-x)f(x)\,dx,$$

exists, for any $\alpha' < \alpha$ and $\beta' > \beta$, we also have

$$g(x) \star f(x) = \left[\int_{\alpha'}^{\beta'} G_+(z-x)\,f(x)\,dx, \int_{\alpha'}^{\beta'} G_-(z-x)\,f(x)\,dx \right]. \tag{2.144}$$

Since both convolutions exists we must have $g(x) \star f(x) = f(x) \star g(x)$. □

Example 2.41. Because the Dirac impulse hyperfunction $\delta(x)$ has the compact support $[\alpha, \beta]$ for any $\alpha < 0$ and any $\beta > 0$, we may write with any $c_1 < \Re z - \beta$ and $c_2 > \Re z - \alpha$,

$$\delta(x) \star g(x) = \left[-\frac{1}{2\pi i} \int_{c_1}^{c_2} \frac{g(x)}{z-x}\,dx \right] = \left[\frac{1}{2\pi i} \int_{c_1}^{c_2} \frac{g(x)}{x-z}\,dx \right]$$

$$= \chi_{(c_1,c_2)}g(x) = \chi_{(-\infty,\infty)}g(x) = g(x).$$

Therefore,

$$\delta(x) \star g(x) = g(x). \tag{2.145}$$

By Proposition 2.52 we also must have $g(x) \star \delta(x) = \delta(x) \star g(x) = g(x)$, thus the Dirac impulse acts as a unity with respect to the convolution product. However, it may be instructive to compute directly $g(x) \star \delta(x)$. If $g(x) = [G_+(z), G_-(z)]$, we have

$$g(x) \star \delta(x) = \left[-\frac{1}{2\pi i} \int_\alpha^\beta \frac{G_+(z-x)}{x}\,dx, -\frac{1}{2\pi i} \int_\alpha^\beta \frac{G_-(z-x)}{x}\,f(x)\,dx \right]$$

$$= \left[\frac{1}{2\pi i} \oint_{(\alpha,\beta)} \frac{G_+(z-\zeta)}{\zeta}\,d\zeta, \frac{1}{2\pi i} \oint_{(\alpha,\beta)} \frac{G_-(z-\zeta)}{\zeta}\,d\zeta \right].$$

By expanding $G_\pm(z - \zeta)$ we obtain

$$\frac{1}{2\pi i} \oint_{(\alpha,\beta)} \frac{G_\pm(z - \zeta)}{\zeta}\, d\zeta = \sum_{k=0}^{\infty} \frac{(-1)^k\, G_\pm^{(k)}(z)}{k!} \frac{1}{2\pi i} \oint_{(\alpha,\beta)} \zeta^{k-1}\, d\zeta = G_\pm(z).$$

Thus,

$$g(x) \star \delta(x) = [G + (z), G_-(z)] = g(x). \tag{2.146}$$

Problem 2.13. Show that

$$\delta(x - a) \star g(x) = g(x) \star \delta(x - a) = g(x - a). \tag{2.147}$$

Proposition 2.54. *Let $f(x)$ and $g(x)$ be two hyperfunctions with compact support $[\alpha, \beta]$ and $[\gamma, \delta]$, respectively. Then, their convolution is again a hyperfunction with compact support, and the support is contained in $[\alpha + \gamma, \beta + \delta]$.*

Proof. The convolution exists by the previous proposition. The functions $F(z - \zeta)$ and $G(\zeta)$, thus their product, are real analytic functions of ζ if $\zeta < \min\{\gamma, \Re z - \beta\}$ and if $\max\{\delta, \Re z - \alpha\} < \zeta$. This means that the convolution integral is reduced to an integral with the finite integration interval $[\min\{\gamma, \Re z - \beta\}, \max\{\delta, \Re z - \alpha\}]$. Thus it exists. $F(z - \zeta)$ is real analytic with respect to z if $\Re z - \zeta < \alpha$ or if $\Re z - \zeta > \beta$. This is fulfilled if $\Re z < \alpha + \gamma$ or if $\Re z > \beta + \delta$. This shows that the support of the convolution is contained in $[\alpha + \gamma, \beta + \delta]$. $\qquad \square$

A *right-sided hyperfunction* has the support contained in an interval $[\alpha, \infty)$, and a *left-sided hyperfunction* in an interval $(-\infty, \beta]$.

Proposition 2.55. *The convolution of two right-sided hyperfunctions $f(x)$ and $g(x)$ with support $[\alpha, \infty)$ and $[\gamma, \infty)$, respectively, exists and is again a right-sided hyperfunction with the support contained in $[\alpha + \gamma, \infty)$. Similarly, the convolution of two left-sided hyperfunctions with support $(-\infty, \beta]$ and $(-\infty, \delta]$, respectively, exists and is again a left-sided hyperfunction with the support contained in $(-\infty, \beta + \delta]$.*

Proof. We prove the case of two right-sided hyperfunctions. $F(z - \zeta)$ and $G(\zeta)$, thus their product, are real analytic functions of the integration variable ζ, if $\Re z - \alpha < \Re \zeta < \gamma$, i.e., if $\Re z < \alpha + \gamma$. This means that the convolution integral as a contour integral over a holomorphic integrand is zero for $\Re z < \alpha + \gamma$. This shows that the convolution is again a right-sided hyperfunction. Let now $\Re z \geq \alpha + \gamma$. Let c_1 be a real point with $c_1 < \alpha + \gamma$. By the foregoing reasoning the convolution integral becomes

$$H(z) := \int_{c_1}^{\infty} F(z - x)g(x)\, dx$$

$$= \int_{\gamma_{c_1,\infty}^+} F(z - \zeta)G(\zeta)\, d\zeta - \int_{\gamma_{c_1,\infty}^-} F(z - \zeta)G(\zeta)\, d\zeta$$

However, if $\Re z - \zeta < \alpha$, i.e., $\zeta > \Re z - \alpha$, $F(z - \zeta)$ is a holomorphic function of z.

If we choose a second point $c_2 > \Re z$, we can write

$$\int_{\gamma_{c_1,\infty}^+} F(z-\zeta)G(\zeta)\,d\zeta - \int_{\gamma_{c_1,\infty}^-} F(z-\zeta)G(\zeta)\,d\zeta$$

$$= \int_{\gamma_{c_1,c_2-\alpha}^+} F(z-\zeta)G(\zeta)\,d\zeta - \int_{\gamma_{c_1,c_2-\alpha}^-} F(z-\zeta)G(\zeta)\,d\zeta$$

$$+ \lim_{b\to\infty} \{ \int_{\gamma_{c_2-\alpha,b}^+} F(z-\zeta)G(\zeta)\,d\zeta - \int_{\gamma_{c_2-\alpha,b}^-} F(z-\zeta)G(\zeta)\,d\zeta.$$

For any finite $b > c_2 - \alpha$,

$$\phi(x) := \{ \int_{\gamma_{c_2-\alpha,b}^+} F(x-\zeta)G(\zeta)\,d\zeta - \int_{\gamma_{c_2-\alpha,b}^-} F(x-\zeta)G(\zeta)\,d\zeta \}$$

is a real analytic function in x. By discarding $\phi(x)$ we obtain an equivalent defining function. Finally, we have

$$f(x) \star g(x) = \left[\int_{c_1}^{c_2} F(z-x)g(x)\,dx \right] \quad (c_1 < \alpha + \gamma, c_2 > \Re z), \tag{2.148}$$

i.e., for any given z, we obtain an integral with a finite integration interval. □

Referring to Proposition 2.52 we know that the convolution product of two right-sided or two left-sided hyperfunctions is commutative.

If both supports are $\mathbb{R}_+$ we may use the handy notation for (2.148)

$$f(x) \star g(x) = \left[\int_{0-}^{\Re z+} F(z-x)g(x)\,dx \right], \tag{2.149}$$

moreover, if the second factor $g(x)$ is an ordinary continuous function, (2.148) simplifies to

$$f(x) \star g(x) = \left[\int_0^{\Re z} F(z-x)g(x)\,dx \right], \tag{2.150}$$

where the integral is now an ordinary integral.

For easy reference when dealing with Fourier transforms later on, we consider the case where $f(x)$ has the compact support $[\alpha, \beta]$, and the hyperfunction $g(x)$ is decomposed as follows:

$$g(x) = g_1(x) + g_2(x), \quad \text{supp } g_1 = (-\infty, 0], \quad \text{supp } g_2 = [0, \infty). \tag{2.151}$$

(take $g_1(x) = \chi_{(-\infty,0)}g(x)$ and $g_2(x) = \chi_{(0,\infty)}g(x)$). Equation (2.143) then is written

$$f(x) \star g(x) = \left[\int_{c_1}^{c_2} F(z-x)g_1(x)\,dx + \int_{c_1}^{c_2} F(z-x)g_2(x)\,dx \right]$$

$$= \left[-\oint_{(c_1,c_2)} F(z-\zeta)G_1(\zeta)\,d\zeta - \oint_{(c_1,c_2)} F(z-\zeta)G_2(\zeta)\,d\zeta \right].$$

Notice that $G_1(z)$ is real analytic for $\Re z > 0$ and $G_2(z)$ is real analytic for $\Re z < 0$. This implies that the last expression can be written more precisely as follows:

$$f(x) \star g(x)$$
$$= \left[-\oint_{(c_1,c_2)\cap(-\infty,0]} F(z - \zeta)G_1(\zeta)\, d\zeta - \oint_{(c_1,c_2)\cap[0,\infty)} F(z - \zeta)G_2(\zeta)\, d\zeta \right].$$
$$(2.152)$$

This gives for $f(x) \star g(x)$, accordingly to $c_1 < \Re z - \beta$ and $c_2 > \Re z - \alpha$,

$$\Re z < \alpha \Longrightarrow \left[-\oint_{(c_1,c_2)} F(z - \zeta)G_1(\zeta)\, d\zeta \right],$$

$$\alpha < \Re z < \beta \Longrightarrow \left[-\oint_{(c_1,0)} F(z - \zeta)G_1(\zeta)\, d\zeta - \oint_{(0,c_2)} F(z - \zeta)G_2(\zeta)\, d\zeta \right],$$

$$\beta < \Re z \Longrightarrow \left[-\oint_{(c_1,c_2)} F(z - \zeta)G_2(\zeta)\, d\zeta \right].$$

2.7.3 Operational Properties

The definition of the convolution implies directly its bilinearity.

Proposition 2.56. *(bilinearity) If a_1, a_2, b_1, b_2 are complex numbers and the four convolutions $f_1(x) \star g_1(x), f_1(x) \star g_2(x), f_2(x) \star g_1(x), f_2(x) \star g_2(x)$ exist, then*

$$(a_1 f_1(x) + a_2 f_2(x)) \star (b_1 g_1(x) + b_2 g_2(x))$$
$$= a_1 b_1 (f_1(x) \star g_1(x)) + a_1 b_2 (f_1(x) \star g_2(x)) \qquad (2.153)$$
$$+ a_2 b_1 (f_2(x) \star g_1(x)) + a_2 b_2 (f_2(x) \star g_2(x)).$$

Proposition 2.57. *Assume that all convolutions $f(x) \star g(x), g(x) \star f(x), Df(x) \star g(x),$ and $f(x) \star Dg(x)$ exist; then*

$$D\{f(x) \star g(x)\} = Df(x) \star g(x) = f(x) \star Dg(x). \qquad (2.154)$$

Proof. By the assumed uniform convergence of the convolution integral we can differentiate with respect to z under the integral sign. From

$$f(x) \star g(x) = \left[\int_{-\infty}^{\infty} F_+(z - x)g(x)\, dx, \int_{-\infty}^{\infty} F_-(z - x)g(x)\, dx \right],$$

and by the definition of the derivative of a hyperfunction, we obtain

$$D\{f(x) \star g(x)\} = \left[\frac{d}{dz} \int_{-\infty}^{\infty} F_+(z - x)g(x)\, dx, \frac{d}{dz} \int_{-\infty}^{\infty} F_-(z - x)g(x)\, dx \right]$$
$$= \left[\int_{-\infty}^{\infty} \frac{dF_+}{dz}(z - x)g(x)\, dx, \int_{-\infty}^{\infty} \frac{dF_-}{dz}(z - x)g(x)\, dx \right]$$
$$= Df(x) \star g(x).$$

By using the commutativity and the same reasoning as above, we also have

$$D\{f(x) \star g(x)\} = D\{g(x) \star f(x)\}f(x) = Dg(x) \star f(x),$$

thus, $Df(x) \star g(x) = Dg(x) \star f(x)$. Since $f(x) \star Dg(x)$ exists too, the statement of the proposition follows.   □

From $\delta(x) \star g(x) = g(x)$, we obtain for any integral $n > 0$,

$$\delta^{(n)}(x) \star g(x) = g^{(n)}(x), \tag{2.155}$$

and from $\delta(x - a) \star g(x) = g(x - a)$,

$$\delta^{(n)}(x - a) \star g(x) = g^{(n)}(x - a). \tag{2.156}$$

Proposition 2.58. *If all involved convolutions exist, then*

$$x\,(f(x) \star g(x)) = (xf(x)) \star g(x) + f(x) \star (x\,g(x)). \tag{2.157}$$

Proof. Let

$$f(x) \star g(x) = \left[\int_{-\infty}^{\infty} F_+(z - x)g(x)\,dx, \int_{-\infty}^{\infty} F_-(z - x)g(x)\,dx\right],$$

then

$$x\,\{f(x) \star g(x)\} = \left[z\int_{-\infty}^{\infty} F_+(z - x)g(x)\,dx, z\int_{-\infty}^{\infty} F_-(z - x)g(x)\,dx\right]$$

$$= \left[\int_{-\infty}^{\infty} \{(z - x)F_+(z - x)\}g(x)\,dx, \int_{-\infty}^{\infty} \{(z - x)F_-(z - x)\}g(x)\,dx\right]$$

$$+ \left[\int_{-\infty}^{\infty} F_+(z - x)\{xg(x)\}\,dx, \int_{-\infty}^{\infty} F_-(z - x)\{xg(x)\}\,dx\right]$$

$$= (xf(x)) \star g(x) + f(x) \star (xg(x)).   □$$

Proposition 2.59. *Let $a \neq 0$, b, and c be real numbers. If $h(x) = f(x)\star g(x)$ exists, then*

$$f(ax + b) \star g(ax + c) = \frac{1}{|a|}\,h(ax + b + c), \tag{2.158}$$

thus, particularly

$$f(x + b) \star g(x + c) = h(x + b + c). \tag{2.159}$$

Proof. We prove the case $a < 0$ and leave the case $a > 0$ to the reader. If $f(x) = [F_+(z), F_-(z)]$ and $g(x) = [G_+(z), G_-(z)]$, we have (see Section 1.3.1),

$$f(ax + b) = [-F_-(az + b), -F_+(az + b)],$$
$$g(ax + b) = [-G_-(az + b), -G_+(az + b)].$$

The upper component of the defining function $H(z; a, b, c)$ of $f(ax + b) \star g(ax + c)$ becomes

$$H_+(z; a, b, c) = \int_{\gamma^+_{-\infty,\infty}} F_-(a(z - \zeta) + b))\,G_-(a\zeta + c)\,d\zeta$$

$$- \int_{\gamma^-_{-\infty,\infty}} F_-(a(z - \zeta) + b)\,G_+(a\zeta + c)\,d\zeta.$$

For the lower component we have

$$H_-(z; a, b, c) = \int_{\gamma_{-\infty,\infty}^+} F_+(a(z - \zeta) + b)) \, G_-(a\zeta + c) \, d\zeta$$

$$- \int_{\gamma_{-\infty,\infty}^-} F_+(a(z - \zeta) + b) \, G_+(a\zeta + c) \, d\zeta.$$

The change of variables $\zeta' := a\zeta + c$ yields

$$H_+(z; a, b, c) = \frac{1}{-a} \{ \int_{\gamma_{-\infty,\infty}^-} F_-((az + b + c) - \zeta')) \, G_-(\zeta') \, d\zeta'$$

$$- \int_{\gamma_{-\infty,\infty}^+} F_-((az + b + c) - \zeta') \, G_+(\zeta') \, d\zeta' \}$$

$$= -\frac{1}{|a|} H_-(az + b + c),$$

and

$$H_-(z; a, b, c) = \frac{1}{-a} \{ \int_{\gamma_{-\infty,\infty}^-} F_+((az + b + c) - \zeta')) \, G_-(\zeta') \, d\zeta'$$

$$- \int_{\gamma_{-\infty,\infty}^+} F_+((az + b + c) - \zeta') \, G_+(\zeta') \, d\zeta' \}$$

$$= -\frac{1}{|a|} H_+(az + b + c),$$

where $h(x) = f(x) \star g(x) = [H_+(z), H_-(z)]$. But for $a < 0$, i.e., $-a = |a|$, we have

$$\frac{1}{|a|} h(ax + b + c) = \left[-\frac{1}{|a|} H_-(az + b + c), -\frac{1}{|a|} H_+(az + b + c) \right]$$

which completes the proof. $\qquad\square$

The above proposition has the following consequence for periodic hyperfunctions.

Proposition 2.60. Let $f(x) = f(x + T)$ be a periodic hyperfunction with period T, and $g(x)$ an arbitrary hyperfunction. If the convolution $f(x) \star g(x)$ exists, then it is a periodic hyperfunction with period T.

Proof. With the assumed existence of $h(x) = f(x) \star g(x)$, and from the above proposition with $a = 1$, $b = T$, $c = 0$, we have $f(x + T) \star g(x) = h(x + T)$. Since $f(x + T) = f(x)$ if follows that $h(x + T) = h(x)$. $\qquad\square$

Without imposing supplementary conditions, the convolution product is not associative. This is shown by the following example, where $1(x) = [1/2, -1/2]$ denotes the constant unity hyperfunction. See also Section 2.7.3.

$$\{1(x) \star \delta'(x)\} \star u(x) = \{1'(x) \star \delta(x)\} \star u(x)$$
$$= \{0 \star \delta(x)\} \star u(x) = 0 \star u(x) = 0,$$
$$1(x) \star \{\delta'(x) \star u(x)\} = 1(x) \star \{\delta(x) \star u'(x)\}$$
$$= 1(x) \star \{\delta(x) \star \delta(x)\} = 1(x) \star \delta(x) = 1(x).$$

However, we can prove

Proposition 2.61. *The convolution product of two right-sided or two left-sided hy-perfunctions is associative.*

Proof. We prove the proposition for the case where both supports are $\mathbb{R}_+$.

$$f_1(x) \star f_2(x) = f_2(x) \star f_1(x) = \left[\int_{0-}^{\Re z+} F_2(z-t) f_1(t)\, dt \right] =: [H(z)],$$

$$
\begin{aligned}
(f_1(x) \star f_2(x)) \star f_3(x) &= \left[\int_{0-}^{\Re z+} H(z-\tau) f_3(\tau)\, d\tau \right] \\
&= \left[\int_{0-}^{\Re z+} \int_{0-}^{\Re z+} F_2(z-\tau-t) f_1(t)\, dt\, f_3(\tau)\, d\tau \right] \\
&= \left[\int_{0-}^{\Re z+} \int_{0-}^{\Re z+} F_2(z-\tau-t) f_1(t)\, f_3(\tau)\, dt\, d\tau \right].
\end{aligned}
$$

On the other hand,

$$f_2(x) \star f_3(x) = \left[\int_{0-}^{\Re z+} F_2(z-\tau) f_3(\tau)\, d\tau \right] =: [G(z)],$$

$$
\begin{aligned}
f_1(x) \star (f_2(x)) \star f_3(x)) &= (f_2(x)) \star f_3(x)) \star f_1(x) \\
&= \left[\int_{0-}^{\Re z+} G(z-t) f_1(t)\, dt \right] \\
&= \left[\int_{0-}^{\Re z+} f_1(t) \int_{0-}^{\Re z+} F_2(z-t-\tau) f_3(\tau)\, d\tau\, dt \right] \\
&= \left[\int_{0-}^{\Re z+} \int_{0-}^{\Re z+} F_2(z-t-\tau) f_1(t)\, f_3(\tau)\, d\tau\, dt \right].
\end{aligned}
$$

An exchange of the order of integration establishes $(f_1(x) \star f_2(x)) \star f_3(x) = f_1(x) \star (f_2(x)) \star f_3(x))$. $\qquad\square$

We have now established that the space of all right-sided hyperfunctions with support in $[0, \infty)$, denoted by $\mathcal{A}(\mathbb{R}_+, \star)$ form a *commutative algebra having a unit element*, where the multiplication is the convolution product. This means that if $f(x), f_1(x), f_2(x), f_3(x) \in \mathcal{A}(\mathbb{R}_+, \star)$ we have

(i) $c_1 f_1(x) + c_2 f_2(x),\ f_1(x) \star f_2(x) \in \mathcal{A}(\mathbb{R}_+, \star)$,

(ii) distributive law (2.153) holds,

(iii) $f_1(x) \star f_2(x) = f_2(x) \star f_1(x)$,

(iv) $(f_1(x) \star f_2(x)) \star f_3(x) = f_1(x) \star (f_2(x)) \star f_3(x))$,

(v) $\delta(x) \star f(x) = f(x)$.

This algebra is called the *convolution algebra*. Given a *convolution equation*

$$f(x) \star y(x) = g(x), \qquad (2.160)$$

where $f(x), g(x) \in \mathcal{A}(\mathbb{R}_+, \star)$ are given hyperfunctions, we can readily solve for the unknown hyperfunction $y(x)$, provided the hyperfunction $f(x)$ has an inverse $f^{\star-1}(x)$ with respect to the convolution product, i.e., $f^{\star-1}(x) \star f(x) = \delta(x)$ holds, because

$$y(x) = \delta(x) \star y(x) = (f^{\star-1}(x) \star f(x)) \star y(x)) = f^{\star-1}(x) \star ((f(x)) \star y(x))$$
$$= f^{\star-1}(x) \star g(x).$$

Not every hyperfunction $f(x)$ has an inverse hyperfunction with respect to $\star$. However, if it has one, it is unique. The reason is simple. Suppose that $h(x)$ would be another inverse of the given $f(x)$, then

$$h(x) \star f(x) = \delta(x), \text{ and } f^{\star-1}(x) \star f(x) = \delta(x).$$

From this we obtain

$$h(x) = h(x) \star \delta(x) = h(x) \star (f(x) \star (f^{\star-1}(x)))$$
$$= (h(x) \star f(x)) \star f^{\star-1}(x) = \delta(x) \star f^{\star-1}(x) = f^{\star-1}(x).$$

Having arrived at this state of affair, we could now construct an operational calculus for solving linear differential equations with constant coefficients because

$$(a_n D^n + a_{n-1} D^{n-1} + \cdots + a_1 D + a_0)y(x) = g(x)$$

can be written in form of a convolution equation

$$(a_n \delta^{(n)}(x) + a_{n-1}\delta^{(n-1)}(x) + \cdots + a_1\delta(x) + a_0) \star y(x) = g(x).$$

We only have to seek the inverse element in $\mathcal{A}(\mathbb{R}_+, \star)$ of the hyperfunction

$$f(x) = a_n \delta^{(n)}(x) + a_{n-1}\delta^{(n-1)}(x) + \cdots + a_1\delta(x) + a_0.$$

This could be achieved directly, however the method of Laplace transformation to be developed in the next chapter is such a powerful and convenient tool for attacking problems of this kind that we postponed further investigations to the next chapter.

Problem 2.14. Show that for $m, n \in \mathbb{N}$ it holds that

$$\mathrm{fp}\frac{1}{(ax+b)^m} \star \mathrm{fp}\frac{1}{(ax+c)^n} = -\frac{\pi^2}{|a|} \frac{(-1)^{m+n}}{(m-1)!\,(n-1)!} \delta^{(m+n-2)}(ax+b+c).$$

Problem 2.15. Prove the following propositions concerning complex conjugation, parity and realness.

(i)

$$\overline{f(x)} \star \overline{g(x)} = \overline{f(x) \star g(x)}, \qquad (2.161)$$

(ii) The convolution of two even or two odd hyperfunctions yields an even hyperfunction. The convolution of an even and an odd hyperfunctions is an odd hyperfunction.

(iii) The convolution of two real or two imaginary hyperfunctions is a real hyperfunction. Convolution of a real and an imaginary hyperfunction gives an imaginary hyperfunction.

2.7.4   Principal Value Convolution

Usually, it is assumed that the improper convolution integral converges independently at both integration limits. If we only demand convergence in the sense of Cauchy's principal value, the range of hyperfunctions for which the convolution exist may be augmented. This, for example, will be assumed for the definition of the Hilbert transformation of two hyperfunctions and also for the convolution of periodic hyperfunctions.

Definition 2.22. Let $f(x) = [F(z)]$ and $g(x) = [G(z)]$ be two hyperfunctions, then, provided the limit exists,

$$f(x) \otimes g(x) := \text{pv} \int_{-\infty}^{\infty} F(z-x)g(x)\,dx = \lim_{R\to\infty} \int_{-R}^{R} F(z-x)g(x)\,dx \quad (2.162)$$

is called the *convolution in the sense of Cauchy's principal-value*, for short the pv-convolution of $f(x)$ and $g(x)$.

It is clear that if the convolution $f(x) \star g(x)$ exists, the pv-convolution $f(x) \otimes g(x)$ exists too, but the latter may exist without the former.

Notice that if

$$f(x) \otimes g(x) := h(x) = [H_+(z), H_-(z)],$$

then we have

$$H_+(z) = -\lim_{R\to\infty} \oint_{(-R,R)} F_+(z-\zeta)\,G(\zeta)\,d\zeta, \quad (2.163)$$

$$H_-(z) = -\lim_{R\to\infty} \oint_{(-R,R)} F_-(z-\zeta)\,G(\zeta)\,d\zeta. \quad (2.164)$$

Proposition 2.62. *Suppose that either $f(x) \otimes g(x)$ or $g(x) \otimes f(x)$ exists. Then the existence of the other, and of the commutativity law $f(x) \otimes g(x) = g(x) \otimes f(x)$ is assured, if and only if*

$$\psi_R(z,a) := \int_{R}^{R+a} F(z-x)g(x)\,dx \quad (2.165)$$

defines for any real a and for all sufficiently large $R > 0$ a real analytic function of z.

Proof. Consider the two functions

$$H_R(z) := \int_{-R}^{R} F(z-x)g(x)\,dx, \quad K_R(z) := \int_{-R}^{R} G(z-x)f(x)\,dx.$$

By (2.141) in the proof of Proposition 2.52 we know that

$$K_R(z) = \int_{\Re z - R}^{\Re z + R} F(z-x)g(x)\,dx + \phi(z)$$

where $\phi(t)$ is a real analytic function for $\Re z - R < t < \Re z + R$. Now write

$$K_R(z) = \int_{-(R-\Re z)}^{R-\Re z} F(z-x)g(x)\,dx + \int_{R-\Re z}^{R+\Re z} F(z-x)g(x)\,dx + \phi(z)$$

$$= \int_{-R'}^{R'} F(z-x)g(x)\,dx + \int_{R'}^{R'+2\Re z} F(z-x)g(x)\,dx + \phi(z)$$

with $R' = R - \Re z$. Observe that for a fixed z, $R \to \infty$ corresponds to $R' \to \infty$. Therefore,

$$K_R(z) = H_{R'}(z) + \int_{R'}^{R'+2\Re z} F(z-x)g(x)\,dx + \phi(z)$$

$$= H_{R'}(z) + \psi_{R'}(z, 2\Re z) + \phi(z).$$

Now for any R as large as you please, it is seen that $K_R(z)$ and $H_{R'}(z)$ are equivalent defining functions. This proves the existence of both $f(x) \otimes g(x)$ and $g(x) \otimes f(x)$, provided one of them exists, and at the same time, $f(x) \otimes g(x) = g(x) \otimes f(x)$. On the other hand, if $f(x) \otimes g(x) = g(x) \otimes f(x)$ holds, we must have that $\psi_R(z, a)$ is a real analytic function for any a and $R > 0$ representing the zero hyperfunction. □

Problem 2.16. Establish the following convolution in the sense of Cauchy's principal value.

$$\text{fp}\frac{1}{x} \otimes 1 = 1 \otimes \text{fp}\frac{1}{x} = 0,$$

$$1 \otimes x = 0, \quad x \otimes 1 \quad \text{does not exist,}$$

$$1 \otimes \text{sgn}\,x = 0, \quad \text{sgn}\,x \otimes 1 = 2x.$$

In Example 2.36 we considered the Dirac impulse comb

$$f(x) = \text{pv} \sum_{k=-\infty}^{\infty} \delta(x - kT) \tag{2.166}$$

which is a periodic hyperfunction with period T, and its projection on the interval $[-T/2, T/2]$,

$$f_0(x) = \chi_{(-T/2, T/2)} f(x) = \delta(x).$$

We can now prove

Proposition 2.63.

$$\left\{ \text{pv} \sum_{k=-\infty}^{\infty} \delta(x - kT) \right\} \otimes f(x) = \text{pv} \sum_{k=-\infty}^{\infty} f(x - kT). \tag{2.167}$$

Proof. If $f(x) = F[z]$, the left-hand side equals

$$\lim_{N \to \infty} \lim_{R \to \infty} -\frac{1}{2\pi i} \oint_{(-R,R)} \left\{ \frac{-1}{z - \zeta} - 2\sum_{k=1}^{N} \frac{z - \zeta}{(z - \zeta)^2 - k^2 T^2} \right\} F(\zeta)\,d\zeta.$$

For a positive integer N and $R > 0$ we then have,

$$\frac{1}{2\pi i} \oint_{(-R,R)} \{\frac{1}{z-\zeta} + 2\sum_{k=1}^{N} \frac{z-\zeta}{(z-\zeta)^2 - k^2 T^2}\} F(\zeta)\, d\zeta$$

$$= f(x) + \sum_{k=1}^{N} \frac{1}{2\pi i} \oint_{(-R,R)} \frac{2(z-\zeta)}{(z-\zeta)^2 - k^2 T^2} \} F(\zeta)\, d\zeta$$

$$= \chi_{(-R,R)} f(x) + \sum_{k=1}^{N} \frac{-1}{2\pi i} \oint_{(-R,R)} \frac{F(\zeta)\, d\zeta}{\zeta - (z - kT)} + \frac{-1}{2\pi i} \oint_{(-R,R)} \frac{F(\zeta)\, d\zeta}{\zeta - (z + kT)}$$

$$= \chi_{(-R,R)} f(x) + \sum_{k=1}^{N} \chi_{(-R,R)} f(x - kT) + \chi_{(-R,R)} f(x + kT)$$

$$= \chi_{(-R,R)} \sum_{k=-N}^{N} f(x - kT).$$

By letting $N \to \infty$ and $R \to \infty$ we obtain

$$\mathrm{pv} \sum_{k=-\infty}^{\infty} f(x - kT). \qquad \square$$

2.8   Integral Equations I

We shall occasionally apply the theory of hyperfunctions to integral equations. We open this track with some general remarks, and we shall come back repeatedly to this subject in the course of this book.

A *Fredholm integral equation of the second kind* is an equation of the type

$$y(x) - \lambda \int_a^b k(x,t) y(t)\, dt = f(x), \tag{2.168}$$

where $y(x)$ is the unknown function or hyperfunction. The second member $f(x)$ and the kernel $k(x,t)$ are given functions or hyperfunctions. λ is a complex or real parameter. The cases with $a = -\infty$ and/or $b = \infty$ are not excluded.

A *Volterra integral equation of the second kind* is an equation of the type

$$y(x) - \lambda \int_a^x k(x,t) y(t)\, dt = f(x). \tag{2.169}$$

When the second member $f(x)$ is identically zero we speak of a *homogeneous* integral equation (Fredholm or Volterra).

Definition 2.23. A complex value λ such that the homogeneous equation (Fredholm or Volterra) has a non-zero solution $y_\lambda(x)$ is called an *eigenvalue*, and the corresponding $y_\lambda(x)$ an *eigenfunction* of the integral equation.

The two types of integral equations are said to be of the *first kind*, if they are of the form

$$\int_a^b k(x,t) y(t)\, dt = f(x), \quad \int_a^x k(x,t) y(t)\, dt = f(x), \tag{2.170}$$

respectively.

A Fredholm integral equation is said to be *non-singular*, if a and b are both finite, and the kernel is a continuous function in both variables. When the interval (a, b) is infinite or if the kernel is not a continuous function, the integral equation is called *singular*.

It is known that a linear non-singular Fredholm integral equation has at most a countable number of eigenvalues, i.e., a so-called discrete spectrum, without an accumulation point in the finite complex plane. We shall see that singular integral equations may have a continuous spectrum.

An important subclass consists of integral equations where the kernel has the special form $k(x, t) = K(x - t)$. In this case, we speak of an integral equation of *convolution type*.

Another known result is that a Volterra integral equation of convolution type

$$y(x) - \lambda \int_a^x K(x - t)y(t)\, dt = 0 \qquad (2.171)$$

with a finite lower limit a has no eigenvalues, i.e., its spectrum is empty. This is a consequence of the famous theorem of Titchmarsh which says that the convolution product of right-sided hyperfunctions has no divisor of zero, i.e., if $k(x) \star y(x) = 0$, and $k(x) \neq 0$ then $y(x) = 0$. (Take $k(x) = \lambda K(x) - \delta(x)$ which is non-zero.)

When working in the framework of generalized functions, the distinction between integral equations of the first and second kind becomes obsolete because

$$y(x) - \lambda \int_{-\infty}^{\infty} K(x - t)y(t)\, dt = f(x)$$
$$\Leftrightarrow\ y(x) - \lambda K(x) \star y(x) = f(x)$$
$$\Leftrightarrow\ \{\delta(x) - \lambda K(x)\} \star y(x) = f(x)$$
$$\Leftrightarrow K_1(x) \star y(x) = f(x)$$
$$\Leftrightarrow \int_{-\infty}^{\infty} K_1(x - t)y(t)\, dt = f(x),$$

where $K_1(x) := \delta(x) - \lambda K(x)$ is just another kernel.

We shall see that the methods of Laplace transforms (Chapter 3), Fourier transforms (Chapter 4) and Hilbert transforms (Chapter 5) are suitable for solving integral equations of convolution type. Here, we shall briefly explain a method based on an assumed associativity of the convolution product (remember that without further restrictions the associativity of the convolution product does generally not hold).

Definition 2.24. For the integral equation

$$\int_{-\infty}^{\infty} K(x - t)y(t)\, dt = f(x) \Leftrightarrow K(x) \star y(x) = f(x) \qquad (2.172)$$

a solution $G(x)$ corresponding to the special second member $f(x) = \delta(x)$ is said to be a *fundamental solution of the integral equation*.

Thus, the fundamental solution is the inverse hyperfunction of the kernel with respect to the convolution product, where the delta hyperfunction takes over the role of the unit element. Suppose now that a fundamental solution of the integral equation is known, then because $G(x) \star K(x) = \delta(x)$ we have

$$K(x) \star y(x) = f(x) \implies G(x) \star (K(x) \star y(x)) = G(x) \star f(x).$$

If associativity of the convolution product is assumed, we can write

$$G(x) \star (K(x) \star y(x)) = (G(x) \star K(x)) \star y(x) = \delta(x) \star y(x) = y(x),$$

thus,

$$y(x) = G(x) \star f(x) = \int_{-\infty}^{\infty} G(x - t) f(t) \, dt. \tag{2.173}$$

If all three hyperfunctions $K(x)$, $f(x)$, $y(x)$ are assumed to be members of the convolution algebra $\mathcal{A}(\mathbb{R}_+, \star)$, associativity of the convolution product is assured and the above method applies.

Example 2.42. In Chapter 5, we shall introduce the Hilbert transform $g(x)$ of $y(x)$ defined by

$$\pi \, g(x) = \int_{-\infty}^{\infty} \frac{y(t)}{t - x} \, dt. \tag{2.174}$$

The inverse Hilbert transform is the solution of the integral equation (2.174) given $g(x)$. The known relation

$$\mathrm{fp}\frac{1}{x} \star \mathrm{fp}\frac{-1}{\pi^2 x} = \delta(x)$$

shows that

$$G(x) = -\mathrm{fp}\frac{1}{\pi^2 x}$$

is a fundamental solution. Thus, the inverse Hilbert transform is given by

$$y(x) = \pi \left(-\mathrm{fp}\frac{1}{\pi^2 x}\right) \star g(x) = -\frac{1}{\pi} \int_{-\infty}^{\infty} \frac{g(t)}{t - x} \, dt. \tag{2.175}$$

There is a final remark to be made about the methods that we shall use subsequently to solve integral equations. If integral transforms are used for solving integral equations, then clearly only solutions possessing the involved integral transform can be found. There may still be other solutions having no such integral transform. Such a solution cannot and will not appear by the method of the used integral transformation. For example, the existence conditions for Laplace and Fourier transforms are not the same, thus the two methods may possibly appear to yield different solutions. Also, when solving an integral equation by any method, it may happen that the solution process produces a hyperfunction which is not a solution of the given problem. So, in principle, the found hyperfunction solution should be checked whether or not it satisfies the given problem.

In this book, we shall deal only with singular integral equations of convolution type.

$$\frac{1}{2\pi i} \int_{-\infty}^{(0+)} e^{-iz\tau} \tau^{-\lambda} \log^2(\tau) \, d\tau$$

$$= \frac{(-iz)^{\lambda-1}}{\Gamma(\lambda)} \{(\log(-iz) - \Psi(\lambda))^2 - \Psi'(\lambda)\} \quad (\Im z > 0). \tag{3.21}$$

3.2 The Two-Sided Laplace Transform

3.2.1 The Classical Laplace Transform

Let $t \mapsto f(t)$ be an ordinary function defined on the entire real axis, and s a complex variable. The two-sided Laplace transformation is defined by

$$\hat{f}(s) := \int_{-\infty}^{\infty} e^{-st} f(t) \, dt, \tag{3.22}$$

provided that the improper integral is convergent for some s. The function $\hat{f}(s)$ is called the two-sided Laplace transform of $f(t)$, or the image function of $f(t)$. We use the notation $\hat{f} = \mathcal{L}[f]$, or more explicitly, $\hat{f}(s) = \mathcal{L}[f(t)](s)$. Moreover, we shall also use the notation

$$f(t) \circ\!\!-\!\!\bullet \hat{f}(s)$$

for the correspondence between the function $f(t)$ and its image function $\hat{f}(s)$. Henceforth, we shall drop the qualifier "two-sided" and simply speak of the Laplace transform when meaning the two-sided Laplace transform. The explicit meaning of (3.22) is

$$\int_{-\infty}^{\infty} e^{-st} f(t) \, dt := \lim_{a \to -\infty, b \to \infty} \int_a^b e^{-st} f(t) \, dt.$$

Concerning the convergence of the improper Laplace integral, we shall see in Section 3.3 below that one of three cases may occur.

(i) There is no s such that (3.22) is convergent, i.e., the Laplace transform does not exist.

(ii) There is a vertical line $\Re s = x_0$ such that (3.22) converges for all complex s lying on this line but for no other s. This case is generally excluded from the framework of Laplace transformation since it is more conveniently covered by the Fourier transformation (see Chapter 4).

(iii) There is a vertical strip $\sigma_- < \Re s < \sigma_+$ such that (3.22) is convergent for all s in the strip. We may then still distinguish the cases of simple and of absolute convergence of the integral (3.22). The latter means

$$\int_{-\infty}^{\infty} |e^{-st} f(t)| \, dt = \int_{-\infty}^{\infty} e^{-\Re s\, t} |f(t)| \, dt < \infty, \quad (\sigma_- < \Re s < \sigma_+).$$

For the classical two-sided Laplace transformation and its operational properties, see, for example, [14, Graf], Section 1.2. Before we introduce the Laplace transform of a hyperfunction, let us emphasize that a merely functional expression of a two-sided Laplace image function does not uniquely determine the original function $f(t)$. Only the functional expression $\hat{f}(s)$, associated with a corresponding vertical strip of convergence, will uniquely define the original function $f(t)$. The same functional expression, but with another strip of convergence, may determine another original $f(t)$. In other words, in order to determine $\mathcal{L}^{-1}[\hat{f}(s)](t)$ the domain of $\hat{f}(s)$, i.e., its vertical strip of convergence matters. Let us emphasize this by two examples.

Example 3.1. Suppose that $\hat{f}(s) = 1/(s-a)^m$, where $m \in \mathbb{N}$. The classical complex inversion formula yields $f(t)$ by

$$f(t) = \mathrm{pv}\,\frac{1}{2\pi i}\int_{c-i\infty}^{c+i\infty} e^{st}\,\hat{f}(s)\,ds.$$

The given expression for the image function admits two vertical strips of convergence where $\hat{f}(s)$ is holomorphic. The first one is $\Re s > \Re a$, i.e., a right-half plane $\Omega_+(a)$, while the second one is $\Re s < \Re a$, i.e., a left-half plane $\Omega_-(a)$. This corresponds to two different holomorphic image functions

$$\hat{f}_1 : \Omega_+(a) \longrightarrow \mathbb{C} \qquad\qquad \hat{f}_2 : \Omega_-(a) \longrightarrow \mathbb{C}$$

$$s \mapsto \frac{1}{(s-a)^m} \qquad\qquad s \mapsto \frac{1}{(s-a)^m}.$$

In the first case the vertical integration contour must lie in $\Omega_+(a)$, i.e., $c > \Re a$. For $t > 0$, and by using Jordan's lemma (see below), the vertical integration contour can be closed by an infinite left semicircle obtaining

$$f_1(t) = \mathrm{pv}\,\frac{1}{2\pi i}\int_{c-i\infty}^{c+i\infty} e^{st}\,\frac{ds}{(s-a)^m}$$

$$= \mathrm{Res}_{s=a}\{\frac{e^{st}}{(s-a)^m}\} = e^{at}\,\mathrm{Res}_{s=a}\{\frac{e^{(s-a)t}}{(s-a)^m}\} = \frac{t^{m-1}}{(m-1)!}\,e^{at}.$$

However, if $t < 0$, we close the vertical contour by a infinite right semicircle, and obtain $f_1(t) = 0$ by Cauchy's theorem. Therefore, we get

$$f_1(t) = Y(t)\,\frac{t^{m-1}}{(m-1)!}\,e^{at}.$$

In the second case, the vertical integration contour must lie in $\Omega_-(a)$, i.e., $c < \Re a$, and and for $t < 0$, the vertical integration contour is closed by an infinite right semicircle.

$$f_2(t) = \mathrm{pv}\,\frac{1}{2\pi i}\int_{c-i\infty}^{c+i\infty} e^{st}\,\frac{ds}{(s-a)^m}$$

$$= -\mathrm{Res}_{s=a}\{\frac{e^{st}}{(s-a)^m}\} = e^{at}\,\mathrm{Res}_{s=a}\{\frac{e^{(s-a)t}}{(s-a)^m}\} = -\frac{t^{m-1}}{(n-1)!}\,e^{at}.$$

The minus sign comes up because the closed contour is traversed in the negative sense. For $t > 0$, and closing the contour by an infinite left semicircle we get $f_2(t) = 0$. Thus, in this case we obtain

$$f_2(t) = -Y(-t)\,\frac{t^{m-1}}{(m-1)!}\,e^{at}.$$

Remark. The technique of closing the vertical contour by an infinite left or right semicircle (or other similar curves) in order to evaluate the complex inversion formula by the calculus of residues uses *Jordan's Lemma* which states (see also the end of Section 2.2):

Let c_0 be a point in the complex s-plane and ℓ the vertical straight line through c_0. Consider a family of semicircles to the left of ℓ,

$$S_n : \; s = c_0 + R_n\, e^{i\tau}, \quad \frac{\pi}{2} \leq \tau \leq \frac{3\pi}{2}$$

with $R_n \to \infty$ as $n \to \infty$. Let $\hat{f}(s)$ be a function defined on these semicircles such that $|\hat{f}(s)|$ tends to zero uniformly with respect to $\frac{\pi}{2} \leq \tau \leq \frac{3\pi}{2}$, then

$$\lim_{n\to\infty} \int_{S_n} e^{t\,s}\hat{f}(s)\,ds = 0,$$

provided $t > 0$.

Similarly, for a family of semicircles to the right of ℓ,

$$S_n : \; s = c_0 + R_n\, e^{i\tau}, \quad -\frac{\pi}{2} \leq \tau \leq \frac{\pi}{2},$$

the same statement holds, provided $t < 0$.

A proof of Jordan's Lemma can be found in [9, Doetsch p.174].

Example 3.2. The expression

$$\hat{f}(s) = \frac{-s + s^2}{(s + 1)^2\,(s^2 + 1)(s - 2)}$$

gives rise to four different image functions, defined by the same functional expression, but with four different domains, the four different vertical strips of convergence where the image function is holomorphic.

$$\hat{f}_1 : s \mapsto \frac{-s + s^2}{(s + 1)^2\,(s^2 + 1)\,(s - 2)}, \quad \Re s < -1,$$

$$\hat{f}_2 : s \mapsto \frac{-s + s^2}{(s + 1)^2\,(s^2 + 1)\,(s - 2)}, \quad -1 < \Re s < 0,$$

$$\hat{f}_3 : s \mapsto \frac{-s + s^2}{(s + 1)^2\,(s^2 + 1)\,(s - 2)}, \quad 0 < \Re s < 2,$$

$$\hat{f}_4 : s \mapsto \frac{-s + s^2}{(s + 1)^2\,(s^2 + 1)\,(s - 2)}, \quad 2 < \Re s.$$

Partial fraction decomposition of $\hat{f}(s)$ gives

$$\hat{f}(s) = -\frac{1}{3}\frac{1}{(s + 1)^2} + \frac{1}{18}\frac{1}{s + 1} - \frac{1 + 3i}{20}\frac{1}{s - i} - \frac{1 - 3i}{20}\frac{1}{s + i} + \frac{2}{45}\frac{1}{s - 2}.$$

In the left-most strip of convergence we choose $c < -1$ for the vertical integration contour, and we obtain, by the method of the previous example,

$$f_1(t) = -Y(-t)\{-\frac{1}{3}te^{-t} + \frac{1}{18}e^{-t} - \frac{1}{10}\cos t + \frac{3}{10}\sin t + \frac{2}{45}e^{2t}\},$$

In the next strip, we choose $-1 < c < 0$, and obtain

$$f_2(t) = Y(t)\{-\frac{1}{3}te^{-t} + \frac{1}{18}e^{-t}\} - Y(-t)\{-\frac{1}{10}\cos t + \frac{3}{10}\sin t + \frac{2}{45}e^{2t}\}.$$

In the following one, we choose $0 < c < 2$ and get

$$f_3(t) = Y(t)\{-\frac{1}{3}te^{-t} + \frac{1}{18}e^{-t} - \frac{1}{10}\cos t + \frac{3}{10}\sin t\} - Y(-t)\frac{2}{45}e^{2t},$$

and finally in the right-most strip of convergence, with $2 < c$, one obtains

$$f_4(t) = Y(t)\{-\frac{1}{3}te^{-t} + \frac{1}{18}e^{-t} - \frac{1}{10}\cos t + \frac{3}{10}\sin t + \frac{2}{45}e^{2t}\}.$$

3.3   Laplace Transforms of Hyperfunctions

For the following let us consider open sets $J := (a, 0) \cup (0, b)$ with some $a < 0$ and some $b > 0$, together with any compact subsets $K := [a', a''] \cup [b', b'']$ with $a < a' \le a'' < 0$ and $0 < b' \le b'' < b$. We then consider the following open neighborhoods $[-\delta, \infty) + iJ$ and $(-\infty, \delta] + iJ$ of $\mathbb{R}_+$ and $\mathbb{R}_-$, respectively, for some $\delta > 0$.

Now let us introduce the subclass $O(\mathbb{R}_+)$ of hyperfunctions $f(t) = [F(z)]$ on $\mathbb{R}$ satisfying

(i) The support supp $f(t)$ is contained in $[0, \infty)$.

(ii) Either the support supp $f(t)$ is bounded on the right by a finite number $\beta > 0$, or we demand that among all equivalent defining functions, there is one, $F(z)$, defined in $[-\delta, \infty) + iJ$ such that for any compact subset $K \subset J$ there exist some real constants $M' > 0$ and σ' such that $|F(z)| \le M' e^{\sigma' \Re z}$ holds uniformly for all $z \in [0, \infty) + iK$.

Because supp$f(t) \subset \mathbb{R}_+$, and since the singular support sing supp$f$ is a subset of the support, we have sing supp$f \subset \mathbb{R}_+$. Therefore $f(t)$ is a holomorphic hyperfunction for all $t < 0$. Moreover, the fact that $F_+(x+i0) - F(x-i0) = 0$, $\forall x < 0$ shows that $F(z)$ is real analytic on the negative part of the real axis. Hence $f(t) \in O(\mathbb{R}_+)$ implies that $\chi_{(-\varepsilon,\infty)}f(t) = f(t)$ for any $\varepsilon > 0$.

Definition 3.1. We call the subclass of hyperfunctions $O(\mathbb{R}_+)$ the *class of right-sided originals*.

In the case of an unbounded support supp $f(t)$ let $\sigma_- := \inf \sigma'$ be the greatest lower bound of all σ', where the infimum is taken over all σ' and all equivalent defining functions satisfying (ii). This number $\sigma_- = \sigma_-(f)$ is called *the growth index* of $f(t) \in O(\mathbb{R}_+)$. It has the properties

(i) $\sigma_- \le \sigma'$.

(ii) For every $\varepsilon > 0$ there is a σ' with $\sigma_- \le \sigma' \le \sigma_- + \varepsilon$, and an equivalent defining function $F(z)$ such that $|F(z)| \le M' e^{\sigma' \Re z}$ uniformly for all $z \in [0, \infty) + iK$.

In the case of a bounded support supp $f(t)$, we set $\sigma_-(f) = -\infty$.

Definition 3.2. The Laplace transform of a right-sided original $f(t) = [F(z)] \in O(\mathbb{R}_+)$ is now defined by

$$\hat{f}(s) = \mathcal{L}[f](s) = -\int_{\infty}^{(0+)} e^{-sz} F(z)\, dz. \qquad (3.23)$$

The contour of the loop integral is shown in Figure 1.12.

Proposition 3.1. *The image function $\hat{f}(s)$ of $f(t) \in O(\mathbb{R}_+)$ is holomorphic in the right half-plane $\Re s > \sigma_-(f)$.*

Proof. **A)** We consider the case of an unbounded support supp $f(t)$. First, we decompose the integration loop $\gamma = \ell(0,0,0)$ into three parts $\gamma = \gamma_0 + \gamma^+ + \gamma^-$. γ_0 is the part of the loop that encircles 0 and is to the left of a vertical line ℓ_0 passing through a point $x_0 > 0$. γ^+ denotes the part of the contour being to the right of ℓ_0 and above the real axis, while γ^- is the part below the real axis and to the right of ℓ_0. Because γ_0 is of finite length the integral

$$-\int_{\gamma_0} e^{-sz} F(z)\, dz$$

represents an entire function of s (the integrand is holomorphic with respect to s and continuous with respect to z). With $z = x + iy$, $|y| < m$, for some $m > 0$, we have the estimate

$$\left| -\int_{\gamma^+} e^{-sz} F(z)\, dz \right| \le \int_{x_0}^{\infty} |e^{-s(x+iy)} F(x + iy)|\, dx$$

$$\le \int_{x_0}^{\infty} e^{-(x\,\Re s + y\,\Im s)} M' e^{\sigma' x}\, dx \le M' e^{m\,\Im s} \int_{x_0}^{\infty} e^{-x\,(\Re s - \sigma')}\, dx.$$

Choose $\epsilon > 0$ and assume $\Re s \ge \sigma_-(f) + \epsilon$. The above σ' can be selected such that $\sigma_-(f) < \sigma' < \sigma_-(f) + \epsilon \le \Re s$, i.e., $\Re s - \sigma' > 0$. Therefore, the integral converges uniformly inside the right half-plane $\Re s > \sigma_-(f)$ and

$$-\int_{\gamma^+} e^{-sz} F(z)\, dz$$

is a holomorphic function of s for $\Re s > \sigma_-(f)$. Likewise, the same result holds for

$$-\int_{\gamma^-} e^{-sz} F(z)\, dz.$$

B) In the case of a bounded support $[a, b] \subset \mathbb{R}_+$, the defining integral becomes

$$\hat{f}(s) = \mathcal{L}[f](s) = -\oint_{(a,b)} e^{-sz} F(z)\, dz$$

which defines an entire function $\hat{f}(s)$, thus $\sigma_-(f) = -\infty$. This proves the proposition. $\qquad \square$

Similarly, we introduce the class $O(\mathbb{R}_-)$ of hyperfunctions specified by

(i) The support supp$f(t)$ is contained in $\mathbb{R}_- = (-\infty, 0]$,

(ii) Either the support supp $f(t)$ is bounded on the left by a finite number $\alpha < 0$, or we demand that among all equivalent defining functions there is one, denoted by $F(z)$, and defined in $(-\infty, \delta] + iJ$, such that for any compact subset $K \subset J$ there are some real constants $M'' > 0$ and σ'' such that $|F(z)| \leq M'' e^{\sigma'' \Re z}$ holds uniformly for $z \in (-\infty, 0] + iK$.

The set $O(\mathbb{R}_-)$ is said to be the *class of left-sided originals*. In the case of an unbounded support let $\sigma_+ := \sup \sigma''$ be the least upper bound of all σ'', where the supremum is taken over all σ'' and all equivalent defining functions satisfying (ii). The number $\sigma_+ = \sigma_+(f)$ is called *the growth index of* $f(t) \in O(\mathbb{R}_-)$. It has the properties

(i) $\sigma'' \leq \sigma_+$.

(ii) For every $\varepsilon > 0$ there is a σ'' such that $\sigma_+ - \varepsilon \leq \sigma'' \leq \sigma_+$, and a defining function $F(z)$ such that $|F(z)| \leq M'' e^{\sigma'' \Re z}$ uniformly for all $z \in (-\infty, 0] + iK$.

If the support $supp f(t)$ is bounded, we set $\sigma_+(f) = +\infty$.

Definition 3.3. The Laplace transform of $f(t) \in O(\mathbb{R}_-)$ is defined by

$$\hat{f}(s) = \mathcal{L}[f](s) = -\int_{-\infty}^{(0+)} e^{-sz} F(z) \, dz. \qquad (3.24)$$

The contour of the loop integral is shown in Figure 1.13.

As before, we have

Proposition 3.2. *The image function* $\hat{f}(s)$ *of* $f(t) \in O(\mathbb{R}_-)$ *is holomorphic in the left half-plane* $\Re s < \sigma_+(f)$.

Example 3.3. We have $\delta(t) \in O(\mathbb{R}_-) \cap O(\mathbb{R}_+)$ and $\sigma_-(\delta) = -\infty$ and $\sigma_+(\delta) = +\infty$. For $f(t) = u(t)e^{-t}$ we have $f(t) \in O(\mathbb{R}_+)$ and $\sigma_-(f) = -1$. Similarly, for $f(t) = u(-t)e^{t}$ we have $f(t) \in O(\mathbb{R}_-)$ and $\sigma_+(f) = +1$. Let $g(t)$ be any polynomial, Then $h_2(t) = u(t) g(t) \in O(\mathbb{R}_+)$ with $\sigma_-(h_2) = 0$, also $h_1(t) = u(-t) g(t) \in O(\mathbb{R}_-)$ with $\sigma_+(h_1) = 0$.

With a left-sided original $g(t) \in O(\mathbb{R}_-)$ with growth index $\sigma_+(g)$ and a right-sided original $f(t) \in O(\mathbb{R}_+)$ with growth index $\sigma_-(f)$ we form the hyperfunction $h(t) := g(t) + f(t)$ whose support is now the entire real axis. If $\hat{g}(s) = \mathcal{L}[g(t)](s)$, $\Re s < \sigma_+(g)$ and $\hat{f}(s) = \mathcal{L}[f(t)](s)$, $\Re s > \sigma_-(f)$ we may add the two image functions, provided they have a common strip of convergence, i.e., that $\sigma_-(f) < \sigma_+(g)$ holds.

Definition 3.4. With $g(t) \in O(\mathbb{R}_-)$, $f(t) \in O(\mathbb{R}_+)$, $h(t) = g(t) + f(t)$,

$$\mathcal{L}[h(t)](s) := \hat{g}(t)(s) + \hat{f}(t)(s), \qquad \sigma_-(f) < \Re s < \sigma_+(g) \qquad (3.25)$$

provided $\sigma_-(f) < \sigma_+(g)$.

Example 3.4. Let us compute the Laplace transform of the Dirac impulse hyperfunction. Because $\delta(t)/2 = [-1/(4\pi i z)] \in O(\mathbb{R}_-) \cap O(\mathbb{R}_+)$ we have

$$\mathcal{L}[\frac{1}{2}\delta(t)](s) = \frac{1}{2}\frac{1}{2\pi i}\int_{-\infty}^{(0+)}\frac{e^{-sz}}{z}\,dz = \frac{1}{2}\operatorname{Res}_{z=0}\{\frac{e^{-sz}}{z}\} = 1/2,$$

$$\mathcal{L}[\frac{1}{2}\delta(t)](s) = \frac{1}{2}\frac{1}{2\pi i}\int_{\infty}^{(0+)}\frac{e^{-sz}}{z}\,dz = \frac{1}{2}\operatorname{Res}_{z=0}\{\frac{e^{-sz}}{z}\} = 1/2.$$

Therefore, $h(t) = \delta(t) = \delta(t)/2 + \delta(t)/2$, and finally,

$$\delta(t) \circ\!\!-\!\!\bullet\, 1. \tag{3.26}$$

For $a > 0$ we have $\delta(t-a) = [-1/(2\pi i(z-a))] \in O(\mathbb{R}_+)$ and

$$\mathcal{L}[\delta(t-a)](s) = \frac{1}{2\pi i}\int_{\infty}^{(0+)}\frac{e^{-sz}}{z-a}\,dz = \operatorname{Res}_{z=a}\{\frac{e^{-sz}}{z-a}\} = e^{-sa}.$$

Finally,

$$\delta(t-a) \circ\!\!-\!\!\bullet\, e^{-sa}. \tag{3.27}$$

Remark. The above example with the Dirac impulse hyperfunction shows that the decomposition (3.25) need not be unique. Instead of $h(t) = \delta(t) = \delta(t)/2 + \delta(t)/2$ another decomposition, for example $h(t) = \delta(t) = \delta(t)/3 + 2\delta(t)/3$ is also possible, and works as well, without altering the final result.

Hyperfunctions of the subclass $O(\mathbb{R}_+)$ are said to be of *bounded exponential growth as* $t \to \infty$, and hyperfunctions of the subclass $O(\mathbb{R}_-)$ are called of bounded exponential growth as $t \to -\infty$. Similarly, an ordinary function $f(t)$ is called of bounded exponential growth as $t \to \infty$, if there are some real constants $M' > 0$ and σ' such that $|f(t)| \leq M' e^{\sigma' t}$, for sufficiently large t. It is called of bounded exponential growth as $t \to -\infty$, if there are some real constants $M'' > 0$ and σ'' such that $|f(t)| \leq M'' e^{\sigma'' t}$, for sufficiently negative large t. A function or hyperfunction is *of bounded exponential growth*, if it is of bounded exponential growth for $t \to -\infty$ as well as for $t \to \infty$. Thus, we may say that a hyperfunction or ordinary function $f(t)$ has a (two-sided) Laplace transform, if it is of bounded exponential growth, and if $\sigma_-(f) < \sigma_+(f)$.

In the next few propositions we show that if there is a point c on the real line where the given hyperfunction is holomorphic, i.e., $c \notin \operatorname{sing\,supp} f$, then certain properties known from the Laplace transformation of ordinary functions are carried over to hyperfunctions.

Proposition 3.3. *If $f(t) = [F(z)]$ is a hyperfunction of bounded exponential growth which is holomorphic at $t = c$, then*

$$-\int_{-\infty}^{(c+)} e^{-sz}F(z)\,dz = \int_{-\infty}^{c} e^{-st}f(t)\,dt,$$

$$-\int_{\infty}^{(c+)} e^{-sz}F(z)\,dz = \int_{c}^{\infty} e^{-st}f(t)\,dt \quad thus,$$

$$-\{\int_{-\infty}^{(c+)} e^{-sz}F(z)\,dz + \int_{\infty}^{(c+)} e^{-sz}F(z)\,dz\} = \int_{-\infty}^{\infty} e^{-st}f(t)\,dt.$$

Note that the integrals on the right-hand sides are integrals over hyperfunctions, and on the left-hand side in the third line we have two semi-infinite u-shaped contours.

Proof. We prove the second statement, the proof of the first one is similar. The contour integral on the left-hand side can be expressed by

$$\int_\infty^{(c+)} e^{-sz} F(z)\, dz = -\int_{c+r}^\infty e^{-sz} F_+(z)\, dz + \int_{S^+(c,r)} e^{-sz} F_+(z)\, dz$$

$$+ \int_{S^-(c,r)} e^{-sz} F_-(z)\, dz + \int_{c+r}^\infty e^{-sz} F_-(z)\, dz$$

where the contour starts at $\infty + i0$, runs on the upper side of the real axis to $c + r$, turns around c on the semicircle $S^+(c,r)$ in the upper half-plane, continues to turn on the semicircle $S^-(c,r)$ in the lower half-plane, and runs back to $\infty - i0$ on the lower side of the real axis. The number r is arbitrarily small and positive. $F_+(z)$ and $F_-(z)$ being holomorphic at c imply

$$\int_{S^+(c,r)} e^{-sz} F_+(z)\, dz \to 0, \quad \int_{S^-(c,r)} e^{-sz} F_-(z)\, dz \to 0,$$

and

$$-\int_{c+r}^\infty e^{-sz} F_+(z)\, dz + \int_{c+r}^\infty e^{-sz} F_-(z)\, dz \to -\int_c^\infty e^{-st} f(t)\, dt$$

as the radius $r \to 0+$. $\square$

The two projections of $f(t) = [F(z)]$ on the intervals (c, ∞) and $(-\infty, c)$ are defined by

$$\chi_{(c,\infty)} f(t) = \left[\frac{e^{\lambda z}}{2\pi i} \int_c^\infty \frac{e^{-\lambda t} f(t)}{t - z}\, dt \right],$$

$$\chi_{(-\infty,c)} f(t) = \left[\frac{e^{\lambda' z}}{2\pi i} \int_{-\infty}^c \frac{e^{-\lambda' t} f(t)}{t - z}\, dt \right],$$

where $e^{\lambda t}$, $e^{\lambda' t}$ are convenient converging factors, possibly needed, in order to force convergence of the integral.

Proposition 3.4. *Let $f(t) = [F(z)]$ be a hyperfunction of bounded exponential growth with an arbitrary support and holomorphic at some point $t = c$. If in addition $\sigma_- = \sigma_-(\chi_{(0,\infty)} f(t)) < \sigma_+ = \sigma_+(\chi_{(-\infty,0)} f(t))$, then its Laplace transform is given by*

$$\mathcal{L}[f(t)](s) = \mathcal{L}[\chi_{(-\infty,c)} f(t)](s) + \mathcal{L}[\chi_{(c,\infty)} f(t)](s)$$

$$= \int_{-\infty}^c e^{-st} f(t)\, dt + \int_c^\infty e^{-st} f(t)\, dt \tag{3.28}$$

$$= \int_{-\infty}^\infty e^{-st} f(t)\, dt.$$

The latter integral is an integral over a hyperfunction and means

$$\int_{\gamma_{-\infty,\infty}^+} e^{-sz} F_+(z)\, dz - \int_{\gamma_{-\infty,\infty}^-} e^{-sz} F_-(z)\, dz.$$

Proof. We take $c = 0$, the proof for another c needs only minor modifications which are left to the reader.

$$\mathcal{L}[\chi_{(0,\infty)} f(t)](s) = -\int_{\infty}^{(0+)} e^{-sz} \frac{e^{\lambda z}}{2\pi i} \int_0^\infty \frac{e^{-\lambda t} f(t)}{t - z}\, dt\, dz$$

where the λ of the converging factor is selected such that $\sigma_- + \delta < \lambda$ for any $\delta > 0$ which assures (absolute and uniform) convergence of the inner integral. The t-contour of the inner integral is located inside the z-contour of the outer integral. If $\Re s > \lambda$, we may interchange the order of integration and obtain

$$\mathcal{L}[\chi_{(0,\infty)} f(t)](s) = \int_0^\infty e^{-\lambda t} f(t) \frac{1}{2\pi i} \int_\infty^{(0+)} \frac{e^{z(\lambda - s)}}{z - t}\, dz\, dt$$

$$= \int_0^\infty e^{-\lambda t} f(t) e^{t(\lambda - s)}\, dt = \int_0^\infty f(t) e^{-ts}\, dt.$$

Similarly, we obtain for $\Re s < \lambda' < \sigma_+ - \delta$,

$$\mathcal{L}[\chi_{(-\infty,0)} f(t)](s) = \int_{-\infty}^0 f(t) e^{-ts}\, dt.$$

Because $F_+(z)$ and $F_-(z)$ are holomorphic at $z = 0$, the separation point $z = 0$ can be removed, and we can write

$$\mathcal{L}[\chi_{(-\infty,0)} f(t)](s) + \mathcal{L}[\chi_{(0,\infty)} f(t)](s) = \int_{-\infty}^\infty f(t) e^{-ts}\, dt$$

$$= \int_{\gamma_{-\infty,\infty}^+} e^{-sz} F_+(z)\, dz - \int_{\gamma_{-\infty,\infty}^-} e^{-sz} F_-(z)\, dz. \qquad \square$$

If, for example, the hyperfunction $f(t)$ is not holomorphic at $t = 0$, i.e., $0 \in \text{sing supp}(f)$ as is the case with $f(t) = \delta(t)$ in the above example, we have to split the given hyperfunctions into components of $O(\mathbb{R}_+)$ and $O(\mathbb{R}_-)$ directly, as we have done in Example 3.4, or by selecting another speration point $c \neq 0$ where $f(t)$ is holomorphic. The above proposition has an obvious generalization which can be stated as follows.

Proposition 3.5. *Let* $f(t) = [F(z)]$ *be a hyperfunction of bounded exponential growth with arbitrary support and which is holomorphic at the points* t_i, $-\infty < t_1 < t_2 < \ldots < t_n$, *and such that* $\hat{f}(s)$ *has the strip of convergence* $\sigma_- < \Re s < \sigma_+$. *Then, its Laplace transform is given by*

$$\mathcal{L}[f(t)](s) = -\oint_{(-\infty,t_1)} e^{-sz}\,F(z)\,dz - \sum_{k=1}^{n-1}\oint_{(t_k,t_{k+1})} e^{-sz}\,F(z)\,dz$$

$$-\oint_{(t_n,\infty)} e^{-sz}\,F(z)\,dz$$

$$= \int_{-\infty}^{t_1} e^{-st}\,f(t)\,dt + \sum_{k=1}^{n-1}\int_{t_k}^{t_{k+1}} e^{-st}\,f(t)\,dt$$

$$+ \int_{t_n}^{\infty} e^{-st}\,f(t)\,dt.$$

(3.29)

Here the first and the last contour is a u-shaped infinite contour (Figure 3.2).

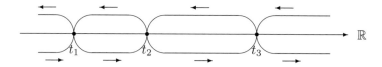

Figure 3.2: Contours in Proposition 3.5

Because there exist large tables of Laplace transforms of right-sided originals, the following proposition is of great practical value. If $f(t) = [F(z)] \in O(\mathbb{R}_-)$ with $\sigma_+ = \sigma_+(f(t))$, then $f(-t) = [-F(-z)] \in O(\mathbb{R}_+)$ and $\sigma_-(f(-t)) = -\sigma_+(f(t)) = -\sigma_+$. Then, for $\Re s < \sigma_+$ or $-\sigma_+ < \Re(-s)$,

$$\mathcal{L}[f(t)](s) = -\int_{-\infty}^{(0+)} e^{-sz}F(z)\,dz = -\int_{\infty}^{(0+)} e^{-s(-z)}F(-z)\,d(-z)$$

$$= -\int_{\infty}^{(0+)} e^{-(-s)z}\{-F(-z)\}\,dz = \mathcal{L}[f(-t)](-s),$$

and we obtain

Proposition 3.6. *Let $f(t) = [F(z)]$ be a hyperfunction with an arbitrary support and which is holomorphic at $t = 0$. Let $f_1(t) := \chi_{(-\infty,0)}f(t) \in O(\mathbb{R}_-)$ and $f_2(t) := \chi_{(0,\infty)}f(t) \in O(\mathbb{R}_+)$ with $\sigma_- = \sigma_-(f_2(t))$ and $\sigma_+ = \sigma_+(f_1(t))$. If $\sigma_- < \sigma_+$ the Laplace transform of $f(t)$ is obtained by*

$$\mathcal{L}[f(t)](s) := \mathcal{L}[f_1(-t)](-s) + \mathcal{L}[f_2(t)](s).$$

(3.30)

with $-\sigma_+ < \Re(-s)$, $\sigma_- < \Re s$.

Example 3.5. Let us compute $\mathcal{L}[\exp(-a\,t^2)](s)$ for $a > 0$. Of course the Laplace transform of $\exp(-at^2)$ can be computed readily by direct evaluation of the conventional Laplace integral; but we will do it by the above decomposition in order to illustrate the method. By using the known correspondence

$$Y(t)\exp(-a\,t^2) \circ\!\!-\!\!\bullet \frac{1}{2}\sqrt{\frac{\pi}{a}}\,\exp(\frac{s^2}{4\,a})\,\mathrm{erfc}(\frac{s}{2\sqrt{a}}),\quad -\infty < \Re s < \infty$$

we can write

$$\mathcal{L}[\chi_{(0,\infty)}\, e^{-a(-t)^2}](-s) = \frac{1}{2}\sqrt{\frac{\pi}{a}}\,\exp(\frac{(-s)^2}{4\,a})\,\mathrm{erfc}(\frac{-s}{2\sqrt{a}}), \quad \Re s < \infty,$$

$$\mathcal{L}[\chi_{(0,\infty)}\, e^{-a t^2}](s) = \frac{1}{2}\sqrt{\frac{\pi}{a}}\,\exp(\frac{s^2}{4\,a})\,\mathrm{erfc}(\frac{s}{2\sqrt{a}}), \quad \Re s > -\infty.$$

By using

$$\mathrm{erfc}(-z) + \mathrm{erfc}(z) = 2,$$

we get

$$\mathcal{L}[\chi_{(0,\infty)}\, e^{-a(-t)^2}](-s) + \mathcal{L}[\chi_{(0,\infty)}\, e^{-a t^2}](s) = \sqrt{\frac{\pi}{a}}\,\exp(\frac{s^2}{4\,a}).$$

Therefore, for $-\infty < \Re s < \infty$,

$$\exp(-a\,t^2) \circ\!\!-\!\!\bullet \sqrt{\frac{\pi}{a}}\,\exp(\frac{s^2}{4\,a}). \tag{3.31}$$

Note that the relation

$$\mathcal{L}[Y(t)\, e^{-a t^2}](s) = \frac{1}{2}\sqrt{\frac{\pi}{a}}\,\exp(\frac{s^2}{4\,a})\,\mathrm{erfc}(\frac{s}{2\sqrt{a}})$$

is valid for $\Re a > 0$ and $-\infty < \Re s < \infty$. The expression on the right-hand side is for a fixed s a holomorphic function in a, where the complex a-plane has an incison along the negative real line. However, if $a = ib$, $b > 0$, the Laplace transform of the left-hand side exists for $\Re s > 0$ because $Y(t)\exp(-ibt^2) = Y(t)\cos(bt^2) - iY(t)\sin(bt^2)$, is a bounded right-sided original. Thus, the validity of the obtained formula can be continued for $\Re s > 0$ from $\Re a > 0$ to $\Re a \geq 0$, $a \neq 0$, and we obtain,

$$Y(t)\exp(-i\,b\,t^2) \circ\!\!-\!\!\bullet \frac{1}{2}\sqrt{\frac{\pi}{2b}}(1-i)\,\exp(-i\,\frac{s^2}{4\,b})\,\mathrm{erfc}(\sqrt{\frac{1}{2b}}\frac{1-i}{2}\,s). \tag{3.32}$$

Example 3.6. Let us compute $\mathcal{L}[\exp(-a|t|)](s)$ for $a > 0$. Because

$$\chi_{(0,\infty)}\, e^{-a|t|} = Y(t)e^{-at} \circ\!\!-\!\!\bullet \frac{1}{s+a}, \quad \Re s > -a,$$

$$\chi_{(-\infty,0)}\, e^{-a|t|} = Y(-t)e^{at} = f(t), \quad f(-t) = u(t)e^{-at},$$

we have

$$\mathcal{L}[e^{-a|t|}](s) = \frac{1}{(-s)+a} + \frac{1}{s+a}$$

$$= -\frac{2a}{s^2 - a^2}$$

and the strip of convergence is $-a < \Re s < a$.

Example 3.7. Let us compute the Laplace transform of

$$f(t) = \sum_{k=-\infty}^{\infty} e^{-|t|}\, \delta(t-k).$$

We decompose $f(t)$ into $f(t) = f_1(t) + f_2(t)$, where

$$f_1(t) = \frac{1}{2}e^t\,\delta(t) + \sum_{k=1}^{\infty} e^t\,\delta(t+k) = \frac{1}{2}\delta(t) + \sum_{k=1}^{\infty} e^{-k}\,\delta(t+k) \in O(\mathbb{R}_-),$$

$$f_2(t) = \frac{1}{2}e^{-t}\,\delta(t) + \sum_{k=1}^{\infty} e^{-t}\,\delta(t-k) = \frac{1}{2}\delta(t) + \sum_{k=1}^{\infty} e^{-k}\,\delta(t-k) \in O(\mathbb{R}_+).$$

By taking the Laplace transform term wise we obtain

$$\hat{f}(s) = 1 + \sum_{k=1}^{\infty} e^{-k} e^{-ks} + \sum_{k=1}^{\infty} e^{-k} e^{ks}.$$

The first geometric series converges for $-1 < \Re s$,

$$\sum_{k=1}^{\infty} e^{-k(s+1)} = \frac{e^{-(s+1)}}{1 - e^{-(s+1)}},$$

while the second geometric series evaluates to

$$\sum_{k=1}^{\infty} e^{k(s-1)} = \frac{e^{s-1}}{1 - e^{s-1}}$$

for $\Re s < 1$. Hence, in the common strip of convergence $-1 < \Re s < 1$, we finally obtain

$$\sum_{k=-\infty}^{\infty} e^{-|t|}\,\delta(t-k) \circ\!\!-\!\!\bullet \frac{1 - e^{-2}}{(1 - e^{-s-1})(1 - e^{s-1})}. \tag{3.33}$$

Example 3.8. Sometimes it is better not to split the Laplace integral in order to compute a Laplace transform. The *Hermite polynomials of degree n* are defined by

$$H_n(x) := (-1)^n\, e^{x^2}\, \frac{d^n}{dx^n} e^{-x^2},$$

and the *Hermite functions* by

$$\mathcal{H}_n(x) := e^{-x^2/2}\, H_n(x) = (-1)^n\, e^{x^2/2}\, \frac{d^n}{dx^n} e^{-x^2}.$$

Let us compute the Laplace transform of a Hermite function. For doing so, we need the following formula which can be established by a simple integration by parts: for a given positive integer n and for $k = 0, 1, 2, \ldots, n-1$ it holds that

$$\int_{-\infty}^{\infty} \frac{d^k}{dx^k}(e^{-sx+x^2/2})\, \frac{d^{n-k}}{dx^{n-k}} e^{-x^2}\, dx$$

$$= -\int_{-\infty}^{\infty} \frac{d^{k+1}}{dx^{k+1}}(e^{-sx+x^2/2})\, \frac{d^{n-k-1}}{dx^{n-k-1}} e^{-x^2}\, dx.$$

Starting with $k = 0$ and successively applying this formula, we obtain

$$(-1)^n \int_{-\infty}^{\infty} e^{-sx+x^2/2} \frac{d^n}{dx^n} e^{-x^2} \, dx = \int_{-\infty}^{\infty} \frac{d^n}{dx^n} \left(e^{-sx+x^2/2}\right) e^{-x^2} \, dx.$$

Now,

$$\hat{\mathcal{H}}_n(s) = \int_{-\infty}^{\infty} e^{-sx} (-1)^n e^{x^2/2} \frac{d^n}{dx^n} e^{-x^2} \, dx$$

$$= (-1)^n \int_{-\infty}^{\infty} e^{-sx+x^2/2} \frac{d^n}{dx^n} e^{-x^2} \, dx$$

$$= \int_{-\infty}^{\infty} \frac{d^n}{dx^n} \left(e^{-sx+x^2/2}\right) e^{-x^2} \, dx$$

$$= e^{-s^2/2} \int_{-\infty}^{\infty} \frac{d^n}{dx^n} \left(e^{(x-s)^2/2}\right) e^{-x^2} \, dx,$$

and this equals

$$\hat{\mathcal{H}}_n(s) = (-1)^n e^{-s^2/2} \int_{-\infty}^{\infty} \frac{d^n}{ds^n} \left(e^{(x-s)^2/2}\right) e^{-x^2} \, dx$$

$$= (-1)^n e^{-s^2/2} \frac{d^n}{ds^n} e^{s^2/2} \int_{-\infty}^{\infty} e^{-sx} e^{-x^2} \, dx.$$

By using (3.31), we finally get

$$\hat{\mathcal{H}}_n(s) = (-1)^n e^{-s^2/2} \frac{d^n}{ds^n} e^{s^2/2} \sqrt{2\pi} e^{s^2/2} = \sqrt{2\pi} (-1)^n e^{-s^2/2} \frac{d^n}{ds^n} e^{s^2}$$

$$= \sqrt{2\pi} i^n (-1)^n e^{(-is)^2/2} (-1)^n \frac{d^n}{d(-is)^n} e^{-(-is)^2},$$

i.e.,

$$\mathcal{L}[\mathcal{H}_n(t)](s) = \sqrt{2\pi} (-i)^n \mathcal{H}_n(-is) \qquad (3.34)$$

with $-\infty < \Re s < \infty$.

In the sequel we shall need the following property.

Proposition 3.7.

$$\mathcal{L}[\overline{f(t)}](s) = \overline{\mathcal{L}[f(t)](\overline{s})}. \qquad (3.35)$$

The proof is left to the reader.

3.4 Transforms of some Familiar Hyperfunctions

3.4.1 Dirac Impulses and their Derivatives

The Dirac impulse hyperfunction at $t = a$ and its derivatives are defined by

$$\delta^{(n)}(t - a) = \left[-\frac{(-1)^n n!}{2\pi i (z - a)^{n+1}} \right].$$

They are right-sided originals if $a \geq 0$. Thus, their Laplace transforms are

$$\frac{(-1)^n n!}{2\pi i} \int_\infty^{(0+)} \frac{e^{-sz}}{(z-a)^{n+1}} \, dz.$$

The integration loop can be replaced by a circle $c(a, r)$ centered at a and with any radius $r > 0$. The integral is evaluated by the calculus of residues. It gives

$$\frac{1}{2\pi i} \int_{c(a,r)} \frac{e^{-sz}}{(z-a)^{(n+1)}} \, dz = \frac{e^{-as}}{2\pi i} \int_{c(a,r)} \frac{e^{-s(z-a)}}{(z-a)^{(n+1)}} \, dz = e^{-as} \frac{(-s)^n}{n!}.$$

Thus, we obtain for $a > 0$ and unrestricted $\Re s$,

$$\delta^{(n)}(t-a) \circ\!\!-\!\!\bullet\, e^{-as} s^n, \qquad\qquad (3.36)$$

$$\delta(t-a) \circ\!\!-\!\!\bullet\, e^{-as}. \qquad\qquad (3.37)$$

For $a = 0$, we get

$$\delta^{(n)}(t) \circ\!\!-\!\!\bullet\, s^n, \qquad\qquad (3.38)$$

$$\delta(t) \circ\!\!-\!\!\bullet\, 1. \qquad\qquad (3.39)$$

Problem 3.2. If $a \leq 0$ then $\delta^{(n)}(x-a)$ is a left-sided original. Show that its Laplace transform, defined by Definition 3.24, is the same as for $a \geq 0$.

Let us now compute the Laplace transform of the Dirac impulse comb

$$\sum_{k=0}^{\infty} \delta^{(n)}(t - kT).$$

Because the corresponding series of the images converges for $\Re s > 0$, we readily obtain

$$\sum_{k=0}^{\infty} e^{-kTs} s^n = s^n \sum_{k=0}^{\infty} e^{-kTs} = \frac{s^n}{1 - e^{-Ts}}.$$

Thus,

$$\sum_{k=0}^{\infty} \delta^{(n)}(t - kT) \circ\!\!-\!\!\bullet\, \frac{s^n}{1 - e^{-Ts}}, \qquad \Re s > 0 \qquad\qquad (3.40)$$

for $n = 0, 1, 2, \ldots$. Likewise, we obtain

$$\sum_{k=1}^{\infty} \delta^{(n)}(t + kT) \circ\!\!-\!\!\bullet\, \frac{s^n e^{Ts}}{1 - e^{Ts}}, \qquad \Re s < 0$$

which shows that the hyperfunction

$$\sum_{k=-\infty}^{\infty} \delta(t - kT) = \frac{1}{T} \sum_{k=-\infty}^{\infty} e^{2\pi i kt/T}$$

has no Laplace transform because the two strips of convergence have no overlap.

Problem 3.3. Show that for $n = 0, 1, 2, \ldots$ and for $\Re s > 0$ we have

$$\sum_{k=0}^{\infty} \delta^{(n)}(t - (2k+1)T) \circ\!\!-\!\!\bullet \frac{s^n}{2\sinh(Ts)}, \qquad (3.41)$$

$$\sum_{k=0}^{\infty} (-1)^k \delta^{(n)}(t - (2k+1)T) \circ\!\!-\!\!\bullet \frac{s^n}{2\cosh(Ts)}, \qquad (3.42)$$

$$\delta^{(n)}(t) + 2\sum_{k=1}^{\infty} \delta^{(n)}(t - 2kT) \circ\!\!-\!\!\bullet s^n \coth(Ts), \qquad (3.43)$$

$$\delta^{(n)}(t) + 2\sum_{k=1}^{\infty} (-1)^k \delta^{(n)}(t - 2kT) \circ\!\!-\!\!\bullet s^n \tanh(Ts). \qquad (3.44)$$

Problem 3.4. In Example 1.38 we have established the hyperfunction

$$f(t) = \sum_{k=0}^{\infty} \frac{(-1)^k}{k!\,(k+1)!} \delta^{(k)}(x) = \left[-\frac{1}{2\pi i} e^{1/z} \right].$$

Establish its Laplace transform by a term by term transposition and show that

$$\hat{f}(s) = \frac{J_1(2\sqrt{s})}{\sqrt{s}}.$$

Verify this result by using our definition of the Laplace transform of a right-sided original, i.e.,

$$\hat{f}(s) = -\int_{\infty}^{(0+)} e^{-sz} \left(-\frac{1}{2\pi i} \right) e^{1/z}\, dz.$$

Hint: Use the integral representation of the Bessel function

$$J_1(z) = -J_{-1}(z) = -\frac{1}{2\pi i} \int_C e^{z/2(h-h^{-1})}\, dh$$

where C is a circle enclosing the origin.

3.4.2 Non-negative Integral Powers

We take the evaluation of the Laplace transform of $f(t) = \chi_{(0,\infty)} t^m = u(t) t^m$ as a test case for our definition of the Laplace transform of a hyperfunction by means of a loop integral. By using

$$u(t)\, t^m = \left[-\frac{1}{2\pi i} z^m \log(-z) \right],$$

its Laplace transform becomes

$$\mathcal{L}[u(t)\, t^m](s) := -\int_{\infty}^{(0+)} e^{-sz} \frac{(-1)}{2\pi i} z^m \log(-z)\, dz$$

$$= \frac{1}{2\pi i} \int_{\infty}^{(0+)} e^{-sz} z^m \log(-z)\, dz.$$

From equation (3.14) we immediately obtain the result

$$u(t)\, t^m \circ\!\!-\!\!\bullet \frac{m!}{s^{m+1}} \quad (\Re s > 0)$$

in agreement with the familiar result.

Problem 3.5. By using (3.15) show that

$$u(-t)\, t^m \circ\!\!-\!\!\bullet -\frac{m!}{s^{m+1}} \quad (\Re s < 0).$$

3.4.3   Negative Integral Powers

The classical Laplace transform of a negative integral power does not exist, but we shall readily obtain the Laplace transform of the hyperfunction

$$\mathrm{fp}\,\frac{u(t)}{t^m} = \left[-\frac{1}{2\pi i}\frac{\log(-z)}{z^m} \right]$$

where m is a positive integer. By our definition we have

$$\begin{aligned}
\mathcal{L}[\mathrm{fp}\frac{u(t)}{t^m}](s) &:= -\int_\infty^{(0+)} e^{-sz}\frac{(-1)}{2\pi i}z^{-m}\log(-z)dz\\
&= \frac{1}{2\pi i}\int_\infty^{(0+)} e^{-sz}\, z^{-m}\log(-z)dz\\
&= \frac{(-1)^m}{2\pi i}\int_\infty^{(0+)} e^{-sz}\,(-z)^{-m}\log(-z)dz\\
&= (-1)^m\frac{s^{m-1}}{\Gamma(m)}\{\log s - \Psi(m)\},
\end{aligned}$$

where we have used (3.10). Euler's Psi-function has an elementary expression when evaluated at integer values:

$$\Psi(m) = -\gamma + \sum_{k=1}^{m-1}\frac{1}{k}$$

where $\gamma = 0.57721\ldots$ is Euler's constant. Hence, we obtain the correspondence

$$\mathrm{fp}\,\frac{u(t)}{t^m} \circ\!\!-\!\!\bullet (-1)^m\frac{s^{m-1}}{(m-1)!}\{\log s + \gamma - \sum_{k=1}^{m-1}\frac{1}{k}\} \quad (\Re s > 0), \tag{3.45}$$

valid for any positive integer m.

Problem 3.6. Derive

$$\mathcal{L}^{-1}[s^m \log s] = (-1)^{m+1}\, m!\, \mathrm{fp}\frac{u(t)}{t^{m+1}} + \Psi(m+1)\,\delta^{(m)}(t). \tag{3.46}$$

Problem 3.7. Establish by the same method as above

$$\mathrm{fp}\,\frac{u(-t)}{t^m} \circ\!\!-\!\!\bullet -(-1)^m\frac{s^{m-1}}{(m-1)!}\{\log(-s) + \gamma - \sum_{k=1}^{m-1}\frac{1}{k}\} \quad (\Re s < 0). \tag{3.47}$$

3.4.4 Non-integral Powers

The classical Laplace transform of a non-integral power $u(t)t^\alpha$ only exists for $\alpha > -1$. However, we shall see that the Laplace transform of the hyperfunction

$$u(t)t^\alpha = \left[-\frac{(-z)^\alpha}{2i\sin(\alpha\pi)} \right]$$

exists for any non-integral α. By using the definition by a loop integral, we have for the image function

$$\mathcal{L}[u(t)\,t^\alpha](s) = -\int_\infty^{(0+)} e^{-sz}(-1)\frac{(-z)^\alpha}{2i\sin(\alpha\pi)}\,dz$$

$$= \frac{\pi}{\sin(\alpha\pi)}\frac{1}{2\pi i}\int_\infty^{(0+)} e^{-sz}(-z)^\alpha\,dz$$

$$= -\frac{\pi}{\sin(\alpha\pi)}\frac{s^{-\alpha-1}}{\Gamma(-\alpha)} = \frac{\pi}{\sin(-\alpha\pi)}\frac{s^{-\alpha-1}}{\Gamma(-\alpha)}, \quad (\Re s > 0).$$

In the last step we used equation (3.2). By using the reflection formula for the gamma function, we deduce the correspondence

$$u(t)\,t^\alpha \circ\!\!-\!\!\bullet \frac{\Gamma(\alpha+1)}{s^{\alpha+1}}, \quad \alpha \in \mathbb{C}\setminus\mathbb{Z}, \quad \Re s > 0. \tag{3.48}$$

Of course this correspondence also holds for a non-negative integer $\alpha = m$.

Problem 3.8. Show by the same method

$$u(-t)\,|t|^\alpha \circ\!\!-\!\!\bullet -\frac{\Gamma(\alpha+1)}{s^{\alpha+1}}, \quad \alpha \in \mathbb{C}\setminus\mathbb{Z}, \quad \Re s < 0. \tag{3.49}$$

3.4.5 Powers with Logarithms

Let us first compute the Laplace transform of the hyperfunction

$$u(t)\frac{\log t}{t^m} = \left[-\frac{1}{4\pi i}\frac{\log^2(-z)}{z^m} \right].$$

By using (3.18) we obtain for $\lambda = m \in \mathbb{N}$,

$$-\int_\infty^{(0+)} e^{-sz}\left(-\frac{1}{4\pi i}\right)\frac{\log^2(-z)}{z^m}\,dz = \frac{1}{2}\frac{1}{2\pi i}\int_\infty^{(0+)} e^{-sz}\,z^{-m}\log^2(-z)\,dz$$

$$= -\frac{(-1)^m}{2}\frac{s^{m-1}}{(m-1)!}\{(\log s - \Psi(m))^2 - \Psi'(m)\}.$$

Finally, for $m \in \mathbb{N}$, we obtain

$$u(t)\frac{\log t}{t^m} \circ\!\!-\!\!\bullet \frac{(-s)^{m-1}}{2(m-1)!}\{(\log s - \Psi(m))^2 - \Psi'(m)\}. \tag{3.50}$$

For Euler's Psi-function we have for $m \in \mathbb{N}$,

$$\Psi(m) = -\gamma + \sum_{k=1}^{m-1}\frac{1}{k}, \quad \Psi'(m) = \sum_{k=0}^{\infty}\frac{1}{(k+m)^2} = \frac{\pi^2}{6} - \sum_{k=1}^{m-1}\frac{1}{k^2}.$$

Thus, for $m = 1$ and $\Re s > 0$,

$$u(t)\frac{\log t}{t} \circ\!\!-\!\!\bullet\, \frac{1}{2}\left\{(\log s + \gamma)^2 - \frac{\pi^2}{6}\right\}. \tag{3.51}$$

Problem 3.9. Establish

$$\mathcal{L}^{-1}[s^m \log^2 s](t) = 2(-1)^m\, m!\left\{\log t - \Psi(m+1)\right\} \mathrm{fp}\frac{u(t)}{t^{m+1}} \tag{3.52}$$
$$+ \left\{\Psi^2(m+1) + \Psi'(m+1)\right\}\delta^{(m)}(t).$$

Problem 3.10. Establish for $\Re s < 0$,

$$u(-t)\frac{\log|t|}{t^m} \circ\!\!-\!\!\bullet\, -\frac{(-s)^{m-1}}{2(m-1)!}\left\{(\log(-s) - \Psi(m))^2 - \Psi'(m)\right\}. \tag{3.53}$$

Next let us compute the Laplace transform of $u(t)|t|^\alpha \log|t| = u(t)t^\alpha \log t$ where α is non-integral. We have (see Section 1.6.4),

$$u(t)t^\alpha\,\log t = \left[-\frac{(-z)^\alpha}{2i\sin(\pi\alpha)}\left\{\log(-z) - \pi\cot(\pi\alpha)\right\}\right].$$

The Laplace transform becomes

$$\frac{1}{2i\sin(\pi\alpha)}\left\{\int_\infty^{(0+)} e^{-sz}(-z)^\alpha\,\log(-z)\,dz - \pi\cot(\pi\alpha)\int_\infty^{(0+)} e^{-sz}(-z)^\alpha\,dz\right\}$$

with $\Re s > 0$. By using (3.10) the first term is evaluated to

$$\frac{2\pi i}{2i\sin(\pi\alpha)}\frac{s^{-\alpha-1}}{\Gamma(-\alpha)}\left\{\log s - \Psi(-\alpha)\right\},$$

and by (3.2) the second term equals

$$-\frac{2\pi i}{2i\sin(\pi\alpha)}\pi\cot(\pi\alpha)(-1)\frac{s^{-\alpha-1}}{\Gamma(-\alpha)}.$$

Adding the two terms yields

$$\frac{\pi}{\sin(\pi\alpha)\Gamma(-\alpha)}\frac{1}{s^{\alpha+1}}\left\{\log s - \Psi(-\alpha) + \pi\cot(\pi\alpha)\right\}.$$

By using the reflection formula for the Gamma function, we obtain

$$-\frac{\Gamma(\alpha+1)}{s^{\alpha+1}}\left\{\log s - \Psi(-\alpha) + \pi\cot(\pi\alpha)\right\}.$$

By the formula $\Psi(z) = \Psi(1-z) - \pi\cot(\pi z)$ we get $\Psi(-\alpha) = \Psi(1+\alpha) + \pi\cot(\pi\alpha)$. Therefore, if α is not an integral number, we have for $\Re s > 0$,

$$u(t)t^\alpha\,\log t \circ\!\!-\!\!\bullet\, \frac{\Gamma(\alpha+1)}{s^{\alpha+1}}\left\{\Psi(\alpha+1) - \log s\right\}. \tag{3.54}$$

The correspondence was derived under the assumption of a non-integral α. It can be shown, by induction on n, and starting from the correspondence $u(t) \log t \circ - \bullet - (\log s + \gamma)/s$, together with the multiplication rule of Section 3.5.3, that it remains valid, if α tends to a non-negative integer n. Thus, for $n \in \mathbb{N}_0$, we have for $\Re s > 0$,

$$u(t)t^n \log t \circ - \bullet \frac{n!}{s^{n+1}} \{\sum_{k=1}^{n} \frac{1}{k} - \gamma - \log s\}. \tag{3.55}$$

Problem 3.11. Establish for $\alpha \notin \mathbb{Z}$,

$$\frac{1}{\Gamma(-\alpha)} u(t) \frac{\Psi(-\alpha) - \log t}{t^{\alpha+1}} \circ - \bullet s^\alpha \log s. \tag{3.56}$$

3.4.6 Exponential Integrals

Exponential integrals come up frequently in the calculation of Laplace transforms of certain hyperfunctions. Hence, let us gather some facts about this subject (see, for example, [29, Olver] for details). This is also necessary because we have the situation that the notation for the various exponential integrals is not uniformly established and may vary from one author to another. We follow the notation used in [1, Abramowitz-Stegun].

The function *Exponential Integral* $E_1(z)$ is defined in the complex plane with a cut along the negative real axis, i.e., for $|\arg z| < \pi$, by

$$E_1(z) := \int_z^\infty \frac{e^{-t}}{t} dt. \tag{3.57}$$

In the defining integral it is assumed that the integration variable t approaches infinity within the sector $-\pi/2 < \arg t < \pi/2$ as $|t|$ becomes sufficiently large. The point $t = 0$ is a pole of first order of the integrand, hence $z = 0$ will be a branch point of $E_1(z)$ analogous to $\log(z)$. The principal branch of $E_1(z)$ is obtained by the mentioned cut along the negative real line. By a change of variables and restricting the independent variable z to the right-half plane, we readily obtain another integral representation with a fixed integration interval

$$E_1(z) = e^{-z} \int_0^\infty \frac{e^{-zt}}{t+1} dt, \quad |\arg z| < \frac{\pi}{2}.$$

Thus, by the definition of the classical Laplace transform, we obtain

$$Y(t) \frac{1}{t+1} \circ - \bullet e^s E_1(s), \quad \Re s > 0,$$

and, by the similarity rule for the right-sided (ordinary) Laplace transform (see, for example, [14, Graf Chapter 1]),

$$Y(t) \frac{1}{t+b} \circ - \bullet e^{bs} E_1(bs), \quad \Re s > 0, b > 0. \tag{3.58}$$

Also, by the t-shift rule,

$$\frac{Y(t-b)}{t} \circ - \bullet E_1(bs), \quad \Re s > 0, b > 0. \tag{3.59}$$

Another change of variables yields the representation

$$E_1(z) = \int_1^\infty \frac{e^{-zt}}{t}\, dt, \quad |\arg z| < \frac{\pi}{2}$$

which leads to the *Generalized Exponential Integral*, defined by

$$E_n(z) := \int_1^\infty \frac{e^{-zt}}{t^n}\, dt, = e^{-z} \int_0^\infty \frac{e^{-zt}}{(t+1)^n}\, dt, \tag{3.60}$$

where $n = 1, 2, 3, \ldots$. Integration py parts gives at once the recursion formula

$$E_{n+1}(z) := \frac{1}{n}\{e^{-z} - z\, E_n(z)\}, \quad \Re z > 0.$$

This shows that

  (i) $E_n(z)$ is defined for $|\arg z| < \pi/2$,

  (ii) $E_n(z)$ has a branch point at $z = 0$.

Problem 3.12. Use the recursion formula to show that for $n = 2, 3, \ldots$ we have

$$E_n(z) = p_{n-2}(z)\, e^{-z} + \frac{(-z)^{n-1}}{(n-1)!} E_1(z), \tag{3.61}$$

where $p_{n-2}(z)$ is a polynomial of degree $n - 2$.

A change of variables leads to the correspondences

$$\frac{u(t)}{(t+b)^n} \circ\!\!-\!\!\bullet \frac{1}{b^{n-1}} e^{bs}\, E_n(bs), \quad \Re s > 0, b > 0, \tag{3.62}$$

and again, by the t-shift rule,

$$\frac{u(t-b)}{t^n} \circ\!\!-\!\!\bullet \frac{1}{b^{n-1}} E_n(bs), \quad \Re s > 0, b > 0. \tag{3.63}$$

We also need the *Complementary Exponential Integral*, defined by

$$Ein(z) := \int_0^z \frac{1 - e^{-t}}{t}\, dt. \tag{3.64}$$

It is an entire function, and integrating term by term gives the power series expansion

$$Ein(z) = \sum_{k=1}^\infty \frac{(-1)^{k-1} z^k}{k\, k!}.$$

The Laplace transform is obtained by using the image integration rule for the right-sided Laplace transform (see, for example, [14, Graf Chapter 1])

$$Y(t)\, Ein(t) \circ\!\!-\!\!\bullet \frac{1}{s} \log\!\left(1 + \frac{1}{s}\right), \quad \Re s > 0. \tag{3.65}$$

By a decomposition of the contour of the defining integral of the complementary exponential integral into the path $0 \to 1 \to z$, and by using the integral representation of Euler's constant

$$\gamma = \int_0^1 \frac{1 - e^{-t}}{t} \, dt - \int_1^\infty \frac{e^{-t}}{t} \, dt$$

(see [29, Olver p.40]) the relation

$$\text{Ein}(z) = E_1(z) + \log z + \gamma \tag{3.66}$$

is established. It is valid for all z of the complex plane with a cut along the negative real axis. As a by-product we obtain the power series expansion of the exponential integral, valid for $|\arg z| < \pi$,

$$E_1(z) = -\log z - \gamma - \sum_{k=1}^\infty \frac{(-1)^k z^k}{k \, k!}. \tag{3.67}$$

By using the result of the above problem we see that the structure of $E_n(z)$ is given by

$$E_n(z) = -\frac{(-z)^{n-1}}{(n-1)!} \log z + h(z) \tag{3.68}$$

where $h(z)$ is an entire function of z. Because $E_1(-z) - (-\log(-z)) = \text{Ein}(-z) - \gamma$ is real analytic the unit-step hyperfunction $u(t)$ can also be represented by

$$u(t) = \left[-\frac{1}{2\pi i} \log(-z) \right] = \left[\frac{1}{2\pi i} E_1(-z) \right].$$

Another Exponential Integral is often used. For $x > 0$ it is defined as a Cauchy principal value integral,

$$\text{Ei}(x) := \text{pv} \int_{-\infty}^x \frac{e^t}{t} \, dt, \quad x > 0. \tag{3.69}$$

By the change of variables $t = s(x - \tau)$, $s > 0$, we can express this integral in the form

$$-e^{-sx} \text{Ei}(sx) = \text{pv} \int_0^\infty e^{-s\tau} \frac{d\tau}{\tau - x} \, d\tau = \mathcal{L}[\frac{Y(t)}{t - x}](s), \tag{3.70}$$

provided that the Laplace integral is taken as a principal-value integral.

Example 3.9. Let us compute the Laplace transform of $f(t) = Y(t)/(t^2 - x^2)$ where x is a positive parameter. Because

$$\frac{1}{t^2 - x^2} = \frac{1}{2x} \{ \frac{1}{t - x} - \frac{1}{t + x} \}$$

we have

$$\mathcal{L}[f(t)](s) = \frac{1}{2x} \{ \mathcal{L}[\frac{u(t)}{t - x}](s) - \mathcal{L}[\frac{u(t)}{t + x}](s) \}.$$

By using (3.70) and (3.58), we obtain

$$Y(t) \frac{1}{t^2 - x^2} \circ\!\!-\!\!\bullet - \frac{1}{2x} \{ e^{-xs} \text{Ei}(xs) + e^{xs} E_1(xs) \}. \tag{3.71}$$

Note that for $x > 0$ we have $E_1(x) = -\text{Ei}(-x)$. By integration along a suitable contour in the complex plane the relation $E_1(-x \pm i0) = -\text{Ei}(x) \mp i\pi$ can be established. This yields

$$\text{Ei}(x) = -\frac{1}{2}\{E_1(-x + i0) + E_1(-x - i0)\}, \quad x > 0. \tag{3.72}$$

Observe that we may write $-\text{Ei}(x) = [E_1(-z)/2, -E_1(-z)/2]$. Using equation (3.67) we can write

$$-E_1(-z) = \log(-z) + \gamma + \sum_{k=1}^{\infty} \frac{z^k}{k\,k!},$$

now valid for z non-positive and real. For $z = x \pm i0, x > 0$, we then have

$$-E_1(-x \mp i0) = \log(-x \mp i0) + \gamma + \sum_{k=1}^{\infty} \frac{x^k}{k\,k!}.$$

In the complex plane, with a cut along the negative real axis, the relation

$$\log(-z) = \log z - i\,\text{sgn}(\Im z)\pi$$

holds, thus

$$\log(-x \mp i0) = \log(x) \mp i\pi.$$

Therefore, we have

$$-E_1(-x \mp i0) \pm i\pi = \log x + \gamma + \sum_{k=1}^{\infty} \frac{x^k}{k\,k!} \quad x > 0,$$

and, by using (3.72), we obtain

$$\text{Ei}(x) = \log x + \gamma + \sum_{k=1}^{\infty} \frac{x^k}{k\,k!},$$

first valid for $x > 0$. By analytic continuation we obtain the power series

$$\text{Ei}(z) = \log z + \gamma + \sum_{k=1}^{\infty} \frac{z^k}{k\,k!} \tag{3.73}$$

valid for $|\arg z| < \pi$. The power series (3.73) defines the *Exponential Integral* Ei in the complex plane with a cut along the negative real axis.

Problem 3.13. By using (3.73) and (3.67) establish

$$-E_1(ze^{\pm i\pi}) = \text{Ei}(z) \pm i\,\pi,$$
$$\text{Ei}(ze^{\pm i\pi}) = -E_1(z) \pm i\,\pi.$$

Let us introduce the *complementary Exponential* Integral* by

$$\text{Ein}^*(n, z) := z^{n-1} \int_0^z \frac{1}{t^n}\left(e^t - \sum_{k=0}^{n-1} \frac{t^k}{k!}\right) dt \quad n = 1, 2, \dots. \tag{3.74}$$

It is an entire function, and $\text{Ein}^*(1, z) = -\text{Ein}(-z)$. The reader may check for himself that

$$\text{Ein}^*(n, z) = \sum_{k=1}^{\infty} \frac{z^{n+k-1}}{k\,(n+k-1)!}. \tag{3.75}$$

Integration by parts yields the recursion formula

$$\text{Ein}^*(n+1, z) = n\left(z^n\,\text{Ein}^*(n, z) + \sum_{k=0}^{n} \frac{z^k}{k!} - e^z\right), \tag{3.76}$$

$$\text{Ein}^*(1, z) = \text{Ei}(z) - \gamma - \log z. \tag{3.77}$$

For example,

$$\text{Ein}^*(2, z) = z(\text{Ei}(z) - \gamma - \log z) + 1 + z - e^z. \tag{3.78}$$

Let us return to the generalized Exponential Integral $E_n(t)$. If we denote the Laplace transform of $E_n(t)$ by $\hat{E}_n(s)$, the recursion formula in the t-domain implies the recursion formula for the Laplace transforms:

$$\hat{E}_{n+1}(s) = \frac{1}{n}\{\frac{1}{s+1} + \frac{d}{ds}\hat{E}_n(s)\}.$$

From the equations (3.65), (3.66) and the known correspondence

$$Y(t)\,\log t \circ\!\!-\!\!\bullet\; -\frac{1}{s}(\gamma + \log s)$$

we obtain the Laplace transform of $E_1(t)$:

$$Y(t)\,E_1(t) \circ\!\!-\!\!\bullet\; \frac{1}{s}\,\log(1+s), \quad \Re s > 0.$$

By using the recursion formula for the image functions, one obtains

$$Y(t)\,E_2(t) \circ\!\!-\!\!\bullet\; \frac{1}{s} - \frac{1}{s^2}\,\log(1+s) \quad \Re s > 0.$$

Iterating the process yields

$$Y(t)\,E_n(t) \circ\!\!-\!\!\bullet\; \frac{(-1)^{n-1}}{s^n}\{\log(1+s) + \sum_{k=1}^{n-1} \frac{(-1)^k\,s^k}{k} \quad \Re s > 0. \tag{3.79}$$

Problem 3.14. Establish

$$Y(t)\,\text{Ei}(t) \circ\!\!-\!\!\bullet\; -\frac{1}{s}\,\log(s-1), \quad \Re s > 1.$$

Another special function is the function erg defined by

$$\text{erg}(z) := \frac{2}{\sqrt{\pi}}\int_0^z e^{t^2}\,dt. \tag{3.80}$$

It is an entire function of the complex variable z. Often used is a variant of it, namely the function $\text{erg}\sqrt{az}$ which has the power series expansion

$$\text{erg}\sqrt{az} = 2\sqrt{\frac{az}{\pi}}\sum_{k=0}^{\infty} \frac{a^k}{(2k+1)k!}\,z^k.$$

Problem 3.15. Establish the inverse Laplace transform

$$\mathcal{L}^{-1}\Big[\frac{\log(1+s)}{s^n}\Big](t) = (-1)^{n-1}\, Y(t)\,\Big\{E_n(t) - \sum_{k=0}^{n-2}\frac{(-1)^k\, t^k}{(n-1-k)\, k!}\Big\}$$

valid for integral $n > 1$, and $\Re s > 0$.

Problem 3.16. Establish the power series expansion of the general exponential integral

$$E_n(z) = \frac{(-z)^{n-1}}{(n-1)!}\left[-\log z + \Psi(n)\right] - \sum_{k=0,\, k\neq n-1}^{\infty}\frac{(-1)^k\, z^k}{(k-n+1)\, k!}.$$

Problem 3.17. Establish the correspondence

$$Y(t)\,\frac{\sinh\sqrt{at}}{t} \quad\circ\!-\!\bullet\quad \pi\,\mathrm{erg}\sqrt{\frac{a}{4s}},\quad \Re s > 0,$$

where $a > 0$. Hint: Use

$$\Gamma\Big(k + \frac{1}{2}\Big) = \sqrt{\pi}\,\frac{(2k)!}{2^{2k}\, k!}$$

and term by term transposition. By the same technique also establish

$$Y(t)\,\frac{\cosh\sqrt{at}}{\sqrt{t}} \quad\circ\!-\!\bullet\quad \sqrt{\frac{\pi}{s}}\, e^{a/(4s)},\quad \Re s > 0.$$

3.4.7   Transforms of Finite Part Hyperfunctions

First, let us remember the discussion in Section 1.6.2, especially the result (1.89), which shows that

$$\mathrm{fp}\,\frac{u(at)}{t^m},\quad \mathrm{fp}\,\frac{u(t)}{t^m},$$

and also

$$\mathrm{fp}\,\frac{u(t)}{(t-b)^m},\quad \mathrm{fp}\,\frac{u(t/b)}{(t-b)^m}\quad (b > 0)$$

are not the same hyperfunctions. They are only homologous for they interpret the same ordinary functions

$$\frac{Y(t)}{t^m},\quad \frac{Y(t)}{(t-b)^m},$$

but differ by a generalized delta-hyperfunction (see Section 2.3.4).

Example 3.10. We compute the Laplace transform of $\mathrm{fp}\,u(t/b)/(t-b)^m$ $(m \in \mathbb{N},\ b > 0)$. With

$$\mathrm{fp}\,\frac{u(t/b)}{(t-b)^m} = \Big[-\frac{1}{2\pi i}\frac{\log(-z/b)}{(z-b)^m}\Big],$$

the image function is defined by

$$\frac{1}{2\pi i}\int_{\infty}^{(0+)} e^{-sz}\,\frac{\log(-z/b)}{(z-b)^m}\, dz.$$

which becomes, by the change of variables $z - b = \zeta$,

$$\frac{1}{2\pi i} \int_\infty^{(-b+)} e^{-s(\zeta+b)} \frac{\log(-\zeta/b - 1)}{\zeta^m} d\zeta.$$

This can be written as

$$\frac{e^{-bs}}{2\pi i} \int_\infty^{(-b+)} e^{-s\zeta} \frac{\log[(-\zeta/b)(1 + b/\zeta)]}{\zeta^m} d\zeta.$$

Separating the log-term we obtain

$$\frac{e^{-bs}}{2\pi i} \int_\infty^{(-b+)} e^{-s\zeta} \frac{\log(-\zeta/b)}{\zeta^m} d\zeta + \frac{e^{-bs}}{2\pi i} \int_\infty^{(-b+)} e^{-s\zeta} \frac{\log(1 + b/\zeta)}{\zeta^m} d\zeta.$$

In the first integral the loop turning around $-b$ can be replaced by a loop turning around 0 due to the fact that the integrand is holomorphic for negative ζ. Thus, the above expression becomes

$$\frac{e^{-bs}}{2\pi i} \int_\infty^{(0+)} e^{-s\zeta} \frac{\log(-\zeta/b)}{\zeta^m} d\zeta + \frac{e^{-bs}}{2\pi i} \int_\infty^{(-b+)} e^{-s\zeta} \frac{\log(1 + b/\zeta)}{\zeta^m} d\zeta.$$

By using (3.10) the first expression evaluates to

$$e^{-bs} (-1)^m \frac{s^{m-1}}{(m-1)!} \{\log(bs) - \Psi(m)\},$$

where $\Psi(m) = \sum_1^{m-1} 1/k - \gamma$. For the second integral, we first replace the infinite contour $(-b, \infty)$ by the finite contour $(-b, 0+)$, then deform it such that $b < |\zeta|$ holds everywhere on it. Expansion of the log term in a series followed by term wise integration yields

$$e^{-bs} \sum_{k=1}^\infty \frac{(-1)^{k-1} b^k}{k} \frac{1}{2\pi i} \int_{b<|\zeta|} e^{-s\zeta} \zeta^{-m-k} d\zeta.$$

The integral is evaluated by the calculus of residues and gives

$$\frac{(-s)^{m+k-1}}{(m+k-1)!}.$$

Thus, referring to (3.75), the second integral becomes

$$(-1)^m s^{m-1} e^{-bs} \sum_{k=1}^\infty \frac{(bs)^k}{k(m+k-1)!} = (-1)^m b^{1-m} e^{-bs} \operatorname{Ein}^*(m, bs).$$

Finally, we obtain

$$\operatorname{fp}\frac{u(t/b)}{(t-b)^m} \circ\!\!-\!\!\bullet (-1)^m e^{-bs}\left\{\frac{s^{m-1}}{(m-1)!} [\log(bs) - \Psi(m)]\right. \tag{3.81}$$
$$\left. + b^{1-m} \operatorname{Ein}^*(m, bs)\right\}.$$

For $m = 1$, using (3.77), we get

$$\text{fp}\frac{u(t/b)}{t-b} \circ\!-\!\bullet \; -e^{-bs}\,\text{Ei}(bs). \tag{3.82}$$

For $m = 2$, we have,

$$\text{fp}\frac{u(t/b)}{(t-b)^2} \circ\!-\!\bullet \; bs\,e^{-bs}\,\{\text{Ei}(bs) + (\tfrac{1}{b} - 1)(\gamma + \log(b\,s) - 1)\} - 1 + e^{-bs}. \tag{3.83}$$

Problem 3.18. Show that

$$\text{fp}\frac{u(t)}{(t-b)^m} \circ\!-\!\bullet \; (-1)^m\,e^{-bs}\{\frac{s^{m-1}}{(m-1)!}\,[\log s - \Psi(m)] \\ + b^{1-m}\,\text{Ein}^*(m,bs)\}, \tag{3.84}$$

$$\text{fp}\frac{u(t)}{t-b} \circ\!-\!\bullet \; -e^{-bs}\,\{\text{Ei}(bs) - \log b\}. \tag{3.85}$$

Example 3.11. Let us now compute the Laplace transform of the right-sided original

$$\chi_{(0,b)}\text{fp}\frac{1}{(t-b)^n} = \left[-\frac{1}{2\pi i\,(z-b)^n}\,\log\frac{z}{z-b}\right].$$

Its Laplace transform is

$$\frac{1}{2\pi i}\int_\infty^{(0+)} e^{-sz}\,\frac{1}{(z-b)^n}\,\log\frac{z}{z-b}\,dz.$$

Because the integrand is holomorphic in the complex plane with a cut from 0 to b, the infinite loop can be replace by a finite contour

$$\frac{1}{2\pi i}\oint_{\gamma_0} e^{-sz}\,\frac{1}{(z-b)^n}\,\log\frac{z}{z-b}\,dz,$$

where the closed contour γ_0 encircles the segment $[0,b]$ in the positive sense. By a change of variables $u = z - b$, the integral becomes

$$\frac{e^{-bs}}{2\pi i}\int_{\gamma_1} e^{-su}\,\frac{1}{u^n}\,\log\frac{u+b}{u}\,du,$$

where the closed contour γ_1 now encircles the segment $[-b,0]$. This contour can be replaced by a circle γ_2 centered at 0 and with a radius greater than $|b|$ implying $|u| > |b|$ everywhere on this contour. The series expansion

$$\log\frac{u+b}{u} = \log(1 + \frac{b}{u}) = -\sum_{k=1}^\infty \frac{(-1)^k\,b^k}{k\,u^k}$$

converges uniformly in u on γ_2. Consequently, the series can be integrated term wise, and the Laplace transform becomes

$$-e^{-bs}\sum_{k=1}^\infty \frac{(-1)^k\,b^k}{k}\,\frac{1}{2\pi i}\int_{\gamma_2}\frac{e^{-su}}{u^{n+k}}\,du.$$

By the calculus of residues we obtain

$$\frac{1}{2\pi i} \int_{\gamma_2} \frac{e^{-su}}{u^{n+k}} \, du = \frac{(-s)^{n+k-1}}{(n+k-1)!},$$

and finally,

$$(-1)^n e^{-bs} b^{1-n} \sum_{k=1}^{\infty} \frac{(bs)^{n+k-1}}{k(n+k-1)!} = (-1)^n e^{-bs} b^{1-n} \operatorname{Ein}^*(n, bs).$$

Therefore, for $\Re s > 0$, we have the result

$$\chi_{(0,b)} \operatorname{fp} \frac{1}{(t-b)^n} \circ\!\!-\!\!\bullet (-1)^n e^{-bs} b^{1-n} \operatorname{Ein}^*(n, bs), \tag{3.86}$$

for $n = 1$,

$$\chi_{(0,b)} \operatorname{fp} \frac{1}{b-t} \circ\!\!-\!\!\bullet e^{-bs} \{\operatorname{Ei}(bs) - \gamma - \log(bs)\}. \tag{3.87}$$

Problem 3.19. Establish

$$\chi_{(0,b)} \operatorname{fp} \frac{1}{(t-b)^2} \circ\!\!-\!\!\bullet se^{-bs} \{\operatorname{Ei}(bs) - \gamma - \log(bs) + 1\} - \frac{1}{b} + \frac{1}{b} e^{-bs}. \tag{3.88}$$

Example 3.12. Next, we compute the Laplace transform of the right-sided original

$$\chi_{(0,b)} \operatorname{fp} \frac{1}{t^n} = \left[-\frac{1}{2\pi i \, z^n} \log \frac{z}{z-b} \right].$$

Its Laplace transform is

$$\frac{1}{2\pi i} \int_{\infty}^{(0+)} e^{-sz} \frac{1}{z^n} \log \frac{z}{z-b} \, dz.$$

As in the last example, the loop integral can be replace by the contour integral

$$\frac{1}{2\pi i} \int_{\gamma_0} e^{-sz} \frac{1}{z^n} \log \frac{z}{z-b} \, dz,$$

where the closed contour γ_0 encircles the segment $[0, b]$ in the positive sense. We replace the contour by a circle centered at 0 and a radius greater than $|b|$ and expand the log term

$$\log \frac{z}{z-b} = \sum_{k=1}^{\infty} \frac{b^k}{k \, z^k}.$$

We then obtain, after an interchange of summation and integration,

$$\sum_{k=1}^{\infty} \frac{b^k}{k} \frac{1}{2\pi i} \int_{\gamma_0} \frac{e^{-sz}}{z^{k+n}} \, dz.$$

By the calculus of residues the integral gives

$$\frac{1}{2\pi i} \int_{\gamma_0} \frac{e^{-sz}}{z^{k+n}} \, dz = \frac{(-s)^{k+n-1}}{(k+n-1)!},$$

and we obtain

$$b^{1-n} \sum_{k=1}^{\infty} \frac{(-bs)^{k+n-1}}{k \, (k+n-1)!}.$$

A change of the summation index: $\ell = k + n - 1$, and replacing ℓ again by k, we get

$$b^{1-n} \sum_{k=n}^{\infty} \frac{(-bs)^k}{(k-n+1)\, k!}.$$

This series can be rearranged and becomes

$$b^{1-n} \{ \sum_{k=1, k \neq n-1}^{\infty} \frac{(-bs)^k}{(k-n+1)\, k!} - \sum_{k=1}^{n-2} \frac{(-bs)^k}{(k-n+1)\, k!} \}.$$

Now we use the power series expansion of the exponential integral $E_n(bs)$ and obtain for $\Re s > 0$, $n = 2, 3, \cdots$,

$$\chi_{(0,b)} \mathrm{fp} \frac{1}{t^n} \circ\!\!-\!\!\bullet \; - \frac{1}{b^{n-1}} \{ E_n(bs) + \sum_{k=1}^{n-2} \frac{(-bs)^k}{(k-n+1)\, k!} \}$$
$$+ \frac{(-s)^{n-1}}{(n-1)!} \{ \Psi(n) - \log(bs) \}. \tag{3.89}$$

For $n = 1$, using $\Psi(1) = -\gamma$, we have

$$\chi_{(0,b)} \mathrm{fp} \frac{1}{t} \circ\!\!-\!\!\bullet \; - \{ E_1(bs) - \log(bs) + \gamma \}. \tag{3.90}$$

Example 3.13. We want to find the Laplace transform of the right-sided hyperfunction

$$f(t) := \mathrm{fp} \frac{u(t) J_0(at)}{t} = \left[-\frac{1}{2\pi i} J_0(az) \frac{\log(-z)}{z} \right], \quad a > 0$$

where $J_0(z)$ is the Bessel function of order 0. By the definition of the Laplace transform, we have

$$\hat{f}(s) = \frac{1}{2\pi i} \int_{\infty}^{(0+)} e^{-sz} \frac{J_0(az)}{z} \log(-z) dz,$$

where $\log$ denotes the principal branch, i.e., $\log$ is the singled-valued function defined on the complex plane with a cut along the negative real axis such that $\log 1 = 0$. We now use the known correspondence

$$Y(t) J_0(at) \circ\!\!-\!\!\bullet \; \frac{1}{\sqrt{a^2 + s^2}}$$

valid for $\Re s > 0$. This can be written as

$$\frac{1}{\sqrt{a^2 + p^2}} = \frac{1}{2\pi i} \int_{\infty}^{(0+)} e^{-pz} J_0(az) \log(-z) dz.$$

Then, from the previous example

$$\mathcal{L}[\text{fp}\frac{u(t)}{t^m}e^{-at}](s) = (-1)^m\frac{(s+a)^{m-1}}{(m-1)!}\{\log(s+a)+\gamma-\sum_{k=1}^{m-1}\frac{1}{k}\}$$

for $\Re s > -\Re a$, and

$$\mathcal{L}[\text{fp}\frac{u(-t)}{t^m}e^{at}](s) = -(-1)^m\frac{(s-a)^{m-1}}{(m-1)!}\{\log(-s+a)+\gamma-\sum_{k=1}^{m-1}\frac{1}{k}\}$$

for $\Re s < \Re a$. Thus, in the vertical strip of convergence $-\Re a < \Re s < \Re a$ we have,

$$\mathcal{L}\left[\text{fp}\frac{u(t)}{t^m}e^{-at}\right](s) + \mathcal{L}\left[\text{fp}\frac{u(-t)}{t^m}e^{at}\right](s)$$

$$= \frac{(-1)^m}{(m-1)!}\{(s+a)^{m-1}\log(s+a)+(s+a)^{m-1}(\gamma-\sum_{k=1}^{m-1}\frac{1}{k}) \qquad (3.100)$$

$$- (s-a)^{m-1}\log(-s+a)-(s-a)^{m-1}(\gamma-\sum_{k=1}^{m-1}\frac{1}{k})\}.$$

For $m = 1, 2$ we have for $-\Re a < \Re s < \Re a$,

$$\text{fp}\frac{e^{-a|t|}}{t} \circ\!\!-\!\!\bullet \log\frac{a-s}{a+s}, \qquad (3.101)$$

$$\text{fp}\frac{e^{-a|t|}}{t^2} \circ\!\!-\!\!\bullet s\log\frac{a+s}{a-s}+a\{\log((a^2-s^2)+2(\gamma-1)\}. \qquad (3.102)$$

The Laplace transform of fp $\exp(-a|t|)/t^m$, $\Re a > 0$ has a unique image function given above. Which is the Laplace transform of $t^{-m}\exp(-a|t|)$, $\Re a > 0$, i.e., without the fp. We could rely on the discussion in Section 1.4 and use (1.64) and write down immediately the result, however it will be instructive to open again the question from the start. From $f(t) = t^{-m}\exp(-a|t|) = [F(z)]$ it follows for $t \ne 0$ that we must have $t^m f(t) = \exp(-a|t|)$. Because

$$e^{-a|t|} := u(-t)e^{at}+u(t)e^{-at} = \left[\frac{e^{az}}{2\pi i}\log z - \frac{e^{-az}}{2\pi i}\log(-z)+\phi(z)\right]$$

where $\phi(x)$ is an arbitrary real analytic function. Thus, we must have

$$z^m F(z) = \frac{e^{az}}{2\pi i}\log z - \frac{e^{-az}}{2\pi i}\log(-z)+\phi(z),$$

hence,

$$F(z) = \frac{e^{az}}{2\pi i\, z^m}\log z - \frac{e^{-az}}{2\pi i\, z^m}\log(-z)+\frac{\phi(z)}{z^m},$$

therefore,

$$f(t) = t^{-m}e^{-a|t|} = e^{at}\,\text{fp}\frac{u(-t)}{t^m}+e^{-at}\,\text{fp}\frac{u(t)}{t^m}+\delta_{(m-1)}(t),$$

where $\delta_{(m-1)}(t)$ is an arbitrary generalized delta-hyperfunction of order $m-1$. Taking the Laplace transform, we arrive at

$$\mathcal{L}[t^{-m}\,e^{-a|t|}](s) = \mathcal{L}[e^{at}\,\mathrm{fp}\frac{u(-t)}{t^m}](s) + \mathcal{L}[e^{-at}\,\mathrm{fp}\frac{u(-t)}{t^m}](s) + p_{m-1}(s), \quad (3.103)$$

where $p_{m-1}(s)$ is now a polynomial in s of degree $m-1$ with m arbitrary coefficients. By the image translation rule, we then have

$$\mathcal{L}[t^{-m}\,e^{-a|t|}](s) = \mathcal{L}[\mathrm{fp}\frac{u(-t)}{t^m}](s-a) + \mathcal{L}[\mathrm{fp}\frac{u(-t)}{t^m}](s+a) + p_{m-1}(s).$$

Finally, we obtain for $-a < \Re s < a$,

$$
\begin{aligned}
t^{-m}\,e^{-a|t|} \;\circ\!\!-\!\!\bullet\; & \frac{(-1)^m}{(m-1)!}\,(\gamma - \sum_{k=1}^{m-1}\frac{1}{k})\,\{(s+a)^{m-1} - (s-a)^{m-1}\} \\
& + \frac{(-1)^m}{(m-1)!}\,\{(s+a)^{m-1}\,\log(s+a) - (s-a)^{m-1}\,\log(s-a)\} \\
& + p_{m-1}(s).
\end{aligned}
\qquad (3.104)
$$

Problem 3.24. Establish the correspondences ($\Re s > \Re a$)

$$\mathrm{fp}\frac{u(t)\,\cosh(at)\,\cos(\omega t)}{t} \;\circ\!\!-\!\!\bullet\; -\frac{1}{4}\,\log[(s^2-a^2)^2 + 2\omega^2\,(s^2+a^2) + \omega^4] - \gamma,$$

$$\mathrm{fp}\frac{u(t)\,\cosh(at)\,\cos(at)}{t} \;\circ\!\!-\!\!\bullet\; -\frac{1}{4}\,\log[(s^4 + 4a^4] - \gamma.$$

Thus,

$$\mathcal{L}^{-1}\left[\log[(s^4 + a^4]\right](t) = -4\gamma\,\delta(t) - 4\,\mathrm{fp}\frac{u(t)\,\cosh(at/\sqrt{2})\,\cos(at/\sqrt{2})}{t}.$$

Problem 3.25. Establish the correspondences ($\Re s > \omega > 0$)

$$\mathrm{fp}\frac{u(t)\,\sin^2(\omega t)}{t} \;\circ\!\!-\!\!\bullet\; \frac{1}{4}\,\log(1 + 4\frac{\omega^2}{s^2}),$$

$$\mathrm{fp}\frac{u(t)\,\cos^2(\omega t)}{t} \;\circ\!\!-\!\!\bullet\; -\frac{1}{4}\,\log[s^2(s^2 + 4\omega^2)] - \gamma.$$

Problem 3.26. Establish the following inverse Laplace transforms, valid for $m = 0, 1, 2, \ldots$.

$$\mathcal{L}^{-1}[(s+a)^m\,\log(s+a)](t)$$

$$= -(-1)^m m!\,\mathrm{fp}\frac{u(t)}{t^{m+1}}e^{-at} + \Phi(m+1)\sum_{k=0}^{m}\binom{m}{k}a^{m-k}\,\delta^{(k)}(t), \quad (\Re s > -\Re a),$$

$$\mathcal{L}^{-1}[(a-s)^m\,\log(a-s)](t)$$

$$= m!\,\mathrm{fp}\frac{u(-t)}{t^{m+1}}e^{at} + \Phi(m+1)\sum_{k=0}^{m}(-1)^k\binom{m}{k}a^{m-k}\,\delta^{(k)}(t), \quad (\Re s < \Re a).$$

3.5.3 The Multiplication or Image Differentiation Rule

Proposition 3.9. *Let $f(t) = [F(z)] \circ\!\!-\!\!\bullet \hat{f}(s)$, with $\sigma_-(f) < \Re s < \sigma_+(f)$, then*

$$t^n f(t) \circ\!\!-\!\!\bullet (-1)^n \frac{d^n \hat{f}}{ds^n}. \tag{3.105}$$

Proof. We prove the rule for $n = 1$, the general case follows by induction on n. By canonical splitting of $f(t)$ into $f_1(t) + f_2(t)$ we arrive at $tf(t) = tf_1(t) + tf_2(t) = [zF(z)] = [zF_1(z)] + [zF_2(z)]$. Then, for $f_1(t) = [F_1(z)]$ we have

$$-\frac{d\hat{f}_1}{ds} = \frac{d}{ds} \int_{-\infty}^{(0+)} e^{-sz} F_1(z)\, dz = \int_{-\infty}^{(0+)} \frac{\partial}{\partial s} e^{-sz} F_1(z)\, dz$$

$$= -\int_{-\infty}^{(0+)} e^{-sz} \left(zF_1(z)\right) dz.$$

By definition, this is the Laplace transform of $tf_1(t)$ (note that the Laplace integral can be differentiated under the integral sign). Similarly, we find

$$-\frac{d\hat{f}_2}{ds} = -\int_{\infty}^{(0+)} e^{-sz} \left(zF_2(z)\right) dz$$

which is the Laplace transform of of $tf_2(t)$. Adding the two results gives

$$t\, f(t) = t\, f_1(t) + t\, f_2(t) \circ\!\!-\!\!\bullet -\frac{d\hat{f}_1}{ds} - \frac{d\hat{f}_2}{ds} = -\frac{d\hat{f}}{ds}. \qquad \square$$

Problem 3.27. Check this rule with (3.101) and (3.102)

3.5.4 Similarity Rule

The similarity rule, known from classical Laplace transformation, is also valid for the Laplace transform of hyperfunctions.

Proposition 3.10. *Let $c \neq 0$ be a constant real factor, and $f(t) = [F(z)] \circ\!\!-\!\!\bullet \hat{f}(s)$ with $\sigma_-(f) < \Re s < \sigma_+(f)$, then*

$$f(ct) \circ\!\!-\!\!\bullet \frac{1}{|c|} \hat{f}\left(\frac{s}{c}\right). \tag{3.106}$$

For $c > 0$, the strip of convergence of $\mathcal{L}[f(ct)](s)$ is $c\,\sigma_-(f) < \Re s < c\,\sigma_+(f)$, and for $c < 0$, it is $c\,\sigma_+(f) < \Re s < c\,\sigma_-(f)$.

Proof. We shall prove the proposition for $c < 0$, the case $c > 0$ is left to the reader. By canonical splitting of $f(t)$ into $f_1(t) + f_2(t)$ we have for $c < 0$ (see Definition 1.5),

$$f(ct) = f_1(ct) + f_2(ct) = [-F_1(cz)] + [-F_2(cz)],$$

where now $f_1(ct) \in O(\mathbb{R}_+)$, $f_2(ct) \in O(\mathbb{R}_-)$.

$$\mathcal{L}[f_2(ct)](s) = -\int_{-\infty}^{(0+)} e^{-sz} (-1)F_2(cz)\, dz, \quad \Re s < c\,\sigma_+(f),$$

$$\mathcal{L}[f_1(ct)](s) = -\int_{\infty}^{(0+)} e^{-sz} (-1)F_1(cz)\, dz, \quad c\,\sigma_-(f) < \Re s.$$

These two loop integrals can be written as

$$\mathcal{L}[f_2(ct)](s) = -\frac{1}{-c} \int_{-\infty}^{(0+)} e^{-(s/c)(cz)} F_2(cz)\, d(cz), \quad \Re(s/c) > \sigma_+(f),$$

$$\mathcal{L}[f_1(ct)](s) = -\frac{1}{-c} \int_{\infty}^{(0+)} e^{-(s/c)(cz)} F_1(cz)\, d(cz), \quad \sigma_-(f) > \Re(s/c).$$

If the change of variables $cz = \zeta$ is made, we have

$$\mathcal{L}[f_2(ct)](s) = -\frac{1}{|c|} \int_{\infty}^{(0+)} e^{-(s/c)\zeta} F_2(\zeta)\, d\zeta, \quad \Re(s/c) > \sigma_+(f),$$

$$\mathcal{L}[f_1(ct)](s) = -\frac{1}{|c|} \int_{-\infty}^{(0+)} e^{-(s/c)\zeta} F_1(\zeta)\, d\zeta, \quad \sigma_-(f) > \Re(s/c).$$

Therefore

$$f(ct) \circ\!\!-\!\!\bullet \frac{1}{|c|} \{-\int_{-\infty}^{(0+)} e^{-(s/c)\zeta} F_1(\zeta)\, d\zeta - \int_{\infty}^{(0+)} e^{-(s/c)\zeta} F_2(\zeta)\, d\zeta\}$$

with $c\sigma_+(f) < \Re s < c\sigma_-(f)$. $\qquad\qquad\square$

Let us verify the rule for the correspondence

$$\mathrm{fp}\frac{u(t)}{t^m} \circ\!\!-\!\!\bullet (-1)^m \frac{s^{m-1}}{(m-1)!} \{\log s + \gamma - \sum_{k=1}^{m-1} \frac{1}{k}\} \quad (\Re s > 0).$$

By the similarity rule we obtain for $c > 0$,

$$\mathrm{fp}\frac{u(ct)}{(ct)^m} \circ\!\!-\!\!\bullet \frac{1}{c} (-1)^m \frac{(s/c)^{m-1}}{(m-1)!} \{\log(s/c) + \gamma - \sum_{k=1}^{m-1} \frac{1}{k}\}$$

i.e.,

$$\frac{1}{c^m}(-1)^m \frac{s^{m-1}}{(m-1)!} \{\log s + \gamma - \sum_{k=1}^{m-1} \frac{1}{k}\} - \frac{(-1)^m\, s^{m-1}}{c^m\,(m-1)!} \log c.$$

By (1.89) we know that

$$\mathrm{fp}\frac{u(ct)}{(ct)^m} = \frac{1}{c^m}\mathrm{fp}\frac{u(t)}{t^m} + \frac{(-1)^{m-1} \log c}{c^m(m-1)!} \delta^{(m-1)}(t)$$

which yields the Laplace transform

$$\frac{1}{c^m}(-1)^m \frac{s^{m-1}}{(m-1)!} \{\log s + \gamma - \sum_{k=1}^{m-1} \frac{1}{k}\} + \frac{(-1)^{m-1} \log c}{c^m(m-1)!} s^{m-1}$$

which is in agreement with the result of the similarity rule. The reader has certainly observed that a proof for an operation rule amounts to proving it for a right-sided and a left-sided original and then patching things together. Henceforth we shall just prove it for a right-sided original and leave it to the reader to prove it for a left-sided one.

3.5.5 Differentiation Rule

The differentiation rule for the classical right-sided Laplace transformation takes into account the initial value of the original at $t = 0+$. If $f(t)$ denotes an ordinary function with its Laplace transform $\hat{f}(s)$, then df/dt has $s\hat{f}(s) - f(0+)$ as its Laplace transform. The differentiation rule for hyperfunctions is simpler because it involves differentiation in a generalized sense.

Proposition 3.11. *Let* $f(t) = [F(z)] \circ - \bullet \hat{f}(s)$, *with* $\sigma_-(f) < \Re s < \sigma_+(f)$, *then* $f^{(n)}(t) = D^n f(t) \circ - \bullet s^n \hat{f}(s)$, *with the strip of convergence unchanged.*

Proof. By canonical splitting of $f(t)$ into $f_1(t) + f_2(t)$, we have for $n = 1$ and the right-sided original $f_2(t)$ that

$$f_2'(t) = \left[\frac{dF_2}{dz}\right] \circ - \bullet - \int_\infty^{(0+)} e^{-sz} \frac{dF_2}{dz} \, dz.$$

Integration by parts yields

$$-\int_\infty^{(0+)} e^{-sz} \frac{dF_2}{dz} \, dz = -e^{-sz} F_2(z) \left| \begin{array}{c} z \to \infty - i0 \\ z \to \infty + i0 \end{array} \right. + \int_\infty^{(0+)} e^{-sz}(-s) F_2(z) \, dz.$$

The first term on the right-hand side is zero which is seen from the following estimate, where we put $z = x \pm iy$,

$$\begin{aligned} |e^{-sz} F_2(z)| &= |\exp[-(\Re s + i\Im s)(x \pm iy)]| \, |F_2(z)| \\ &= \exp(-x\Re s \pm y\Im s) |F_2(z)| \\ &\le \exp(-x\Re s) M_1 \exp(\sigma' x) \\ &= M_1 e^{-x(\Re s - \sigma')}. \end{aligned}$$

Note that $e^{\pm y\Im s}$ remains bounded by M_1, and $|F_2(z)|$ by $e^{\sigma' x}$ since $F_2(z)$ is supposed to be of bounded exponential growth. Thus, as $x \to \infty$ and $\Re s > \sigma'$, the term tends to zero. Going on we have

$$-\int_\infty^{(0+)} e^{-sz} \frac{dF_2}{dz} \, dz = -s \int_\infty^{(0+)} e^{-sz} F_2(z) \, dz, \quad \Re s > \sigma_-(f).$$

In the same way one proves

$$-\int_{-\infty}^{(0+)} e^{-sz} \frac{dF_1}{dz} \, dz = -s \int_{-\infty}^{(0+)} e^{-sz} F_1(z) \, dz \quad \Re s < \sigma_+(f).$$

By adding the last two relations, we obtain the assertion for $n = 1$. The case $n \in \mathbb{R}$ follows by induction on n. $\square$

Example 3.18. We apply the proposition to the correspondence $(a > 0)$

$$f(t) = \mathrm{fp}\frac{u(t)J_0(at)}{t} \circ - \bullet \hat{f}(s) = -\log\left(\frac{s + \sqrt{a^2 + s^2}}{2}\right) - \gamma,$$

where $\Re s > 0$. Observe that $J_0(t)$ is a real analytic function. The derivative of the right-sided original gives:

$$f'(t) = D\,\mathrm{fp}\frac{u(t)J_0(at)}{t}$$

$$= \left[-\frac{1}{2\pi i}\frac{d}{dz}\{\frac{J_0(az)}{z}\log(-z)\}\right]$$

$$= a\left[-\frac{1}{2\pi i}\frac{J_0'(az)}{z}\log(-z)\right] - \left[-\frac{1}{2\pi i}\frac{J_0(az)}{z^2}\log(-z)\right] + \left[-\frac{1}{2\pi i}\frac{J_0(az)}{z^2}\right]$$

$$= -a\frac{u(t)J_1(at)}{t} - \mathrm{fp}\frac{u(t)J_0(at)}{t^2} - \delta'(t),$$

where the result $J_0'(z) = -J_1(z)$ was used. By the differentiation rule the Laplace transform of this hyperfunction must be

$$-s\log(\frac{s+\sqrt{a^2+s^2}}{2}) - s\gamma.$$

We use the known correspondence

$$u(t)\frac{J_n(at)}{t}\;\circ\!\!-\!\!\bullet\;\frac{(\sqrt{s^2+s^2}-s)^n}{n\,a^n},\quad n = 1,2,\dots.$$

Gathering these facts yields the result

$$\mathrm{fp}\frac{u(t)J_0(at)}{t^2}\;\circ\!\!-\!\!\bullet\;s\log(\frac{s+\sqrt{a^2+s^2}}{2}) + s\gamma - \sqrt{s^2+a^2}\quad(\Re s > 0)\qquad(3.107)$$

which is in agreement with the result of Problem 3.21.

Problem 3.28. Establish for $\Re s > 0$ and $a > 0$,

$$\mathrm{fp}\frac{u(t)J_1(at)}{t^2}\;\circ\!\!-\!\!\bullet\;-\frac{a}{2}\{\log(\frac{s+\sqrt{a^2+s^2}}{2})$$

$$+\gamma - \frac{1}{2} + \frac{s}{\sqrt{s^2+a^2}+s}\},$$

$$\mathrm{fp}\frac{u(t)J_0(at)}{t^3}\;\circ\!\!-\!\!\bullet\;\frac{a^2}{4}\{\log(\frac{s+\sqrt{a^2+s^2}}{2})$$

$$+\gamma - 1 + \frac{s}{\sqrt{s^2+a^2}+s}\} + \frac{s^2}{2}.$$

Problem 3.29. Establish, by using the known correspondences

$$u(t)\frac{\cosh\sqrt{at}}{\sqrt{t}}\;\circ\!\!-\!\!\bullet\;\sqrt{\frac{\pi}{s}}\,e^{a/(4s)},\quad u(t)\frac{\sinh\sqrt{at}}{t}\;\circ\!\!-\!\!\bullet\;\pi\,\mathrm{erg}\sqrt{\frac{a}{4s}},$$

and the differentiation rule, that

$$u(t)\,\mathrm{fp}\frac{\cosh\sqrt{at}}{t^{3/2}}\;\circ\!\!-\!\!\bullet\;\pi\sqrt{a}\,\mathrm{erg}\sqrt{\frac{a}{4s}} - 2\sqrt{\pi s}\,e^{a/(4s)}.\qquad(3.108)$$

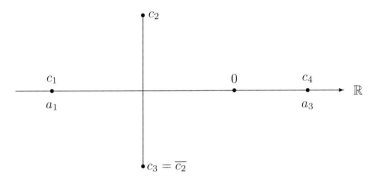

Figure 3.3: Roots and partition of $\mathbb{C}$ in Example 3.19

Example 3.19. Let $P(s)$ be a polynomial with real coefficients and r different roots $c_k \in \mathbb{C}$. Consider the associated linear differential operator

$$P(D) := (D - c_1)^{m_1}(D - c_2)^{m_2}\cdots(D - c_r)^{m_r}. \tag{3.109}$$

Denote by $a_1, a_2, \ldots, a_p$ the distinct real parts of the roots $(1 \leq p \leq r)$, arranged such that $a_1 < a_2 < \ldots < a_p$ (see Figure 3.3). Non-real roots occur in pairs $(c_k, \overline{c_k})$ having the same multiplicity. The complex plane partitions into $p+1$ vertical strips $a_{\ell-1} < \Re s < a_\ell$. For $\ell = 1, \ldots p+1$ we propose to solve the boundary value problem

$$P(D)y(t) = f(t), \quad -\infty < t < \infty$$
$$y(t) = o(e^{(a_\ell - \epsilon)t}) \text{ as } t \to -\infty, \tag{3.110}$$
$$y(t) = o(e^{(a_{\ell-1} + \epsilon)t}) \text{ as } t \to \infty,$$

where $\epsilon > 0$, and $a_0 = -\infty$, $a_{p+1} = \infty$. The case $y(t) = o(e^{-\infty t})$ is to be understood as $y(t) = o(e^{-kt}) \; \forall k > 0$ as $t \to -\infty$, and the case $y(t) = o(e^{\infty t})$ as $y(t) = o(e^{kt}) \; \forall k > 0$ as $t \to \infty$. We seek Green's function $G_\ell(t - x)$ for every problem. $G_\ell(t)$ is the solution of (3.110) where $f(t)$ replaced by $\delta(t)$. Once Green's function is determined, the solution of (3.110) is given by

$$y(t) = \int_{-\infty}^{\infty} G_\ell(t - x)f(x)\, dx \tag{3.111}$$

provided the integral converges. Suppose that $G_\ell(t) \circ\!\!-\!\!\bullet \hat{G}(s)$, then $\hat{G}(s) = 1/P(s)$. Partial fraction decomposition yields

$$\hat{G}(s) = \frac{1}{P(s)} = \sum_{k=1}^{r}\left\{\frac{c_{k,1}}{(s - c_k)^{m_k}} + \frac{c_{k,2}}{(s - c_k)^{m_k - 1}} + \cdots + \frac{c_{k,m_k}}{s - c_k}\right\}.$$

In every selected strip $a_{\ell-1} < \Re s < a_\ell$, $\ell = 1, \ldots, p+1$, where $\hat{G}(s)$ is a holomorphic function, the corresponding $G_\ell(t)$ is an expression of type

$$G_\ell(t) = u(-t)\sum_{k=\ell}^{p} p_k(t) f_k(t) + u(t)\sum_{k=1}^{\ell-1} p_k(t) f_k(t),$$

where $f_k(t) = e^{a_k t}$, if $c_k = a_k$ is a real root, and $f_k(t) = e^{a_k t}(A_k \cos \omega_k t + B_k \sin \omega_k t)$, if we have a pair of non-real roots, respectively. The $p_k(t)$ are polynomials of degree $m_k - 1$ at most. With an $\epsilon > 0$, $G_\ell(t)$ can be written in the form

$$G_\ell(t) = u(-t)\, e^{(a_\ell - \epsilon)t}\, o(1) + u(t)\, e^{(a_{\ell - 1} + \epsilon)t} o(1),$$

which shows the pretended limit behavior. From an application point of view, the most interesting case is the one where we have

$$\Re a_1 < \Re a_2 < \cdots < \Re a_{n-1} < 0 < \Re a_n < \cdots < \Re a_p$$

i.e., where the first $n - 1$ zeros have strictly negative real part and following ones strictly positive real parts. Then Green's function $G_n(t)$ will be bounded on the entire real axis, and

$$y(t) = \int_{-\infty}^{\infty} G_n(t - x) f(x)\, dx,$$

$$y(t) = o(e^{(a_n - \epsilon)t}) \text{ as } t \to -\infty,$$

$$y(t) = o(e^{(a_{n-1} + \epsilon)t}) \text{ as } t \to \infty \quad (\epsilon > 0)$$

will solve the boundary value problem for any continuous $f(t)$ of slow growth. Computations are readily done with the *Mathematica* package of the author (see [14, Graf]). For example, with

$$P(D) = \frac{1}{25}\{D^5 + D^4 + 8D^3 + 2D^2 - 8D - 8\}$$

$$= \frac{1}{25}(D+2)^2(D+1-i))(D+1+i))(D-1)$$

and $f(t) = t^2$ we find

$$G_3(t) = \frac{5}{9} u(-t)e^t + u(t)\left\{(\frac{50}{9} + \frac{25}{6}t)e^{-2t} + 2e^{-t}(-\frac{5}{2}\cos t + \frac{5}{4}\sin t)\right\},$$

and the solution of the boundary problem becomes

$$y(t) = -\frac{25}{4}\left(\frac{5}{4} - t + \frac{1}{2}t^2\right).$$

3.5.6 Integration Rule

Primitives of a Hyperfunction

So far, we only have defined the definite integral of a hyperfunction $f(x) = [F(z)]$ by

$$\int_a^b f(x)\, dx := -\oint_{(a,b)} F(z)\, dz.$$

We now define the converse process of differentiation of a hyperfunction.

Definition 3.5. Let $f(x) = [F_+(z), F_-(z)]$ be a given hyperfunction holomorphic at the point $x = a$. The *primitive or antiderivative of* $f(x)$ is the hyperfunction which is defined and denoted by

$$_aD^{-1}f(x) := \left[\int_a^z F_+(\zeta)\, d\zeta, \int_a^z F_-(\zeta)\, d\zeta\right]. \tag{3.112}$$

For $n \in \mathbb{N}$, *the n-th antiderivative of* $f(x)$ is specified by

$$_aD^{-n}f(x) := \frac{1}{(n-1)!}\left[\int_a^z (z-\zeta)^{n-1} F_+(\zeta)\, d\zeta, \int_a^z (z-\zeta)^{n-1} F_-(\zeta)\, d\zeta\right]. \tag{3.113}$$

Here the contour for the integrals is any path from the real point a to the point z in the upper and lower half-plane, respectively. The initial point a may also be $-\infty$ or ∞, provided the integrals converge,

$$_{-\infty}D^{-1}f(x) := \left[\int_{-\infty}^z F_+(\zeta)\, d\zeta, \int_{-\infty}^z F_-(\zeta)\, d\zeta\right], \tag{3.114}$$

$$_{\infty}D^{-1}f(x) := \left[-\int_z^{\infty} F_+(\zeta)\, d\zeta, -\int_z^{\infty} F_-(\zeta)\, d\zeta\right]. \tag{3.115}$$

By induction on n, it can be shown that

$$_aD^{-1}(_aD^{-n}f(x)) = {}_aD^{-(n+1)}f(x), \tag{3.116}$$

$$D(_aD^{-n}f(x)) = {}_aD^{-n+1}f(x), \quad (n > 1). \tag{3.117}$$

If $\phi(x) = [\phi(z)/2, -\phi(z)/2]$ is a real analytic hyperfunction, and $\Phi(x)$ is any primitive of $\phi(x)$, i.e., $\Phi'(x) = \phi(x)$, then

$$\int_a^z \phi(\zeta)\, d\zeta = \Phi(z) - \Phi(a),$$

and

$$_aD^{-1}\phi(x) = \left[\frac{\Phi(z) - \Phi(a)}{2}, -\frac{\Phi(z) - \Phi(a)}{2}\right]$$
$$= [\Phi(z)/2, -\Phi(z)/2] - [\Phi(a)/2, -\Phi(a)/2]$$
$$= \Phi(x) - \Phi(a).$$

For example,

$$_aD^{-1}x^n = \frac{x^{n+1}}{n+1} - \frac{a^{n+1}}{n+1}.$$

Example 3.20. Let $f(x) = \delta(x) = [-1/(2\pi i z)]$. For $a < 0$ we have

$$\int_a^z -\frac{1}{2\pi i}\frac{d\zeta}{\zeta} = -\frac{\log(-\zeta)}{2\pi i}\Big|_a^z = -\frac{1}{2\pi i}\{\log(-z) - \log(-a)\},$$

and for $a > 0$,

$$\int_a^z -\frac{1}{2\pi i}\frac{d\zeta}{\zeta} = -\frac{\log\zeta}{2\pi i}\Big|_a^z = -\frac{1}{2\pi i}\{\log z - \log a\}.$$

The constant term on the right-hand side can be discarded giving an equivalent defining function, and we finally have,

$$_aD^{-1}\delta(x) = \begin{cases} u(x), & a < 0 \\ -u(-x), & a > 0. \end{cases}$$

Example 3.21. Let $f(x) = \mathrm{fp}(u(x)/x) = [-\log(-z)/(2\pi i z)]$. For $a < 0$ we have

$$\int_a^z \frac{-\log(-\zeta)}{2\pi i \zeta} d\zeta = -\frac{1}{2\pi i} \int_{-a}^{-z} \frac{\log \zeta}{\zeta} d\zeta$$

$$= -\frac{1}{4\pi i} \log^2(-z) + \frac{1}{4\pi i} \log^2(-a).$$

The last term on the right-hand side can be discarded giving an equivalent defining function. The result is

$$_aD^{-1}\mathrm{fp}\frac{u(x)}{x} = u(x)\log x \quad (a < 0). \tag{3.118}$$

Problem 3.30. Show that

$$_aD^{-1}u(x) = u(x)x \quad (a < 0),$$
$$_aD^{-1}u(-x) = u(-x)x \quad (a > 0).$$

Let us return to the subject of Laplace transformations. If $f(t) \in O(\mathbb{R}_+)$ has the growth index σ_-, and if $\sigma_0 > \max[\sigma_-, 0]$, we have for $\Re s > \sigma_0$,

$$e^{-sz} \int_a^z (z - \zeta)^{n-1} F(\zeta) d\zeta \to 0, \quad \text{as } \Re z \to \infty$$

which is a consequence of the bounded exponential growth of $f(x) = [F(z)]$. Integration by parts yields

$$-\int_\infty^{(0+)} e^{-sz} \frac{1}{(n-1)!} \int_a^z (z - \zeta)^{n-1} F(\zeta) d\zeta \, dz$$

$$= \frac{1}{s} \int_\infty^{(0+)} e^{-sz} \frac{1}{(n-2)!} \int_a^z (z - \zeta)^{n-1} F(\zeta) d\zeta \, dz$$

giving the result

$$\mathcal{L}[_aD^{-n}f(t)](s) = \frac{1}{s} \mathcal{L}[_aD^{-n+1}f(t)](s), \tag{3.119}$$

thus,

Proposition 3.12. If $f(t) \in O(\mathbb{R}_+)$ with growth index σ_-, then, for $\sigma_0 > \max[\sigma_-, 0]$, we have for $\Re s > \sigma_0$

$$\mathcal{L}[_aD^{-n}f(t)](s) = \frac{1}{s^n} \mathcal{L}[f(t)](s), \tag{3.120}$$

where $a < 0$ is any negative real number.

If $f(t) = [F_+(z), F_-(z)]$ is a given hyperfunction of bounded exponential growth having a Laplace transform $\hat{f}(s)$ for $\sigma_- < \Re s < \sigma_+$, the integration rule is

$$\mathcal{L}[_{-\infty}D^{-1}f(t)](s) = \frac{1}{s}\mathcal{L}[f(t)](s), \quad \max[\sigma_-, 0] < \Re s < \sigma_+ \tag{3.121}$$

provided $\sigma_+ > 0$, and

$$\mathcal{L}[_{\infty}D^{-1}f(t)](s) = -\frac{1}{s}\mathcal{L}[f(t)](s), \quad \sigma_- < \Re s < \min[\sigma_+, 0] \tag{3.122}$$

provided $\sigma_- < 0$. Details are left to the reader.

3.5.7 Original Translation Rule

Proposition 3.13. *Let $f(t) = [F(z)] \circ\!-\!\bullet \hat{f}(s)$, with $\sigma_-(f) < \Re s < \sigma_+(f)$, then for $c \in \mathbb{R}$ we have*

$$f(t - c) \circ\!-\!\bullet e^{-cs}\hat{f}(s) \tag{3.123}$$

with $\sigma_-(f) < \Re s < \sigma_+(f)$.

Proof. By canonical splitting of $f(t)$ into $f_1(t) + f_2(t)$ we arrive at $f(t - c) = f_1(t - c) + f_2(t - c)$ and the two integration loops now turn around the point c:

$$f_2(t - c) \circ\!-\!\bullet -\int_{\infty}^{(c+)} e^{-sz} F_2(z - c)\, dz = -\int_{\infty}^{(0+)} e^{-s(\zeta+c)} F_2(\zeta)\, d\zeta$$

$$= e^{-cs}(-1)\int_{\infty}^{(0+)} e^{-s\zeta} F_2(\zeta)\, d\zeta = e^{-cs}\,\hat{f}_2(s).$$

Similarly,

$$f_1(t - c) \circ\!-\!\bullet -\int_{-\infty}^{(c+)} e^{-sz} F_1(z - c)\, dz = -\int_{-\infty}^{(0+)} e^{-s(\zeta+c)} F_1(\zeta)\, d\zeta$$

$$= e^{-cs}(-1)\int_{-\infty}^{(0+)} e^{-s\zeta} F_1(\zeta)\, d\zeta = e^{-cs}\,\hat{f}_1(s).$$

Thus, we obtain

$$f(t - c) = f_1(t - c) + f_2(t - c) \circ\!-\!\bullet e^{-cs}\{\hat{f}_1(s) + \hat{f}_2(s)\} = e^{-cs}\,\hat{f}(s). \qquad \square$$

3.5.8 Linear Substitution Rules

If we combine the similarity rule and original translation rule we obtain

Proposition 3.14. *Let $f(t) = [F(z)] \circ\!-\!\bullet \hat{f}(s)$, with $\sigma_-(f) < \Re s < \sigma_+(f)$, then for $a, b \in \mathbb{R}$, $a \neq 0$, we have*

$$f(a\,t + b) \circ\!-\!\bullet \frac{1}{|a|} e^{(b/a)\,s} \hat{f}(\frac{s}{a}), \tag{3.124}$$

with $a\,\sigma_-(f) < \Re s < a\,\sigma_+(f)$ for $a > 0$, and $a\,\sigma_+(f) < \Re s < a\,\sigma_-(f)$ for $a < 0$.

We conclude this section with the special case where a linear change of variables is made in a right-sided original. Assume the correspondence $f(t) = [F(z)] \in O(\mathbb{R}_+) \circ - \bullet \hat{f}(s)$, $\Re s > \sigma_-(f)$. We are looking for the Laplace transform of $f(a\,t - b)$ where both a and b are assumed to be positive. The defining function $F(z)$ is now real analytic on the negative real axis. Because $f(a t - b) = [F(az - b)]$, we have that $G(z) = F(az - b)$ is real analytic on $(-\infty, b/a)$. That means that $g(t) = f(a\,t - b)$ is a hyperfunction that vanishes on $(-\infty, b/a)$ and therefore is in $O(\mathbb{R}_+)$. Thus, its Laplace transform is

$$\hat{g}(s) = -\int_{\infty}^{(0+)} e^{-sz} F(az - b)\, dz.$$

A change of variables leads to

$$\hat{g}(s) = -\frac{e^{-sb/a}}{a} \int_{\infty}^{(-b+)} e^{-s/a\,z} F(z)\, dz$$

where the integration loop now turns around the point $-b$. Since $F(z)$ is real analytic for $x < 0$ we can replace the new loop by the old one and obtain

$$\hat{g}(s) = -\frac{e^{-sb/a}}{a} \int_{\infty}^{(0+)} e^{-s/a\,z} F(z)\, dz$$

which equals $e^{-sb/a} \hat{f}(s/a)/a$. By keeping track of the growth index $\sigma_-(f(t))$ we see that $\sigma_-(f(a\,t - b)) = a\,\sigma_-(f(t))$. Because $f(at - b)$ vanishes on $(-\infty, b/a)$ we may write $u(t - b/a)f(at - b)$ for it. We have proved

Proposition 3.15. *Let $f(t) \in O(\mathbb{R}_+)$ and a and b positive constants. If $f(t) \circ - \bullet \hat{f}(s)$, for $\Re s > \sigma_-$, then $f(a\,t - b) \circ - \bullet e^{-sb/a} \hat{f}(s/a)/a$, for $\Re s > a\,\sigma_-$, where the original $f(a\,t - b)$ vanishes on $(-\infty, b/a)$. For $a = 1$, we obtain the translation rule that can be written as*

$$f(t) \circ - \bullet \hat{f}(s) \implies u(t - b)f(t - b) \circ - \bullet e^{-sb} \hat{f}(s). \qquad (3.125)$$

A similar statement holds if $f(t) \in O(\mathbb{R}_-)$, $a > 0$, $b > 0$, and $f(t) \circ - \bullet \hat{f}(s)$ for $\Re s < \sigma_+$, then

$$f(a\,t + b)\, u(-b/a - t) \circ - \bullet \frac{e^{sb/a}}{a}\, \hat{f}(\frac{s}{a}), \quad \Re s < a\,\sigma_+, \qquad (3.126)$$

i.e., $f(at + b)$ vanishes on $(-b/a, \infty)$.

Example 3.22. Because

$$\text{fp}\frac{u(t)J_0(t)}{t} \circ - \bullet - \log(\frac{s + \sqrt{1 + s^2}}{2}) - \gamma$$

we obtain

$$\text{fp}\frac{u(t - b)J_0(t - b)}{t - b} \circ - \bullet - e^{-bs} \{\log(\frac{s + \sqrt{1 + s^2}}{2}) - \gamma\}.$$

Let now $f(t) = [F(z)] \in O(\mathbb{R}_+)$. If $b > 0$, $f(t + b) = [F(z + b)]$ is no more necessarily in $O(\mathbb{R}_+)$, the projection on $(0, \infty)$, however, is,

$$\chi_{(0,\infty)} f(t + b) = \left[-\frac{\phi(z)}{2\pi i} \int_\infty^{(0+)} \frac{F(t + b)}{\phi(t)\,(t - z)}\, dt \right] \in O(\mathbb{R}_+).$$

Here $\phi(t) = e^{\sigma' t}$ with $\sigma' > \sigma_-(f)$ is a appropriate convergence factor. Let us compute its Laplace transform. By definition, the Laplace transform is given by

$$\chi_{(0,\infty)} f(t + b) \circ\!-\!\bullet - \int_\infty^{(0+)} e^{-sz}\left\{ -\frac{\phi(z)}{2\pi i} \int_\infty^{(0+)} \frac{F(t + b)}{\phi(t)\,(t - z)}\, dt \right\} dz$$

where the z-integration loop lies outside the t-integration loop. Exchange of the order of integration yields

$$\chi_{(0,\infty)} f(t + b) \circ\!-\!\bullet - \int_\infty^{(0+)} \frac{F(t + b)}{\phi(t)} \left\{ \frac{1}{2\pi i} \int_\infty^{(0+)} \frac{e^{-sz}\,\phi(z)}{t - z}\, dz \right\} dt.$$

The infinite loop of the inner integral can be replaced by a finite one. Evaluating the inner integral by Cauchy's integral theorem yields $-e^{-st}\,\phi(t)$ because $e^{-sz}\,\phi(z)$ is holomorphic in z. Thus,

$$\chi_{(0,\infty)} f(t + b) \circ\!-\!\bullet - \int_\infty^{(0+)} \frac{F(t + b)}{\phi(t)}\, e^{-st}\,\phi(t)\,dt = -\int_\infty^{(0+)} F(t + b)\, e^{-st}\, dt$$

$$= -e^{bs} \int_\infty^{(0+)} F(t + b)\, e^{-s(t+b)}\, d(t + b) = -e^{bs} \int_\infty^{(b+)} F(\zeta)\, e^{-s\zeta}\, d\zeta$$

$$= -e^{bs}\left\{ \int_\infty^{(0+)} F(\zeta)\, e^{-s\zeta}\, d\zeta - \int_{C_b} F(\zeta)\, e^{-s\zeta}\, d\zeta \right\}$$

$$= e^{bs}\left\{ \hat{f}(s) + \int_{C_b} F(\zeta)\, e^{-s\zeta}\, d\zeta \right\}.$$

The contour C_b starts at $b > 0$ rounds 0 in the positive sense and returns to b (see Figure 2.12). The last but one step is only justified if we suppose that $F(z)$ is holomorphic at $z = b > 0$. The right-half plane of convergence of $f(t)$ and of $\chi_{(0,\infty)} f(t + b)$ is the same. This gives the following translation rule.

Proposition 3.16. *Let* $f(t) = [F(z)] \in O(\mathbb{R}_+) \circ\!-\!\bullet \hat{f}(s)$, *and suppose that* $F(z)$ *is holomorphic at* $z = b > 0$. *Then,*

$$\chi_{(0,\infty)} f(t + b) \circ\!-\!\bullet e^{bs}\left\{ \hat{f}(s) + \int_{C_b} e^{-sz}\, F(z)\, dz \right\}. \qquad (3.127)$$

Problem 3.31. Use Lemma 2.45, and Problem 3.16 to establish

$$-\frac{1}{2\pi i} \int_{C_b} \frac{e^{-sz}\,\log(-z)}{z^m}\, dz = -\frac{E_m(bs)}{b^{m-1}} + \frac{(-s)^{m-1}}{(m - 1)!}\,[\log s - \Psi(m)].$$

Problem 3.32. Use

$$\mathrm{fp}\,\frac{u(t)}{t^m} \circ\!-\!\bullet (-1)^m \frac{s^{m-1}}{(m - 1)!}\left\{ \log s + \gamma - \sum_{k=1}^{m-1} \frac{1}{k} \right\}$$

and Proposition 3.16 to verify the correspondence

$$\frac{u(t)}{(t+b)^n} \circ\!\!-\!\!\bullet \frac{1}{b^{n-1}} e^{bs} E_n(bs).$$

3.6   Inverse Laplace Transforms and Convolutions

3.6.1   Inverse Laplace Transforms

For easy reference we restate below the operational rules regarded from the point of view of the inverse Laplace transformation.

Proposition 3.17. *If the image function $\hat{f}(s)$, defined and holomorphic in the vertical strip $\sigma_- < \Re s < \sigma_+$, has as the inverse Laplace transform the hyperfunction $f(t)$, then*

(i) *the translated image $\hat{f}(s-c)$ with $(\sigma_- + \Re c) < \Re s < (\sigma_+ + \Re c)$ has $e^{ct} f(t)$,*

(ii) $s^n \, \hat{f}(s)$ *has $D^n f(t) = f^{(n)}(t)$,*

(iii) $e^{-as} \, \hat{f}(s)$ *has $f(t-a)$,*

(iv) $\hat{f}(as)$ *with $\sigma_-/a < \Re s < \sigma_+/a$ for $a > 0$, and $\sigma_+/a < \Re s < \sigma_-/a$ for $a < 0$ has $f(t/a)/|a|$,*

(v) $d^n \hat{f}(s)/ds^n$ *has $(-t)^n \, f(t)$ as inverse Laplace transform.*

Example 3.23.

$$\mathcal{L}^{-1}[(s-c)^n](t) = e^{ct} \, \delta^{(n)}(t)$$

$$= \sum_{k=0}^{n} (-1)^k \begin{pmatrix} n \\ k \end{pmatrix} c^k \, \delta^{(n-k)}(t),$$

in accordance with the formula

$$\phi(t)\delta^{(n)}(t-a) = \sum_{k=0}^{n} (-1)^k \begin{pmatrix} n \\ k \end{pmatrix} \phi^{(k)}(a) \, \delta^{(n-k)}(t-a)$$

valid for any real analytic function $\phi(t)$.

The inverse (two-sided) Laplace transform of any rational expression in s can now be computed.

Example 3.24. Consider the rational expression

$$\hat{f}(s) = \frac{s^6 + s^5 + 2s^4 - 2s^3 - 12s^2 - 7s - 5}{(s+1)^2(s^2+1)(s-2)}.$$

Long division first yields a sum of a proper rational function and a polynomial in s.

$$\hat{f}(s) = \frac{4s^4 + 2s^3 - 7s^2 - 2s - 3}{(s+1)^2(s^2+1)(s-2)} + s + 1.$$

Partial fraction decomposition of the first term results in

$$\hat{f}(s) = \frac{1}{(s+1)^2} + \frac{1}{(s+1)} + \frac{1}{(s+i)} + \frac{1}{(s-i)} + \frac{1}{(s-2)} + s + 1.$$

This gives rise to four different image functions, defined by the same functional expression, but with different vertical strips of convergence where the image function is holomorphic. Note that the Laplace transforms of Dirac Impulses and their derivatives have the entire complex plane as strip of convergence:

$$\hat{f}_1 : s \mapsto \hat{f}(s), \ \Re s < -1,$$
$$\hat{f}_2 : s \mapsto \hat{f}(s), \ -1 < \Re s < 0,$$
$$\hat{f}_3 : s \mapsto \hat{f}(s), \ 0 < \Re s < 2,$$
$$\hat{f}_4 : s \mapsto \hat{f}(s), \ 2 < \Re s.$$

For the left-most strip of convergence, we obtain

$$f_1(t) = -u(-t)\left\{ te^{-t} + e^{-t} + 2\cos t + e^{2t} \right\} + \delta(t) + \delta'(t).$$

In the next strip of convergence, we have

$$f_2(t) = u(t)\left\{ te^{-t} + e^{-t} \right\} - u(-t)\left\{ 2\cos t + e^{2t} \right\} + \delta(t) + \delta'(t).$$

For the following one, we get

$$f_3(t) = u(t)\left\{ te^{-t} + e^{-t} + 2\cos t \right\} - u(-t)\, e^{2t} + \delta(t) + \delta'(t),$$

and finally for the right-most strip of convergence, for $2 < \Re s$, one obtains

$$f_4(t) = u(t)\left\{ te^{-t} + e^{-t} + 2\cos t + e^{2t} \right\} + \delta(t) + \delta'(t).$$

B-Transforms

If $f(t)$ is an ordinary function of locally bounded variation, and the Laplace integral is absolutely convergent in the vertical strip $\sigma_- < \Re s < \sigma_+$, then the complex inversion formula

$$\frac{f(t+) + f(t-)}{2} = \lim_{R \to \infty} \frac{1}{2\pi i} \int_{c-iR}^{c+iR} e^{st}\, \hat{f}(s)\, ds$$

holds, where the contour is a vertical line $\sigma_- < \Re s = c < \sigma_+$. If $t = x$ is a point of continuity of $f(t)$ we may write

$$f(x) = \frac{1}{2\pi i} \int_{c-i\infty}^{c+i\infty} e^{sx}\, \hat{f}(s)\, ds = \frac{1}{2\pi i} \int_{c-i\infty}^{c} e^{sx}\, \hat{f}(s)\, ds + \frac{1}{2\pi i} \int_{c}^{c+i\infty} e^{sx}\, \hat{f}(s)\, ds$$

$$= \frac{1}{2\pi i} \int_{c}^{c+i\infty} e^{sx}\, \hat{f}(s)\, ds - \left\{ -\frac{1}{2\pi i} \int_{c-i\infty}^{c} e^{sx}\, \hat{f}(s)\, ds \right\}$$

$$:= F_+(x + i0) - F_-(x - i0).$$

This suggests defining the inverse Laplace transform of a hyperfunction $f(t) = [F_+(z), F_-(z)]$ by

$$F_+(z) := \frac{1}{2\pi i} \int_c^{c+i\infty} e^{sz} \hat{f}(s) \, ds, \quad \Im z > 0, \tag{3.128}$$

$$F_-(z) := -\frac{1}{2\pi i} \int_{c-i\infty}^c e^{sz} \hat{f}(s) \, ds, \quad \Im z < 0. \tag{3.129}$$

Let us take a closer look at this idea. For $z = x + iy$ and $s = c + i\omega$ we consider the expression

$$|e^{sz}| = e^{\Re((c+i\omega)(x+iy))} = e^{(cx-\omega y)}$$

which for $y > 0$ exponentially tends to zero as $\omega \to \infty$. Likewise, for $y < 0$ it decays exponentially to zero as $\omega \to -\infty$. This means that for $y = \Im z > 0$ and $\hat{f}(s) = O(s^m)$ as $s \to \infty$ the function $e^{sz} \hat{f}(s)$ is absolutely integrable on $\Im s = \omega \in [0, \infty)$. Similarly, for $y = \Im z < 0$, the function is absolutely integrable on $\Im s = \omega \in (-\infty, 0]$. This leads us to introduce a new integral transformation, called the B-transform (for "back transformation").

Definition 3.6. Let $\hat{f}(s)$ be holomorphic in a vertical strip $\sigma_- < \Re s < \sigma_+$, and let for some $m \in \mathbb{N}$, $\hat{f}(s) = O(s^m)$ as $|\Im s| \to \infty$. Then, the hyperfunction

$$f(t) = [F_+(z), F_-(z)] \tag{3.130}$$

defined by

$$F_+(z) := \frac{1}{2\pi i} \int_c^{c+i\infty} e^{sz} \hat{f}(s) \, ds, \quad \Im z > 0, \tag{3.131}$$

$$F_-(z) := -\frac{1}{2\pi i} \int_{c-i\infty}^c e^{sz} \hat{f}(s) \, ds, \quad \Im z < 0 \tag{3.132}$$

is called the B-transform of $\hat{f}(s)$ denoted by $B[\hat{f}(s)](t)$.

If $\hat{f}(s)$ is such that the two defining integrals above converge for $z = t \in \mathbb{R}$ (this is the case if $\hat{f}(c + i\omega) = O(|\omega|^\mu)$, $\mu < -1$ as $|\omega| \to \infty$), then $f(t) = F_+(t) - F_-(t)$ and the B-transform realizes the converse of the Laplace transform.

Note that for the B-transform any c with $\sigma_- < c < \sigma_+$ can be selected and will yield the same $f(t)$. This follows from the holomorphicity of the integrand in the vertical strip $\sigma_- < \Re s < \sigma_+$ in the two integrals. To see this, consider for $\sigma_- < c_1 < c_2 < \sigma_+$ the two following rectangular contours in the s-plane. In the upper half-plane: $c_1 \to c_2 \to c_2 + iR \to c_1 + iR \to c_1$. Similarly, in the lower half-plane: $c_2 \to c_1 \to c_1 - iR \to c_2 - iR \to c_2$. Applying Cauchy's theorem and letting $R \to \infty$ then yields

$$G_+(z) = F_+(z) + \phi(z), \quad G_-(z) = F_-(z) + \phi(z)$$

where

$$G_+(z) = \frac{1}{2\pi i} \int_{c_2}^{c_2+i\infty} e^{sz} \hat{f}(s)\, ds, \quad \Im z > 0,$$

$$G_-(z) = -\frac{1}{2\pi i} \int_{c_2-i\infty}^{c_2} e^{sz} \hat{f}(s)\, ds, \quad \Im z < 0,$$

$$\phi(x) = -\frac{1}{2\pi i} \int_{c_1}^{c_2} e^{sx} \hat{f}(s)\, ds.$$

But $\phi(x)$ is a real analytic function ensuring that $F(z)$ and $G(z)$ are equivalent defining functions determining the same hyperfunction $f(t)$.

Also, the integration limit $c = c \pm i0$ in the above integrals can be replaced by $c + ia$ for any finite and real a, i.e., one may also take

$$F_+(z) := \frac{1}{2\pi i} \int_{c+ia}^{c+i\infty} e^{sz} \hat{f}(s)\, ds, \quad \Im z > 0, \tag{3.133}$$

$$F_-(z) := -\frac{1}{2\pi i} \int_{c-i\infty}^{c+ia} e^{sz} \hat{f}(s)\, ds. \quad \Im z < 0, \tag{3.134}$$

because this amounts to adding the entire function

$$\phi(z) = -\frac{1}{2\pi i} \int_{c}^{c+ia} e^{sz} \hat{f}(s)\, ds$$

to the upper and lower component of the defining function which results in an equivalent defining function. The next proposition states conditions upon the image function such that the B-transform becomes real- or imaginary-valued.

Proposition 3.18. *An image function satisfying $\hat{f}(s) = \overline{\hat{f}(\bar{s})}$ produces a real-valued B-transform. Similarly, if $\hat{f}(s) = -\overline{\hat{f}(\bar{s})}$ then the B-transform is imaginary-valued.*

Proof. We prove the first part of the statement. Writing $z = x + iy$, $s = c + i\omega$, we have

$$F_+(z) = \frac{e^{c(x+iy)}}{2\pi} \int_0^\infty e^{-\omega y} e^{i\omega x} \hat{f}(c + i\omega)\, d\omega, \quad y > 0,$$

$$F_-(z) = -\frac{e^{c(x+iy)}}{2\pi} \int_{-\infty}^0 e^{-\omega y} e^{i\omega x} \hat{f}(c + i\omega)\, d\omega, \quad y < 0.$$

Thus, for $\Im z < 0$, and by using $\hat{f}(s) = \overline{\hat{f}(\bar{s})}$ we have

$$\overline{F_+(\bar{z})} = \frac{e^{c(x-iy)}}{2\pi} \overline{\int_0^\infty e^{\omega y} e^{i\omega x} \hat{f}(c + i\omega)\, d\omega}$$

$$= \frac{e^{c(x+iy)}}{2\pi} \int_0^\infty e^{\omega y} e^{-i\omega x} \overline{\hat{f}(c + i\omega)}\, d\omega$$

$$= \frac{e^{cz}}{2\pi} \int_{-\infty}^0 e^{-\omega y} e^{i\omega x} \hat{f}(c + i\omega)\, d\omega = -F_-(z),$$

and, for $\Im z > 0$, similarly,

$$\overline{F_-(\bar{z})} = -\frac{e^{cz}}{2\pi} \int_0^\infty e^{-\omega y} e^{i\omega x} \, \hat{f}(c + i\omega) \, d\omega = -F_+(z).$$

Hence, $\overline{f(t)} = \left[-\overline{F_-(\bar{z})}, -\overline{F_+(\bar{z})}\right] = [F_+(z), F_-(z)] = f(t).$ □

Proposition 3.19. *Let* $\hat{f}(s) = O(s^m)$, *for some* $m \in \mathbb{N}$, *as* $|\Im s| \to \infty$. *If* $\hat{f}(s)$ *is holomorphic in a right half-plane* $\sigma_- < \Re s$, *then* $B[\hat{f}(s)](t) \in O(\mathbb{R}_+)$. *Similarly, if* $\hat{f}(s)$ *is holomorphic in a left half-plane* $\Re s < \sigma_+$, *then* $B[\hat{f}(s)](t) \in O(\mathbb{R}_-)$. *In both cases the upper and lower component of the defining function of the B-transform* $f(t) = [F(z)]$ *are analytic continuations of each other, i.e.,* $F_+(z) = F_-(z) = F(z)$.

Proof. We prove (i), the proof of (ii) is similar and is left to the reader. The upper and lower components of the defining functions of the B-transform may be written in the form

$$F_+(z) = \frac{e^{cx}}{2\pi} \int_0^\infty e^{-\omega y} e^{i(cy + \omega x)} \, \hat{f}(c + i\omega) \, d\omega, \quad y > 0,$$

$$F_-(z) = -\frac{e^{cx}}{2\pi} \int_{-\infty}^0 e^{-\omega y} e^{i(cy + \omega x)} \, \hat{f}(c + i\omega) \, d\omega, \quad y < 0.$$

The two integrals are convergent Laplace integrals. First, this shows that $F_+(z)$ and $F_-(z)$ are holomorphic functions in the upper and lower half-plane, respectively, and secondly, that the B-transform is of bounded exponential growth. If now $\Re z = x < 0$ and by the fact that $c > 0$ can be chosen as large as you wish, we conclude that

$$F_+(z) = \lim_{c \to \infty} \frac{e^{cx}}{2\pi} \int_0^\infty e^{-\omega y} e^{i(cy + \omega x)} \, \hat{f}(c + i\omega) \, d\omega = 0,$$

$$F_-(z) = -\lim_{c \to \infty} \frac{e^{cx}}{2\pi} \int_{-\infty}^0 e^{-\omega y} e^{i(cy + \omega x)} \, \hat{f}(c + i\omega) \, d\omega = 0.$$

This shows that the support of $f(t)$ is contained in $\mathbb{R}_+$. The fact that $F_+(x) = F_-(x) = 0$ for negative x also shows that one function is the analytic continuation (across the negative real axis) of the other. □

Example 3.25. Let $\hat{f}(s) = s^n$, with a non-negative integer n. Here the vertical strip of convergence is the entire complex plane, thus we may take $c = 0$. We set $s = i\omega$, and $z = x + iy$. First let $y > 0$. Then

$$F_+(z) = \frac{1}{2\pi i} \int_0^{i\infty} e^{sz} s^n \, ds = \frac{i^n}{2\pi} \int_0^\infty e^{-\omega y} \, e^{ix\omega} \omega^n \, d\omega.$$

This is a Laplace integral and the image translation rule gives immediately

$$F_+(z) = \frac{i^n}{2\pi} \frac{n!}{(y - ix)^{n+1}} = -\frac{1}{2\pi i} \frac{(-1)^n n!}{z^{n+1}}.$$

Now let $y < 0$, then

$$F_-(z) = -\frac{1}{2\pi i} \int_{i\infty}^0 e^{sz} s^n \, ds$$

$$= -\frac{i^n(-1)^n}{2\pi} \int_0^\infty e^{-\omega(-y)} e^{-ix\omega} \omega^n \, d\omega$$

and we have again a Laplace integral. The image translation rule gives

$$F_-(z) = -\frac{i^n(-1)^n}{2\pi} \frac{n!}{(-y+ix)^{n+1}} = -\frac{1}{2\pi i} \frac{(-1)^n n!}{z^{n+1}}.$$

Therefore, we obtain $f(t) = B[s^n](t) = \delta^{(n)}(t)$ and the B-transform has yielded the inverse Laplace transform.

In the practical important case where $\hat{f}(s)$ is holomorphic in the right half-plane $\sigma_- = 0 < \Re s$ with $c > 0$, and $\hat{f}(s)$ is integrable at $s = 0$, it can be shown that the integrals defining $F_+(z)$ and $F_-(z)$ can be replaced by

$$\frac{1}{2\pi i} \int_0^{i\infty} e^{sz} \hat{f}(s) \, ds - \frac{1}{2\pi i} \int_0^c e^{sz} \hat{f}(s) \, ds$$

$$- \frac{1}{2\pi i} \int_{-i\infty}^0 e^{sz} \hat{f}(s) \, ds - \frac{1}{2\pi i} \int_0^c e^{sz} \hat{f}(s) \, ds.$$

This can be established by considering the two following rectangular contours in the s-plane. In the upper half-plane: $0 \to c \to c + iR \to iR \to 0$ with a small indentation in the form of a quater-circle of radius ϵ to avoid the origin. Similarly, in the lower half-plane: $0 \to -iR \to c-iR \to c \to 0$ with an analogous indentation to avoid 0. The above relations then follows by Cauchy's integral theorem and by letting $\epsilon \to 0$ and $R \to \infty$. Now the integral

$$\phi(z) := \frac{1}{2\pi i} \int_0^c e^{sz} \hat{f}(s) \, ds$$

is an entire function of z, thus can be discarded from the lower and upper compo-nent of the defining function, eventually giving the equivalent defining function

$$F_+(z) = \frac{1}{2\pi i} \int_0^{i\infty} e^{sz} \hat{f}(s) \, ds, \quad (\Im z > 0),$$

$$F_-(z) = \frac{1}{2\pi i} \int_{-\infty}^0 e^{sz} \hat{f}(s) \, ds, \quad (\Im z < 0).$$

Since one of these functions is the analytic continuation of the other, it is sufficient (as in the above example) to compute one of the two integrals in order to find the analytic expression of $F(z)$.

Example 3.26. Let $\hat{f}(s) = s^m \log s$, $\Re s > 0$. We have $\sigma_- = 0$ and $\hat{f}(s)$ is integrable at $s = 0$. Thus, we have

$$F(z) = \frac{1}{2\pi i} \int_0^{i\infty} e^{sz} \hat{f}(s) \, ds, \quad \Im z > 0.,$$

With $z = x + iy$, $y > 0$ we obtain

$$F(z) = \frac{1}{2\pi} \int_0^\infty e^{i\tau(x+iy)} (i\tau)^m \log(i\tau)\, d\tau$$

$$= \frac{i^m}{2\pi} \int_0^\infty e^{-y\tau} e^{i\tau x} \tau^m \{\log\tau + i\frac{\pi}{2}\}\, d\tau$$

$$= \frac{i^{m+1}}{4} \int_0^\infty e^{-y\tau} e^{i\tau x} \tau^m\, d\tau + \frac{i^m}{2\pi} \int_0^\infty e^{-y\tau} e^{i\tau x} \tau^m \log\tau\, d\tau.$$

These are two Laplace integrals. The evaluation of the first one yields

$$\frac{(-1)^{m+1} m!}{4\, z^{m+1}}.$$

The evaluation of the second one gives

$$\frac{(-1)^{m+1} m!}{2\pi i\, z^{m+1}} \{\Psi(m+1) - \log(-z)\} - \frac{(-1)^{m+1} m!}{4\, z^{m+1}}.$$

Therefore,

$$F(z) = -\frac{\Psi(m+1)}{2\pi i} \frac{(-1)^m m!}{z^{m+1}} + (-1)^{m+1} m! \left(-\frac{1}{2\pi i} \frac{\log(-z)}{z^{m+1}}\right)$$

giving the B-transform

$$f(t) = (-1)^{m+1} m!\, \mathrm{fp}\frac{u(t)}{t^{m+1}} + \Psi(m+1)\, \delta^{(m)}(t) \tag{3.135}$$

which, compared with the result of Problem 3.6, shows that in this case too the B-transform has given the inverse Laplace transform.

Example 3.27. Let us consider $\hat{f}(s) = s^\alpha$ in the domain $\Re s > 0$, for $\Re\alpha > -1, \alpha \notin \mathbb{Z}$. By Proposition 3.19 we have $f(t) = [F(z)] \in O(\mathbb{R}_+)$, and

$$f(t) = \mathcal{L}^{-1}[\hat{f}](t) = [F(z) = F_+(z)] = \left[\frac{1}{2\pi i} \int_c^{c+i\infty} e^{sz}\, \hat{f}(s)\, ds\right], \tag{3.136}$$

where, for computing the integral, we assume $\Im z = y > 0$. For $\Re\alpha > -1$, $\hat{f}(s)$ is integrable at $s = 0$ and we may select $c = 0$, thus,

$$F(z) = \frac{1}{2\pi i} \int_0^{i\infty} e^{sz} s^\alpha\, ds.$$

By putting $z = t + iy, y > 0$ and $s = i\omega$ we obtain

$$F(z) = \frac{i^\alpha}{2\pi} \int_0^\infty e^{-y\omega} e^{i\omega t} \omega^\alpha\, d\omega$$

which is an ordinary Laplace integral. By using $u(t)\omega^\alpha \circ\!\!-\!\!\bullet \Gamma(\alpha+1)/y^{\alpha+1}$ and the image shift rule, we obtain

$$F(z) = \frac{i^\alpha}{2\pi} \frac{\Gamma(\alpha+1)}{i^{\alpha+1}(-t-iy)^{\alpha+1}} = \frac{1}{2\pi i} \frac{\Gamma(\alpha+1)}{(-z)^{\alpha+1}}$$

$$= -\frac{\Gamma(\alpha+1)\sin(\pi(-\alpha-1))}{\pi} \{-\frac{(-z)^{-(\alpha+1)}}{2i\sin(\pi(-\alpha-1))}\}$$

$$= -\frac{\Gamma(\alpha+1)\sin(\pi\alpha)}{\pi} \{-\frac{(-z)^{-(\alpha+1)}}{2i\sin(\pi(-\alpha-1))}\}.$$

By the reflection principle for the gamma function, we have

$$\Gamma(1+\alpha) = \Gamma(1-(-\alpha)) = \frac{\pi}{\sin(\pi(-\alpha))\Gamma(-\alpha)} = \frac{\pi}{-\sin(\alpha\pi)\Gamma(-\alpha)}$$

which finally results in

$$F(z) = \frac{1}{\Gamma(-\alpha)}\left\{-\frac{(-z)^{-(\alpha+1)}}{2i\sin(\pi(-\alpha-1))}\right\}.$$

Therefore, we have found for $\Re\alpha > -1, \alpha \notin \mathbb{Z}$, and $\Re s > 0$,

$$B[s^\alpha](t) = u(t)\frac{1}{\Gamma(-\alpha)}t^{-\alpha-1} \qquad (3.137)$$

and this B-transform is again the inverse Laplace transform of s^α. This result generalizes the classical correspondence which is only valid for $\Re\alpha < 0$. Thus, by the Identity Principle for holomorphic functions (3.137) holds for $\alpha \in \mathbb{C} \setminus \mathbb{N}_0$. For $\alpha = n \in \mathbb{N}_0$, we know that

$$\mathcal{L}^{-1}[s^n](t) = \delta^{(n)}(t). \qquad (3.138)$$

Problem 3.33. Find again the inverse Laplace transform of $\hat{f}(s) = s^m \log s$ by differentiating the integral

$$\frac{1}{2\pi i}\int_c^{c+i\infty} e^{sz}\, s^\alpha\, ds$$

with respect to the parameter α. Establish then again

$$B[(s+c)^m \log(s+c)](t) = \mathcal{L}^{-1}[(s+c)^m \log(s+c)](t)$$

$$= \Psi(m+1)\sum_{k=0}^m \binom{m}{k} c^k \delta^{(m-k)}(t) + (-1)^{m+1}\, m!\, e^{-ct}\, \mathrm{fp}\frac{u(t)}{t^{m+1}}.$$

Let us look again at the formula for the B-transform in the case where $\hat{f}(s)$ is holomorphic in a right half-plane:

$$B[\hat{f}(s)](t) = [F(z)], \quad F(z) = \frac{1}{2\pi i}\int_c^{c+i\infty} e^{sz}\, \hat{f}(s)\, ds.$$

Consider the integrand on the quarter circle $s = c + Re^{i\theta}$, $\pi/2 \geq \theta \geq 0$. If $\hat{f}(s) = O(s^m)$ we have the estimate

$$\left|e^{sz}\, \hat{f}(s)\, ds\right| < Ke^{cx}\, e^{xR\cos\theta - yR\sin\theta}\, R^{m+1}\, d\theta,$$

and nothing prevents us from placing $z = x + iy$ in the left half-plane, i.e., with $\Re z = x < 0$ when computing the expression for $F(z)$. The above estimate now indicates that the integrand tends exponentially to zero as $R \to \infty$ and $\pi/2 \geq \theta \geq 0$. Hence, the vertical contour of the integral may be bound to the right, and we obtain

$$F(z) = \frac{1}{2\pi i}\int_c^\infty e^{sz}\, \hat{f}(s)\, ds \qquad (3.139)$$

where the contour is any path starting at c and going to $\infty\, e^{i\theta}, 0 \leq \theta \leq \pi/2$. Thus, it was not at random that in the computing of the B-transform in the above examples absolutely convergent Laplace integrals came up.

We do not provide a formal proof which establishes that the introduced B-transform yields always the converse of the Laplace transform in the general case where the hyperfunction has a support spread over the entire real line. See, however, the end of the next section.

The Inversion Formulas

We shall now prove that for an $f(t) \in O(\mathbb{R}_+)$ the deduced formula from the B-transform provides the converse of the Laplace transform.

Proposition 3.20. *Let $\hat{f}(s)$ be the Laplace transform of the hyperfunction $f(t) = [F(z)] \in O(\mathbb{R}_+)$ assumed to be a holomorphic for $\Re s > \sigma_-$ and satisfying $\hat{f}(s) = O(s^m)$ as $s \to \infty$ for some $m \in \mathbb{N}$. Moreover, let*

$$G(z) = \frac{1}{2\pi i} \int_c^\infty e^{sz}\, \hat{f}(s)\, ds \tag{3.140}$$

with $c > \sigma_-$ and where the contour is any path from c to $\infty\, e^{i\theta}, 0 \leq \theta \leq \pi/2$. Then $G(z) \sim F(z)$.

Proof. We have

$$\hat{f}(s) = -\int_\infty^{(0+)} e^{-s\zeta}\, F(\zeta)\, d\zeta.$$

Inserting this expression into the the integral of $G(z)$, yields

$$G(z) := -\frac{1}{2\pi i} \int_c^\infty e^{sz} \int_\infty^{(0+)} e^{-s\zeta}\, F(\zeta)\, d\zeta\, ds,$$

where we have assumed that z is placed in such a way that $\Re z$ lies to the left of the integration loop of the integral for $\hat{f}(s)$, i.e., we have $\Re(z - \zeta) < 0$ for all ζ. Interchanging the order of integration results in

$$G(z) := -\frac{1}{2\pi i} \int_\infty^{(0+)} F(\zeta) \int_c^\infty e^{s(z-\zeta)}\, ds\, d\zeta.$$

The interchange of the order of integration can be justified since $F(\zeta)$ is of bounded exponential growth yielding an absolute converging double integral. The inner integral can now be evaluated due to $\Re(z - \zeta) < 0$ and gives

$$\int_c^\infty e^{s(z-\zeta)}\, ds = -\frac{e^{c(z-\zeta)}}{z - \zeta}$$

such that

$$G(z) := -\frac{1}{2\pi i} \int_\infty^{(0+)} \frac{e^{c(z-\zeta)}\, F(\zeta)}{\zeta - z}\, d\zeta.$$

We now extend the contour of the integral defining $\hat{f}(s)$ to the left until the point z is enclosed within it, and call the new loop C_z (see Figure 3.4). We may then write

$$G(z) := -\frac{e^{cz}}{2\pi i} \int_{C_z} \frac{e^{-c\zeta}\, F(\zeta)}{\zeta - z}\, d\zeta + \frac{e^{cz}}{2\pi i} \oint_{|\zeta - z| = r} \frac{e^{-c\zeta}\, F(\zeta)}{\zeta - z}\, d\zeta.$$

Inside the small circle $|\zeta - z| = r$ the function $e^{-c\zeta} F(\zeta)$ is holomorphic, such that the second term becomes $F(z)$ due to Cauchy's theorem. We then have

$$G(z) = F(z) + \phi(z),$$

where

$$\phi(z) = -\frac{e^{cz}}{2\pi i} \int_{\gamma_{a,\infty}^+} \frac{e^{-c\zeta} F(\zeta)}{\zeta - z} \, d\zeta + \frac{e^{cz}}{2\pi i} \int_{\gamma_{a,\infty}^-} \frac{e^{-c\zeta} F(\zeta)}{\zeta - z} \, d\zeta$$

is a holomorphic function of z in each compact subset K which does not hit the contour C_z and having an overlap with the real axis (see Figure 3.5 and also Appendix A.3). This establishes that $G(z)$ is an equivalent defining function for the hyperfunction $f(t) = [F(z)]$. $\qquad\square$

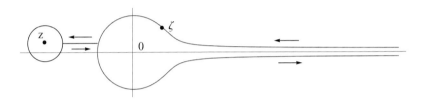

Figure 3.4: Integration loop C_z

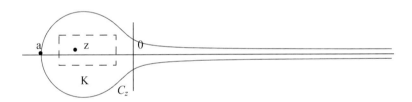

Figure 3.5: Integration loop C_z deformed

The inversion formula (3.140) is the same as formula (A.24) given at the end of Appendix B by the theory of Laplace hyperfunctions.

Example 3.28. Let $\hat{f}(s) = 1/s^n$. We compute the inverse Laplace transform by using (3.139). We may select $c = 1$ and obtain

$$F(z) = \frac{1}{2\pi i} \int_1^\infty e^{-s(-z)} \frac{1}{s^n} \, ds = \frac{1}{2\pi i} E_n(-z).$$

Using (3.68) we find the equivalent defining function

$$-\frac{1}{2\pi i} \frac{z^{n-1}}{(n-1)!} \log(-z)$$

which is a defining function of $u(t) t^{n-1}/(n-1)!$.

Also note that while the integral

$$\mathrm{pv}\frac{1}{2\pi i}\int_{c-i\infty}^{c+i\infty} e^{st}\,\hat{f}(s)\,ds,$$

being essentially a Fourier integral, cannot be differentiated under the integral sign without further restrictions, the integral

$$F(z) = \frac{1}{2\pi i}\int_c^{\infty} e^{sz}\,\hat{f}(s)\,ds$$

is a Laplace integral and may be differentiated under the integral sign as many times as you want, i.e.,

$$\frac{d^n F}{dz^n} = \frac{1}{2\pi i}\int_c^{\infty} e^{sz}\,s^n\,\hat{f}(s)\,ds.$$

In a similiar way the following proposition may be proved, where we leave the details to the reader.

Proposition 3.21. *Let $\hat{f}(s)$ be the Laplace transform of the hyperfunction $f(t) = [F(z)] \in O(\mathbb{R}_-)$ assumed to be holomorphic for $\Re s < \sigma_+$ and satisfying $\hat{f}(s) = O(s^m)$ as $s \to \infty$ for some $m \in \mathbb{N}$. Moreover, let*

$$G(z) = -\frac{1}{2\pi i}\int_{-\infty}^c e^{sz}\,\hat{f}(s)\,ds \qquad\qquad (3.141)$$

with $c < \sigma_+$ and where the contour is any path from $\infty\,e^{-i\theta}$, $-\pi \le \theta \le -\pi/2$ to c. Then $G(z) \sim F(z)$.

Problem 3.34. Use (3.141) to establish that for $\hat{f}(s) = 1/s^n$ and $\hat{f}(s) = s^n$, and for $\Re s < 0$, respectively, we obtain

$$f(t) = -u(-t)\frac{t^n}{(n-1)!} \in O(\mathbb{R}_-),$$

$$f(t) = \delta^{(n)}(t) \in O(\mathbb{R}_-),$$

respectively.

The two above inversion formulas can be used to convert a given image function $\hat{f}(s)$ which is only holomorphic in a vertical strip $\sigma_- < \Re s < \sigma_+$ of finite width, provided the image function can be decomposed into a sum

$$\hat{f}(s) = \hat{f}_1(s) + \hat{f}_2(s), \qquad\qquad (3.142)$$

where $\hat{f}_1(s)$ is holomorphic in the left half-plane $\Re s < \sigma_+$ and $\hat{f}_2(s)$ is holomorphic in the right half-plane $\Re s > \sigma_-$. The following example illustrates this case.

Example 3.29. Let

$$\hat{f}(s) = \log\frac{1-s}{1+s}, \qquad -1 < \Re s < 1.$$

Then

$$\hat{f}(s) = \log \frac{1-s}{1+s} = \log(1-s) - \log(1+s)$$
$$=: \hat{f}_1(s) + \hat{f}_2(s)$$

where the first term is now holomorphic in the left half-plane $\Re s < 1$ and the second one in the right half-plane $\Re s > -1$. There is no need to compute the integrals of the inversion formulas, because Problem 3.26 yields

$$\mathcal{L}^{-1}[\log(s+1)](t)$$
$$= -\mathrm{fp}\frac{u(t)}{t}e^{-t} + \Phi(1)\,\delta(t), \quad (\Re s > -1),$$
$$\mathcal{L}^{-1}[\log(1-s)](t)$$
$$= \mathrm{fp}\frac{u(-t)}{t}e^t + \Phi(1)\,\delta(t), \quad (\Re s < 1).$$

We then obtain

$$\mathcal{L}^{-1}[\log \frac{1-s}{1+s}](t) = \mathrm{fp}\frac{u(t)}{t}e^{-t} + \mathrm{fp}\frac{u(-t)}{t}e^t = \mathrm{fp}\frac{1}{t}e^{-|t|} \tag{3.143}$$

where the the two Dirac impulses have been canceled out.

Laplace hyperfunctions

We have introduced the Laplace and inverse Laplace transform of hyperfunctions in a straightforward and intuitive way with a minimum of abstract mathematical prerequisites. There is another approach due to [22, Komatsu] which is from a pure mathematico-logical point of view perhaps preferable. This approach is based on the theory of Laplace transforms of holomorphic functions of exponential type in a sector of the complex plane, a theory first developed by G. Pólya and A. J. Macintyre (see [24, Macintyre]).

   In the paper of [22, Komatsu] , the so-called Laplace hyperfunctions are defined by

$$\mathcal{B}_L(\mathbb{R}_+) := \mathcal{O}_{exp}(\mathbb{C} \setminus [0, \infty))/\mathcal{O}_{exp}(\mathbb{C}), \tag{3.144}$$

i.e., by the quotient space of the set of holomorphic functions $F(z)$ of exponential type in $\mathbb{C} \setminus [0, \infty)$, over the set of holomorphic functions of exponential type in the entire plane.

   The Laplace transform of a Laplace hyperfunction is then defined in a slightly more special way. Instead of a general loop which encloses the positive part of the real axis an angle is used. The advantage of this approach is that by using the mentioned theory of Pólya and Macintyre, the inversion formula for the inverse Laplace transform of Laplace hyperfunctions directly results. It is the same as (3.140).

   The mathematics of Komatsu's approach is highly sophisticated and about the level used throughout this book, so we have presented for the mathematically interested reader a slightly simplified version of Komatsu's theory in Appendix B.

3.6.2   The Convolution Rule

Let $f(t) = [F(z)]$ and $g(t) = [G(z)]$ be two given hyperfunctions such that their convolution

$$h(t) = f(t) \star g(t) = [H(z)]$$

exists, where

$$H_+(z) = \int_{-\infty}^{\infty} F_+(z-x)g(x)\,dx, \quad \Im z > 0,$$

$$H_-(z) = \int_{-\infty}^{\infty} F_-(z-x)g(x)\,dx, \quad \Im z < 0.$$

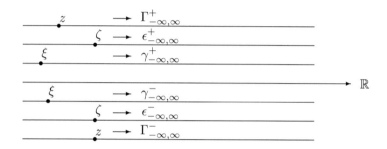

Figure 3.6: Contours for the proof of Proposition 3.22

Let us assume further that there is a real point c where both $f(t)$ and $g(t)$ are holomorphic, then we can write for the Laplace transform of the convolution

$$\hat{h}(s) = \int_{-\infty}^{\infty} e^{-sz} H(z)\,dz$$

$$= \int_{\Gamma_{-\infty,\infty}^{+}} e^{-sz} \int_{-\infty}^{\infty} F_+(z-x)g(x)\,dx\,dz - \int_{\Gamma_{-\infty,\infty}^{-}} e^{-sz} \int_{-\infty}^{\infty} F_-(z-x)g(x)\,dx\,dz$$

$$= \int_{\Gamma_{-\infty,\infty}^{+}} e^{-sz}\Big[\int_{\gamma_{-\infty,\infty}^{+}} F_+(z-\xi)G_+(\xi)\,d\xi - \int_{\gamma_{-\infty,\infty}^{-}} F_+(z-\xi)G_-(\xi)\,d\xi\Big]\,dz$$

$$- \int_{\Gamma_{-\infty,\infty}^{-}} e^{-sz}\Big[\int_{\gamma_{-\infty,\infty}^{+}} F_-(z-\xi)G_+(\xi)\,d\xi - \int_{\gamma_{-\infty,\infty}^{-}} F_-(z-\xi)G_-(\xi)\,d\xi\Big]\,dz.$$

The various contours are indicated in Figure 3.6. Interchanging the order of integration yields

$$\int_{\gamma_{-\infty,\infty}^{+}} G_+(\xi)\Big[\int_{\Gamma_{-\infty,\infty}^{+}} e^{-sz} F_+(z-\xi)\,dz - \int_{\Gamma_{-\infty,\infty}^{-}} e^{-sz} F_-(z-\xi)\,dz\Big]\,d\xi$$

$$- \int_{\gamma_{-\infty,\infty}^{-}} G_-(\xi)\Big[\int_{\Gamma_{-\infty,\infty}^{+}} e^{-sz} F_+(z-\xi)\,dz + \int_{\Gamma_{-\infty,\infty}^{-}} e^{-sz} F_-(z-\xi)\,dz\Big]\,d\xi.$$

The substitution $z - \xi = \zeta$ in the inner integrals results in

$$\int_{\gamma^+_{-\infty,\infty}} e^{-s\xi} G_+(\xi)\, d\xi\, [\int_{\epsilon^+_{-\infty,\infty}} e^{-s\zeta} F_+(\zeta)\, d\zeta - \int_{\epsilon^-_{-\infty,\infty}} e^{-s\zeta} F_-(\zeta)\, d\zeta]$$

$$- \int_{\gamma^-_{-\infty,\infty}} e^{-s\xi} G_-(\xi)\, d\xi\, [\int_{\epsilon^+_{-\infty,\infty}} e^{-s\zeta} F_+(\zeta)\, d\zeta + \int_{\epsilon^-_{-\infty,\infty}} e^{-sz} F_-(\zeta)\, d\zeta],$$

which eventually becomes

$$\int_{\gamma^+_{-\infty,\infty}} e^{-s\xi} G_+(\xi)\, d\xi\, \hat{f}(s) - \int_{\gamma^-_{-\infty,\infty}} e^{-s\xi} G_-(\xi)\, d\xi\, \hat{f}(s)$$

$$= \hat{f}(s)\, \hat{g}(s) = \hat{h}(s).$$

We have established the *Convolution Rule for Laplace transformation.*

Proposition 3.22. *Let the Laplace transforms $\hat{f}(s)$ and $\hat{g}(s)$ of the hyperfunctions $f(t)$ and $g(t)$ have a non-void common strip of convergence $\sigma_- < \Re s < \sigma_+$, and assume further that the convolution $h(t) := (f \star g)(t)$ exists and has the Laplace transform $\hat{h}(s)$, then*

$$\hat{h}(s) = \hat{f}(s)\, \hat{g}(s) \tag{3.145}$$

holds.

In the event of the properties

$$\delta(t - a) \star \delta(t - b) = \delta(t - a - b),$$
$$\delta^{(n)}(t - a) \star g(t) = D^n g(t - a)$$

the reader may readily verify the convolution rule.

The convolution of arbitrary hyperfunctions need not exist. However, we know that if $f(t), g(t) \in \mathcal{A}(\mathbb{R}_+, \star)$ (convolution algebra), the convolution product always exists, and is commutative and associative. Moreover, in this case we have

$$(f \star g)(t) = \left[-\oint_{(c,c_z)} F(z - \zeta)\, G(\zeta)\, d\zeta \right], \tag{3.146}$$

where the contour (c, c_z) (Figure 3.7) encircles the point 0 and has the right real vertex c_z with $c_z > \Re z$.

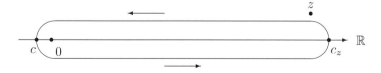

Figure 3.7: Contour (c, c_z)

In order to compute the above contour integral, note that if $G(\zeta)$ and $F(\zeta)$ are integrable at $\zeta = 0$ the contour can be shrinked to the upper and lower side of the segment $(0, c_z)$, where z of the upper half-plane is identified with $c_z + i0$ and z of the lower half-plane with $c_z - i0$. Thus, we write

$$(f \star g)(t) = \left[\int_0^z F(z - \zeta)\, g(\zeta)\, d\zeta \right]. \tag{3.147}$$

The following examples illustrate this feature.

Example 3.30. Let us compute the convolution of the hyperfunctions $f(t) = t^n u(t) = \left[-\frac{\zeta^n}{2\pi i} \log(-\zeta) \right]$ and $g(t) = t^m u(t) = \left[-\frac{\zeta^m}{2\pi i} \log(-\zeta) \right]$. Here n, m are positive integral exponents.

$$(f \star g)(t) = \left[(-\frac{1}{2\pi i})^2 \oint_{(c, c_z)} (z - \zeta)^n\, \zeta^m \log(\zeta - z) \log(-\zeta) d\zeta \right].$$

Since the integrand is integrable at $x = 0$ and at $x = z$, we may shrink the contour to the upper and lower side of the interval $(0, z)$ which results in

$$(f \star g)(t) = \left[-\frac{1}{2\pi i} \int_0^z (z - \zeta)^n\, \zeta^m \log(\zeta - z) d\zeta \right].$$

By using the binomial theorem, we obtain

$$-\frac{1}{2\pi i} \int_0^z (z - \zeta)^n\, \zeta^m \log(\zeta - z) d\zeta = -\frac{1}{2\pi i} \int_0^z \zeta^n\, (z - \zeta)^m \log(-\zeta) d\zeta$$

$$= -\sum_{k=0}^m (-1)^k \binom{m}{k} z^{m-k} \frac{1}{2\pi i} \int_0^z \zeta^{n+k} \log(-\zeta)\, d\zeta$$

$$= -(-1)^n \sum_{k=0}^m \binom{m}{k} z^{m-k} \frac{1}{2\pi i} \int_0^z (-\zeta)^{n+k} \log(-\zeta)\, d\zeta.$$

Integration by parts yields

$$-\frac{1}{2\pi i} \int_0^z (z - \zeta)^n\, \zeta^m \log(\zeta - z) d\zeta$$

$$= -(-1)^n \sum_{k=0}^m \binom{m}{k} z^{m-k} \frac{-1}{2\pi i} \{ \frac{(-z)^{n+k+1}}{n+k+1} [\log(-z) - \frac{1}{n+k+1}] \}.$$

The real analytic term can be discarded and we obtain

$$\sum_{k=0}^m (-1)^k \frac{1}{n+k+1} \binom{m}{k} \frac{-1}{2\pi i} z^{n+m+1} \log(-z).$$

Now, it can be shown (hint: use the binomial theorem in the defining integral for Euler's Beta function) that

$$\sum_{k=0}^m (-1)^k \frac{1}{n+k+1} \binom{m}{k} = \frac{n!\, m!}{(n+m+1)!} = B(n+1, m+1).$$

So, we finally obtain

$$u(t)\, t^n \star u(t)\, t^m u(t) = \frac{n!\, m!}{(n+m+1)!}\, u(t)\, t^{n+m+1}$$

in agreement with a computation using the convolution rule.

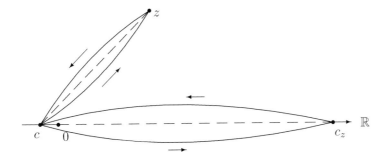

Figure 3.8: Contours in the proof of (3.152)

The first integral can be written as

$$\frac{1}{2\pi i} \oint_{(c,z)} (z - \zeta)^{n-1} \log(\zeta - z) F(\zeta) \, d\zeta$$

$$= \frac{1}{2\pi i} \int_c^z (z - \zeta)^{n-1} \log(\zeta - z) F(\zeta) \, d\zeta$$

$$- \frac{1}{2\pi i} \int_c^z (z - \zeta)^{n-1} \{\log(\zeta - z) + 2\pi i\} F(\zeta) \, d\zeta$$

$$= - \int_c^z (z - \zeta)^{n-1} F(\zeta) \, d\zeta.$$

Thus,

$$\int_c^z (z - \zeta)^{n-1} F(\zeta) \, d\zeta = \frac{1}{2\pi i} \oint_{(c,c_z)} (z - \zeta)^{n-1} \log(\zeta - z) F(\zeta) \, d\zeta. \qquad \Box$$

3.6.3 Fractional Integrals and Derivatives

In Section 3.5.6 we have defined the n-th anti-derivative of $f(x) = F[z] \in O(\mathbb{R}_+)$ as

$$_cD^{-n} f(t) := \frac{1}{(n-1)!} \int_c^z (z - \zeta)^{n-1} F(\zeta) \, d\zeta,$$

where $n \in \mathbb{N}$, $c < 0$. At the end of the previous section we have proved that we can also write

$$_cD^{-n} f(t) := \frac{1}{(n-1)!} \frac{1}{2\pi i} \oint_{(c,c_z)} (z - \zeta)^{n-1} \log(\zeta - z) F(\zeta) \, d\zeta$$

$$= \frac{1}{(n-1)!} t^{n-1} u(t) \star f(t) = \frac{1}{\Gamma(n)} t^{n-1} u(t) \star f(t).$$

The last expression on the right-hand side has still a meaning if the positive integer n is replaced by any real number α different from zero or a negative integer. This leads to the following

Definition 3.7. Let $\alpha \in \mathbb{R}$ be different from zero or any negative integer, then the *fractional integral of order α of the hyperfunction* $f(t) = F[z] \in O(\mathbb{R}_+)$ *is defined*

and denoted by

$$I^\alpha f(t) := \frac{1}{\Gamma(\alpha)} t^{\alpha-1} u(t) \star f(t). \tag{3.153}$$

The *fractional derivative of order* $\alpha \notin \mathbb{N}_0$ is defined by

$$D^\alpha f(t) := I^{-\alpha} f(t) := \frac{1}{\Gamma(-\alpha)} t^{-\alpha-1} u(t) \star f(t). \tag{3.154}$$

The convolution rule immediately gives

$$I^\alpha f(t) = \mathcal{L}^{-1}[s^{-\alpha} \hat{f}(s)](t), \tag{3.155}$$

$$D^\alpha f(t) = \mathcal{L}^{-1}[s^\alpha \hat{f}(s)](t). \tag{3.156}$$

Example 3.33.

$$D^{1/2}\delta(t) = \mathcal{L}^{-1}[\sqrt{s}\,1](t) = \frac{1}{2\sqrt{\pi}} \operatorname{fp} \frac{u(t)}{t^{3/2}},$$

$$D^{1/2}\delta^{(m)}(t-a) = \mathcal{L}^{-1}[s^{m+1/2} e^{-as}](t)$$

$$= \frac{1}{\Gamma(-m-3/2)} \operatorname{fp} \frac{u(t-a)}{(t-a)^{m+3/2}}.$$

If $\beta - \alpha + 1$ and $\beta + 1$ are neither zero nor a negative integer,

$$D^\alpha u(t)t^\beta = \mathcal{L}^{-1}[\frac{\Gamma(\beta+1)}{s^{\beta-\alpha+1}}](t) = \frac{\Gamma(\beta+1)}{\Gamma(\beta-\alpha+1)} u(t)t^{\beta-\alpha},$$

$$D^{1/2}u(t)e^{at} = \mathcal{L}^{-1}[\frac{\sqrt{s}}{s-a}](t) = u(t)\left\{\frac{1}{\sqrt{\pi t}} + \sqrt{a}\, e^{at} \operatorname{erf}(\sqrt{at})\right\}.$$

3.7   Right-sided Laplace Transforms

The *right-sided (one-sided or ordinary) Laplace transform* of an ordinary function $f(t)$, $t \in \mathbb{R}$, and of bounded exponential growth as $t \to \infty$ is defined by

$$\mathcal{L}_+[f(t)](s) := \int_0^\infty e^{-st} f(t)\, dt. \tag{3.157}$$

Let us assume that in this section *all hyperfunctions are*

(i) *defined on the entire real line and are of bounded exponential growth as* $t \to$ ∞,

(ii) *holomorphic at* $t = 0$.

If $f(t) = [F_-(z), F_+(z)]$ is a hyperfunction, then

$$\exp(-st)\, f(t) = [\exp(-sz)F_-(z), \exp(-sz)F_+(z)]$$

is also a hyperfunction due to the real analyticity of the exponential factor, and its integral

$$\int_0^\infty e^{-st} f(t)\, dt := \int_{\gamma_{0,\infty}^+} e^{-sz} F_+(z)\, dz - \int_{\gamma_{0,\infty}^-} e^{-sz} F_-(z)\, dz$$

make sense. We show now

Proposition 3.25. *We have*

$$\mathcal{L}[\chi_{(0,\infty)} f(t)](s) = \mathcal{L}_+[f(t)](s). \tag{3.158}$$

Proof. We have

$$\chi_{(0,\infty)} f(t) = \left[\frac{e^{\lambda z}}{2\pi i} \int_0^\infty \frac{e^{-\lambda t} f(t)}{t - z} \, dt \right] \in O(\mathbb{R}_+)$$

where $\exp(-\lambda t)$ is a convenient converging factor in order to assure convergence of the integral. Then, by the definition of the Laplace transform, and by interchanging the order of integration (see Appendix A.3), we have

$$\mathcal{L}[\chi_{(0,\infty)} f(t)](s) = - \int_\infty^{(0+)} e^{-sz} \frac{e^{\lambda z}}{2\pi i} \int_0^\infty \frac{e^{-\lambda t} f(t)}{t - z} \, dt \, dz$$

$$= \int_0^\infty e^{-\lambda t} f(t) \frac{1}{2\pi i} \int_\infty^{(0+)} \frac{e^{-z(s-\lambda)}}{z - t} \, dz \, dt$$

$$= \int_0^\infty e^{-\lambda t} f(t) \operatorname{Res}_{z=t} \left\{ \frac{e^{-z(s-\lambda)}}{z - t} \right\} dt$$

$$= \int_0^\infty e^{-\lambda t} f(t) e^{-t(s-\lambda)} \, dt = \int_0^\infty e^{-st} f(t) \, dt$$

$$= \mathcal{L}_+[f(t)](s). \qquad \square$$

Corollary 3.26. *If $\alpha \notin \mathbb{Z}$ we have for the fractional derivative*

$$D^\alpha \left(\chi_{(0,\infty)} f(t) \right) \circ - \bullet s^\alpha \, \mathcal{L}_+[f(t)](s). \tag{3.159}$$

Proof. By (3.156) we have

$$D^\alpha \left(\chi_{(0,\infty)} f(t) \right) \circ - \bullet s^\alpha \, \mathcal{L}[\chi_{(0,\infty)} f(t)](s) = s^\alpha \, \mathcal{L}_+[f(t)](s). \qquad \square$$

We know that $\mathcal{L}[D\chi_{(0,\infty)} f(t)](s) = s \, \mathcal{L}[\chi_{(0,\infty)} f(t)](s)$. Consider now $\mathcal{L}[\chi_{(0,\infty)} Df(t)](s)$. We have then for $f(t) = [F_-(z), F_+(z)]$,

$$\chi_{(0,\infty)} Df(t) = \left[\frac{e^{\lambda z}}{2\pi i} \left\{ \int_{\gamma_{(0,\infty)}^+} \frac{e^{-\lambda \zeta} F_+'(\zeta)}{\zeta - z} \, d\zeta - \int_{\gamma_{(0,\infty)}^-} \frac{e^{-\lambda \zeta} F_-'(\zeta)}{\zeta - z} \, d\zeta \right\} \right]$$

$$= \left[-\frac{e^{\lambda z}}{2\pi i} \oint_{(0,\infty)} \frac{e^{-\lambda \zeta} F'(\zeta)}{\zeta - z} \, d\zeta \right]$$

with

$$F'(z) := \begin{cases} F_+'(z), & \Im z > 0 \\ F_-'(z), & \Im z < 0. \end{cases}$$

Integration by parts yields

$$\chi_{(0,\infty)}Df(t) = \left[\frac{e^{\lambda z}}{2\pi i} \left\{ \frac{F_+(0) - F_-(0)}{z} - \lambda \frac{e^{\lambda z}}{2\pi i} \oint_{(0,\infty)} \frac{e^{-\lambda \zeta}F(\zeta)}{z - \zeta} d\zeta \right. \right.$$

$$+ \left[\frac{e^{\lambda z}}{2\pi i} \oint_{(0,\infty)} \frac{e^{-\lambda \zeta}F(\zeta)}{(z - \zeta)^2} d\zeta \right]$$

$$= -f(0)\,\delta(t)$$

$$+ \left[\frac{e^{\lambda z}}{2\pi i} \left\{ -\lambda \oint_{(0,\infty)} \frac{e^{-\lambda \zeta}F(\zeta)}{z - \zeta} d\zeta + \oint_{(0,\infty)} \frac{e^{-\lambda \zeta}F(\zeta)}{(z - \zeta)^2} d\zeta \right\} \right]$$

where in the first expression on the right-hand side we used $e^{\lambda z} = 1 + \lambda z + \dots$ invoking a simplification by a passage to an equivalent defining function. Now,

$$\mathcal{L}[\chi_{(0,\infty)}Df(t)](s) = -f(0) + \int_\infty^{(0+)} e^{-sz} \frac{e^{\lambda z}}{2\pi i} \{ -\lambda \oint_{(0,\infty)} \frac{e^{-\lambda \zeta}F(\zeta)}{z - \zeta} d\zeta$$

$$+ \oint_{(0,\infty)} \frac{e^{-\lambda \zeta}F(\zeta)}{(z - \zeta)^2} d\zeta \} dz.$$

Interchanging the order of integration yields

$$- f(0) + \oint_{(0,\infty)} e^{-\lambda \zeta} F(\zeta) \{ -\frac{\lambda}{2\pi i} \int_\infty^{(0+)} \frac{e^{-z(s-\lambda)}}{z - \zeta} dz$$

$$+ \frac{1}{2\pi i} \int_\infty^{(0+)} \frac{e^{-z(s-\lambda)}}{(z - \zeta)^2} dz \} d\zeta$$

$$= -f(0) + \oint_{(0,\infty)} e^{-\lambda \zeta} F(\zeta) \{ -\lambda e^{-\zeta(s-\lambda)} - (s - \lambda) e^{-\zeta(s-\lambda)} \} d\zeta$$

$$= -f(0) + \lambda \mathcal{L}_+[f(t)](s) + (s - \lambda) \mathcal{L}_+[f(t)](s) = -f(0) + s\mathcal{L}_+[f(t)](s).$$

We have proved

Proposition 3.27. *If $f(t) = [F_+(z), F_-(z)]$, then*

$$\mathcal{L}[\chi_{(0,\infty)}Df(t)](s) = \mathcal{L}_+[Df(t)](s) = s\mathcal{L}_+[f(t)](s) - f(0)$$
$$= s\mathcal{L}[\chi_{(0,\infty)}f(t)](s) - f(0). \tag{3.160}$$

Replacing $f(t)$ by $Df(t)$, we obtain

$$\mathcal{L}_+[D^2 f(t)](s) = s^2 \mathcal{L}_+[f(t)](s) - sf(0) - f'(0) \tag{3.161}$$

where $f'(0) = F'_+(0) - F'_-(0)$, and so on.

Example 3.34. Let $\phi(t)$ be a real analytic function defined on $\mathbb{R}$. We consider the hyperfunction $f(t) = [\phi(z)/2, -\phi(z)/2]$, then $Df(t) = [\phi'(z)/2, -\phi'(z)/2]$. Clearly,

$$\mathcal{L}_+[\phi(t)](s) = \int_0^\infty e^{-st} \phi(t)\, dt = \hat{f}(s), \quad \mathcal{L}_+[\frac{d\phi(t)}{dt}](s) = \int_0^\infty e^{-st} \frac{d\phi(t)}{dt}\, dt$$

provided the integrals are convergent. Then,

$$\mathcal{L}_+[\frac{\phi(t)}{dt}](s) = s\,\hat{f}(s) - \phi(0).$$

Example 3.35. Consider the following fractional ordinary differential equation (see also [7, Debnath - Bhatta Section 6.3])

$$D y(t) + b\, D^\alpha y(t) + a\, y(t) = 0, \quad t > 0, \tag{3.162}$$

under the initial data $y(0) = y_0 \neq 0$. By taking the right-sided Laplace transform, we obtain the image system

$$s\hat{y}(s) + b\, s^\alpha \hat{y}(s) + a\, \hat{y}(s) = y_0$$

i.e.,

$$\hat{y}(s) = \frac{y_0}{s + b\, s^\alpha + a}$$

where $\hat{y}(s) = \mathcal{L}_+[y(t)](s)$. The case $a = 0$ can be inverted by using the *Mittag-Leffler function*

$$E_{\mu,\nu}(z) := \sum_{k=0}^{\infty} \frac{z^k}{\Gamma(k\,\mu + \nu)}, \tag{3.163}$$

(for example $E_{1,1}(z) = \exp(z)$). By writing

$$\frac{1}{s + b\, s^\alpha} = \frac{1}{s}\, \frac{1}{1 + \frac{b}{s^{1-\alpha}}},$$

developing the second factor into a series, using the Integration Rule for the right-sided Laplace transformation, and finally some elementary properties of the Gamma function, we obtain for $0 < \alpha < 1$,

$$Y(t)\, E_{1-\alpha,1}(-b\, t^{1-\alpha}) \circ\!\!-\!\!\bullet \frac{1}{s + b\, s^\alpha}. \tag{3.164}$$

The case $\alpha = 1/2$ can be handled by writing

$$\hat{y}(s) = \frac{y_0}{s + b\, \sqrt{s} + a} = \frac{y_0'}{(\sqrt{s})^2 + b\, \sqrt{s} + a}$$

and applying partial fraction decomposition to the rational function

$$g(s) := \frac{1}{s^2 + b\, s + a} = \frac{c_1}{s - \gamma_1} + \frac{c_2}{s - \gamma_2},$$

and finally using the correspondences

$$\mathcal{L}^{-1}[\frac{1}{\sqrt{s} - \gamma}](t) = Y(t)\,\{\frac{1}{\sqrt{\pi t}} + \gamma\, e^{\gamma^2 t}\,(1 + \mathrm{erf}(\gamma\,\sqrt{t}))\},$$

$$\mathcal{L}^{-1}[\frac{1}{\sqrt{s} + \gamma}](t) = Y(t)\,\{\frac{1}{\sqrt{\pi t}} - \gamma\, e^{\gamma^2 t}\,(1 + \mathrm{erfc}(\gamma\,\sqrt{t}))\}.$$

3.8 Integral Equations II

3.8.1 Volterra Integral Equations of Convolution Type

A *Volterra integral equation of the first kind* is

$$\int_0^t k(t - x)\, y(x)\, dx = f(t), \tag{3.165}$$

where $y(t)$ is the unknown function or hyperfunction, and the kernel $k(t)$ and the second member $f(t)$ are given functions or hyperfunctions. A *Volterra integral equation of the second kind* is an equation of the form

$$y(t) + \int_0^t k(t-x)\, y(x)\, dx = f(t). \tag{3.166}$$

By assuming that all involved functions or hyperfunctions are right-sided, i.e., their support is contained in $\mathbb{R}_+$, we can write

$$k(t) \star y(t) = f(t), \quad y(t) + k(t) \star y(t) = f(t) \tag{3.167}$$

for these equations. Because $\delta(t)\star y(t) = y(t)$, we may write for the second equation

$$y(t) + k(t) \star y(t) = \delta(t) \star y(t) + k(t) \star y(t) = (\delta(t) + k(t)) \star y(t)$$
$$=: k_1(t) \star y(t) = f(t),$$

and the difference between the two types has disappeared. Volterra integral equations of convolution type are most easily solved by means of Laplace transforms.

Example 3.36.

$$\int_0^t k(t-x)\, y(x)\, dx = k^{(n)}(t).$$

The solution is given by $y(t) = \delta^{(n)}(t)$, reflecting $k(t) * \delta^{(n)}(t) = D^n k(t)$. Note that the involved derivative is in the generalized sense.

Example 3.37. For $t > 0$ let

$$\int_0^t \cos(t-x)\, y(x)\, dx = 1 + t + t^2. \tag{3.168}$$

We write this as

$$u(t) \cos t \star y(t) = (1 + t + t^2) u(t),$$

where $y(t)$ is supposed to be in $O(\mathbb{R}_+)$. By taking Laplace transforms on both sides, we obtain the image equation

$$\frac{s}{s^2 + 1}\, \hat{y}(s) = \frac{1}{s} + \frac{1}{s^2} + \frac{2}{s^3}$$

due to the convolution rule. Solving the algebraic equation for the unknown $\hat{y}(s)$ yields for $\Re s > 0$,

$$\hat{y}(s) = \frac{(s^2 + 1)(s^2 + s + 2)}{s^4} = 1 + \frac{1}{s} + \frac{3}{s^2} + \frac{1}{s^3} + \frac{2}{s^4}.$$

Taking the inverse Laplace transform gives

$$y(t) = \delta(t) + u(t)\,(1 + 3t + t^2/2 + t^3/3).$$

The occurrence of delta impulses and/or their derivatives are typical for solutions of Volterra integral equations of the first kind. Instead of inserting $u(t)$, and working with the two-sided Laplace transform, we can also do it with the one-sided Laplace transform from the outset for problems of this kind.

Example 3.38. For $t > 0$ let

$$y(t) + \int_0^t \sin(t - x)\, y(x)\, dx = \cos t. \tag{3.169}$$

By taking the one-sided Laplace transforms on both sides, we obtain the image equation

$$\hat{y}(s) + \frac{1}{s^2 + 1}\, \hat{y}(s) = \frac{s}{s^2 + 1}$$

due to the convolution rule. Solving the algebraic equation for the unknown $\hat{y}(s)$ yields

$$\hat{y}(s) = \frac{s}{s^2 + 2},$$

thus, for $t > 0$, we have

$$y(t) = \cos(\sqrt{2}\, t).$$

The *Mathematica* package `LaplaceAndzTransforms` of the author is a useful help for doing calculations of this kind (see [14, Graf]).

Expansion of the Kernel

Consider the integral equation

$$y(t) + \lambda \int_0^t k(t - x)\, y(x)\, dx = f(t), \tag{3.170}$$

where λ is a real or complex parameter. Assume that all involved functions or hyperfunctions are right-sided, i.e., we consider the problem in the convolution algebra $\mathcal{A}(\mathbb{R}_+, \star)$. We have

$$\{\delta(t) + \lambda\, k(t)\} \star y(t) = f(t), \tag{3.171}$$

and we want to find the inverse element of $\delta(t) + \lambda\, k(t)$ with respect to the convolution product. Consider

$$\{\delta - \lambda\, k + \lambda^2\, k \star k - \lambda^3\, k \star k \star k\} \star \{\delta + \lambda k\}$$
$$= \delta \star \delta + \lambda \delta \star k - \lambda\, k \star \delta - \lambda^2\, k \star k + \lambda^2\, k \star k \star \delta + \lambda^3\, k \star k \star k - \lambda^3\, k \star k \star k \star \delta$$
$$- \lambda^4\, k \star k \star k \star k = \delta - \lambda^4\, k \star k \star k \star k,$$

and we see that for any finite m we have

$$\left\{\delta(t) + \sum_{n=1}^{m} (-\lambda)^n\, k(t)^{\star n}\right\} \star \{\delta(t) + \lambda\, k(t)\} = \delta(t) - \lambda^{m+1}\, k(t)^{\star(m+1)}$$

which shows, provided $\lambda^m\, k(t)^{\star m} \to 0$ as $m \to \infty$, that the inverse of $\delta(t) + \lambda\, k(t)$ with respect to the convolution product is given by

$$K(t) := \delta(t) + \sum_{n=1}^{\infty} (-\lambda)^n\, k(t)^{\star n}, \tag{3.172}$$

and the solution of (3.171) becomes

$$y(t) = K(t) \star f(t) \tag{3.173}$$

or,

$$y(t) = \int_0^t K(t - x) f(x) \, dx. \tag{3.174}$$

Assume now, that the kernel $k(t)$ has a one-sided Laplace transform, then by taking the Laplace transform of equation (3.171), where $f(t)$ is replaced by $\delta(t)$, we obtain

$$(1 + \lambda \hat{k}(s)) \hat{K}(s) = 1,$$

and the inverse element $K(t)$ is given for $t > 0$ by

$$K(t) = \mathcal{L}^{-1} \Big[\frac{1}{1 + \lambda \hat{k}(s)} \Big](t). \tag{3.175}$$

Abel's Integral equation

The integral equation

$$\int_0^t \frac{y(x)}{(t - x)^\alpha} \, dx = f(t) \quad (0 < \alpha < 1) \tag{3.176}$$

is known as Abel's integral equation. If we take Hadamard's finite part of the integral, i.e.,

$$\mathrm{FP} \int_0^t \frac{y(x)}{(t - x)^\alpha} \, dx = f(t), \tag{3.177}$$

we can lessen the restriction $0 < \alpha < 1$ to $\alpha \notin \mathbb{Z}$. We shall solve the convolution equation

$$y(t) \star u(t) t^{-\alpha} = f(t).$$

with $y(t), f(t) \in O(\mathbb{R}_+)$. Taking Laplace transforms yields the image equation

$$\hat{y}(s) \frac{\Gamma(-\alpha + 1)}{s^{-\alpha+1}} = \hat{f}(s),$$

thus

$$\hat{y}(s) = \frac{s}{\Gamma(1 - \alpha)} \frac{\hat{f}(s)}{s^\alpha} = \frac{s}{\Gamma(1 - \alpha)\Gamma(\alpha)} \frac{\Gamma(\alpha)}{s^\alpha} \hat{f}(s).$$

Taking the inverse Laplace transform and using the convolution and differentiation rules yields

$$y(t) = \frac{1}{\Gamma(1 - \alpha)\Gamma(\alpha)} D(u(t) t^{\alpha - 1} \star f(t)) \tag{3.178}$$

or, if we assume that the convolution integral exists,

$$\begin{aligned}
y(t) &= \frac{1}{\Gamma(1 - \alpha)\Gamma(\alpha)} D \int_0^t f(x) (t - x)^{\alpha - 1} \, dx \\
&= \frac{\sin(\alpha\pi)}{\pi} D \int_0^t \frac{f(x)}{(t - x)^{1 - \alpha}} \, dx.
\end{aligned} \tag{3.179}$$

Note that the differentiation with respect to t of the integral is in the generalized sense. Let us check the above formula with $f(t) = u(t)t^{-\alpha}$, i.e., we attempt to solve the integral equation

$$\int_0^t \frac{y(x)}{(t-x)^\alpha}\, dx = u(t)t^{-\alpha}.$$

The solution is given by

$$y(t) = \frac{\sin(\alpha\pi)}{\pi} D(u(t)t^{\alpha-1} \star u(t)t^{-\alpha}).$$

By using the result (3.150), we obtain

$$y(t) = \frac{\sin(\alpha\pi)}{\pi} B(\alpha, -\alpha+1)Du(t)$$

$$= \frac{1}{\Gamma(1-\alpha)\Gamma(\alpha)}\, \Gamma(\alpha)\Gamma(-\alpha+1)\, Du(t) = \delta(t)$$

as it is to be expected.

Problem 3.35. Show that the solution of the integral equation

$$\int_0^t \frac{y(x)}{(t-x)^\alpha}\, dx = u(t)t^\beta,$$

where α, β are non-integral, and $\alpha + \beta \notin -\mathbb{N}_0$ is given by

$$y(t) = \frac{\sin(\pi\alpha)\, \Gamma(\alpha)\, \Gamma(1+\beta)}{\pi\, \Gamma(\alpha+\beta)}\, u(t)t^{\alpha+\beta-1}.$$

From (3.178) and the fact that $Du(t)t^{\alpha-1} = (\alpha-1)u(t)t^{\alpha-2}$ we obtain

$$y(t) = \frac{\alpha-1}{\Gamma(1-\alpha)\Gamma(\alpha)} u(t)t^{\alpha-2} \star f(t). \qquad (3.180)$$

By using Proposition 3.23 we obtain the solution of Abel's integral equation for $f(t) = \text{fp}\, u(t)t^{-m}$, m a positive integer, by

$$\frac{\alpha-1}{\Gamma(1-\alpha)\Gamma(\alpha)}\, u(t)\, t^{\alpha-2} \star u(t)\, t^{-m} = \frac{(-1)^{m-1}(\alpha-1)\,\Gamma(\alpha-1)}{\Gamma(m)\,\Gamma(\alpha)\,\Gamma(1-\alpha)\,\Gamma(\alpha-m)}$$

$$\times \{\Psi(m) - \Psi(\alpha-m) + \log t\}\, u(t)\, t^{\alpha-m-1},$$

thus

$$y(t) = \frac{(-1)^{m-1}\sin(\pi\alpha)\,\Gamma(\alpha)}{\pi\,\Gamma(m)\,\Gamma(\alpha-m)}\{\Psi(m) - \Psi(\alpha-m) + \log t\}\, u(t)\, t^{\alpha-m-1}. \qquad (3.181)$$

Problem 3.36. Show that if $f(t) = u(t)t^{-\alpha-m}$ we obtain

$$y(t) = \frac{(-1)^m \Gamma(\alpha)}{\Gamma(\alpha+m)}\, \delta^{(m)}(t). \qquad (3.182)$$

If in (3.178) we substitute the hyperfunction $f(t) = u(t)\phi(t)$, where $\phi(t)$ is a real analytic function, we obtain

$$y(t) = \frac{1}{\Gamma(1-\alpha)\Gamma(\alpha)} u(t) t^{\alpha-1} \star D(u(t)\phi(t)), \qquad (3.183)$$

where we have used the rule $D(f(t) \star g(t)) = f'(t) \star g(t) = f(t) \star g'(t)$. Because

$$D(u(t)\phi(t)) = \phi(0)\,\delta(t) + u(t)\frac{d\phi}{dt}$$

we eventually get

$$y(t) = \frac{\sin(\alpha\pi)}{\pi}\left\{\phi(0)u(t)t^{\alpha-1} + \int_0^t \frac{1}{(t-x)^{1-\alpha}}\frac{d\phi}{dx}\,dx\right. \qquad (3.184)$$

as solution of the integral equation

$$\int_0^t \frac{y(x)}{(t-x)^\alpha}\,dx = u(t)\phi(t) \quad (0 < \alpha < 1). \qquad (3.185)$$

Consider now Abel's integral equation where the parameter $\alpha = m$ is a positive integer, i.e.,

$$\mathrm{FP}\int_0^t \frac{y(x)}{(t-x)^m}\,dx = f(t). \qquad (3.186)$$

Notice that without taking the finite part of the integral the equation makes no sense. We interpret it as the convolution equation

$$\mathrm{fp}\frac{u(t)}{t^m} \star y(t) = f(t) \qquad (3.187)$$

where all involved hyperfunctions are assumed to be in the convolution algebra $\mathcal{A}(\mathbb{R}_+, \star)$. For $f(t) = \delta(t)$, and by taking Laplace transforms, we obtain, by using (3.45) for the inverse element $k(t)$ with respect to the convolution product,

$$(-1)^m \frac{s^{m-1}}{(m-1)!}\left\{\log s + \gamma - \sum_{k=1}^{m-1}\frac{1}{k}\right\}\hat{k}(s) = 1.$$

Thus

$$\hat{k}(s) = \frac{1}{c_m^{m-1}}\frac{(-1)^m\,(m-1)!}{(s/c_m)^{m-1}\,\log(s/c_m)}$$

with $c_m = \log\Psi(m)$. If $m \in \{4, 5, \ldots\}$, $c_m > 0$, and we can apply the Scale Change Rule and obtain

$$k(t) = u(t)\frac{(-1)^m\,(m-1)!}{c_m^{m-2}}\nu(c_m t, m-2)$$

where $\nu(t, \alpha)$ denotes the Volterra type function

$$\nu(t, \alpha) := \int_0^\infty \frac{t^{x+\alpha}}{\Gamma(x+\alpha+1)}\,dx. \qquad (3.188)$$

3.8.2 Convolution Integral Equations over an Infinite Range

Example 3.39. Let us study the following Fredholm singular integral equation by using the method of Laplace transformation. In Section 4.7 we shall revisit the same equation but by using Fourier transformation.

$$y(t) - \lambda \int_{-\infty}^{\infty} e^{-|t-x|} y(x) \, dx = f(t). \tag{3.189}$$

The integral equation can be written as

$$y(t) - \lambda e^{-|t|} \star y(t) = f(t),$$

or as

$$\{\delta(t) - \lambda e^{-|t|}\} \star y(t) = f(t).$$

The new kernel has the Laplace transform

$$k(t) = \delta(t) - \lambda e^{-|t|} \circ - \bullet \hat{k}(s) = 1 + \frac{2\lambda}{s^2 - 1} = \frac{s^2 + 2\lambda - 1}{s^2 - 1}$$

valid in the vertical strip $-1 < \Re s < 1$. We now assume that $y(t)$ and $f(t)$ have Laplace transform $\hat{y}(s)$ and $\hat{f}(s)$, respectively. In order that the Laplace transform of the convolution equation makes sense, the strip of convergence of $\hat{f}(s)$ must have a non-void intersection with $-1 < \Re s < 1$. We then obtain the image equation

$$\frac{s^2 + 2\lambda - 1}{s^2 - 1} \hat{y}(s) = \hat{f}(s),$$

or

$$\hat{y}(s) = \frac{s^2 - 1}{s^2 + 2\lambda - 1} \hat{f}(s) = \hat{f}(s) - \lambda \frac{2}{s^2 + 2\lambda - 1} \hat{f}(s).$$

Hence,

$$y(t) = f(t) - 2\lambda \mathcal{L}^{-1}[\frac{1}{s^2 + 2\lambda - 1} \hat{f}(s)](t). \tag{3.190}$$

Let us continue with the second member

$$f(t) = \delta(t - \tau) \tag{3.191}$$

having the Laplace transform $\hat{f}(s) = e^{-\tau s}$ and being an entire function. Three cases have to be distinguished.
  A) $2\lambda - 1 = \Omega^2 > 0$, i.e., $\lambda > 1/2$:

$$\frac{1}{s^2 + 2\lambda - 1} = \frac{1}{s^2 + \Omega^2}.$$

We have two strips of holomorphicity $-1 < \Re s < 0$ and $0 < \Re s < 1$. The first one corresponds to

$$\mathcal{L}^{-1}[\frac{1}{s^2 + \Omega^2}] = -u(-t) \frac{\sin \Omega t}{\Omega},$$

and the second one to

$$\mathcal{L}^{-1}[\frac{1}{s^2 + \Omega^2}] = u(t) \frac{\sin \Omega t}{\Omega}.$$

By the Original Translation Rule, we then obtain the two solutions

$$y_1(t) = \delta(t - \tau) - 2\lambda\, u(\tau - t)\frac{\sin(\Omega(\tau - t))}{\Omega}, \tag{3.192}$$

$$y_2(t) = \delta(t - \tau) - 2\lambda\, u(t - \tau)\frac{\sin(\Omega(t - \tau))}{\Omega} \tag{3.193}$$

with $\Omega = \sqrt{2\lambda - 1}$. Note that $y(t) = y_1(t)/2 + y_2(t)/2$ is also a solution of (3.189), which can be written as

$$y(t) = \delta(t - \tau) - \lambda\frac{\sin(\Omega|t - \tau|)}{\Omega}. \tag{3.194}$$

This solution cannot directly be found by Laplace transformation.

Problem 3.37. Show that for
  B) $\lambda = 1/2$: we have the solutions

$$y_1(t) = \delta(t - \tau) - 2\lambda\, u(\tau - t)(\tau - t), \tag{3.195}$$
$$y_2(t) = \delta(t - \tau) - 2\lambda\, u(t - \tau)(t - \tau), \tag{3.196}$$

and

$$y(t) = \delta(t - \tau) - 2\lambda\, |t - \tau|. \tag{3.197}$$

  C) $1 - 2\lambda = \Omega^2 > 0$, i.e., $\lambda < 1/2$:

$$y_1(t) = \delta(t - \tau) - 2\lambda\, u(\tau - t)\frac{\sinh(\Omega(\tau - t))}{\Omega}, \tag{3.198}$$

$$y_2(t) = \delta(t - \tau) - 2\lambda\, u(t - \tau)\frac{\sinh(\Omega(t - \tau))}{\Omega}, y(t) \quad = \delta(t - \tau) - \lambda\frac{\sinh(\Omega|t - \tau|)}{\Omega}.$$
$$\tag{3.199}$$

We shall now discuss some integral equations over a semi-infinite range, sometimes called *Wiener-Hopf Equations* due to the famous paper where these authors had discussed the integral equation

$$y(t) = -\frac{1}{2}\int_0^\infty E_1(-|t - x|)y(x)\,dx.$$

Here $E_1(x)$ is the exponential integral. We shall present two simple examples which exhibit the *Wiener-Hopf technique*. For the same example we also show how the Wiener-Hopf procedure can be avoided (see also [6, Davies], [30, Pipkin]). An integral equation of the type

$$y(t) = \lambda\int_0^\infty k(t - x)y(x)\,dx + f(t)$$

is not a convolution equation unless the kernel $k(x)$ is a right-sided function or hyperfunction, in which case the equation becomes a Volterra integral equation.

Example 3.40. Consider the integral equation

$$f(t) = \lambda\int_0^\infty e^{-\alpha|t - x|}y(x)\,dx, \quad t > 0, \tag{3.200}$$

by the Wiener-Hopf technique. Again we assume that $y(t) \in O(\mathbb{R}_+)$ has the Laplace transform $\hat{y}(s)$, $\Re s > \sigma_-(y)$, while we interpret the right-hand side as $u(t)$ with its Laplace transform $1/s$, $\Re s > 0$. Also, we introduce an unknown function $g(t) \in O(\mathbb{R}_-)$ with its Laplace transform $\hat{g}(s)$, $\Re s < \sigma_+(g)$. The Laplace transform of the kernel is

$$|t|\, e^{-\alpha|t|} \circ - \bullet \frac{2(s^2 + \alpha^2)}{(s^2 - \alpha^2)^2}, \qquad -\alpha < \Re s < \alpha.$$

We make the assumption that there is a real number $0 < \sigma_0 < \alpha$ such that both Laplace transforms $\hat{y}(s)$ and $\hat{g}(s)$ are absolutely convergent for $\Re s = \sigma_0$. This implies that we must have $\sigma_0 > \sigma_-(y)$ and $\sigma_0 < \sigma_+(g)$. The given integral equation can now be written as

$$\lambda\, |t| e^{-\alpha|t|} \star y(t) = u(t) + g(t), \tag{3.216}$$

where the unknown $g(t)$ takes into account the values of the convolution for $t < 0$. The image equation becomes

$$\lambda \frac{2(s^2 + \alpha^2)}{(s^2 - \alpha^2)^2}\, \hat{y}(s) = \frac{1}{s} + \hat{g}(s),$$

and this equation is valid on the vertical line $\Re s = \sigma_0$. The above equation is now separated as

$$E_+(s) := \lambda\, s\, \frac{2(s^2 + \alpha^2)}{(s + \alpha)^2}\, \hat{y}(s) = (s - \alpha)^2 \{1 + s\, \hat{g}(s)\} =: E_-(s).$$

The function $E_+(s)$ is holomorphic in the right half-plane $\Re s > \sigma_1 := \max[\sigma_-(y), -\alpha]$, and $E_-(s)$ is holomorphic in the left half-plane $\Re s < \sigma_2 := \min[\sigma_+(g), \alpha]$. We now have a vertical strip $\sigma_1 < \Re s < \sigma_2$ containing the vertical line $\Re s = \sigma_0$ in it where $E_-(s) = E_+(s)$ holds. Therefore, there is one entire function $E(s) = E_+(s) = E_-(s)$ with the representations $E_+(s)$ for $\Re s > \sigma_1$ and $E_-(s)$ for $\Re s < \sigma_2$. By the same reasoning as in the previous example we arrive at the conclusion that $E(s) = O(s^2)$ as $s \to \infty$. This implies (as a consequence of Liouville's theorem) that $E(s)$ is a polynomial of second degree at most. But then the representation $E_+(s)$ shows immediately, that it can only be a polynomial of degree 0, i.e., a constant, say A, if $\hat{y}(s)$ should be an ordinary functions. We then find

$$\hat{y}(s) = A\, \frac{1}{2\lambda} \frac{(s + \alpha)^2}{s(s^2 + \alpha^2)} = \frac{A}{\lambda} \{ \frac{1}{2s} + \frac{\alpha}{s^2 + \alpha^2} \},$$

thus

$$y(t) = \frac{A}{\lambda} \{ (\frac{1}{2} + \sin(\alpha t) \}.$$

It we substitute this solution into the given integral equation and set $t = 0$ we obtain

$$A\, \mathcal{L}[x\, (\frac{1}{2} + \sin(\alpha x))](\alpha) = \frac{A}{\alpha^2} = 1$$

which yields $A = \alpha^2$, finally giving

$$y(t) = \frac{\alpha^2}{\lambda} \{ \frac{1}{2} + \sin(\alpha t) \}. \tag{3.217}$$

Chapter 4

Fourier Transforms

From the outset we shall explore the relation between Laplace and Fourier transforms. The subclass of hyperfunctions of slow growth $\mathcal{S}(\mathbb{R})$ and their Fourier transforms are introduced. It will be shown that the Fourier transform of a hyperfunction can be computed by evaluating the Laplace transform of two right-sided hyperfunctions. This fact is exploited to the hilt in the sequel, almost all Fourier transforms in this book are computed via Laplace transforms. The inverse Fourier transformation and the important Reciprocity Rule are formulated for hyperfunctions. All operational rules for the Fourier transform of hyperfunctions are carefully established. Conditions for the validity of the convolution property of Fourier transformation are stated. Many concrete examples of Fourier transforms of hyperfunctions are presented. The chapter terminates with Poisson's summation formula and some applications to differential and integral equations.

4.1 Fourier Transforms of Hyperfunctions

4.1.1 Basic Definitions

Let $f(x)$ be an ordinary function defined on the real line, and ω a real variable. The classical Fourier transformation is defined by

$$\hat{f}(\omega) := \int_{-\infty}^{\infty} e^{-i\omega x} f(x)\, dx, \tag{4.1}$$

provided that the integral converges. The function $\hat{f}(\omega)$ is called the Fourier transform of $f(x)$. We use the notation $\hat{f} = \mathcal{F}[f]$, or more explicitly, $\hat{f}(\omega) = \mathcal{F}[f(x)](\omega)$. The symbol $f(x) \triangleleft\!-\!\blacktriangleright \hat{f}(\omega)$ for the correspondence between a function and its Fourier transform is also used. Strong restrictions upon $f(x)$ are needed in order that the Fourier integral converges. For example, familiar functions such as the constant function $f(x) = 1$ or the power $f(x) = x^n$ do not have Fourier transforms in the classical sense because the defining integral fails to be convergent. Conditions such as

$$\int_{-\infty}^{\infty} |f(x)|\, dx < \infty, \text{ or } \int_{-\infty}^{\infty} |f(x)|^2\, dx < \infty,$$

i.e., $f \in L^1(\mathbb{R})$ or $f \in L^2(\mathbb{R})$, are usually imposed upon $f(x)$ to assure convergence and to build a firm theory (see [5, Chandrasekharan]). The introduction of generalized functions in mathematics has greatly enhanced the applicability of the Fourier transformation. It can be said that the Fourier transformation is one of the greatest beneficiary of the theory of generalized functions. In order to convey the Fourier transformation to hyperfunctions, we shall proceed in a similar manner as we have done with the Laplace transformation.

Now let us consider hyperfunctions of the form $f(x) = f_1(x) + f_2(x)$ where we assume $f_1(x) \in O(\mathbb{R}_-)$, and $f_2(x) \in O(\mathbb{R}_+)$ (canonical splitting of $f(x)$). Let $\sigma_+(f_1)$ and $\sigma_-(f_2)$ be the growth index of $f_1(x)$ and $f_2(x)$, as $x \to -\infty$, and $x \to \infty$, respectively. Then the Laplace transform $\mathcal{L}[f_1(x)](s)$ is a holomorphic function in the left half-plane $\Re s < \sigma_+(f_1)$ and $\mathcal{L}[f_2(x)](s)$ is holomorphic in the right half-plane $\Re s > \sigma_-(f_2)$. If $\sigma_-(f_2) < 0 < \sigma_+(f_1)$ holds, the two image functions have a common vertical strip of convergence $\sigma_-(f_2) < \Re s < \sigma_+(f_1)$ containing the imaginary axis, and the two image functions can be added. By definition, we have $\mathcal{L}[f(x)](s) := \mathcal{L}[f_1(x)](s) + \mathcal{L}[f_2(x)](s)$ with $\sigma_-(f_2) < \Re s < \sigma_+(f_1)$. In the event of hyperfunctions $f_1(x) = [F_1(z)]$, $f_2(x) = [F_2(z)]$, we have

$$\mathcal{L}[f_1(x)](s) = -\int_{-\infty}^{(0+)} e^{-sz} F_1(z)\, dz, \quad \mathcal{L}[f_2(x)](s) = -\int_{\infty}^{(0+)} e^{-sz} F_2(z)\, dz.$$

We now make a change of variables $s = i\zeta$, i.e., $\zeta = -is$. The vertical strip of convergence containing the imaginary axis in the s-plane becomes a horizontal strip of convergence $-\sigma_+(f_1) \le \Im\zeta \le -\sigma_-(f_2)$ in the ζ-plane containing the real axis. With

$$H_+(\zeta) := \mathcal{L}[f_1(x)](i\zeta) = -\int_{-\infty}^{(0+)} e^{-i\zeta z} F_1(z)\, dz, \qquad (4.2)$$

and

$$H_-(\zeta) := -\mathcal{L}[f_2(x)](i\zeta) = \int_{\infty}^{(0+)} e^{-i\zeta z} F_2(z)\, dz, \qquad (4.3)$$

we define

$$\mathcal{F}[f(x)](\zeta) := H_+(\zeta) - H_-(\zeta), \quad -\sigma_+(f_1) < \Im\zeta < -\sigma_-(f_2) \qquad (4.4)$$

and call $\hat{f}(\zeta)) := \mathcal{F}[f(x)](\zeta)$ the Fourier transform of the hyperfunction $f(x)$. So far the Fourier transform is only a hidden Laplace transform and nothing is really new.

Example 4.1. From the known Laplace transform correspondences

$$\exp(-a\,t^2) \circ\!\!-\!\!\bullet \sqrt{\frac{\pi}{a}}\, \exp(\frac{s^2}{4\,a}), \quad -\infty < \Re s < \infty,$$

$$\delta(t) \circ\!\!-\!\!\bullet 1, \quad \delta(t-a) \circ\!\!-\!\!\bullet e^{-sa}, \quad \delta^{(n)}(t-a) \circ\!\!-\!\!\bullet e^{-as}\, s^n, \quad -\infty < \Re s < \infty,$$

$$e^{-a|t|} \circ\!\!-\!\!\bullet -\frac{2a}{s^2 - a^2}, \quad -a < \Re s < a,$$

$$\sum_{k=-\infty}^{\infty} e^{-|t|}\, \delta(t-k) \circ\!\!-\!\!\bullet \frac{1 - e^{-2}}{(1 - e^{-s-1})\,(1 - e^{s-1})}, \quad -1 < \Re s < 1,$$

$$\mathcal{H}_n(t) \circ\!\!-\!\!\bullet \sqrt{2\pi}\,(-i)^n\, \mathcal{H}_n(-is), \quad -\infty < \Re s < \infty,$$

follows immediately

$$\exp(-a\,x^2) \lhd - \blacktriangleright \sqrt{\frac{\pi}{a}}\,\exp(-\frac{\zeta^2}{4\,a}), \quad -\infty < \Im\zeta < \infty,$$

$$\delta(x) \lhd - \blacktriangleright 1, \quad \delta(x-a) \circ - \bullet\, e^{-i\zeta a}, \quad -\infty < \Im\zeta < \infty,$$

$$\delta^{(n)}(x-a) \lhd - \blacktriangleright e^{-ia\zeta}\,(i\zeta)^n, \quad -\infty < \Im\zeta < \infty,$$

$$e^{-a|x|} \lhd - \blacktriangleright \frac{2a}{\zeta^2 + a^2}, \quad -a < \Im\zeta < a,$$

$$\sum_{k=-\infty}^{\infty} e^{-|x|}\,\delta(x-k) \lhd - \blacktriangleright \frac{1 - e^{-2}}{(1 - e^{-i\zeta-1})\,(1 - e^{i\zeta-1})}, \quad -1 < \Im\zeta < 1,$$

$$\mathcal{H}_n(x) \lhd - \blacktriangleright \sqrt{2\pi}\,(-i)^n\,\mathcal{H}_n(\zeta), \quad -\infty < \Im\zeta < \infty.$$

For the last correspondence see also [5, Chandrasekharan], [37, Vladimirov]. It shows that the Fourier transformation viewed as an operator has the Hermite functions $\mathcal{H}_n(x)$ (see Example 3.8) as eigenfunctions with the corresponding eigenvalues $\sqrt{2\pi}\,(-i)^n$.

The Fourier transform viewed as a function of a complex variable ζ, varying in a horizontal strip containing the real axis, is sometimes called the generalized Fourier transform. Notice that the above introduced functions $H_+(\zeta)$ and $H_-(\zeta)$ are holomorphic in the upper half-plane $\Im\zeta > -\sigma_+(f_1)$ and the lower half-plane $\Im\zeta < -\sigma_-(f_2)$, respectively. We have required $-\sigma_+(f_1) < 0 < \sigma_-(f_2)$ giving a horizontal strip of convergence of positive width. We may now lessen this requirement and demand only that $-\sigma_+(f_1) \leq 0 \leq -\sigma_-(f_2)$ holds. Equation (4.4) then suggests the following

Definition 4.1. Let $f(x) = f_1(x) + f_2(x)$ be a hyperfunction with $f_1(x) = [F_1(z)] \in \mathrm{O}(\mathbb{R}_-)$, $f_2(x) = [F_2(z)] \in \mathrm{O}(\mathbb{R}_+)$. Moreover, assume $-\sigma_+(f_1) \leq 0 \leq -\sigma_-(f_2)$ holds, then, the *Fourier transform of* $f(x)$ is defined as being the hyperfunction $\hat{f}(\omega) := [H_+(\zeta), H_-(\zeta)]$, where the two components of the defining function are given by (4.2), (4.3). We write $\hat{f}(\omega) = \mathcal{F}[f(x)](\omega)$ and use the correspondence sign $f(x) \lhd - \blacktriangleright \hat{f}(\omega)$.

The ω in $\hat{f}(\omega)$ is now a real variable. We have at once

Proposition 4.1. *If it strictly holds that* $-\sigma_+(f_1) < 0 < -\sigma_-(f_2)$, *then the Fourier transform is a holomorphic hyperfunction on* $\mathbb{R}$.

Also, immediately,

Corollary 4.2. *Let* $f(x)$ *be an ordinary function or a hyperfunction with compact support, then its Laplace and its Fourier transform are entire, respectively.*

Definition 4.2. A hyperfunction $f(x)$ which can be written as $f(x) = f_1(x) + f_2(x)$ with $f_1(x) \in \mathrm{O}(\mathbb{R}_-)$, $f_2(x) \in \mathrm{O}(\mathbb{R}_+)$, and satisfies $\sigma_+(f_1) \leq 0 \leq \sigma_-(f_2)$ is said to be of *slow growth* and is denoted by $f(x) \in \mathcal{S}(\mathbb{R})$.

Remember that $\sigma_-(f_2) = \inf\sigma'$, where the infimum is taken over all σ' satisfying $|F_2(z)| \leq M'\,e^{\sigma'\,\Re z}$, as $\Re z \to \infty$. Likewise, $\sigma_+(f_1) = \sup\sigma''$, where the supremum is taken over all σ'' satisfying $|F_1(z)| \leq M''\,e^{\sigma''\,\Re z}$, as $\Re z \to -\infty$. Thus, we may state

Proposition 4.3. *A hyperfunction $f(x)$ is of slow growth, i.e., $f(x) \in \mathcal{S}(\mathbb{R})$, if there is a positive integer N, and a defining function $F(z)$ for $f(x)$ such that in a full neighborhood of the real axis $F(x + iy) = O(|x|^N)$ as $|x| \to \infty$.*

Polynomials in x are of slow growth. Thus, hyperfunctions of slow growth have Fourier transforms. We also get from the above definition

Proposition 4.4. *The Fourier transform of a right-sided hyperfunction of slow growth is a lower hyperfunction, while the Fourier transform of a left-sided hyperfunction of slow growth is an upper hyperfunction.*

These properties are important because the product of two lower or two upper hyperfunctions is always defined. Also remember that the product of two hyperfunctions is well defined if one factor is a holomorphic hyperfunction.

Analogous to the case of Laplace transformation the formulas (4.2), (4.3) can be replaced by the following ones if the hyperfunction $f(x) = [F(z)]$ is holomorphic at $x = c$.

$$H_+(\zeta) := - \oint_{(-\infty,c)} e^{-i\zeta z} F(z) \, dz, \tag{4.5}$$

and

$$H_-(\zeta) := \int_{(c,\infty)} e^{-i\zeta z} F(z) \, dz \tag{4.6}$$

with the two u-shaped semi-infinite contours having the point c as common vertex. We have also the following generalization analogous to Proposition 3.5.

Proposition 4.5. *Let $f(x) = [F(z)]$ be a hyperfunction of slow growth with an arbitrary support and which is holomorphic at the finite points $x_1 < x_2 < \ldots < x_n$. Then, its Fourier transform is given by*

$$\begin{aligned}
\mathcal{F}[f(x)](\omega) = [&- \oint_{(-\infty,x_1)} e^{-i\zeta z} F(z) \, dz - \sum_{k=1}^{n-1} \oint_{(x_k,x_{k+1})} e^{-i\zeta z} F(z) \, dz \\
&- \oint_{(x_n,\infty)} e^{-\zeta z} F(z) \, dz] \\
= &\int_{-\infty}^{x_1} e^{-i\omega x} f(x) \, dx + \sum_{k=1}^{n-1} \int_{x_k}^{x_{k+1}} e^{-i\omega x} f(x) \, dx \\
&+ \int_{x_n}^{\infty} e^{-i\omega x} f(x) \, dx.
\end{aligned} \tag{4.7}$$

4.1.2 Connection to Laplace Transformation

It is clear from our definition of the Fourier transform that we may benefit from known Laplace transform correspondences to compute Fourier transforms. For a hyperfunction $f(x) = f_1(x) + f_2(x)$ with $f_1(x) \in O(\mathbb{R}_-), \sigma_+(f_1) \geq 0$, and $f_2(x) \in O(\mathbb{R}_+), \sigma_-(f_2) \leq 0$, i.e., of slow growth, we have at once,

$$\mathcal{F}[f(x)](\omega) = \hat{f}(\omega) = [\mathcal{L}[f_1(x)](i\zeta), -\mathcal{L}[f_2(x)](i\zeta)] . \tag{4.8}$$

If $\sigma_+(f_1) > 0$ and $\sigma_-(f_2) < 0$ the Fourier transform is the ordinary function $\hat{f}(\zeta) = \mathcal{L}[f_1(x)](i\zeta) + \mathcal{L}[f_2(x)](i\zeta)$, holomorphic in the horizontal strip $-\sigma_+(f_1) <$

$\Im\zeta < -\sigma_-(f_2)$. Moreover, since large tables of right-sided Laplace transforms are available, it is useful to pass at $g_1(x) := f_1(-x) \in O(\mathbb{R}_+)$ with $\sigma_+(f_1) = -\sigma_-(g_1)$. Then we have

$$\mathcal{L}[f_1(x)](s) = \mathcal{L}[f_1(-x)](-s)] = \mathcal{L}[g_1(x)](-s) = \mathcal{L}[g_1(x)](-i\zeta).$$

Hence, computing a Fourier transform is reduced to computing Laplace transforms of two right-sided hyperfunctions:

$$\mathcal{F}[f(x)](\omega) = \hat{f}(\omega) = [\mathcal{L}[f_1(-x)](-i\zeta), -\mathcal{L}[f_2(x)](i\zeta)], \tag{4.9}$$

provided that $\sigma_-(g_1) = -\sigma_+(f_1(-x)) \le 0$ and $\sigma_-(f_2) \le 0$.

Example 4.2. Let us compute the Fourier transform of $f(x) = \exp(-i\,b\,x^2)$. We write

$$f(x) = Y(-x)\,e^{-ibx^2} + Y(x)\,e^{-ibx^2} = f_1(x) + f_2(x).$$

From (3.32) we obtain

$$\mathcal{L}[f_1(-x)](-i\zeta) = \frac{1}{2}\sqrt{\frac{\pi}{2b}}(1-i)\exp(i\frac{\zeta^2}{4\,b})\operatorname{erfc}(-\sqrt{\frac{1}{2b}}\frac{1-i}{2}i\zeta)$$

$$-\mathcal{L}[f_2(x)](i\zeta) = -\frac{1}{2}\sqrt{\frac{\pi}{2b}}(1-i)\exp(i\frac{\zeta^2}{4\,b})\operatorname{erfc}(\sqrt{\frac{1}{2b}}\frac{1-i}{2}i\zeta).$$

Thus,

$$\hat{f}(\omega) = \frac{1}{2}\sqrt{\frac{\pi}{2b}}(1-i)\exp(i\frac{\zeta^2}{4\,b})\left[\operatorname{erfc}(-\sqrt{\frac{1}{2b}}\frac{1-i}{2}i\zeta), -\operatorname{erfc}(\sqrt{\frac{1}{2b}}\frac{1-i}{2}i\zeta)\right],$$

and, by using $\operatorname{erfc}(-z) = 2 - \operatorname{erfc}(z)$ and the fact that $\operatorname{erfc}(z)$ is an entire function, we eventually obtain, for $b > 0$,

$$e^{-ibx^2} \; \vartriangleleft\!-\!\blacktriangleright \; \sqrt{\frac{\pi}{2b}}(1-i)\exp(i\frac{\omega^2}{4\,b}). \tag{4.10}$$

Taking real and imaginary parts on both sides yields

$$\cos(b\,x^2) \; \vartriangleleft\!-\!\blacktriangleright \; \sqrt{\frac{\pi}{2b}}\left(\cos\frac{\omega^2}{4b} + \sin\frac{\omega^2}{4b}\right),$$

$$\sin(b\,x^2) \; \vartriangleleft\!-\!\blacktriangleright \; \sqrt{\frac{\pi}{2b}}\left(\cos\frac{\omega^2}{4b} - \sin\frac{\omega^2}{4b}\right).$$

Example 4.3. From known Laplace transform correspondences, with $a > 0$, let us compute the Fourier transform of

$$\frac{e^{-a|x|}}{x^m} := \operatorname{fp}\frac{u(-x)}{x^m}\,e^{ax} + \operatorname{fp}\frac{u(x)}{x^m}\,e^{-ax} = f_1(x) + f_2(x).$$

By using Example 3.17 we have

$$\operatorname{fp}\frac{e^{-a|t|}}{t} \; \circ\!-\!\bullet \; \log\frac{a-s}{a+s},$$

$$\text{fp}\frac{e^{-a|t|}}{t^2} \circ\!\!-\!\!\bullet\, s\log\frac{a+s}{a-s} + a\,\{\log((a^2-s^2)) + 2(\gamma-1)\}.$$

Therefore,

$$\text{fp}\frac{e^{-a|x|}}{x} \lhd\!-\!\blacktriangleright\, \log\frac{a-i\zeta}{a+i\zeta} = \frac{2}{i}\arctan(\frac{\zeta}{a}), \tag{4.11}$$

$$\text{fp}\frac{e^{-a|x|}}{x^2} \lhd\!-\!\blacktriangleright\, i\zeta\log\frac{a+i\zeta}{a-i\zeta} + a\,\{\log((a^2+\zeta^2)) + 2(\gamma-1)\}. \tag{4.12}$$

The Fourier transform is an ordinary function, holomorphic in the horizontal strip $-a < \Im\zeta < a$.

Example 4.4. Let us compute the Fourier transform of the Bessel function $J_0(ax)$. We write $J_0(ax) = u(-x)J_0(ax) + u(x)J_0(ax) = u(-x)J_0(-ax) + u(x)J_0(ax)$, because $J_0(x)$ is even, hence $f_1(-x) = f_2(x) = u(x)J_0(ax)$. Because

$$\mathcal{L}[u(t)J_0(at)(s)] = \frac{1}{\sqrt{a^2+s^2}}$$

we obtain

$$\hat{f}(\omega) = \mathcal{F}[J_0(ax)](\omega) = \left[\frac{1}{\sqrt{a^2-\zeta^2}}, -\frac{1}{\sqrt{a^2-\zeta^2}}\right].$$

Of the function $F(\zeta) = 1/\sqrt{a^2-\zeta^2}$ holomorphic in the ζ-plane with an incision from $-\infty$ to $-a$ and from a to ∞, we select the branch which is positive for $-a < \zeta < a$. $F(\omega)$ is real analytic for $-a < \omega < a$ and the hyperfunction $\hat{f}(\omega) = 2/\sqrt{a^2-\omega^2}\,[1/2, -1/2]$ is equal to $2/\sqrt{a^2-\omega^2}$ on this interval. Above and below $I := (-\infty, -a)\cup(a, \infty)$ of the ω-axis, $F(\zeta)$ takes values that distinguish themselves by a factor -1. Thus, $F(\omega+i0)$ and $-F(\omega-i0)$ are the same for $\omega \in I$. This means that $\hat{f}(\omega) = 0$ for $\omega \in I$. Eventually, we obtain

$$\mathcal{F}[J_0(ax)](\omega) = \chi_{(-a,a)}\frac{2}{\sqrt{a^2-\omega^2}}. \tag{4.13}$$

4.2   Fourier Transforms of Some Familiar Hyperfunctions

All computations of Fourier transforms below will be based on Laplace transformation.

Example 4.5. $f(x) = x^m$, $m = 0, 1, 2, \ldots$. We write

$$x^m = u(-t)x^m + u(t)x^m = f_1(x) + f_2(x).$$
$$H_+(\zeta) := \mathcal{L}[f_1(x)](i\zeta) = \mathcal{L}[u(t)(-x)^m](-i\zeta)$$
$$= \frac{(-1)^m\, m!}{(-i\zeta)^{m+1}} = -\frac{m!}{(i\zeta)^{m+1}}, \quad \Im\zeta > 0,$$
$$H_-(\zeta) := -\mathcal{L}[f_2(x)](i\zeta) = -\mathcal{L}[u(t)x^m](i\zeta)$$
$$= -\frac{m!}{(i\zeta)^{m+1}}, \quad \Im\zeta < 0.$$

Thus, we find that the defining function of the Fourier transform is

$$H_+(\zeta) = H_-(\zeta) = 2\pi\, i^m \left(-\frac{1}{2\pi i}\right) \frac{(-1)^m\, m!}{\zeta^{m+1}}$$

which is the defining function of $2\pi\, i^m \delta^{(m)}(\omega)$. Therefore, the Fourier transform of a power function is

$$x^m \lhd\!-\!\blacktriangleright 2\pi\, i^m \delta^{(m)}(\omega) \quad m = 0, 1, 2, \ldots. \tag{4.14}$$

Example 4.6. Let us compute the Fourier transform of the unit-step function $u(x) = 0 + u(x) = f_1(x) + f_2(x)$.

$$H_+(\zeta) := \mathcal{L}[f_1(t)](-i\zeta) = 0,$$
$$H_-(\zeta) := -\mathcal{L}[f_2(x)](i\zeta) = -\mathcal{L}[u(x)](i\zeta)$$
$$= -\frac{1}{i\zeta}.$$

Then,

$$[0, -1/(i\zeta)] = [1/(2i\zeta), -1/(2i\zeta)] - [1/(2i\zeta), 1/(2i\zeta)]$$
$$= \frac{1}{i}\, [1/(2\zeta), -1/(2\zeta)] + \pi\, [-1/(2\pi i\zeta)]$$

and finally,

$$u(x) \lhd\!-\!\blacktriangleright \frac{1}{i}\, \mathrm{fp}\frac{1}{\omega} + \pi\, \delta(\omega) = \frac{1}{i}\frac{1}{\omega - i0}. \tag{4.15}$$

If we return to the definition of the Laplace transform we can write with $[0, -1/(i\zeta)] = [1/(i\zeta)\mathbf{1}_-(z)$, and $\Im\zeta < 0$,

$$1/(i\zeta)\mathbf{1}_-(z) = \int_0^\infty e^{-(i\zeta)x}\, dx\, \mathbf{1}_-(z)$$
$$= \{\int_0^\infty \cos(\zeta x)\, dx - i \int_0^\infty \sin(\zeta x)\, dx\}\mathbf{1}_-(z).$$

Comparing real and imaginary parts with the above result yields

$$\int_0^\infty \cos(\omega(x - i0))\, dx = \pi\, \delta(\omega), \quad \int_0^\infty \sin(\omega(x - i0))\, dx = \mathrm{fp}\frac{1}{\omega}. \tag{4.16}$$

These expressions are definite integrals over the lower hyperfunctions $\cos(\omega(x-i0))$ and $\sin(\omega(x - i0))$.

Problem 4.1. Establish, by the same method,

$$u(-x) \lhd\!-\!\blacktriangleright -\frac{1}{i}\, \mathrm{fp}\frac{1}{\omega} + \pi\, \delta(\omega) = \frac{i}{\omega + i0}, \tag{4.17}$$

$$\mathrm{sgn}(x) \lhd\!-\!\blacktriangleright \frac{2}{i}\, \mathrm{fp}\frac{1}{\omega}, \tag{4.18}$$

$$u(x)\, x^n \lhd\!-\!\blacktriangleright i^{n-1}\, (-1)^n n!\, \mathrm{fp}\frac{1}{\omega^{n+1}} + i^n\, \pi\, \delta^{(n)}(\omega). \tag{4.19}$$

Example 4.7. Let us compute the Fourier transform of $f(x) = u(x)\sin(\omega_0 x)$. We have

$$\mathcal{F}[f(x)](\omega) = \hat{f}(\omega) = [0, -\mathcal{L}[u(t)\sin(\omega_0 t)](i\zeta)],$$

hence

$$\hat{f}(\omega) = \left[0, -\frac{\omega_0}{-\zeta^2 + \omega_0^2}\right].$$

The defining function can be transformed as follows:

$$\left[0, \frac{\omega_0}{\zeta^2 - \omega_0^2}\right] = \left[-\frac{1}{2}\frac{\omega_0}{\zeta^2 - \omega_0^2}, \frac{1}{2}\frac{\omega_0}{\zeta^2 - \omega_0^2}\right] + \left[\frac{1}{2}\frac{\omega_0}{\zeta^2 - \omega_0^2}, \frac{1}{2}\frac{\omega_0}{\zeta^2 - \omega_0^2}\right]$$

$$= \frac{1}{4}\left[\frac{1}{\zeta - \omega_0} - \frac{1}{\zeta - \omega_0}, \frac{1}{\zeta - \omega_0} - \frac{1}{\zeta - \omega_0}\right] + \left[-\frac{1}{2}\frac{\omega_0}{\zeta^2 - \omega_0^2}, \frac{1}{2}\frac{\omega_0}{\zeta^2 - \omega_0^2}\right]$$

$$= -i\frac{\pi}{2}\left\{\left[-\frac{1}{2\pi i}\frac{1}{\zeta - \omega_0}\right] - \left[-\frac{1}{2\pi i}\frac{1}{\zeta + \omega_0}\right]\right\} - \left[\frac{1}{2}\frac{\omega_0}{\zeta^2 - \omega_0^2}, -\frac{1}{2}\frac{\omega_0}{\zeta^2 - \omega_0^2}\right].$$

Finally, we obtain

$$u(t)\sin(\omega_0 x) \vartriangleleft - \blacktriangleright -i\frac{\pi}{2}\{\delta(\omega - \omega_0) - \delta(\omega + \omega_0)\} - \mathrm{fp}\frac{\omega_0}{\omega^2 - \omega_0^2}. \qquad (4.20)$$

Example 4.8. We compute the Fourier Transform of $f(x) = \log|x|$. We write $\log|x| = u(-x)\log(-x) + u(x)\log(x) = f_1(x) + f_2(x)$. The Laplace transform of $u(t)\log t$ is $-(\gamma + \log s)/s$. Therefore,

$$\hat{f}(\omega) = [H_+(\zeta), H_-(\zeta)] = [\mathcal{L}[u(t)\log t](-i\zeta), -\mathcal{L}[u(t)\log t](i\zeta)]$$

$$= \left[-\frac{1}{-i\zeta}(\gamma + \log(-i\zeta)), \frac{1}{i\zeta}(\gamma + \log(i\zeta))\right]$$

$$= \left[\frac{1}{i\zeta}(\gamma + \log(-i\zeta)), \frac{1}{i\zeta}(\gamma + \log(i\zeta))\right]$$

$$= -2\pi\gamma\left[-\frac{1}{2\pi i}\frac{1}{\zeta}\right] + \left[\frac{\log(-i\zeta)}{i\zeta}, \frac{\log(i\zeta)}{i\zeta}\right]$$

$$= -2\pi\gamma\,\delta(\omega) - \pi\frac{\mathrm{sgn}\,\omega}{\omega}.$$

Therefore,

$$\log|x| \vartriangleleft - \blacktriangleright -2\pi\gamma\,\delta(\omega) - \pi\frac{1}{|\omega|}. \qquad (4.21)$$

Problem 4.2. Show for $m \in \mathbb{N}$,

$$x^m\,\log|x| \vartriangleleft - \blacktriangleright 2\pi\,i^m\,\Psi(m+1)\,\delta^{(m)}(\omega) - \frac{\pi m!}{i^m}\frac{\mathrm{sgn}\,\omega}{\omega^{m+1}}, \qquad (4.22)$$

and also

$$u(x)x^m\,\log|x| \vartriangleleft - \blacktriangleright \pi\,i^m\,\Psi(m+1)\,\delta^{(m)}(\omega)$$
$$+ \frac{m!}{(i\omega)^{m+1}}\{\Psi(m+1) - \log|\omega| - \frac{\pi}{2}i\,\mathrm{sgn}\,\omega\}, \qquad (4.23)$$

$$u(-x)x^m\,\log|x| \vartriangleleft - \blacktriangleright \pi\,i^m\,\Psi(m+1)\,\delta^{(m)}(\omega)$$
$$- \frac{m!}{(i\omega)^{m+1}}\{\Psi(m+1) - \log|\omega| + \frac{\pi}{2}i\,\mathrm{sgn}\,\omega\}. \qquad (4.24)$$

Problem 4.3. Show that

$$u(x)\cos(\omega_0 x) \lhd - \blacktriangleright [0, \frac{i\zeta}{\zeta^2 - \omega_0^2}]$$

$$= \frac{\pi}{2}\{\delta(\omega - \omega_0) + \delta(\omega + \omega_0)\} - i\,\mathrm{fp}\frac{\omega}{\omega^2 - \omega_0^2},$$

(4.25)

$$u(-x)\cos(\omega_0 x) \lhd - \blacktriangleright [\frac{i\zeta}{\zeta^2 - \omega_0^2}, 0]$$

$$= \frac{\pi}{2}\{\delta(\omega - \omega_0) + \delta(\omega + \omega_0)\} + i\,\mathrm{fp}\frac{\omega}{\omega^2 - \omega_0^2},$$

(4.26)

$$\cos(\omega_0 x) \lhd - \blacktriangleright \pi[\delta(\omega - \omega_0) + \delta(\omega + \omega_0)].$$

(4.27)

Problem 4.4. Establish by the same method the correspondences

$$u(x)x^m e^{-\gamma x} \lhd - \blacktriangleright \frac{m!}{(i\omega + \gamma)^{m+1}} \quad (\Re\gamma > 0),$$

(4.28)

$$u(-x)x^m e^{\gamma x} \lhd - \blacktriangleright -\frac{m!}{(i\omega - \gamma)^{m+1}} \quad (\Re\gamma > 0),$$

(4.29)

$$u(x)e^{i\omega_0 x} \lhd - \blacktriangleright \pi\delta(\omega - \omega_0) - i\,\mathrm{fp}\frac{1}{\omega - \omega_0}.$$

(4.30)

Example 4.9. Let us compute the Fourier transform of the hyperfunction $f(x) = \mathrm{fp}(1/x^m)$, where m is a positive integer. We write $\mathrm{fp}(1/x^m) = \mathrm{fp}\,u(-x)/x^m + \mathrm{fp}\,u(x)/x^m = f_1(x) + f_2(x)$. We have to compute

$$\left[\mathcal{L}[\mathrm{fp}\frac{u(t)}{(-t)^m}](-i\zeta), -\mathcal{L}[\mathrm{fp}\frac{u(t)}{t^m}](i\zeta)\right].$$

By using (3.45), we get for $\Re(-i\zeta) = \Im(\zeta) > 0$,

$$\mathcal{L}[\mathrm{fp}\frac{u(t)}{(-t)^m}](-i\zeta) = (-1)^m\,\mathcal{L}[\mathrm{fp}\frac{u(t)}{t^m}](-i\zeta)$$

$$= \frac{(-i)^{m-1}\zeta^{m-1}}{(m-1)!}\{\log(-i\zeta) + \gamma - \sum_{k=1}^{m-1}\frac{1}{k}\}.$$

For $\Re(i\zeta) = -\Im(\zeta) > 0$, we obtain

$$-\mathcal{L}[\mathrm{fp}\frac{u(t)}{t^m}](i\zeta) = -\frac{(-1)^m\,s^{m-1}}{(m-1)!}\{\log s + \gamma - \sum_{k=1}^{m-1}\frac{1}{k}\}|_{s=i\zeta}$$

$$= \frac{(-i)^{m-1}\zeta^{m-1}}{(m-1)!}\{\log(i\zeta) + \gamma - \sum_{k=1}^{m-1}\frac{1}{k}\},$$

thus

$$\left[\mathcal{L}[\mathrm{fp}\frac{u(t)}{(-t)^m}](-i\zeta), -\mathcal{L}[\mathrm{fp}\frac{u(t)}{t^m}](i\zeta)\right] = \frac{(-i)^{m-1}\zeta^{m-1}}{(m-1)!}[\log(-i\zeta), \log(i\zeta)]$$

$$+ \frac{(-i)^{m-1}\zeta^{m-1}}{(m-1)!}\{\gamma - \sum_{k=1}^{m-1}\frac{1}{k}\}.$$

The last term is a polynomial, thus a real analytic function that can be discarded. Finally, by using the result of Problem 1.25,

$$\text{fp}\frac{1}{x^m} \lhd - \blacktriangleright \frac{\pi}{i^m\,(m-1)!}\,\omega^{m-1}\,\text{sgn}\,\omega. \tag{4.31}$$

Example 4.10. Let us now compute the Fourier transform of the hyperfunction $f(x) = u(x)/x^m$, where m is again a positive integer. Now, we have only to evaluate

$$\left[0, -\mathcal{L}[\text{fp}\frac{u(t)}{t^m}](i\zeta)\right].$$

Using the known Laplace transform we obtain

$$\frac{(-i)^{m-1}\zeta^{m-1}}{(m-1)!}\,[0,\log(i\zeta)] + [0, \frac{(-i)^{m-1}\zeta^{m-1}}{(m-1)!}\,\{\gamma - \sum_{k=1}^{m-1}\frac{1}{k}\}].$$

By the result of Problem 1.25 we obtain for the first term

$$\frac{(-i)^{m-1}\zeta^{m-1}}{(m-1)!}\,[0,\log(i\zeta)] = -\frac{(-i)^{m-1}\omega^{m-1}}{(m-1)!}\,\{\log|\omega| + i\frac{\pi}{2}\,\text{sgn}\,\omega\},$$

and for the second term

$$[0, \frac{(-i)^{m-1}\zeta^{m-1}}{(m-1)!}\,\{\gamma - \sum_{k=1}^{m-1}\frac{1}{k}\}] = -\frac{(-i)^{m-1}\omega^{m-1}}{(m-1)!}\,\{\gamma - \sum_{k=1}^{m-1}\frac{1}{k}\}.$$

Finally,

$$\frac{u(x)}{x^m} \lhd - \blacktriangleright -\frac{(-i\omega)^{m-1}}{(m-1)!}\,\{\log|\omega| + i\frac{\pi}{2}\,\text{sgn}\,\omega + \gamma - \sum_{k=1}^{m-1}\frac{1}{k}\}. \tag{4.32}$$

Problem 4.5. Show that

$$\frac{u(-x)}{x^m} \lhd - \blacktriangleright \frac{(-i\omega)^{m-1}}{(m-1)!}\,\{\log|\omega| - i\frac{\pi}{2}\,\text{sgn}\,\omega + \gamma - \sum_{k=1}^{m-1}\frac{1}{k}\}. \tag{4.33}$$

Example 4.11. Let us now compute the Fourier transform of the hyperfunction $f(x) = |x|^\alpha\,\text{sgn}\,x$, where α is not an integer. We write $f(x) = -|x|^\alpha u(-x) + x^\alpha u(x) = f_1(x) + f_2(x)$. Then we have

$$[-\mathcal{L}[|-t|^\alpha u(t)](-i\zeta), -\mathcal{L}[|t|^\alpha u(t)](i\zeta)] = \left[-\frac{\Gamma(\alpha+1)}{(-i\zeta)^{\alpha+1}}, -\frac{\Gamma(\alpha+1)}{(i\zeta)^{\alpha+1}}\right]$$

$$= -\Gamma(\alpha+1)\left[(-i\zeta)^{-\alpha-1}, (i\zeta)^{-\alpha-1}\right].$$

We finally arrive at

$$|x|^\alpha\,\text{sgn}\,x \lhd - \blacktriangleright -2i\,\Gamma(\alpha+1)\,\sin(\frac{\pi}{2}(\alpha+1))\,\frac{\text{sgn}\,\omega}{|\omega|^{\alpha+1}}. \tag{4.34}$$

Problem 4.6. Show that for $\alpha \in \mathbb{R} \setminus \mathbb{Z}$,

$$|x|^{\alpha} \lhd\!-\!\blacktriangleright 2\,\Gamma(\alpha+1)\,\cos(\tfrac{\pi}{2}(\alpha+1))\,\frac{1}{|w|^{\alpha+1}}. \tag{4.35}$$

Problem 4.7. Show that for $\alpha \in \mathbb{R} \setminus \mathbb{Z}$,

$$|x|^{\alpha} u(x) \lhd\!-\!\blacktriangleright \Gamma(\alpha+1)\,e^{i\frac{\pi}{2}(\alpha+1)\,\operatorname{sgn} w}\,\frac{1}{|w|^{\alpha+1}}. \tag{4.36}$$

Problem 4.8. Show that for a positive integer m it holds that

$$\mathrm{fp}\frac{1}{|x|^m} \lhd\!-\!\blacktriangleright -\frac{2\,w^{m-1}}{i^{m-1}\,(m-1)!}\,\{\log|w| + \gamma - \sum_{k=1}^{m-1}\frac{1}{k}\}, \tag{4.37}$$

and in particular

$$\mathrm{fp}\frac{1}{|x|} \lhd\!-\!\blacktriangleright -2\,\{\log|w| + \gamma\}. \tag{4.38}$$

4.3 Inverse Fourier Transforms

If $f(x) = f_1(x) + f_2(x)$ is an ordinary function which is locally of bounded variation satisfying $f_1(x) \in O(\mathbb{R}_-)$, $f_2(x) \in O(\mathbb{R}_+)$, and $\sigma_-(f_2) < 0 < \sigma_+(f_1)$, the Laplace integral is absolutely convergent in the vertical strip $\sigma_-(f_2) < \Re s < \sigma_+(f_1)$, and the complex inversion formula

$$\frac{f(x+) + f(x-)}{2} = \mathrm{pv}\frac{1}{2\pi i}\int_{c-i\infty}^{c+\infty} e^{sx}\,\mathcal{L}[f](s)\,ds$$

holds. The contour is a vertical line $\Re s = c$ with $\sigma_-(f_2) < c < \sigma_+(f_1)$. If we normalize $f(x)$ at every point of discontinuity (being of the first kind) to be

$$f(x) := \frac{f(x+0) + f(x-0)}{2},$$

we may write

$$f(x) = \frac{1}{2\pi i}\int_{-\infty-ic}^{\infty-ic} e^{i\zeta x}\,\mathcal{L}[f](i\zeta)\,d(i\zeta).$$

By replacing $-c$ by γ, we obtain

$$f(x) = \frac{1}{2\pi}\int_{-\infty+i\gamma}^{\infty+i\gamma} e^{i\zeta x}\,\mathcal{F}[f(x)](\zeta)\,d\zeta, \tag{4.39}$$

where the contour is now the horizontal line $\Im\zeta = \gamma$ lying inside the horizontal strip of convergence $-\sigma_+(f_1) < \Im\zeta < -\sigma_-(f_2)$ containing the real axis. Formula (4.39) is the inversion formula for the Fourier transformation when the Fourier transform is a holomorphic function in the horizontal strip $-\sigma_+(f_1) < \Im\zeta < -\sigma_-(f_2)$. We may choose $\gamma = 0$ and (4.39) becomes

$$f(x) = \frac{1}{2\pi}\int_0^{\infty} e^{i\zeta x}\,\hat{f}(\zeta)\,d\zeta - \{-\frac{1}{2\pi}\int_{-\infty}^{0} e^{i\zeta x}\,\hat{f}(\zeta)\,d\zeta\}$$
$$:= F_+(x) - F_-(x),$$

where $\hat{f}(\zeta) = \mathcal{F}[f(x)](\zeta)$.

Replace the real variable x by the complex variable $z = x + iy$, and write $\zeta = w + i\eta$. Take into account

$$\left| e^{iz\zeta} \right| = \left| e^{i(x+iy)(w+i\eta)} \right| = e^{-(x\eta+yw)},$$

and integrate along the real axis. Then it is seen that the integrals

$$F_-(z) = -\frac{1}{2\pi} \int_{-\infty}^{0} e^{iwz}\, \hat{f}(w)\, dw, \quad \Im z = y < 0, \tag{4.40}$$

and

$$F_+(z) = \frac{1}{2\pi} \int_{0}^{\infty} e^{iwz}\, \hat{f}(w)\, dw, \quad \Im z = y > 0, \tag{4.41}$$

become convergent Laplace integrals, provided there exist positive numbers M, n such that $|\hat{f}(w)| \le M\,|w|^n$, as $\Re\zeta = w \to \pm\infty$, which is met if the Fourier transform $\hat{f}(w) \in \mathcal{S}(\mathbb{R})$. Also, the functions $F_-(z)$ and $F_+(z)$ are holomorphic in the lower and upper half-plane, respectively. This suggests the following definition of the inverse Fourier transform.

Definition 4.3. If the Fourier transform $\hat{f}(w)$ is an ordinary function of slow growth, its inverse Fourier transform is the hyperfunction $f(x) = [F_+(z), F_-(z)]$, where the lower and upper components of the defining function are given by (4.40) and (4.41).

Let now $\hat{f}(w) = [H(z)] \in \mathcal{S}(\mathbb{R})$ be a given hyperfunction of slow growth. By canonical splitting, we write it as

$$\hat{f}(w) = \hat{f}_1(w) + \hat{f}_2(w) = [H_1(\zeta)] + [H_2(\zeta)] \tag{4.42}$$

where $H_1(\zeta)$ is real analytic on $(0, \infty)$ and $H_2(\zeta)$ is real analytic on $(-\infty, 0)$. If $\hat{f}(w)$ is holomorphic at $w = 0$, we take $\hat{f}_1(w) = \chi_{(-\infty,0)}\hat{f}(w)$ and $\hat{f}_2(w) = \chi_{(0,\infty)}\hat{f}(w)$. If $\hat{f}(w)$ is not holomorphic at $w = 0$, we must split it from the outset, as for example with

$$\hat{f}(w) = \delta(w) = \left[-\frac{1}{4\pi i\,\zeta} \right] + \left[-\frac{1}{4\pi i\,\zeta} \right] = H_1(\zeta) + H_2(\zeta).$$

The integrals (4.40) and (4.41) are replaced by the loop integrals

$$F_-(z) = \frac{1}{2\pi} \int_{-\infty}^{(0+)} e^{i\zeta z}\, H_1(\zeta)\, d\zeta, \tag{4.43}$$

and

$$F_+(z) = -\frac{1}{2\pi} \int_{\infty}^{(0+)} e^{i\zeta z}\, H_2(\zeta)\, d\zeta, \tag{4.44}$$

and the inverse Fourier transform is defined as follows.

Definition 4.4. If the Fourier transform $\hat{f}(w)$ is a hyperfunction of slow growth that is written by a canonical splitting (4.42) as $\hat{f}(w) = \hat{f}_1(w) + \hat{f}_2(w) \in \mathcal{S}(\mathbb{R})$, then its inverse Fourier transform is the hyperfunction

$$f(x) = [F_+(z), F_-(z)], \tag{4.45}$$

where the lower and upper components of the defining function are given by (4.43) and (4.44).

Note that we may also express (4.43), (4.44), by Laplace transforms. Indeed, we have

$$F_+(z) = \frac{1}{2\pi} \mathcal{L}[\hat{f}_2(\omega)](-iz),$$

$$F_-(z) = \frac{1}{2\pi} \int_{-\infty}^{(0+)} e^{i\zeta z}\, H_1(\zeta)\, d\zeta$$

$$= \frac{1}{2\pi} \int_{\infty}^{(0+)} e^{-i\zeta z}\, H_1(-\zeta)\, d(-\zeta)$$

$$= \frac{1}{2\pi} \int_{\infty}^{(0+)} e^{-(iz)\zeta}\, \{-H_1(-\zeta)\}\, d\zeta = -\mathcal{L}[u(\omega)\hat{f}_1(-\omega)](iz),$$

such that we may also write for the inverse Fourier transform

$$f(x) = \frac{1}{2\pi}\left[\mathcal{L}[\hat{f}_2(\omega)](-iz), -\mathcal{L}[\hat{f}_1(-\omega)](iz)\right]. \tag{4.46}$$

Example 4.12. Let us compute the inverse Fourier transform of $\hat{f}(\omega) = \delta^{(n)}(\omega - a)$. The defining function of this hyperfunction is

$$H(\zeta) = -\frac{1}{2\pi i}\frac{(-1)^n\, n!}{(\zeta - a)^{n+1}}.$$

Suppose $a > 0$, then we write $H(\zeta) = 0 + H_2(\zeta)$, i.e., $H_1(\zeta) = 0$. Then,

$$F_+(z) = -\frac{1}{2\pi}\int_{\infty}^{(0+)} e^{i\zeta z} H_2(\zeta)\, d\zeta = \frac{1}{2\pi}\frac{(-1)^n\, n!}{2\pi i}\int_{\infty}^{(0+)} \frac{e^{i\zeta z}}{(\zeta - a)^{n+1}}\, d\zeta$$

$$= \frac{(-1)^n}{2\pi}\frac{d^n}{d\zeta^n}\left(e^{i\zeta z}\right)\Big|_{\zeta = a} = \frac{(-1)^n}{2\pi}(iz)^n\, e^{iaz},$$

where Cauchy's integral formula was used. $F_-(z) = 0$ since $H_1(\zeta) = 0$ Hence,

$$f(x) = \left[\frac{(-iz)^n\, e^{iaz}}{2\pi}, 0\right]$$

and

$$\mathcal{F}^{-1}[\delta^{(n)}(\omega - a)](x) = \frac{(-ix)^n\, e^{iax}}{2\pi}. \tag{4.47}$$

If $a = 0$, we write $H(\zeta) = H(\zeta)/2 + H(\zeta)/2 = H_1(\zeta) + H_2(\zeta)$.

$$F_+(z) = -\frac{1}{2\pi}\int_{\infty}^{(0+)} e^{i\zeta z} H_2(\zeta)\, d\zeta = \frac{1}{4\pi}\frac{(-1)^n\, n!}{2\pi i}\int_{\infty}^{(0+)} \frac{e^{i\zeta z}}{\zeta^{n+1}}\, d\zeta$$

$$= \frac{(-1)^n}{4\pi}\frac{d^n}{d\zeta^n}\left(e^{i\zeta z}\right)\Big|_{\zeta = 0} = \frac{(-1)^n}{4\pi}(iz)^n.$$

Likewise,

$$F_-(z) = \frac{1}{2\pi}\int_{-\infty}^{(0+)} e^{i\zeta z} H_1(\zeta)\, d\zeta = \frac{1}{4\pi}\frac{(-1)^n\, n!}{2\pi i}\int_{-\infty}^{(0+)} \frac{e^{i\zeta z}}{\zeta^{n+1}}\, d\zeta$$

$$= -\frac{(-1)^n}{4\pi}(iz)^n.$$

Then,

$$\mathcal{F}^{-1}[\delta^{(n)}(w)](z) = \frac{(-1)^n}{2\pi}(iz)^n \, [1/2, -1/2] = \frac{1}{2\pi}(-ix)^n. \tag{4.48}$$

In order to save space and in view of the Reciprocity Rule of the next section, we do not present a detailed proof that the so-defined inverse Fourier transform is really the inverse of the direct Fourier transform.

4.3.1 Reciprocity

The Fourier transform of a hyperfunction $f(x) \in \mathcal{S}(\mathbb{R})$ assumed in the split form $f(x) = f_1(x) + f_2(x)$ is the hyperfunction

$$\mathcal{F}[f(x)](w) = \hat{f}(w) = [\mathcal{L}[f_1(-x)](-i\zeta), -\mathcal{L}[f_2(x)](i\zeta)]. \tag{4.49}$$

Similarly, the inverse Fourier transform of a hyperfunction $\hat{f}(w) \in \mathcal{S}(\mathbb{R})$ with $\hat{f}(w) = \hat{f}_1(w) + \hat{f}_2(w)$ is

$$f(x) = \frac{1}{2\pi} \left[\mathcal{L}[\hat{f}_2(w)](-iz), -\mathcal{L}[\hat{f}_1(-w)](iz) \right]. \tag{4.50}$$

Comparing (4.49) with (4.50) and using the fact that

$$f(x) \triangleleft\!-\!\blacktriangleright \hat{f}(w) \Longleftrightarrow f(-x) \triangleleft\!-\!\blacktriangleright \hat{f}(-w)$$

where we have anticipated (4.65), we see that we must have $\hat{\hat{f}}(x) = 2\pi f(-x)$. Thus, we have established

Proposition 4.6. *If the Fourier transform of the hyperfunction of slow growth* $f(x) = [F_+(z), F_-(z)]$ *is again of slow growth, we have the reciprocity relation*

$$\hat{\hat{f}}(x) = 2\pi \, f(-x). \tag{4.51}$$

This relation may also be expressed as

$$f(x) \triangleleft\!-\!\blacktriangleright \hat{f}(w) \Longrightarrow \hat{f}(x) \triangleleft\!-\!\blacktriangleright 2\pi \, f(-w).$$

There are numerous applications of this rule.

Example 4.13. By using

$$x^n \, e^{iax} \triangleleft\!-\!\blacktriangleright 2\pi \, i^n \delta^{(n)}(w - a),$$

we obtain

$$2\pi \, i^n \delta^{(n)}(x - a) \triangleleft\!-\!\blacktriangleright 2\pi \, (-w)^n \, e^{-iaw},$$

or

$$\delta^{(n)}(x - a) \triangleleft\!-\!\blacktriangleright (iw)^n \, e^{-iaw}. \tag{4.52}$$

Example 4.14. We know that

$$u(x) \triangleleft\!-\!\blacktriangleright \frac{1}{i} \, \mathrm{fp}\frac{1}{w} + \pi \, \delta(w),$$

thus

$$\frac{1}{i}\,\mathrm{fp}\frac{1}{x} + \pi\,\delta(x)\,\triangleleft\!-\!\blacktriangleright\,2\pi\,u(-\omega),$$

or

$$\mathrm{fp}\frac{1}{x}\,\triangleleft\!-\!\blacktriangleright\,2\pi i\,u(-\omega) - i\pi,$$

which can be written as

$$\mathrm{fp}\frac{1}{x}\,\triangleleft\!-\!\blacktriangleright\,-\,\pi i\,\mathrm{sgn}(\omega). \tag{4.53}$$

Example 4.15. For $\Re\gamma > 0$ we have

$$u(x)x^m e^{-\gamma x}\,\triangleleft\!-\!\blacktriangleright\,\frac{m!}{(i\omega+\gamma)^{m+1}},$$

$$u(-x)x^m e^{\gamma x}\,\triangleleft\!-\!\blacktriangleright\,-\,\frac{m!}{(i\omega-\gamma)^{m+1}},$$

$$\exp(-a\,x^2)\,\triangleleft\!-\!\blacktriangleright\,\sqrt{\frac{\pi}{a}}\,\exp(-\frac{\omega^2}{4\,a}).$$

Therefore,

$$\sqrt{\frac{1}{4\pi a}}\,\exp(-\frac{x^2}{4\,a})\,\triangleleft\!-\!\blacktriangleright\,\exp(-a\,\omega^2), \tag{4.54}$$

and

$$\frac{m!}{(ix+\gamma)^{m+1}}\,\triangleleft\!-\!\blacktriangleright\,2\pi\,u(-\omega)(-\omega)^m e^{\gamma\omega},$$

or

$$\frac{1}{(ix+\gamma)^{m+1}}\,\triangleleft\!-\!\blacktriangleright\,\frac{2\pi}{m!}\,u(-\omega)(-\omega)^m e^{\gamma\omega}.$$

Replacing $m+1$ by m finally gives

$$\frac{1}{(x-i\gamma)^m}\,\triangleleft\!-\!\blacktriangleright\,\frac{2\pi i^m}{(m-1)!}\,u(-\omega)(-\omega)^{m-1}e^{\gamma\omega}\quad(\Re\gamma>0). \tag{4.55}$$

Similarly,

$$\frac{1}{(x+i\gamma)^m}\,\triangleleft\!-\!\blacktriangleright\,-\,\frac{2\pi i^m}{(m-1)!}\,u(\omega)(-\omega)^{m-1}e^{-\gamma\omega}\quad(\Re\gamma>0). \tag{4.56}$$

Problem 4.9. Use the reciprocity rule and (4.22) to establish for $m \in \mathbb{N}$

$$\frac{\mathrm{sgn}\,x}{x^m}\,\triangleleft\!-\!\blacktriangleright\,-\,\frac{2}{i^{m-1}\,(m-1)!}\,\{\log|\omega| - \Psi(m)\}. \tag{4.57}$$

4.4 Operational Properties

Most operational rules for the Fourier transformation follow from the corresponding rules of the Laplace transformation. We only have to use

$$\mathcal{F}[f(x)](\omega) = \hat{f}(\omega) = [\mathcal{L}[f_1(x)](s), -\mathcal{L}[f_2(x)](s)]\,,\quad s = i\zeta$$

for the split hyperfunction $f(x) = f_1(x) + f_2(x) \in \mathcal{S}(\mathbb{R})$, $f_1(x) \in O(\mathbb{R}_-)$, $f_2(x) \in O(\mathbb{R}_+)$. Moreover, the reciprocity rule of the previous section produces, for a rule of the Fourier transformation, always a corresponding rule for the inverse Fourier transformation. Clearly, the Fourier transformation and inverse Fourier transformation of a hyperfunction are linear operators, i.e.,

$$\mathcal{F}[c_1 f_1 + c_2 f_2] = c_1 \mathcal{F}[f_1] + c_2 \mathcal{F}[f_2],$$
$$\mathcal{F}^{-1}[c_1 f_1 + c_2 f_2] = c_1 \mathcal{F}^{-1}[f_1] + c_2 \mathcal{F}^{-1}[f_2]. \tag{4.58}$$

This follows immediately from the linearity of the defining integrals.

Example 4.16. From

$$\mathcal{F}^{-1}[\delta^{(n)}(\omega - a)](x) = \frac{(-ix)^n e^{iax}}{2\pi},$$

or

$$\frac{(-ix)^n e^{iax}}{2\pi} \vartriangleleft - \blacktriangleright \delta^{(n)}(\omega - a)],$$

we obtain

$$x^n e^{iax} \vartriangleleft - \blacktriangleright 2\pi i^n \delta^{(n)}(\omega - a), \tag{4.59}$$
$$x^n \vartriangleleft - \blacktriangleright 2\pi i^n \delta^{(n)}(\omega), \tag{4.60}$$
$$x^n \cos ax \vartriangleleft - \blacktriangleright \pi i^n \{\delta^{(n)}(\omega - a) + \delta^{(n)}(\omega + a)\}, \tag{4.61}$$
$$x^n \sin ax \vartriangleleft - \blacktriangleright \pi i^{n-1} \{\delta^{(n)}(\omega - a) - \delta^{(n)}(\omega + a)\}. \tag{4.62}$$

4.4.1 Linear Substitution Rule

Proposition 4.7. If $f(x) \vartriangleleft - \blacktriangleright \hat{f}(\omega)$, then, for real $a \neq 0$ and b, we have

$$f(ax + b) \vartriangleleft - \blacktriangleright \frac{1}{|a|} e^{i(b/a)\omega} \hat{f}(\frac{\omega}{a}). \tag{4.63}$$

Particularly,

$$f(x - b) \vartriangleleft - \blacktriangleright e^{-bi\omega} \hat{f}(\omega) \tag{4.64}$$

and

$$f(ax) \vartriangleleft - \blacktriangleright \frac{1}{|a|} \hat{f}(\frac{\omega}{a}), \quad f(-x) \vartriangleleft - \blacktriangleright \hat{f}(-\omega). \tag{4.65}$$

Proof. We have with $s = i\zeta$, and by using the linear substitution rule for the Laplace transformation,

$$\mathcal{F}[f(ax + b)](\omega) = [\mathcal{L}[f_1(ax + b)](s), -\mathcal{L}[f_2(ax + b)](s)]$$

$$= \left[\frac{1}{|a|} e^{(b/a)s} \mathcal{L}[f_1(x)](\frac{s}{a}), -\frac{1}{|a|} e^{(b/a)s} \mathcal{L}[f_2(x)](\frac{s}{a})\right]$$

$$= \frac{1}{|a|} e^{i(b/a)\omega} \left[\mathcal{L}[f_1(x)](\frac{i\zeta}{a}), -\mathcal{L}[f_2(x)](\frac{i\zeta}{a})\right]$$

$$= \frac{1}{|a|} e^{i(b/a)\omega} \hat{f}(\frac{\omega}{a}). \qquad \square$$

Example 4.17. From $\delta(x) \vartriangleleft - \blacktriangleright 1$ we conclude

$$\delta(ax + b) \vartriangleleft - \blacktriangleright \frac{1}{|a|} e^{i(b/a)\omega}, \quad \delta(x - b) \vartriangleleft - \blacktriangleright e^{-ib\omega}. \tag{4.66}$$

A consequence of (4.65) is

Proposition 4.8. *The Fourier Transformation maintains the parity of a hyperfunction, i.e., if $f(x)$ is an even/odd hyperfunction, then its Fourier transform $\hat{f}(\omega)$ is an even/odd hyperfunction.*

4.4.2 Shift-Rules

By using the image shift-rule of Laplace transformation we have

$$
\begin{aligned}
\mathcal{F}[e^{iax} f(x)](\omega) &= \left[\mathcal{L}[e^{iax} f_1(x)](s), -\mathcal{L}[e^{iax} f_2(x)](s) \right] \\
&= \left[\mathcal{L}[f_1(x)](i\zeta - ia), -\mathcal{L}[f_2(x)](i\zeta - ia) \right] \\
&= \left[\mathcal{L}[f_1(x)](i(\zeta - a)), -\mathcal{L}[f_2(x)](i(\zeta - a)) \right] \\
&= \mathcal{F}[f(x)](\omega - a).
\end{aligned}
$$

Thus, if $f(x) \lhd\!-\!\blacktriangleright \hat{f}(\omega)$, we have the two Shift-Rules

$$
e^{iax} f(x) \lhd\!-\!\blacktriangleright \hat{f}(\omega - a), \qquad f(x - b) \lhd\!-\!\blacktriangleright e^{-biw} \hat{f}(\omega). \tag{4.67}
$$

The latter relation is from the previous section.

Example 4.18. From this formula and the linearity we conclude

$$
f(x) \cos bx \lhd\!-\!\blacktriangleright \frac{1}{2} \{ \hat{f}(\omega - b) + \hat{f}(\omega + b) \}, \tag{4.68}
$$

$$
f(x) \sin bx \lhd\!-\!\blacktriangleright \frac{1}{2i} \{ \hat{f}(\omega - b) - \hat{f}(\omega + b) \}. \tag{4.69}
$$

Problem 4.10. Show for $a < b$ that

$$
\mathrm{sgn}(x; a, b) = u(x - b) - u(a - x) \lhd\!-\!\blacktriangleright \frac{e^{-ia\omega} + e^{-bi\omega}}{i} \, \mathrm{fp} \frac{1}{\omega} \tag{4.70}
$$

thus,

$$
\mathrm{sgn}(x; -b, b) \lhd\!-\!\blacktriangleright 2 \frac{\cos(b\omega)}{i} \, \mathrm{fp} \frac{1}{\omega}. \tag{4.71}
$$

4.4.3 Complex Conjugation and Realness

Let $f(x) \lhd\!-\!\blacktriangleright \hat{f}(\omega)$; then what is the Fourier transform of $\overline{f(x)}$? If $\hat{f}(\omega) = [H_+(\zeta), H_-(\zeta)]$ is a given hyperfunction, the complex-conjugate hyperfunction of $\hat{f}(\omega)$ is defined as

$$
\overline{\hat{f}(\omega)} := - \left[\overline{H_-(\overline{\zeta})}, \overline{H_+(\overline{\zeta})} \right],
$$

and, from this,

$$
\overline{\hat{f}(-\omega)} := \left[\overline{H_+(-\overline{\zeta})}, \overline{H_-(-\overline{\zeta})} \right].
$$

Here, $H_+(\zeta) = \mathcal{L}[f_1(x)](i\zeta)$, and $H_-(\zeta) = -\mathcal{L}[f_2(x)](i\zeta)$. On the other hand $f(x) = f_1(x) + f_2(x)$ implies $\overline{f(x)} = \overline{f_1(x)} + \overline{f_2(x)}$, thus

$$
\mathcal{F}[\overline{f(x)}](\omega) = \left[\mathcal{L}[\overline{f_1(x)}](s), -\mathcal{L}[\overline{f_2(x)}](s) \right].
$$

By using (3.35) this gives

$$\mathcal{F}[\overline{f(x)}](\omega) = \left[\overline{\mathcal{L}[f_1(x)](\overline{s})}, -\overline{\mathcal{L}[f_2(x)](\overline{s})}\right]$$

$$= \left[\overline{\mathcal{L}[f_1(x)](-i\overline{\zeta})}, -\overline{\mathcal{L}[f_2(x)](-i\overline{\zeta})}\right]$$

$$= \left[\overline{H_+(-\overline{\zeta})}, \overline{H_-(-\overline{\zeta})}\right] = \overline{\hat{f}(-\omega)}.$$

We have proved

Proposition 4.9. *If* $f(x) \lhd - \blacktriangleright \hat{f}(\omega)$, *then*

$$\overline{f(x)} \lhd - \blacktriangleright \overline{\hat{f}(-\omega)}, \quad \overline{f(-x)} \lhd - \blacktriangleright \overline{\hat{f}(\omega)}. \tag{4.72}$$

The latter rule follows from the former one by Proposition 4.7.
A hyperfunction is said to be *real* if $f(x) = \overline{f(x)}$; it is called *imaginary*, if $f(x) = -\overline{f(x)}$. Suppose that a hyperfunction $f(x)$ of slow growth is even and real. Because the parity is invariant under Fourier transformation, we have $f(x) \lhd - \blacktriangleright \hat{f}(\omega)$ and $\overline{f(x)} \lhd - \blacktriangleright \overline{\hat{f}(-\omega)} = \overline{\hat{f}(\omega)}$. Thus, $f(x) = \overline{f(x)}$ implies $\hat{f}(\omega) = \overline{\hat{f}(\omega)}$. If $f(x)$ is odd and real, then $f(x) = -f(-x) \lhd - \blacktriangleright - \hat{f}(-\omega)$, and $-f(-x)$ $\lhd - \blacktriangleright - \overline{\hat{f}(\omega)}$, i.e., we have $\hat{f}(\omega) = -\overline{\hat{f}(\omega)}$. Therefore,

Proposition 4.10. *The Fourier transform of a real and even hyperfunction is a real hyperfunction. The Fourier transform of a real and odd hyperfunction is an imaginary hyperfunction.*

4.4.4 Differentiation and Multiplication Rule

If $f(x) = [F_+(z), F_-(z)]$, the n-th derivative of the hyperfunction is defined by

$$D^n f(x) = f^{(n)}(x) := \left[\frac{d^n F_+}{dz^n}, \frac{d^n F_-}{dz^n}\right].$$

If $f(x)$ is a hyperfunction of slow growth, its n-th derivative is of slow growth too. Let again $f(x) = f_1(x) + f_2(x) \in \mathcal{S}(\mathbb{R})$ be a canonical decomposition, then $D^n f(x) = D^n f_1(x) + D^n f_2(x) \in \mathcal{S}(\mathbb{R})$, and with $s = i\zeta$, we have

$$\mathcal{F}[D^n f(x)](\omega) = [\mathcal{L}[D^n f_1(x)](s), -\mathcal{L}[D^n f_2(x)](s)]$$

$$= [s^n \mathcal{L}[f_1(x)](s), -s^n \mathcal{L}[f_2(x)](s)]$$

$$= (i\omega)^n [\mathcal{L}[f_1(x)](i\zeta), -\mathcal{L}[f_2(x)](i\zeta)] = (i\omega)^n \mathcal{F}[f(x)](\omega).$$

Therefore,

Proposition 4.11. *If* $f(x) = [F_+(z), F_-(z)]$ *is a hyperfunction of slow growth with its Fourier transform* $\hat{f}(\omega)$, *then its (generalized) nth derivative* $D^n f(x) = f^{(n)}(x)$ *has the Fourier transform* $(i\,\omega)^n\,\hat{f}(\omega)$.

Example 4.19.

$$\delta(ax + b) \lhd - \blacktriangleright \frac{1}{|a|} e^{i(b/a)\,\omega} \text{ implies } \delta^{(n)}(ax + b) \lhd - \blacktriangleright \frac{(i\omega)^n}{|a|\,a^n} e^{i(b/a)\,\omega}.$$

By the same technique or by using the Reciprocity Rule the following Multiplication Rule may be proved.

Proposition 4.12. *If* $f(x) = [F_+(z), F_-(z)]$ *is a hyperfunction of slow growth with its Fourier transform* $\hat{f}(\omega)$, *then* $x^n f(x)$ *is again of slow growth, and has the Fourier transform* $i^n D^n \hat{f}(\omega)$.

Example 4.20. From $\mathrm{sgn}(x) \vartriangleleft - \blacktriangleright (2/i)\mathrm{fp}(1/\omega)$ we find for $n \in \mathbb{N}$,

$$x^n \, \mathrm{sgn}(x) \vartriangleleft - \blacktriangleright i^n \, \frac{2}{i} \, D^n \mathrm{fp}\frac{1}{\omega} = \frac{2n!}{i^{n+1}} \, \mathrm{fp}\frac{1}{\omega^{n+1}}, \tag{4.73}$$

$$|x| = x \, \mathrm{sgn}(x) \vartriangleleft - \blacktriangleright - 2 \, \mathrm{fp}\frac{1}{\omega^2}. \tag{4.74}$$

By the Shift-Rule, for a real b, we obtain

$$\frac{i^n}{2(n-1)!} \, x^{n-1} \, e^{ibx} \, \mathrm{sgn}(x) \vartriangleleft - \blacktriangleright \mathrm{fp}\frac{1}{(\omega - b)^n}. \tag{4.75}$$

Problem 4.11. By using Problem 4.4 establish, for a non-real Ω,

$$\frac{i^n}{(n-1)!} \, x^{n-1} \, u(x) \, e^{i\Omega x} \vartriangleleft - \blacktriangleright \mathrm{fp}\frac{1}{(\omega - \Omega)^n}, \quad (\Im\Omega > 0), \tag{4.76}$$

$$-\frac{i^n}{(n-1)!} \, x^{n-1} \, u(-x) \, e^{i\Omega x} \vartriangleleft - \blacktriangleright \mathrm{fp}\frac{1}{(\omega - \Omega)^n}, \quad (\Im\Omega < 0). \tag{4.77}$$

Problem 4.12. Establish for $\alpha = a + bi$, $c = A + Bi$,

$$\mathcal{F}^{-1}[\mathrm{fp}\frac{c}{(i\omega - \alpha)^n} + \mathrm{fp}\frac{\bar{c}}{(i\omega - \bar{\alpha})^n}](x)$$

$$= \frac{2}{(n-1)!} \, u(-x) \, x^{n-1} \, e^{ax} \, \{A\cos(bx) - B\sin(bx)\}, \quad (\Re\alpha = a > 0),$$

$$\mathcal{F}^{-1}[\mathrm{fp}\frac{c}{(i\omega - \alpha)^n} + \mathrm{fp}\frac{\bar{c}}{(i\omega - \bar{\alpha})^n}](x)$$

$$= \frac{2}{(n-1)!} \, u(x) \, x^{n-1} \, e^{ax} \, \{A\cos(bx) - B\sin(bx)\}, \quad (\Re\alpha = a < 0).$$

These correspondences allow us, by using partial fraction decomposition, to find the inverse Fourier transform of any rational function of ω.

Example 4.21. Green's function $G(x, x')$ of the boundary value problem

$$(D^2 + \omega_0^2) \, y(x) = f(x), \quad -\infty < x < \infty$$

is the solution of $y(x)$ of

$$(D^2 + \omega_0^2) \, y(x) = \delta(x - x').$$

The parameter x' is fixed, and we are looking for a solution $y(x) \in S(\mathbb{R})$, i.e., for one having a Fourier transform. If $y(x) \vartriangleleft - \blacktriangleright \hat{y}(\omega)$, the equation in the frequency domain becomes

$$(-\omega^2 + \omega_0^2) \, \hat{y}(\omega) = e^{-ix'\omega},$$

thus

$$\hat{y}(\omega) = -e^{-ix'\omega} \, \text{fp} \frac{1}{\omega^2 - \omega_0^2} = \frac{e^{-ix'\omega}}{2\omega_0} \left[\text{fp} \frac{1}{\omega + \omega_0} - \text{fp} \frac{1}{\omega - \omega_0} \right].$$

Using (4.75) we get

$$\mathcal{F}^{-1}[\text{fp} \frac{1}{\omega + \omega_0}](x) = \frac{i}{2} e^{-i\omega_0 x} \, \text{sgn} \, x, \quad \mathcal{F}^{-1}[\text{fp} \frac{1}{\omega - \omega_0}](x) = \frac{i}{2} e^{i\omega_0 x} \, \text{sgn} \, x,$$

thus

$$\frac{i}{2} \text{sgn}(x) e^{-i\omega_0 x} - \frac{i}{2} \text{sgn}(x) e^{i\omega_0 x} = \text{sgn}(x) \sin(\omega_0 x).$$

By taking into account the shift rule for the exponential factor, we obtain

$$G(x, x') = \frac{1}{2\omega_0} \text{sgn}(x - x') \sin(\omega_0(x - x')) = \frac{\sin(\omega_0 |x - x'|)}{2\omega_0}.$$

The solution of the boundary value problem for a given right-hand side $f(x)$ can then be written as

$$y(x) = \frac{1}{2\omega_0} \int_{-\infty}^{\infty} \sin(\omega_0 |x - x'|) f(x') \, dx'$$

$$= \frac{1}{2\omega_0} \left\{ \int_{-\infty}^{x} \sin(\omega_0 (x - x')) f(x') \, dx' + \int_{x}^{\infty} \sin(\omega_0 (x' - x)) f(x') \, dx' \right\},$$

provided the infinite integrals exist.

Problem 4.13. Show that Green's function over $(-\infty, \infty)$ of the differential operator $-D^2 + a^2$, i.e., the solution of

$$(-D^2 + a^2)y(x) = \delta(x - x') \quad (a > 0),$$

is given by

$$G(x, x') = \frac{1}{2a} \{ u(x - x') e^{-a(x - x')} + u(x' - x) e^{-a(x' - x)} \}.$$

Problem 4.14. Use Problem 4.12 to show that the solution of the differential equation

$$(D^3 - 3D^2 + D + 5)y(x) = \delta(x)$$

having a Fourier transform is given by

$$y(x) = u(-x) \frac{e^{2x}}{10} (\cos(x) - 3\sin(x)) + u(x) \frac{e^{-x}}{10}.$$

4.4.5 Convolution Rules

For two ordinary functions having both Fourier transforms, the Fourier transform of their convolution, provided it exists, equals the ordinary product of the two Fourier transforms. In the case of hyperfunctions the situation is more complicated because the Fourier transform of a hyperfunction may again be a hyperfunction and the product of hyperfunctions does not always exist. However, we have two special cases where the product of the two Fourier transforms can always be formed.

(i) If one of the two Fourier transforms is a holomorphic hyperfunction, i.e., it has a horizontal strip of convergence of positive width containing the real axis. This is the case if one of the two hyperfunctions has a compact support.

(ii) If both Fourier transforms $\hat{f}(\omega), \hat{g}(\omega)$ are either lower or upper hyperfunctions. This is the case if both $f(x)$ and $g(x)$ are either right-sided or left-sided hyperfunctions.

See Propositions 4.1 and 4.4.

Definition 4.5. We say that for two hyperfunctions $f(x), g(x) \in S(\mathbb{R})$ with their Fourier transforms $\hat{f}(\omega)$ and $\hat{g}(\omega)$, the convolution property holds, if

$$f(x) \star g(x) \vartriangleleft - \blacktriangleright \hat{f}(\omega) \cdot \hat{g}(\omega). \tag{4.78}$$

Let us first consider some examples.

Example 4.22. Suppose the hyperfunction $g(x)$ is of slow growth. We know that

$$\delta(x - a) \star g(x) = g(x) \star \delta(x - a) = g(x - a).$$

With $g(x) \vartriangleleft - \blacktriangleright \hat{g}(\omega)$ and $\delta(x - a) \vartriangleleft - \blacktriangleright e^{-ia\omega}$ we have $g(x - a) \vartriangleleft - \blacktriangleright e^{-ia\omega} \hat{g}(\omega)$ by the linear substitution rule, and the convolution property is verified.

Example 4.23. For integrals m, n we have

$$\text{fp} \frac{1}{(ax + b)^m} \star \text{fp} \frac{1}{(ax + c)^n} = -\frac{\pi^2}{|a|} \frac{(-1)^{m+n-1}}{(m-1)!\,(n-1)!} \delta^{(m+n-2)}(ax + b + c). \tag{4.79}$$

We know the correspondence

$$\text{fp} \frac{1}{x^m} \vartriangleleft - \blacktriangleright \frac{\pi}{i^m (m-1)!} \omega^{m-1} \text{sgn}\,\omega,$$

hence, by the linear substitution rule we obtain

$$\text{fp} \frac{1}{(ax + b)^m} \vartriangleleft - \blacktriangleright \frac{\pi}{i^m (m-1)!\,|a|} e^{i\,(b/a)\omega} \left(\frac{\omega}{a}\right)^{m-1} \text{sgn}\left(\frac{\omega}{a}\right),$$

$$\text{fp} \frac{1}{(ax + b)^n} \vartriangleleft - \blacktriangleright \frac{\pi}{i^n (n-1)!\,|a|} e^{i\,(c/a)\omega} \left(\frac{\omega}{a}\right)^{n-1} \text{sgn}\left(\frac{\omega}{a}\right).$$

Because the square of a sign-function equals 1, we obtain for the product of the two Fourier transforms

$$\frac{\pi^2}{i^{m+n} (m-1)!\,(n-1)!\,|a|^2\, a^{m+n-2}} e^{i\,((b+c)/a)\omega} \omega^{m+n-2}. \tag{4.80}$$

On the other hand, from $\delta^{(m+n-2)}(x) \vartriangleleft - \blacktriangleright (i\,\omega)^{m+n-2}$ and by the linear substitution rule, we have

$$\delta^{(m+n-2)}(ax + b + c) \vartriangleleft - \blacktriangleright \frac{1}{|a|} e^{i\,((b+c)/a)\omega} \left(i\,\frac{\omega}{a}\right)^{m+n-2}.$$

The Fourier transform of the right-hand side of (4.79) then gives

$$\frac{\pi^2}{i^{m+n} |a|^2 (m-1)!\,(n-1)!\,a^{m+n-2}} e^{i\,((b+c)/a)\omega} \omega^{m+n-2}$$

in agreement with (4.80). Therefore, the convolution property is verified.

The next example shows some possible difficulties.

Example 4.24. Consider the two hyperfunctions

$$f(x) = u(x) \triangleleft - \blacktriangleright - i \operatorname{fp}\frac{1}{\omega} + \pi\,\delta(\omega), \quad g(x) = x^{-2}\triangleleft - \blacktriangleright - \pi\,\omega\,\operatorname{sgn}\omega.$$

In the sense of ordinary functions the convolution exists and equals

$$u(x) \star \frac{1}{x^2} = \int_{-\infty}^{x} \frac{1}{t^2}\,dt = -\frac{1}{x},$$

which, interpreted as a hyperfunction, has the Fourier transform $i\pi\,\operatorname{sgn}\omega$. If we form the product of the Fourier transforms, we obtain

$$\hat{f}(\omega) \cdot \hat{g}(\omega) = \{-i\operatorname{fp}\frac{1}{\omega} + \pi\,\delta(\omega)\} \cdot \{-\pi\,\omega\,\operatorname{sgn}\omega\}.$$

On the right-hand side we have the product of two hyperfunctions for which sing supp$f(x) \cap$ sing supp$g(x) = \{0\} \neq \phi$, i.e., whose existence is not established by our theory. However, if we assume associativity and write the product in the form

$$i\pi\,(\operatorname{fp}\frac{1}{\omega}\,\omega)\,\operatorname{sgn}\omega - \pi^2\,(\delta(\omega)\,\omega)\,\operatorname{sgn}\omega = i\pi\,\operatorname{sgn}\omega,$$

because $\omega\,\delta(\omega) = 0$, the result would be in accordance with the convolution property.

We can prove the following properties.

Proposition 4.13. *Let $f(x) = [F(z)]$ be a hyperfunction with compact support $[\alpha, \eta]$, and $g(x) = [G(z)] \in \mathcal{S}(\mathbb{R})$ a hyperfunction of slow growth, then the convolution property holds.*

Proof. By Proposition 2.53 the convolution exists, and by Corollary 4.2 we know that $\hat{f}(\omega)$ is a real analytic function given by

$$\hat{f}(\omega) = - \oint_{(a',b')} e^{-i\omega z}\,F(z)\,dz$$

with $a' < \alpha$ and $b' > \beta$. Thus $\hat{f}(\omega)\hat{g}(\omega)$ make sense as the product of a real analytic function with a hyperfunction. If we write $g(x) = g_1(x) + g_2(x)$, where $g_1(x) = [G_1(z)] \in O(\mathbb{R}_-)$, and $g_2(x) = [G_2(z)] \in O(\mathbb{R}_+)$, we have with

$$\hat{G}_1(\zeta) = - \int_{-\infty}^{(0+)} e^{-i\zeta z}\,G_1(z)\,dz, \quad \hat{G}_2(\zeta) = \int_{\infty}^{(0+)} e^{-i\zeta z}\,G_2(z)\,dz,$$

$$g_1(x) \triangleleft - \blacktriangleright \hat{g}_1(\omega) = \left[\hat{G}_1(\zeta), 0\right], \quad g_2(x) \triangleleft - \blacktriangleright \hat{g}_2(\omega) = \left[0, \hat{G}_2(\zeta)\right],$$

thus

$$\hat{f}(\omega)\,\hat{g}_1(\omega) = \left[\hat{f}(\zeta)\,\hat{G}_1(\zeta), 0\right], \quad \hat{f}(\omega)\,\hat{g}_2(\omega) = \left[0, \hat{f}(\zeta)\,\hat{G}_2(\zeta)\right],$$

or

$$\hat{f}(\omega)\,\hat{g}(\omega) = \hat{f}(\omega)\,\hat{g}_1(\omega) + \hat{f}(\omega)\,\hat{g}_2(\omega) = \left[\hat{f}(\zeta)\,\hat{G}_1(\zeta), \hat{f}(\zeta)\,\hat{G}_2(\zeta)\right].$$

Consider the upper component, where $\Im \zeta > 0$,

$$\hat{f}(\zeta)\,\hat{G}_1(\zeta) = -\oint_{(a',b')} e^{-i\zeta\,z'}\,F(z')\,dz' \times (-1)\int_{-\infty}^{(0+)} e^{-i\zeta\,z}\,G_1(z)\,dz$$

$$= \int_{-\infty}^{(0+)} \oint_{(a',b')} e^{-i\zeta\,(z+z')}\,F(z')G_1(z)\,dz'\,dz.$$

We make the change of variables $Z = z + z'$, where z is held fixed. If the infinite contour $(-\infty, 0+)$ is chosen sufficiently near the real axis, the contour (a', b') is transformed into a similar contour $(c_1 = \Re z - b', c_2 = \Re z - a')$. Thus,

$$\hat{f}(\zeta)\,\hat{G}_1(\zeta) = \int_{-\infty}^{(0+)} \oint_{(c_1,c_2)} e^{-i\zeta\,Z}\,F(Z-z)G_1(z)\,dz\,dZ$$

$$= -\int_{-\infty}^{(0+)} e^{-i\zeta\,Z}\,\{-\oint_{(c_1,c_2)\cap(-\infty,0]} F(Z-z)G_1(z)\,dz\}\,dZ$$

because $\operatorname{supp} g_1(x) = (-\infty, 0)$. Likewise, we obtain for the second expression $\Im \zeta < 0$,

$$\hat{f}(\zeta)\,\hat{G}_2(\zeta) = \oint_{\infty}^{(0+)} e^{-i\zeta\,Z}\,\{-\int_{(c_1,c_2)\cap[0,\infty)} F(Z-\zeta)G_2(\zeta)\,d\zeta\}\,dZ.$$

The last two expressions mean that the hyperfunction $\left[\hat{f}(\zeta)\,\hat{G}_1(\zeta), \hat{f}(\zeta)\,\hat{G}_2(\zeta)\right]$ is the Fourier transform of the hyperfunction

$$-\oint_{(c_1,c_2)\cap(-\infty,0]} F(Z-z)G_1(z)\,dz - \int_{(c_1,c_2)\cap[0,\infty)} F(Z-\zeta)G_2(\zeta)\,d\zeta\,dZ$$

$$= f(x) \star g(x).$$

as it is shown by comparison with (2.152). $\qquad\qquad\square$

Proposition 4.14. *For two right-sided or two left-sided hyperfunctions of slow growth the convolution property holds.*

Proof. Since we have already proved this theorem in the framework of Laplace transformation, we shall use that result. Assume that $f(x), g(x) \in O(\mathbb{R}_+)$ with $\sigma_-(f) \le 0$ and $\sigma_-(g) \le 0$. Then their Laplace transforms $\mathcal{L}[f(t)](s), \mathcal{L}[g(t)](s)$ are holomorphic functions in the right half-plane $\Re s > 0$, and their Fourier transforms are lower hyperfunctions

$$f(x) \triangleleft\!-\!\blacktriangleright \hat{f}(\omega) = [0, -\mathcal{L}[f(x)](i\zeta)], \quad g(x) \triangleleft\!-\!\blacktriangleright \hat{g}(\omega) = [0, -\mathcal{L}[g(x)](i\zeta)].$$

The convolution theorem of Laplace transforms says

$$\mathcal{L}[f(x) \star g(x)](s) = \mathcal{L}[f(t)](s)\mathcal{L}[g(t)](s).$$

The product of two lower hyperfunctions is always defined, and we have

$$\hat{f}(\omega) \cdot \hat{g}(\omega) = [0, -\mathcal{L}[f(x)](i\zeta)\,\mathcal{L}[g(x)](i\zeta)]$$

$$= [0, -\mathcal{L}[f(x) \star g(x)](i\zeta)].$$

Because $h(x) = f(x) \star g(x) \in O(\mathbb{R}_+)$ with $\sigma_-(h) \le 0$, we can write

$$\mathcal{F}[f(x) \star g(x)] = [0, -\mathcal{L}[f(x) \star g(x)](i\zeta] = \hat{f}(\omega) \cdot \hat{g}(\omega)$$

which proves the theorem. The proof for two left-sided hyperfunctions is quite similar. The case where the finite endpoint of the support is not 0 needs some technical modifications but is essentially again the same and is omitted. □

Example 4.25. The following example is instructive. We have

$$u(x) \triangleleft - \blacktriangleright \frac{1}{i}\, \mathrm{fp}\frac{1}{\omega} + \pi\, \delta(\omega).$$

If we want to compute the convolution $u(x) \star u(x)$ by using Fourier transformation, we are confronted with the problem of forming the product of the two hyperfunctions

$$\{\frac{1}{i}\, \mathrm{fp}\frac{1}{\omega} + \pi\, \delta(\omega)\} \cdot \{\frac{1}{i}\, \mathrm{fp}\frac{1}{\omega} + \pi\, \delta(\omega)\}.$$

Regarded in this form, it seems that the product is not defined because the singular support of the two factors is not empty. But the hyperfunction in question is in fact a hidden lower hyperfunction,

$$u(x) \triangleleft - \blacktriangleright [0, -1/(i\zeta)],$$

and the product of two lower hyperfunctions is defined. Therefore by the convolution property we have

$$u(x) \star u(x) \triangleleft - \blacktriangleright [0, -\{-1/(i\zeta)\}^2]$$
$$= [0, 1/\zeta^2] = -[1/(2\zeta^2), -1/(2\zeta^2)] + [1/(2\zeta^2), 1/(2\zeta^2)]$$
$$= -\mathrm{fp}\frac{1}{\omega^2} + \pi i\, \delta'(\omega).$$

The last expression is the Fourier transform of $u(x)x$.

Example 4.26. The Fourier transform of $u(x)|x|^\alpha$, where α is not integral, is

$$\Gamma(\alpha+1)\, e^{i\frac{\pi}{2}(\alpha+1)\,\mathrm{sgn}\,\omega}\, \frac{1}{|\omega|^{\alpha+1}}.$$

Similarly for $u(x)|x|^\beta$. Since the Fourier transform of a right-sided hyperfunctions is always a lower hyperfunction, and the product of two lower hyperfunctions can be formed, we may abbreviate the procedure and simply multiply formally the two Fourier transforms. Thus the Fourier transform of $u(x)|x|^\alpha \star u(x)|x|^\beta$, α, β not integral, is given by

$$\{\Gamma(\alpha+1)\, e^{i\frac{\pi}{2}(\alpha+1)\,\mathrm{sgn}\,\omega}\, \frac{1}{|\omega|^{\alpha+1}}\}\{\Gamma(\beta+1)\, e^{i\frac{\pi}{2}(\beta+1)\,\mathrm{sgn}\,\omega}\, \frac{1}{|\omega|^{\beta+1}}\}$$
$$= \Gamma(\alpha+1)\Gamma(\beta+1)\, e^{i\frac{\pi}{2}(\alpha+\beta+2)\,\mathrm{sgn}\,\omega}\, \frac{1}{|\omega|^{\alpha+\beta+2}}$$
$$= \frac{\Gamma(\alpha+1)\Gamma(\beta+1)}{\Gamma(\alpha+\beta+2)}\, \Gamma(\alpha+\beta+2)\, e^{i\frac{\pi}{2}(\alpha+\beta+2)\,\mathrm{sgn}\,\omega}\, \frac{1}{|\omega|^{\alpha+\beta+2}}.$$

The last expression is the Fourier transform of

$$B(\alpha + 1, \beta + 1) |x|^{\alpha+\beta+1} u(x),$$

therefore,

$$u(x)|x|^{\alpha} \star u(x)|x|^{\beta} = B(\alpha + 1, \beta + 1) |x|^{\alpha+\beta+1} u(x), \tag{4.81}$$

where B denote Euler's beta function.

In order to compute convolutions of right-sided hyperfunctions Laplace transformation often suits better than Fourier transformation.

4.5 Further Examples

The following simple rule is a consequence of the linearity of the Fourier transformation.

Proposition 4.15. *If the hyperfunction $f(x, \alpha)$ depends smoothly on the parameter α, and $f(x, \alpha) \lhd - \blacktriangleright \hat{f}(\omega, \alpha)$, then*

$$\frac{\partial^n}{\partial \alpha^n} f(x, \alpha) \lhd - \blacktriangleright \frac{\partial^n}{\partial \alpha^n} \hat{f}(\omega, \alpha). \tag{4.82}$$

Example 4.27. By differentiating with respect to α, the correspondence

$$|x|^{\alpha} u(x) \lhd - \blacktriangleright \Gamma(\alpha + 1) e^{i\frac{\pi}{2}(\alpha+1)\,\text{sgn}\,\omega} \frac{1}{|\omega|^{\alpha+1}},$$

where α ranges over an interval containing no integral number, yields

$$|x|^{\alpha} \log |x| \, u(x) \lhd - \blacktriangleright$$
$$\frac{\Gamma(\alpha + 1)}{|\omega|^{\alpha+1}} e^{-i\pi(\alpha+1)\,\text{sgn}(\omega)/2} \{\Psi(\alpha + 1) - i\frac{\pi}{2} \text{sgn}(\omega) - \log |\omega|\}. \tag{4.83}$$

Example 4.28. By differentiating the correspondence

$$|x|^{\alpha} \lhd - \blacktriangleright - 2\,\Gamma(\alpha + 1) \sin(\frac{\pi}{2}\alpha) \frac{1}{|\omega|^{\alpha+1}}$$

with respect to the non-integral parameter α, we obtain

$$|x|^{\alpha} \log |x| \lhd - \blacktriangleright$$
$$\frac{2\Gamma(\alpha + 1)}{|\omega|^{\alpha+1}} \{[\Psi(\alpha + 1) + \log |\omega|] \sin(\frac{\pi}{2}\alpha) - \frac{\pi}{2} \cos(\frac{\pi}{2}\alpha)\}. \tag{4.84}$$

Similarly, by differentiating the correspondence

$$|x|^{\alpha} \,\text{sgn}\,x \lhd - \blacktriangleright - 2i\,\Gamma(\alpha + 1) \sin(\frac{\pi}{2}(\alpha + 1)) \frac{\text{sgn}\,\omega}{|\omega|^{\alpha+1}}$$

with respect to α, we get

$$|x|^{\alpha} \,\text{sgn}\,x \log |x| \lhd - \blacktriangleright - \frac{2i\,\Gamma(\alpha + 1)\,\text{sgn}(\omega)}{|\omega|^{\alpha+1}}$$
$$\times \{[\Psi(\alpha + 1) - \log |\omega|] \cos(\frac{\pi}{2}\alpha) - \frac{\pi}{2} \sin(\frac{\pi}{2}\alpha)\}. \tag{4.85}$$

Example 4.29. Here we compute the Fourier transform of the right and left-sided functions $u(x) \exp(-ax^2)$ and $u(-x) \exp(-ax^2)$ with $a > 0$, respectively. We use the Laplace correspondence

$$u(t)e^{-at^2} \circ\!\!-\!\!\bullet \frac{1}{2}\sqrt{\frac{\pi}{a}} e^{s^2/(4a)} \operatorname{erfc}(\frac{s}{2\sqrt{a}})$$

and

$$\mathcal{F}[u(x)f(x)](\omega) = [0, -\mathcal{L}[f(t)](i\zeta)], \quad \mathcal{F}[u(-x)f(x)](\omega) = [\mathcal{L}[f(-t)](-i\zeta), 0].$$

Thus

$$\mathcal{F}[u(x)\exp(-ax^2)](\omega) = \left[0, -\frac{1}{2}\sqrt{\frac{\pi}{a}} e^{-\zeta^2/(4a)} \operatorname{erfc}(\frac{i\zeta}{2\sqrt{a}})\right],$$

$$\mathcal{F}[u(-x)\exp(-ax^2)](\omega) = \left[\frac{1}{2}\sqrt{\frac{\pi}{a}} e^{-\zeta^2/(4a)} \operatorname{erfc}(\frac{-i\zeta}{2\sqrt{a}}), 0\right]$$

yielding

$$u(x)e^{-ax^2} \triangleleft\!-\!\blacktriangleright \frac{1}{2}\sqrt{\frac{\pi}{a}} e^{-\omega^2/(4a)} \operatorname{erfc}(\frac{i\omega}{2\sqrt{a}}),$$

$$u(-x)e^{-ax^2} \triangleleft\!-\!\blacktriangleright \frac{1}{2}\sqrt{\frac{\pi}{a}} e^{-\omega^2/(4a)} \operatorname{erfc}(\frac{-i\omega}{2\sqrt{a}}).$$

We now use *Dawson's integral*

$$\operatorname{Daw}(z) := e^{-z^2} \int_0^z e^{t^2} dt \tag{4.86}$$

(see [29, Olver p.44]), and the relations

$$\operatorname{erfc}(iz) = 1 - \frac{2i}{\sqrt{\pi}} \int_0^z e^{t^2} dt, \quad \operatorname{erfc}(-iz) = 1 + \frac{2i}{\sqrt{\pi}} \int_0^z e^{t^2} dt \tag{4.87}$$

(see [1, Abramowitz - Stegun p.84]) to obtain

$$u(x)e^{-ax^2} \triangleleft\!-\!\blacktriangleright \frac{1}{2}\sqrt{\frac{\pi}{a}} e^{-\omega^2/(4a)} - \frac{i}{\sqrt{a}} \operatorname{Daw}\left(\frac{\omega}{2\sqrt{a}}\right), \tag{4.88}$$

$$u(-x)e^{-ax^2} \triangleleft\!-\!\blacktriangleright \frac{1}{2}\sqrt{\frac{\pi}{a}} e^{-\omega^2/(4a)} + \frac{i}{\sqrt{a}} \operatorname{Daw}\left(\frac{\omega}{2\sqrt{a}}\right). \tag{4.89}$$

Adding the two results gives

$$e^{-ax^2} \triangleleft\!-\!\blacktriangleright \sqrt{\frac{\pi}{a}} e^{-\omega^2/(4a)}, \tag{4.90}$$

and forming the difference yields

$$\operatorname{sgn}(x) e^{-ax^2} \triangleleft\!-\!\blacktriangleright -\frac{2i}{\sqrt{a}} \operatorname{Daw}\left(\frac{\omega}{2\sqrt{a}}\right). \tag{4.91}$$

Applying the Reciprocity Rule to the last result gives

$$\mathcal{F}^{-1}[\operatorname{sgn}(\omega) e^{-a\omega^2}](x) = i \frac{1}{\pi\sqrt{a}} \operatorname{Daw}\left(\frac{x}{2\sqrt{a}}\right). \tag{4.92}$$

4.6 Poisson's Summation Formula

In Section 2.6 we have established the relation (2.123)

$$\text{pv} \sum_{k=-\infty}^{\infty} \delta(x - kT) = \frac{1}{T} \, \text{pv} \sum_{k=-\infty}^{\infty} e^{2\pi i k x / T} = \frac{1}{T} \, \text{pv} \sum_{n=-\infty}^{\infty} e^{-2\pi i n x / T}.$$

We can form the product of a hyperfunction $f(x)$ and the Dirac hyperfunction $\delta(x - kT)$ provided the points $x_k = kT$, $k \in \mathbb{Z}$, are not contained in sing supp$f(x)$, i.e., $f(x)$ is holomorphic at the points x_k (see Example 2.32). In this event we have $f(x) \cdot \delta(x - kT) = f(kT)\delta(x - kT)$. Hence, in the above relation, we form the termwise product on both sides with the hyperfunction $f(x)$ and integrate the resulting hyperfunction from $-\infty$ to ∞. The integral is taken in the Cauchy principal sense in order to extend the applicability of the method.

$$\sum_{k=-\infty}^{\infty} f(kT) \, \text{pv} \int_{-\infty}^{\infty} \delta(x - kT) \, dx = \frac{1}{T} \sum_{n=-\infty}^{\infty} \text{pv} \int_{-\infty}^{\infty} e^{-2\pi i n x / T} f(x) \, dx$$

$$= \frac{1}{T} \, \text{pv} \int_{-\infty}^{\infty} f(x) \, dx$$

$$+ \frac{1}{T} \sum_{n=1}^{\infty} \{ \text{pv} \int_{-\infty}^{\infty} e^{-2\pi i n x / T} f(x) \, dx + \text{pv} \int_{-\infty}^{\infty} e^{2\pi i n x / T} f(x) \, dx \}.$$

This can be rewritten as

$$\text{pv} \sum_{k=-\infty}^{\infty} f(kT) = \frac{1}{T} \, \text{pv} \int_{-\infty}^{\infty} f(x) \, dx$$

$$+ \frac{1}{T} \sum_{n=1}^{\infty} \left\{ \hat{f}(-\frac{2\pi}{T} n) + \hat{f}(\frac{2\pi}{T} n) \right\} \qquad (4.93)$$

provided the series are convergent, the pv-integral over $f(x)$ exists, and the Fourier transform $\hat{f}(\omega)$ on the right-hand side has values at the points $\omega_n = 2\pi n / T$, $\in \mathbb{Z} \backslash 0$. (4.93) is known as *Poisson's summation formula*.

Normally the case $n = 0$ in the sum of the right-hand side is not singled out and the classical Poisson summation formula for ordinary functions is stated as

$$\text{pv} \sum_{k=-\infty}^{\infty} f(kT) = \frac{1}{T} \, \text{pv} \sum_{n=-\infty}^{\infty} \hat{f}(\frac{2\pi}{T} n). \qquad (4.94)$$

In this case let us replace the hyperfunction $f(x)$ by $g(x + t)$, where t is a real parameter such that the hyperfunction $g(x)$ is holomorphic at all points $x_k = kT + t$, $k \in \mathbb{Z}$. By the Shift Rule we have

$$\mathcal{F}[f(x)](\omega) = \mathcal{F}[g(x + t)](\omega) = e^{-it\omega} \, \hat{g}(\omega),$$

therefore

$$\text{pv} \sum_{k=-\infty}^{\infty} g(kT + t) = \frac{1}{T} \, \text{pv} \sum_{n=-\infty}^{\infty} \hat{g}(\frac{2\pi}{T} n) \, e^{-it(2\pi/T)n},$$

or

$$\text{pv} \sum_{k=-\infty}^{\infty} g(t - kT) = \frac{1}{T} \text{pv} \sum_{n=-\infty}^{\infty} \hat{g}(\frac{2\pi}{T}n) \, e^{i\frac{2\pi}{T}tn}. \tag{4.95}$$

Formula (4.95) is only a relation between two infinite series of values of a hyperfunction and not a relation of two infinite series of hyperfunctions (t is a fixed parameter and not the variable of the hyperfunction). In the case where the hyperfunction is an ordinary function the parameter t can be considered as a continuous variable x and (4.95) becomes a functional relation

$$\text{pv} \sum_{k=-\infty}^{\infty} f(x - kT) = \frac{1}{T} \text{pv} \sum_{n=-\infty}^{\infty} \hat{f}(n\omega) \, e^{in\omega x} \tag{4.96}$$

where $\omega = 2\pi/T$, and $\hat{f} = \mathcal{F}[f]$.

Example 4.30. Take $f(x) = \exp(-ax^2)$. Its Fourier transform is $\hat{f}(\omega) = \sqrt{\pi/a} \times \exp(-\omega^2/4a)$. Thus,

$$\text{pv} \sum_{k=-\infty}^{\infty} e^{-a(x-kT)^2} = \frac{1}{T} \sqrt{\frac{\pi}{a}} \text{pv} \sum_{n=-\infty}^{\infty} e^{-\frac{n^2\omega^2}{4a}} e^{in\omega x}. \tag{4.97}$$

With $T = 1$, and replacing k by $-k$ on the left-hand side and writing $a = 1/t$, $t > 0$, we have

$$\sum_{k=-\infty}^{\infty} e^{-(x+k)^2/t} = \sqrt{\pi t} \, \text{pv} \sum_{n=-\infty}^{\infty} e^{-\frac{n^2\omega^2}{4}t} e^{in\omega x}. \tag{4.98}$$

Thus,

$$\frac{1}{\sqrt{\pi t}} \sum_{k=-\infty}^{\infty} e^{-(x+k)^2/t} = 1 + 2 \sum_{n=1}^{\infty} e^{-n^2\pi^2 t} \cos(2\pi nx) \tag{4.99}$$

which is Jacobi's elliptic theta-3 function, denoted by $\vartheta_3(x|t)$.

Example 4.31. Take $f(x) = 1/(x + ia)$, $\Re a > 0$. By (4.56) we have $\hat{f}(\omega) = -2\pi i \, u(\omega) \exp(-a\omega)$; then,

$$\text{pv} \sum_{k=-\infty}^{\infty} f(kT) = -\frac{1}{ia} + 2ia \sum_{k=1}^{\infty} \frac{1}{a^2 + k^2 T^2}.$$

The right-hand side of (4.93) becomes

$$\text{pv} \int_{-\infty}^{\infty} \frac{dx}{x + ia} + \sum_{n=1}^{\infty} e^{-2an\pi/T} = \frac{1}{2} + \frac{e^{-2a\pi/T}}{1 - e^{-2a\pi/T}}$$

$$= \frac{1}{2} + \frac{e^{-a\pi/T}}{e^{a\pi/T} - e^{-a\pi/T}} = \frac{1}{2} \coth\left(\frac{\pi}{T}a\right).$$

Eventually, we obtain

$$\frac{\pi}{T} \coth\left(\frac{\pi}{T}a\right) = \frac{1}{a} + 2a \sum_{k=1}^{\infty} \frac{1}{a^2 + k^2 T^2}$$

valid for $\Re a > 0$. Analytic continuation extends the relation to all complex z and yields the partial fraction decomposition of the coth-function.

$$\frac{\pi}{T} \coth\left(\frac{\pi}{T}z\right) = \frac{1}{z} + 2z \sum_{k=1}^{\infty} \frac{1}{z^2 + k^2 T^2}. \tag{4.100}$$

Example 4.32. We take $f(x) = J_0(ax)$, $a > 0$ and $T = 1$. By (4.13) we get

$$\hat{f}(\omega) = \chi_{(-a,a)} \frac{2}{\sqrt{a^2 - \omega^2}}.$$

By (4.94) we obtain

$$\mathrm{pv} \sum_{k=-\infty}^{\infty} J_0(ak) = \mathrm{pv} \sum_{n=-\infty}^{\infty} \chi_{(-a,a)} \frac{2}{\sqrt{a^2 - 4\pi^2 n^2}}.$$

If $0 < a < 2\pi$ only the term corresponding to $n = 0$ remains. If $a \in (2m\pi, 2(m+1)\pi)$ the symmetric sums contains $2m + 1$ terms. Eventually, we obtain

$$1 + 2 \sum_{k=1}^{\infty} J_0(ak) = \begin{cases} 2/a, & 0 < a < 2\pi \\ 2/a + 4\sum_{n=1}^{m} 1/\sqrt{a^2 - 4\pi^2 n^2}, & 2m\pi < a < 2(m+1)\pi. \end{cases}$$

Example 4.33. We take $f(x) = J_0(x) \cos(ax)$ and $T = \pi$. From (4.68) we find

$$\hat{f}(\omega) = \frac{1}{2} \left\{ \chi_{(-1-a,1-a)} \frac{2}{\sqrt{1 - (\omega - a)^2}} + \chi_{(-1+a,1+a)} \frac{2}{\sqrt{1 - (\omega + a)^2}} \right\}.$$

By (4.94) we obtain

$$\mathrm{pv} \sum_{k=-\infty}^{\infty} J_0(\pi k) \cos(a\pi k) = \frac{1}{\pi} \mathrm{pv} \sum_{n=-\infty}^{\infty} \left\{ \chi_{(-1-a,1-a)} \frac{1}{\sqrt{1 - (2n - a)^2}} \right.$$

$$\left. + \chi_{(-1+a,1+a)} \frac{1}{\sqrt{1 - (2n + a)^2}} \right\}.$$

For $-1 < a < 1$ only the term corresponding to $n = 0$ remains of the sum. Thus,

$$\mathrm{pv} \sum_{k=-\infty}^{\infty} J_0(\pi k) \cos(a\pi k) = \frac{2}{\pi} \frac{1}{\sqrt{1 - a^2}}. \tag{4.101}$$

This reflects that the Fourier series of the periodic function $f(x) = f(x + 2)$, obtained by periodic repetition to the right and to the left of

$$f_0(x) = \frac{1}{\pi} \frac{1}{\sqrt{1 - x^2}}, \quad -1 < x < 1,$$

is

$$\frac{1}{2} + J_0(\pi) \cos(\pi x) + J_0(2\pi) \cos(2\pi x) + J_0(3\pi) \cos(3\pi x) + \cdots.$$

4.7 Application to Integral and Differential Equations

In this section we continue to present some simple applications to differential and integral equations.

4.7.1 Integral Equations III

Example 4.34. Let us consider the following singular integral equation ([21, Kanwal]) of Cauchy's type

$$a\,y(x) + \frac{b}{\pi}\,\mathrm{pv}\int_{-\infty}^{\infty}\frac{y(t)}{t-x}\,dt = f(x) \tag{4.102}$$

with $a^2 + b^2 \neq 0$. We interpret this equation as

$$a\,y(x) - \frac{b}{\pi}\,y(x)\star\mathrm{fp}\frac{1}{x} = f(x).$$

By taking the Fourier transform on both sides, we obtain the image equation

$$a\,\hat{y}(\omega) - \frac{b}{\pi}\,\hat{y}(\omega)\,\frac{\pi}{i}\,\mathrm{sgn}(\omega) = (a + ib\,\mathrm{sgn}(\omega))\,\hat{y}(\omega) = \hat{f}(\omega).$$

Multiplying with $a - ib\,\mathrm{sgn}(\omega)$ yields

$$(a^2 + b^2)\,\hat{y}(\omega) = (a - ib\,\mathrm{sgn}(\omega))\,\hat{f}(\omega),$$

thus

$$\hat{y}(\omega) = \frac{a}{a^2 + b^2}\,\hat{f}(\omega) + \frac{b}{a^2 + b^2}\,\frac{1}{\pi}\,\hat{f}(\omega)\,\frac{\pi}{i}\,\mathrm{sgn}(\omega).$$

By taking the inverse Fourier transform, we obtain

$$y(x) = \frac{a}{a^2 + b^2}\,f(x) + \frac{b}{a^2 + b^2}\,\frac{1}{\pi}\,\mathrm{pv}\int_{-\infty}^{\infty}\frac{f(t)}{x-t}\,dt. \tag{4.103}$$

Example 4.35. We continue the previous example and take

$$f(x) = \sum_{k=1}^{\infty}\delta(x - k^2).$$

We form the convolution

$$\mathrm{fp}\frac{1}{x}\star f(x) = \sum_{k=1}^{\infty}\mathrm{fp}\frac{1}{x}\star\delta(x - k^2) = \sum_{k=1}^{\infty}\frac{1}{x - k^2}$$

$$= -\frac{1}{2x} + \frac{\pi}{2\sqrt{x}}\,\cot(\pi\sqrt{x}),$$

where we have used (2.114). Therefore, we obtain the solution of the integral equation

$$a\,y(x) + \frac{b}{\pi}\,\mathrm{pv}\int_{-\infty}^{\infty}\frac{y(t)}{t-x}\,dt = \sum_{k=1}^{\infty}\delta(x - k^2) \tag{4.104}$$

as

$$y(x) = \frac{a}{a^2 + b^2}\sum_{k=1}^{\infty}\delta(x - k^2) + \frac{b}{a^2 + b^2}\left\{-\frac{1}{2\pi x} + \frac{1}{2\sqrt{x}}\,\cot(\pi\sqrt{x})\right\}. \tag{4.105}$$

Chapter 5

Hilbert Transforms

Because the convolution of two hyperfunctions is an established concept, we may use it to define the Hilbert transform of a hyperfunction. On the other hand, the classical Hilbert transform is also linked to the Cauchy-type integral. These two parents have unfortunately led to the fact that there is no adopted standard definition of the Hilbert transform. After some hesitation I have chosen the definition which relays on the Cauchy-type integral.

First, it is shown that the Hilbert transform of a hyperfunction exists, if and only if the hyperfunction has a strong defining function (see Section 2.4.2). After presenting several concrete examples of Hilbert transforms of familiar hyperfunctions, the operational rules for the Hilbert transformation are established. The connection between Hilbert transforms and Fourier transforms is then exploited. The section on analytic signals and conjugate hyperfunctions is important for signal theory. The chapter closes with some applications to integral equations. The presentation of this chapters follows mostly [19, Imai].

The Hilbert transform of an ordinary function $f(t)$ is defined and denoted by

$$\mathcal{H}[f](t) := \frac{1}{\pi} \, \mathrm{pv} \int_{-\infty}^{\infty} \frac{f(x)}{x - t} \, dx, \qquad (5.1)$$

where the integral is a principal-value integral in a double sense:

$$\mathrm{pv} \int_{-\infty}^{\infty} \frac{f(x)}{x - t} \, dx := \lim_{b \to \infty} \lim_{\varepsilon \to 0+} \{ \int_{-b}^{t-\varepsilon} \frac{f(x)}{x - t} \, dx + \int_{t+\varepsilon}^{b} \frac{f(x)}{x - t} \, dx \}.$$

Since the Hilbert transformation is widely used in signal theory, we denote the independent variable by t, thinking of it as time. Also, in the notation of a signal $f(t)$ and its Hilbert transform $\mathcal{H}[f](t)$ or $\mathcal{H}[f(t)](t)$, both independent real variables are often denoted by the same symbol t, We also adopt this convention if there is no danger of confusion.

5.1 Hilbert Transforms of Hyperfunctions

5.1.1 Definition and Basic Properties

The definition of the Hilbert transform can be regarded as a convolution integral in the sense of Cauchy's principal value. Because the convolution of hyperfunctions is now a well-defined concept, we may use it to define the Hilbert transform of a hyperfunction.

Definition 5.1. The *Hilbert transform of the hyperfunction* $f(t) = [F(z)]$ is defined by the pv-convolution (see Section 2.7.4)

$$\mathcal{H}[f](t) := -\frac{1}{\pi} \, \text{fp} \frac{1}{t} \otimes f(t), \tag{5.2}$$

provided the pv-convolution exists.

It would be convenient to benefit from the commutative relation

$$\text{fp} \frac{1}{t} \otimes f(t) = f(t) \otimes \text{fp} \frac{1}{t}. \tag{5.3}$$

By Proposition 2.62, the pv-convolution is commutative if

$$\psi_R(x, a) := \int_R^{R+a} F(x - t) \, \text{fp} \frac{1}{t} \, dt \tag{5.4}$$

defines for any a and sufficiently large $R > 0$ a real analytic function of x. If sing supp f is contained in a finite interval, and R is sufficiently large, the above integral over the hyperfunction becomes

$$\psi_R(x, a) := \int_R^{R+a} \frac{F(x - t)}{t} \, dt, \tag{5.5}$$

and the condition is satisfied if $F(z) = O(z^{-\alpha})$, $\alpha > 0$, as $|\Re z| \to \infty$, due to the uniform convergence.

Under this assumption, familiar properties of convolution then implies

(i)

$$f(t) \in \mathcal{B}_+(\mathbb{R}) \Longrightarrow \mathcal{H}[f](t) \in \mathcal{B}_+(\mathbb{R}),$$

$$f(t) \in \mathcal{B}_-(\mathbb{R}) \Longrightarrow \mathcal{H}[f](t) \in \mathcal{B}_-(\mathbb{R}),$$

i.e., upper and lower hyperfunctions are transformed into upper and lower hyperfunctions, respectively.

(ii) If $f(t)$ is an even/odd hyperfunction then $\mathcal{H}[f](t)$ is an odd/even hyperfunction, i.e., the parity changes by Hilbert transformation.

The definitions of the concepts of pv-convolution and strong defining function $\tilde{F}(z)$, together with

$$\text{fp} \frac{1}{t} = \left[\frac{1}{2z}, -\frac{1}{2z} \right] = \left[\frac{1}{z} \mathbf{1}(z) \right],$$

imply

$$\mathcal{H}[f](t) = -\frac{1}{\pi} \operatorname{fp}\frac{1}{t} \otimes f(t)$$

$$= \left[\frac{1}{\pi} \operatorname{pv} \int_{-\infty}^{\infty} \frac{1}{x-z} \mathbf{1}(z) f(x) \, dx\right] = \left[\frac{1}{\pi} \operatorname{pv} \int_{-\infty}^{\infty} \frac{1}{x-z} f(x) \, dx \, \mathbf{1}(z)\right]$$

$$= 2i \left[\frac{1}{2\pi i} \operatorname{pv} \int_{-\infty}^{\infty} \frac{f(x)}{x-z} \, dx \, \mathbf{1}(z)\right] = 2i \left[\tilde{F}(z) \mathbf{1}(z)\right].$$

This shows that the existence of a Hilbert transform is intimately related to the existence of a strong defining function (see Section 2.4.2).

Proposition 5.1. *The Hilbert transform of a hyperfunction* $f(t) = [F(z)]$ *exists, if and only if* $f(t)$ *has a strong defining function* $\tilde{F}(z)$. *In the affirmative case, we have*

$$\mathcal{H}[f](t) = 2i \left[\tilde{F}(z) \mathbf{1}(z)\right]. \tag{5.6}$$

Since for a hyperfunction $f(t)$ with compact support the two concepts of standard defining function and of strong defining function coincide, and $\tilde{F}(z)$ then satisfies $\tilde{F}(z) = O(1/z)$ as $|z| \to \infty$, we have

Corollary 5.2. *If* $f(t)$ *is a hyperfunction with compact support, then*

$$\mathcal{H}[f](t) = O(\frac{1}{t}), \ as \ |t| \to \infty. \tag{5.7}$$

Example 5.1. Because the strong defining function of the constant hyperfunction $1 = [\mathbf{1}_+(z)]$ is $\tilde{F}(z) = \mathbf{1}(z)$, we have

$$\mathcal{H}[1](t) = 2i \left[\mathbf{1}^2(z)\right] = \left[\frac{1}{4}\right] = 0,$$

i.e., *the Hilbert transform of a constant hyperfunction is zero.* Since a polynomial of positive degree has no strong defining function, the Hilbert transform of a polynomial of positive degree does not exist.

The next example shows that Dirac impulses are mapped on finite part hyperfunctions, and vice versa.

Example 5.2. The defining function $-(-1)^n n!/(2\pi i(z-a)^{n+1})$ of $\delta^{(n)}(t-a)$ equals its strong defining function $\tilde{F}(z)$. Thus, for a non-negative integer n, we obtain

$$\mathcal{H}[\delta^{(n)}(t-a)](t) = 2i \left[-\frac{(-1)^n n!}{2\pi i (z-a)^{n+1}} \mathbf{1}(z)\right]$$

$$= -\frac{(-1)^n n!}{\pi} \left[\frac{1}{(z-a)^{n+1}} \mathbf{1}(z)\right] = \frac{(-1)^{n+1} n!}{\pi} \operatorname{fp}\frac{1}{(t-a)^{n+1}},$$

i.e.,

$$\delta^{(n)}(t-a) \circ\!\!\xrightarrow{\mathcal{H}}\!\!\bullet \frac{(-1)^{n+1} n!}{\pi} \operatorname{fp}\frac{1}{(t-a)^{n+1}}. \tag{5.8}$$

On the other hand, for $n \in \mathbb{N}$,

$$\operatorname{fp}\frac{1}{(t-a)^n} = \left[\frac{1}{(z-a)^n} \mathbf{1}(z)\right] = [\tilde{F}(z)],$$

where the indicated defining function again equals the strong defining function, we have,

$$\mathcal{H}[\mathrm{fp}\frac{1}{(t-a)^n}](t) = 2i\left[\frac{1}{(z-a)^n}\mathbf{1}^2(z)\right] = \frac{i}{2}\left[\frac{1}{(z-a)^n}\right]$$

$$= \frac{(-1)^{n-1}\pi}{(n-1)!}\left[-\frac{1}{2\pi i}\frac{(-1)^{n-1}(n-1)!}{(z-a)^n}\right]$$

$$= \frac{(-1)^{n-1}\pi}{(n-1)!}\delta^{(n-1)}(t-a),$$

where we used again that $[\mathbf{1}^2(z)] = [1/4, 1/4] = [1/4]$. Thus

$$\mathrm{fp}\frac{1}{(t-a)^n} \overset{\mathcal{H}}{\underset{\bullet}{\circ}} \frac{(-1)^{n-1}\pi}{(n-1)!}\delta^{(n-1)}(t-a). \tag{5.9}$$

Note that $(\mathcal{H}\circ\mathcal{H})[f(t)] = -f(t)$ is verified for these two hyperfunctions.

Example 5.3. In (2.81) we have established the strong defining function of e^{-t^2}, and using it, we can write

$$e^{-t^2} = \left[\frac{1}{2}e^{-z^2}\,\mathrm{erfc}(-iz), -\frac{1}{2}e^{-z^2}\,\mathrm{erfc}(iz)\right].$$

Therefore,

$$[2i\tilde{F}(z)\mathbf{1}(z)] = \left[\frac{i}{2}e^{-z^2}\,\mathrm{erfc}(-iz), \frac{i}{2}e^{-z^2}\,\mathrm{erfc}(iz)\right]$$

is its Hilbert transform (as a hyperfunction). By using (4.86), (4.87) we obtain,

$$[2i\tilde{F}(z)\mathbf{1}(z)] = \left[i\,e^{-z^2}/2 - \frac{1}{\sqrt{\pi}}\mathrm{Daw}(z), i\,e^{-z^2}/2 + \frac{1}{\sqrt{\pi}}\mathrm{Daw}(z)\right]$$

$$\sim -\frac{2}{\sqrt{\pi}}\left[\frac{1}{2}\mathrm{Daw}(z), -\frac{1}{2}\mathrm{Daw}(z)\right] = -\frac{2}{\sqrt{\pi}}\mathrm{Daw}(t),$$

since Dawson's function is an entire function. Eventually, we have

$$e^{-t^2} \overset{\mathcal{H}}{\underset{\bullet}{\circ}} -\frac{2}{\sqrt{\pi}}\mathrm{Daw}(t). \tag{5.10}$$

Example 5.4. Let $P_n(t)$ be the Legendre polynomial of the first kind. In Example 2.21 we have established the standard defining function of $\chi_{(-1,1)}P_n(t)$ (which is the same as its strong defining function) as $-Q_n(z)/(\pi i)$, where $Q_n(z)$ denotes the Legendre functions of the second kind. Thus, by using this result, we have for the defining function of the Hilbert transform of $\chi_{(-1,1)}P_n(t)$,

$$\tilde{F}(z) = -2\frac{Q_n(z)}{\pi} = -\frac{P_n(z)}{\pi}\log\frac{z+1}{z-1}\mathbf{1}(z)$$

$$+ \frac{2}{\pi}\sum_{k=0}^{n-1}\frac{(n+k)!}{(n-k)!\,(k!)^2\,2^k}\{\Psi(n+1) - \Psi(k+1)\}(z-1)^k\,\mathbf{1}(z).$$

The first term on the right-hand side gives the zero hyperfunction. It remains the polynomial. Thus,

$$\chi_{(-1,1)} P_n(t) \circ\!\!-\!\!\bullet \frac{\mathcal{H}}{} \frac{2}{\pi} \sum_{k=0}^{n-1} \frac{(n+k)!}{(n-k)!\,(k!)^2\,2^k} \{\Psi(n+1) - \Psi(k+1)\}(t-1)^k. \quad (5.11)$$

If we know that the Hilbert transform of a hyperfunction $f(t)$ is the zero hyperfunction, what can be said about $f(t)$?

Proposition 5.3. *Suppose that the hyperfunction $f(t)$ has a strong defining function $\tilde{F}(z)$ which satisfies, for some positive integer m,*

$$z^{-m}\,\tilde{F}(z) \to 0 \ as \ z \to \infty, \quad (5.12)$$

then $\mathcal{H}[f](t) = 0$ implies $f(t) = const.$

Proof. We have $\mathcal{H}[f](t) = 2i\left[\tilde{F}(z)\mathbf{1}(z)\right]$. $\mathcal{H}[f](t) = 0$ implies that $\tilde{F}(z)\mathbf{1}(z)$ $=: \phi(z)$ must be a real analytic function. Multiplying this equation with $\mathbf{1}(z)$ yields $\tilde{F}(z) = 4\phi(z)\mathbf{1}(z)$. The fact that $\tilde{F}(z)$ is holomorphic in the upper and lower half-plane implies that $\phi(z)$ can analytically be continued into the entire complex plane, i.e., becomes an entire function. For this entire function it holds that $z^{-m}\phi(z) \to 0$ as $z \to \infty$, too. Liouville's theorem on entire functions (see for example [4, Behnke-Sommer p.155]) then implies that $\phi(z)$ must be a polynomial of degree $m-1$ at most. Because we have $f(t) = 4\phi(t)$ we see that $f(t)$ is a polynomial of degree $m-1$ at most. However, the Hilbert transform of a polynomial of degree 1 or higher does not exist. Thus, $f(t)$ must be a constant. □

Most familiar hyperfunctions encountered in applications satisfy the hypotheses of the above proposition, so it is worth to singling out this class.

Definition 5.2. We denote the subspace of $\mathcal{B}(\mathbb{R})$ of all hyperfunctions which

(i) have a strong defining function $\tilde{F}(z)$,

(ii) satisfy condition (5.12) for some positive integer m,

by $\mathcal{B}_1(\mathbb{R})$.

In Section 2.4.2 we have computed several strong defining functions of given hyperfunctions. By using those results, we have immediately the following Hilbert transforms.

Example 5.5. Because the strong defining function of the hyperfunction $1/(t-c)^m$, $\Im c \neq 0$, is

$$\tilde{F}(z) := \begin{cases} \frac{1}{(z-c)^m}\,\mathbf{1}_+(z), & \Im c < 0 \\ \frac{1}{(z-c)^m}\,\mathbf{1}_-(z), & \Im c > 0 \end{cases}$$

and $\mathbf{1}_+(z)\mathbf{1}(z) = \mathbf{1}_+(z)/2$, and $\mathbf{1}_-(z)\mathbf{1}(z) = -\mathbf{1}_-(z)/2$, we obtain

$$\frac{1}{(t-c)^m} \circ\!\!-\!\!\bullet \frac{\mathcal{H}}{} -i\,\mathrm{sgn}(\Im c)\,\frac{1}{(t-c)^m}, \quad (\Im c \neq 0, m \in \mathbb{N}). \quad (5.13)$$

Problem 5.1. Establish the following Hilbert transforms.

$$\exp(i\Omega t) \circ \!\!\overset{\mathcal{H}}{-\!\!\bullet}\, i \,\mathrm{sgn}(\Omega) \exp(i\Omega t),$$

$$\cos(\Omega t) \circ \!\!\overset{\mathcal{H}}{-\!\!\bullet}\, -\,\mathrm{sgn}\,\Omega \,\sin(\Omega t),$$

$$\sin(\Omega t) \circ \!\!\overset{\mathcal{H}}{-\!\!\bullet}\, \mathrm{sgn}\,\Omega \,\cos(\Omega t).$$

Example 5.6. The strong defining function of the Dirac comb

$$f(t) = \mathrm{pv} \sum_{k=-\infty}^{\infty} \delta(t - kT)$$

is (see Example 2.36)

$$\tilde{F}(z) = -\frac{1}{2Ti} \cot\left(\frac{\pi z}{T}\right).$$

Thus, the defining function of the Hilbert transform becomes

$$-\frac{1}{T} \cot\left(\frac{\pi z}{T}\right) \mathbf{1}(z)$$

which results in

$$\mathrm{pv} \sum_{k=-\infty}^{\infty} \delta(t - kT) \circ \!\!\overset{\mathcal{H}}{-\!\!\bullet}\, -\mathrm{fp}\, \frac{1}{T} \cot\left(\frac{\pi t}{T}\right). \qquad (5.14)$$

We have introduced the subclass $\mathcal{B}_0(\mathbb{R})$ of hyperfunctions where the upper and lower component of their defining functions are upper and lower half-plane functions, respectively (see Definition 2.14). For this subclass of hyperfunctions the standard defining functions and the strong defining functions are the same.

Proposition 5.4. *For hyperfunctions $f(t) \in \mathcal{B}_0(\mathbb{R})$, we have*

$$\mathcal{H}[\mathcal{H}[f]](t) = -f(t), \qquad (5.15)$$

i.e., for $f(t) \in \mathcal{B}_0(\mathbb{R})$ we may write, for any constant c,

$$f(t) \circ \!\!\overset{\mathcal{H}}{-\!\!\bullet}\, g(t) \Longrightarrow g(t) + c \circ \!\!\overset{\mathcal{H}}{-\!\!\bullet}\, -f(t). \qquad (5.16)$$

Proof. The defining function of $\mathcal{H}[f](t)$ is given by $F_1(z) := 2i\tilde{F}(z)\mathbf{1}(z)$ where $\tilde{F}(z)$ is the strong defining function of $f(t)$. $F_1(z)$ satisfies the conditions of Proposition 2.33 too, hence equals the standard defining function of $\mathcal{H}[f](t)$. Therefore, the defining function of $\mathcal{H}[\mathcal{H}[f]](t)$ becomes $2i\, F_1(z)\mathbf{1}(z) = -4\tilde{F}(z)(1/4)$ since $\mathbf{1}^2(z) = 1/4$. This implies $\mathcal{H}[\mathcal{H}[f]](t) = -f(t)$. The last statement follows because the Hilbert transform of a constant is zero. $\qquad \square$

The next two examples illustrate the pitfalls, when the premise of the above proposition is not met.

Example 5.7. Take $f(t) = 1$.

$$\mathcal{H}[\mathcal{H}[1]](t) = \mathcal{H}[0] = 0 \neq -1.$$

Example 5.8. The Dirac comb

$$\text{pv} \sum_{k=-\infty}^{\infty} \delta(t - kT) = \left[-\frac{1}{2Ti} \cot(\frac{\pi z}{T}) \right]$$

is not a member of $\mathcal{B}_0(\mathbb{R})$ because

$$-\frac{1}{2Ti} \cot(\frac{\pi z}{T}) = -\frac{1}{2T} \frac{1 + e^{-2i\pi z/T}}{1 - e^{-2i\pi z/T}} \rightarrow -\frac{1}{2T}$$

as $\Im z \rightarrow -\infty$, and

$$-\frac{1}{2Ti} \cot(\frac{\pi z}{T}) = \frac{1}{2T} \frac{e^{2i\pi z/T} + 1}{e^{2i\pi z/T} - 1} \rightarrow \frac{1}{2T}$$

as $\Im z \rightarrow \infty$. But

$$F_1(z) = -\frac{1}{2Ti} \cot(\frac{\pi z}{T}) - \frac{1}{T} \mathbf{1}(z)$$

is the defining function of a hyperfunction $f_1(t) \in \mathcal{B}_0(\mathbb{R})$. We have

$$f_1(t) = \text{pv} \sum_{k=-\infty}^{\infty} \delta(t - kT) - \frac{1}{T}$$

and its Hilbert transform is

$$-\frac{1}{T} \text{fp} \cot(\frac{\pi t}{T})$$

since the image of a constant is zero. By the above proposition, we then obtain

$$\text{fp} \cot(\frac{\pi t}{T}) \circ \stackrel{\mathcal{H}}{-\!\!-\!\!\bullet} T \text{pv} \sum_{k=-\infty}^{\infty} \delta(t - kT) - 1. \tag{5.17}$$

Let us now use the fact that any hyperfunction $f(t) = [F_+(z), F_-(z)]$ can be decomposed into an upper and a lower hyperfunction. Indeed, we may write

$$f(t) = [F_+(z), F_-(z)] = [F_+(z), 0] - [0, -F_-(z)] \tag{5.18}$$
$$= [F_+(z) \mathbf{1}_+(z)] - [F_-(z) \mathbf{1}_-(z)] = F_+(t + i0) - F_-(t - i0).$$

Proposition 5.5. *If the hyperfunction $f(t)$ has a strong defining function $\tilde{F}(z)$, then*

$$f(t) = \tilde{F}_+(t + i0) - \tilde{F}_-(t - i0),$$
$$\mathcal{H}[f](t) = i\{\tilde{F}_+(t + i0) + \tilde{F}_-(t - i0)\},$$
$$f(t) - i\mathcal{H}[f](t) = 2\tilde{F}_+(t + i0),$$
$$f(t) + i\mathcal{H}[f](t) = -2\tilde{F}_-(t - i0)\}.$$

Proof. The first statement follows from (5.18) where the (arbitrary) defining function is replaced by the strong defining function of $f(t)$. The second statement follows from the fact $\mathbf{1}(z) = (\mathbf{1}_+(z) + \mathbf{1}_-(z))/2$, and

$$\mathcal{H}[f](t) = 2i[\tilde{F}(z) \mathbf{1}(z)] = i \left[\tilde{F}(z)(\mathbf{1}_+(z) + \mathbf{1}_-(z)) \right]$$
$$= i \left[\tilde{F}(z)\mathbf{1}_+(z) \right] + i \left[\tilde{F}(z)\mathbf{1}_-(z) \right] = i\{\tilde{F}(t + i0) + \tilde{F}(t - i0)\}.$$

The last two statements are obtained by adding and subtracting the first and the second statement. $\qquad\square$

The first two of the above statements can be used to generate Hilbert transform correspondences from any defining function which qualifies as a strong defining function.

Example 5.9. The standard defining function of the characteristic function $\chi_{a,b}(t)$, which is also its strong defining function, is

$$\tilde{F}(z) = \frac{1}{2\pi i} \int_a^b \frac{1}{t-z}\, dt = \frac{1}{2\pi i} \log \frac{b-z}{a-z}.$$

Therefore,

$$\tilde{F}(t+i0) = \frac{1}{2\pi i} \log \left|\frac{b-t}{a-t}\right| - \frac{1}{2}$$

and

$$\tilde{F}(t-i0) = \frac{1}{2\pi i} \log \left|\frac{b-t}{a-t}\right| + \frac{1}{2}.$$

Thus,

$$\mathcal{H}[\chi_{a,b}(t)](t) = i\{\tilde{F}(t+i0) + \tilde{F}(t-i0)\} = \frac{1}{\pi} \log \left|\frac{b-t}{a-t}\right|,$$

or

$$\chi_{a,b}(x) \circ\!\!\!\!\overset{\mathcal{H}}{-\!\!\!-}\!\!\!\bullet\ \frac{1}{\pi} \log \left|\frac{b-t}{a-t}\right|. \tag{5.19}$$

From this result we may find again

$$\mathcal{H}[1](t) := \lim_{R\to\infty} \mathcal{H}[\chi_{-R,R}](t) = \lim_{R\to\infty} \frac{1}{\pi} \log \left|\frac{R-t}{-R-t}\right| = 0,$$

i.e.,

$$1 \circ\!\!\!\!\overset{\mathcal{H}}{-\!\!\!-}\!\!\!\bullet\ 0. \tag{5.20}$$

Example 5.10. Again, by using the strong defining function of $f(t) = 1/(t-c)^m$, $m \in \mathbb{N}$, given by

$$\tilde{F}(z) := \begin{cases} \frac{1}{(z-c)^m}\, \mathbf{1}_+(z), & \Im c < 0 \\ \frac{1}{(z-c)^m}\, \mathbf{1}_-(z), & \Im c > 0, \end{cases}$$

we obtain, by putting $c = -ia$, $a > 0$, that $\tilde{F}(z) = 1/(z+ia)^m\, \mathbf{1}_+(z)$ is the strong defining function of $1/(t+ia)^m$. Proposition 5.5 and the taking of the real part in both members yields

$$\frac{\sum_{k=0}^{[m/2]}(-1)^k \binom{m}{2k} a^{2k}\, t^{m-2k}}{(t^2+a^2)^m}$$

$$\circ\!\!\!\!\overset{\mathcal{H}}{-\!\!\!-}\!\!\!\bullet\ \frac{\sum_{k=1}^{[(m+1)/2]}(-1)^{k-1} \binom{m}{2k-1} a^{2k-1}\, t^{m+1-2k}}{(t^2+a^2)^m}.$$

For $m = 1, 2, 3$ we obtain the correspondences $(a > 0)$

$$\frac{t}{t^2 + a^2} \overset{\mathcal{H}}{\circ\!-\!\bullet} \frac{a}{t^2 + a^2},$$

$$\frac{t^2 - a^2}{(t^2 + a^2)^2} \overset{\mathcal{H}}{\circ\!-\!\bullet} \frac{2at}{(t^2 + a^2)^2},$$

$$\frac{t^3 - 3a^2 t}{(t^2 + a^2)^3} \overset{\mathcal{H}}{\circ\!-\!\bullet} \frac{3at^2 - a^3}{(t^2 + a^2)^3}.$$

If we take the imaginary part in both members, we would obtain the duals of the above relations corresponding to $\mathcal{H}[\mathcal{H}[f]](t) = -f(t)$.

Upper or lower hyperfunctions behave in an a simple manner under Hilbert transformation.

Proposition 5.6. *Let $f(t) := F(t + i0) = [F(z)\mathbf{1}_+(z)] \in \mathcal{B}_+(\mathbb{R})$ be an upper hyperfunction with the defining function $F(z) \in \mathcal{O}_0(\mathbb{C}_+)$ being an upper half-plane function. Then,*

$$\mathcal{H}[f(t)](t) = i\,f(t), \ \mathcal{H}[\Re f(t)](t) = -\Im f(t), \ \mathcal{H}[\Im f(t)](t) = \Re f(t).$$

Similarly, let $F(t - i0) = [F(z)\mathbf{1}_-(z)] \in \mathcal{B}_-(\mathbb{R})$ be a lower hyperfunction with $F(z) \in \mathcal{O}_0(\mathbb{C}_-)$, then

$$\mathcal{H}[f(t)](t) = -i\,f(t), \ \mathcal{H}[\Re f(t)](t) = \Im f(t), \ \mathcal{H}[\Im f(t)](t) = -\Re f(t).$$

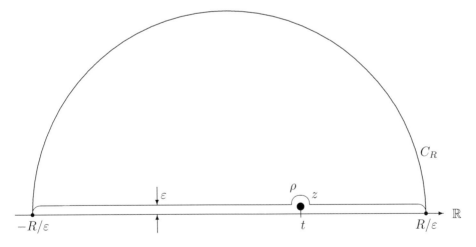

Figure 5.1: Contour in the proof of Proposition 5.6

Proof. We prove the first part of the statement. Consider the closed contour $C(R, \varepsilon)$ of Figure 5.1 starting at $-R/\varepsilon$ running at distance $\varepsilon > 0$ to $t - \varepsilon + i\varepsilon$ turning around $z = t + i\varepsilon$ in an upper semicircle ρ of radius ε, continuing to R/ε, and going back by a large upper semicircle C_R to $-R/\varepsilon$. By Cauchy's integral theorem we have

$$0 = \oint_{C(R,\varepsilon)} \frac{F(\zeta)}{\zeta - z}\, d\zeta = \int_{-R/\varepsilon}^{t-\varepsilon+i\varepsilon} \frac{F(\zeta)}{\zeta - z}\, d\zeta + \int_{\rho} \frac{F(\zeta)}{\zeta - z}\, d\zeta$$

$$+ \int_{t+\varepsilon+i\varepsilon}^{R/\varepsilon} \frac{F(\zeta)}{\zeta - z}\, d\zeta + \int_{C_R} \frac{F(\zeta)}{\zeta - z}\, d\zeta.$$

For a small $\varepsilon > 0$ we have

$$\int_{\rho} \frac{F(\zeta)}{\zeta - z}\, d\zeta \approx F(z) \int_{\rho} \frac{1}{\zeta - z}\, d\zeta = F(z)(-i\pi).$$

By letting $\varepsilon \to 0+$, the radius of the large upper semicircle tends to ∞, and by the hypothesis that $F(z)$ is an upper half-plane function, the integral over the upper semicircle C_R tends to zero. Therefore, letting $\varepsilon \to 0+$ we shall have in the limiting case

$$\mathrm{pv}\frac{1}{\pi} \int_{-\infty}^{\infty} \frac{1}{x - t} f(x)\, dx = iF(t + i0) = if(t),$$

or

$$\mathcal{H}[f(t)](t) = -\mathrm{fp}\frac{1}{x} \otimes f(t) = if(t)$$

which proves the first relation. We have

$$\mathcal{H}[f(t)](t) = 2\,i\,[F(z)\,\mathbf{1}_+(z)\,\mathbf{1}(z)]\,.$$

Because $\mathbf{1}_+(z)\,\mathbf{1}(z) = 1/2\,\mathbf{1}_+(z)$ we obtain

$$\mathcal{H}[f(t)](t) = i\,[F(z)\,\mathbf{1}_+(z)] = i\,f(t).$$

Next, $f(t) = \Re f(t) + i\,\Im f(t)$, thus

$$\mathcal{H}[f(t)](t) = \mathcal{H}[\Re f(t)](t) + i\,\mathcal{H}[\Im f(t)](t)$$
$$= i\,\Re f(t)](t) + i^2\,\Im f(t)](t) = -\Im f(t)](t) + i\,\Re f(t)](t).$$

Thus,

$$\mathcal{H}[\Re f(t)](t) = -\Im f(t)](t),\quad \mathcal{H}[\Im f(t)](t) = \Re f(t)](t). \qquad \square$$

Corollary 5.7. *An upper and a lower hyperfunction where the defining functions is an upper and a lower half-plane function, respectively, is completely determined by its real or its imaginary part alone.*

Proof. Let $\Re f(t)$ be given. Taking its Hilbert transform yields $-\Im f(t)$, hence $f(t) = \Re f(t) + i\,\Im f(t)$ is determined. Similarly, if $\Im f(t)$ is given. $\qquad \square$

Example 5.11.

$$f(t) = \frac{1}{t + i0} = [\frac{1}{z}, 0]$$

implies

$$\mathcal{H}[\frac{1}{t + i0}](t) = \frac{i}{t + i0}. \qquad (5.21)$$

For $a > 0$,

$$f(t) = \frac{1}{t + ia} = [\frac{1}{z + ia}, 0]$$

implies

$$\mathcal{H}[\frac{1}{t + ia}](t) = \frac{i}{t + ia}, \tag{5.22}$$

and, by taking real and imaginary parts, we find again

$$\mathcal{H}[\frac{t}{t^2 + a^2}](t) = \frac{a}{t^2 + a^2}, \quad \mathcal{H}[\frac{1}{t^2 + a^2}](t) = -\frac{1}{a}\frac{t}{t^2 + a^2}. \tag{5.23}$$

Definition 5.3. A complex-valued function $F(z)$ defined in a neighborhood of the real axis is called of *real type*, if $F(z) = \overline{F(\overline{z})}$. We call a hyperfunction $f(t) = [F(z)]$ to be of *real type*, if its defining function is of real type.

Problem 5.2. Show that if $F(z)$ is of real type, then $\Re F(t + i0) = \Re F(t - i0)$, and $\Im F(t + i0) = -\Im F(t - i0)$. Furthermore, $F(z)$ may be decomposed as $F(z) = F_{\Re}(z) + i F_{\Im}(z)$, where

$$F_{\Re}(z) := \frac{F(z) + \overline{F(\overline{z})}}{2}, \quad F_{\Im}(z) := \frac{F(z) - \overline{F(\overline{z})}}{2i}.$$

Show that both $F_{\Re}(z)$ and $F_{\Im}(z)$ are of real type.

Problem 5.3. Suppose that if the hyperfunction $f(t)$ has a defining function of real type, then it may be represented as

$$f(t) = 2i\{\Im F_{\Re}(t + i0) + i\Im F_{\Im}(t + i0)\}$$
$$= -2i\{\Im F_{\Re}(t - i0) + i\Im F_{\Im}(t - i0)\}.$$

Problem 5.4. With $\tilde{F}(z) = \tilde{F}_{\Re}(z) + i\tilde{F}_{\Im}(z)$ and the fact that $\tilde{F}_{\Re}(z)$ and $\tilde{F}_{\Im}(z)$ are of real type, show that

$$\mathcal{H}[f](t) = 2i\{\Re\tilde{F}_{\Re}(t + i0) + i\Re\tilde{F}_{\Im}(t + i0)\}$$
$$= 2i\{\Re\tilde{F}_{\Re}(t - i0) + i\Re\tilde{F}_{\Im}(t - i0)\},$$

provided that $f(t)$ has a strong defining function $\tilde{F}(z)$.

The last two problems show that a real type hyperfunction $f(t)$ and its Hilbert transform are completely determined by either the upper or the lower component of $\tilde{F}(z) = \tilde{F}_{\Re}(z) + i\tilde{F}_{\Im}(z)$.

Proposition 5.5 may be used to produce Hilbert transform pairs by starting with an appropriate defining function that qualifies as a strong defining function. This is illustrated by the next two examples.

Example 5.12. The function $Q(z) := \sqrt{(z - a)(z - b)}$, $a < b$ is a holomorphic function in $\mathbb{C} \setminus [a, b]$. Let us choose the branch of $Q(z)$ that takes positive values for $t = z > b$. Because $Q(z) \approx z$ as $z \to \infty$,

$$\tilde{F}(z) := \frac{i}{2\,Q(z)}$$

qualifies as a strong defining function of a certain hyperfunction $f(t)$. Let us find this hyperfunction and its Hilbert transform by applying Proposition 5.5. By using

$$\tilde{F}(t \pm i0) = \frac{1}{2 \,|Q(t)|} \exp[i \,(\pi/2 - \arg(Q(t \pm i0)))]$$

we find:

$$b < t \Longrightarrow \tilde{F}(t \pm i0) = \frac{i}{2 \,|Q(t)|}, \qquad a < t < b \Longrightarrow \tilde{F}(t \pm i0) = \frac{\pm 1}{2 \,|Q(t)|},$$

$$t < a \Longrightarrow \tilde{F}(t \pm i0) = \frac{-i}{2 \,|Q(t)|}.$$

Therefore,

$$f(t) = \tilde{F}(t + i0) - \tilde{F}(t - i0) = \chi_{(a,b)} \frac{1}{|Q(t)|},$$

$$\mathcal{H}[f](t) = i \,\{\tilde{F}(t + i0) + \tilde{F}(t - i0)\} = \frac{\mathrm{sgn}(t, a, b)}{|Q(t)|},$$

or

$$\chi_{(a,b)} \frac{1}{\sqrt{(t - a)(b - t)}} \quad \overset{\mathcal{H}}{\circ\!\!-\!\!\bullet} \quad - \frac{\mathrm{sgn}(t; a, b)}{\sqrt{(t - a)(t - b)}}. \tag{5.24}$$

For $a = -1, b = 1$ we obtain

$$\chi_{(-1,1)} \frac{1}{\sqrt{1 - t^2}} \quad \overset{\mathcal{H}}{\circ\!\!-\!\!\bullet} \quad - \frac{\mathrm{sgn}(t; -1, 1)}{\sqrt{t^2 - 1}}. \tag{5.25}$$

Problem 5.5. By using the same method, result (5.24) may be generalized. Show that for any non-negative integer n,

$$\chi_{(a,b)} \frac{1}{(t - a)^n \,\sqrt{(t - a)(b - t)}} \quad \overset{\mathcal{H}}{\circ\!\!-\!\!\bullet} \quad - \frac{\mathrm{sgn}(t; a, b)}{(t - a)^n \,\sqrt{(t - a)(t - b)}} \tag{5.26}$$

holds.

$Q(z)$ itself cannot be a strong defining function, but because

$$\lim_{z \to \infty} Q(z) - z = -\frac{a + b}{2},$$

$\tilde{G}(z) := Q(z) - z + (a + b)/2$ qualifies as a strong defining function, for $\tilde{G}(z) = O(1/|z|)$ as $|z| \to \infty$. From the above relations, we then have

$$t < a \Longrightarrow \tilde{G}(t \pm i0) = -|Q(t)| - t + (a + b)/2,$$

$$a < t < b \Longrightarrow \tilde{G}(t \pm i0) = \pm i \,|Q(t)| - t + (a + b)/2,$$

$$b < t \Longrightarrow \tilde{G}(t \pm i0) = |Q(t)| - t + (a + b)/2.$$

By using $g(t) = \tilde{G}(t + i0) - \tilde{G}(t - i0)$ and $\mathcal{H}[g](t) = i \,\{\tilde{G}(t + i0) + \tilde{G}(t - i0)\}$, we finally get the correspondences

$$\chi_{(a,b)} \sqrt{(t - a)(b - t)} \quad \overset{\mathcal{H}}{\circ\!\!-\!\!\bullet} \quad \mathrm{sgn}(t; a, b) \sqrt{(t - a)(t - b)} + t - \frac{a + b}{2}. \tag{5.27}$$

$$\chi_{(-1,1)} \sqrt{1 - t^2} \quad \overset{\mathcal{H}}{\circ\!\!-\!\!\bullet} \quad \mathrm{sgn}(t; -1, 1) \sqrt{t^2 - 1} + t. \tag{5.28}$$

Proposition 5.8. *If the hyperfunction $f(t)$ has finite moments up to order $n - 1$ at $t = a$, i.e.,*

$$m_k(a) := \int_{-\infty}^{\infty} (t - a)^k f(t)\, dt < \infty \quad (k = 0, 1, \ldots, n - 1),$$

we have

$$\mathcal{H}[(t - a)^n f(t)](t) = (t - a)^n \mathcal{H}[f(t)](t) + \frac{1}{\pi} \sum_{k=1}^{n} m_{k-1}(a)\, (t - a)^{n-k}. \quad (5.29)$$

Particularly, for $a = 0$, $n = 1$,

$$\mathcal{H}[(t\, f(t)](t) = t\, \mathcal{H}[f(t)](t) + \frac{1}{\pi} \int_{-\infty}^{\infty} f(t)\, dt \quad (5.30)$$

holds.

Proof. Let us start with the relation

$$t\, (g(t) \star f(t)) = (t\, g(t)) \star f(t) + g(t) \star (t\, f(t))$$

with $g(t) = (1/\pi)\, \mathrm{fp}(1/t)$. Because $t\, g(t) = 1/\pi$, the fact that

$$\frac{1}{\pi} \star f(t) = \frac{1}{\pi} \int_{-\infty}^{\infty} f(t)\, dx,$$

and by using the definition of the Hilbert transform, we obtain

$$t\, \mathcal{H}[f(t)](t) = -\frac{1}{\pi} \int_{-\infty}^{\infty} f(x)\, dx + \mathcal{H}[t\, f(t)](t). \quad (5.31)$$

Replacing $f(t)$ by $t\, f(t)$ in this relation yields

$$\mathcal{H}[t^2\, f(t)](t) = t^2\, \mathcal{H}[t^2\, f(t)](t) + \frac{1}{\pi} \int_{-\infty}^{\infty} x\, f(x)\, dx + \frac{t}{\pi} \int_{-\infty}^{\infty} f(x)\, dx.$$

By induction on n, we finally get

$$\mathcal{H}[t^n\, f(t)](t) = t^n\, \mathcal{H}[f(t)](t) + \frac{1}{\pi} \sum_{k=1}^{n} t^{n-k} \int_{-\infty}^{\infty} x^{k-1}\, f(x)\, dx.$$

This proves the proposition for $a = 0$. By anticipating Proposition 5.11 of the next section we have

$$\mathcal{H}[(t - a)^n\, f(t - a)](t) = \mathcal{H}[t^n\, f(t)](t - a),$$

and the right-hand side transforms into

$$(t - a)^n\, \mathcal{H}[f(t)](t - a) + \frac{1}{\pi} \sum_{k=1}^{n} (t - a)^{n-k} \int_{-\infty}^{\infty} x^{k-1}\, f(x)\, dx.$$

In the relation

$$\mathcal{H}[(t-a)^n f(t-a)](t) = (t-a)^n \mathcal{H}[f(t)](t-a)$$
$$+ \frac{1}{\pi} \sum_{k=1}^{n} (t-a)^{n-k} \int_{-\infty}^{\infty} x^{k-1} f(x)\, dx,$$

we replace $f(t-a)$ by $f(t)$ and obtain

$$\mathcal{H}[(t-a)^n f(t)](t) = (t-a)^n \mathcal{H}[f(t+a)](t-a)$$
$$+ \frac{1}{\pi} \sum_{k=1}^{n} (t-a)^{n-k} \int_{-\infty}^{\infty} x^{k-1} f(x+a)\, dx.$$

Again by Proposition 5.11 of the next section, we have

$$\mathcal{H}[f(t+a)](t-a) = \mathcal{H}[f(t)](t+a-a) = \mathcal{H}[f(t)](t),$$

and finally obtain

$$\mathcal{H}[(t-a)^n f(t)](t) = (t-a)^n \mathcal{H}[f(t)](t)$$
$$+ \frac{1}{\pi} \sum_{k=1}^{n} (t-a)^{n-k} \int_{-\infty}^{\infty} (t-a)^{k-1} f(t)\, dt$$

which proves the proposition. $\square$

Example 5.13. In (5.31) we take $f(t) = \sin t/t$. Then,

$$t\,\mathcal{H}[\frac{\sin t}{t}](t) = -\frac{1}{\pi} \int_{-\infty}^{\infty} \frac{\sin x}{x}\, dx + \mathcal{H}[\sin t](t) = -1 + \cos t.$$

Thus,

$$\mathcal{H}[\frac{\sin t}{t}](t) = \frac{\cos t - 1}{t}, \tag{5.32}$$

and from this result,

$$\mathcal{H}[\frac{1-\cos t}{t}](t) = \frac{\sin t}{t}.$$

By linearity and from (5.9),

$$\mathcal{H}[\frac{\cos t}{t}](t) = \pi\,\delta(t) - \frac{\sin t}{t}, \tag{5.33}$$

where $1/t$ is interpreted as $\mathrm{fp}(1/t)$.

Example 5.14. Take $f(t) = \chi_{a,b}(t)$. Because

$$\mathcal{H}[\chi_{a,b}(t)](y) = \frac{1}{\pi} \log\left|\frac{b-y}{a-y}\right|,$$

we have

$$\mathcal{H}[\chi_{(a,b)} t^n](t) = \frac{t^n}{\pi} \log\left|\frac{b-t}{a-t}\right| + \frac{1}{\pi} \sum_{k=1}^{n} \frac{b^k - a^k}{k}\, t^{n-k}. \tag{5.34}$$

Example 5.15. We start from correspondence (5.24).

$$f(t) = \chi_{(a,b)} \frac{1}{\sqrt{(t-a)(b-t)}}, \quad \mathcal{H}[f](t) = -\frac{\mathrm{sgn}(t;a,b)}{\sqrt{(t-a)(t-b)}}.$$

At a and b, $f(t)$ has moments of all order,

$$m_k(a) = \int_a^b \frac{(t-a)^k}{\sqrt{(t-a)(b-t)}} \, dt$$

$$= (b-a)^k \frac{\Gamma(k+1/2)\,\Gamma(1/2)}{\Gamma(k+1)} = (b-a)^k \frac{\Gamma(k+1/2)\,\sqrt{\pi}}{k!}.$$

Thus, by the above proposition,

$$\chi_{(a,b)} \frac{(t-a)^n}{\sqrt{(t-a)(b-t)}} \circ\!\!-\!\!\bullet^{\mathcal{H}} -\frac{\mathrm{sgn}(t;a,b)\,(t-a)^n}{\sqrt{(t-a)(t-b)}}$$

$$+ \frac{1}{\sqrt{\pi}} \sum_{k=1}^{n} \frac{(b-a)^{k-1}\,\Gamma(k-1/2)}{(k-1)!} (t-a)^{n-k}. \tag{5.35}$$

Similarly, we find

$$\chi_{(a,b)} \frac{(t-b)^n}{\sqrt{(t-a)(b-t)}} \circ\!\!-\!\!\bullet^{\mathcal{H}} -\frac{\mathrm{sgn}(t;a,b)\,(t-b)^n}{\sqrt{(t-a)(t-b)}}$$

$$+ \frac{1}{\sqrt{\pi}} \sum_{k=1}^{n} \frac{(a-b)^{k-1}\,\Gamma(k-1/2)}{(k-1)!} (t-a)^{n-k}. \tag{5.36}$$

Problem 5.6. Show that

$$\chi_{(-1,1)} \frac{t^n}{\sqrt{1-t^2}} \circ\!\!-\!\!\bullet^{\mathcal{H}} -\frac{\mathrm{sgn}(t;-1,1)\,t^n}{\sqrt{t^2-1}}$$

$$+ \frac{1}{\sqrt{\pi}} \sum_{k=1}^{[(n+1)/2]} \frac{\Gamma(k-1/2)}{(k-1)!} t^{n+1-2k}. \tag{5.37}$$

Problem 5.7. Show that

$$\chi_{(a,b)} \frac{t}{\sqrt{(t-a)(b-t)}} \circ\!\!-\!\!\bullet^{\mathcal{H}} -\frac{t\,\mathrm{sgn}(t;a,b)}{\sqrt{(t-a)(t-b)}} + 1. \tag{5.38}$$

Problem 5.8. The triangle function is defined by

$$\Delta_a(t) := \begin{cases} 1 - |t/a|, & |t| \le a \\ 0, & |t| > a \end{cases}$$

where $a > 0$. By writing

$$\Delta_a(t) = \chi_{(-a,0)}\left(1 + \frac{t}{a}\right) + \chi_{(0,a)}\left(1 - \frac{t}{a}\right),$$

establish

$$\Delta_a(t) \circ\!\!-\!\!\bullet^{\mathcal{H}} \frac{1}{\pi a} \left\{ (a+t) \log\left|\frac{t}{a+t}\right| - (a-t) \log\left|\frac{t}{a-t}\right| \right\}. \tag{5.39}$$

Problem 5.9. Let $f(t)$ be a given hyperfunction, and let $f_0(t) := \chi_{(a,b)} f(t)$ be its projection on the interval (a, b). Let $q(t) := (t - a)(b - t)$ (notice that $q(t) > 0$ for $a < t < b$). Show first that

$$\mathcal{H}[q(t) \, f_0(t)](t) = q(t) \, \mathcal{H}[f_0(t)](t) - \frac{1}{\pi} \{m_0(0)t + m_1(0) - (a + b)m_0(0)\},$$

with

$$m_k(0) := \int_a^b t^k \, f(t) \, dt.$$

Replace then $f_0(t)$ by $f_0(t)/\sqrt{q(t)}$, and obtain

$$\sqrt{q(t)} \, \mathcal{H}[\frac{f_0(t)}{\sqrt{q(t)}}](t) = \frac{1}{\sqrt{q(t)}} \, \mathcal{H}[\sqrt{q(t)} \, f_0(t)](t) + \frac{c_0 \, t + c_1}{\sqrt{q(t)}} \tag{5.40}$$

with

$$c_0 = \frac{1}{\pi} \int_a^b \frac{f(x)}{\sqrt{q(x)}} \, dx, \quad c_1 = \frac{1}{\pi} \int_a^b \frac{x f(x)}{\sqrt{q(x)}} \, dx - (a + b) \, c_0.$$

5.1.2 Operational Properties

From the definition of the Hilbert transformation

$$\mathcal{H}[f](t) := -\frac{1}{\pi} \, \mathrm{fp}\frac{1}{t} \otimes f(t),$$

and the operational rules of the convolution product, the following properties can be deduced.

Proposition 5.9. *The Hilbert transformation is a linear operator, i.e.,*

$$\mathcal{H}[c_1 \, f_1(t) + c_2 \, f_2(t)](t) = c_1 \, \mathcal{H}[f_1(t)](t) + c_2 \, \mathcal{H}[f_2(t)](t). \tag{5.41}$$

Proof. This follows readily from the fact that the convolution is linear with respect to each of its factors. □

Example 5.16. Let $a \neq 0$ be real.

$$\mathcal{H}[\sin(at)](t) = \mathcal{H}[\frac{e^{iat} - e^{-iat}}{2i}](t) = \frac{1}{2i} \{\mathcal{H}[e^{iat}](t) - \mathcal{H}[e^{-iat}](t)\}$$

$$= \frac{1}{2i} \{i \operatorname{sgn}(a) \, e^{iat} - i \operatorname{sgn}(-a) e^{-iat}\} = \operatorname{sgn}(a) \cos(at).$$

In the same way we get

$$\mathcal{H}[\cos(at)](t) = - \operatorname{sgn}(a) \sin(at).$$

For the next proposition see also (5.4).

Proposition 5.10. *If*

$$\mathrm{fp}\frac{1}{t} \otimes f(t) = f(t) \otimes \mathrm{fp}\frac{1}{t}, \quad \mathrm{fp}\frac{1}{t} \otimes f'(t) = f'(t) \otimes \mathrm{fp}\frac{1}{t} \tag{5.42}$$

hold, which particularly means that the Hilbert transforms of $f(t)$ and $f'(t)$ exist, then

$$f'(t) \circ \!\!\!\xrightarrow{\mathcal{H}}\!\!\!\bullet \, g'(t). \tag{5.43}$$

Proof. This follows from Proposition 2.57, for

$$D(\mathcal{H}[f])(t) := -\frac{1}{\pi} D(\text{fp}\frac{1}{t} \otimes f(t)) = -\frac{1}{\pi} \text{fp}\frac{1}{t} \otimes f'(t) = \mathcal{H}[f'(t)])(t). \qquad \square$$

Example 5.17. From (5.33) we obtain by differentiation

$$\mathcal{H}[\frac{-t \sin t + \cos t}{t^2}](t) = \pi \delta'(t) - \frac{t \cos t - \sin t}{t^2},$$

thus

$$\frac{-\cos t + 1}{t} + \mathcal{H}[\frac{\cos t}{t^2}](t) = \pi \delta'(t) - \frac{\cos t}{t} + \frac{\sin t}{t^2},$$

and finally,

$$\mathcal{H}[\frac{\cos t}{t^2}](t) = \pi \delta'(t) + \frac{\sin t - t}{t^2}. \qquad (5.44)$$

Proposition 5.11. *If the Hilbert transform of the hyperfunction $f(t)$ is $g(t)$, then, for real $a \neq 0$, b, we have*

$$f(at + b) \circ \overset{\mathcal{H}}{-\!\!\bullet} \text{sgn } a\, g(at + b), \qquad (5.45)$$

particularly,

$$f(-t) \circ \overset{\mathcal{H}}{-\!\!\bullet} -g(-t), \quad f(t - b) \circ \overset{\mathcal{H}}{-\!\!\bullet} g(t - b).$$

This implies that a hyperfunction and its Hilbert transform have opposite parities.

Proof. This follows from Proposition 2.59, for

$$\mathcal{H}[f(at + b)](t) = -\frac{1}{\pi} \text{fp}\frac{1}{t} \otimes f(at + b)) = -\frac{a}{\pi} \text{fp}\frac{1}{at} \otimes f(at + b))$$

$$= -\frac{a}{|a|\pi} (\text{fp}\frac{1}{t} \otimes f(t))(at + 0 + b) = \text{sgn}(a) \mathcal{H}[f(t)](at + b). \qquad \square$$

According to Proposition 2.36 a given hyperfunction $f(x) = [F_+(z), F_-(z)]$ satisfying
(i) $F_+(z) \in \mathcal{O}(\mathbb{C}_+), F_-(z) \in \mathcal{O}(\mathbb{C}_-)$,
(ii) $\lim_{y\to\infty} F_+(x + iy) - c_+ = O(1/z)$ as z tends to infinity in the upper half-plane, and $\lim_{y\to\infty} F_-(x - iy) - c_- = O(1/z)$ as z tends to infinity in the lower half-plane,
has a strong defining function given by

$$\tilde{F}(z) = F(z) - \frac{c_+ + c_-}{2}.$$

Hence, its Hilbert transform is given by

$$\mathcal{H}[f](t) = 2i \left[\{F(z) - \frac{c_+ + c_-}{2}\} \mathbf{1}(z) \right].$$

We can now prove

Proposition 5.12. *Let $f(x) = [F(z)] = [F_+(z), F_-(z)]$ be a hyperfunction satisfying the above hypotheses* (i) *and* (ii), *and let $g(t)$ be another hyperfunction with compact support $[a, b]$. Then, we have*

$$f(t) \star g(t) \circ \overset{\mathcal{H}}{-\!\!\!-\!\!\bullet} \mathcal{H}[f](t) \star g(t). \tag{5.46}$$

Proof. We first prove the special case when $c_+ = c_- = 0$. Let $f(t) \star g(t) = [H(z)]$, then

$$H(z) = \int_a^b F(z - x)g(x)dx$$

is holomorphic in the upper and lower half-plane, and $H(z) \to 0$ as $|\Im z| \to \infty$. Thus $H(z)$ qualifies as a strong defining function (see Corollary 2.34), and we have

$$\mathcal{H}[f \star g](t) = 2i\, [H(z)\,\mathbf{1}(z)]\,.$$

Now

$$\mathcal{H}[f](t) \star g(t) = \left[2i \int_a^b F(z - x)\,\mathbf{1}(z - x)g(x)dx\right]$$

$$= \left[2i \int_a^b F(z - x)\,\mathbf{1}(z)g(x)dx\right]$$

$$= \left[2i \int_a^b F(z - x)\,g(x)dx\,\mathbf{1}(z)\right]$$

$$= 2i\, [H(z)\,\mathbf{1}(z)] = \mathcal{H}[f \star g](t)$$

which proves the proposition in this special case. Now let

$$F_1(z) := F(z) - [c_+, c_-],$$
$$f_1(t) = [F_1(z)] = f(t) - (c_+ - c_-).$$

$F_1(z)$ may be taken for the above special case, and we have

$$\mathcal{H}[f_1 \star g](t) = \mathcal{H}[f_1](t) \star g(t).$$

Since the Hilbert transform of a constant is zero we have

$$\mathcal{H}[f](t) \star g(t) = \mathcal{H}[f_1](t) \star g(t)$$
$$= \mathcal{H}[(f(t) - (c_+ - c_-)) \star g](t)$$
$$= \mathcal{H}[(f \star g](t) - (c_+ - c_-)\,\mathcal{H}[(1 \star g](t)$$
$$= \mathcal{H}[f \star g](t),$$

because

$$\mathcal{H}[1 \star g](t) = \mathcal{H}[\int_a^b g(x)\,dx](t) = 0. \qquad \square$$

Problem 5.10. Verify this rule for $f(t) = \delta(t - a)$, $g(t) = \delta(t - b)$, $a \neq b$.

5.1.3 Using Fourier Transforms

Fourier transformation may also be used to find Hilbert transform pairs. Because the Fourier transform of the hyperfunction $\mathrm{fp}(1/t)$ is $\pi \operatorname{sgn}\omega/i$, we have, applying the convolution theorem, with $f(t) \lhd\!-\!\blacktriangleright \hat{f}(\omega)$,

$$\mathcal{H}[f](t) = i\,\mathcal{F}^{-1}[\operatorname{sgn}(\omega)\,\hat{f}(\omega)](t). \tag{5.47}$$

This requires at least that the product of $\operatorname{sgn}(\omega)$ and the hyperfunction $\hat{f}(\omega)$ is defined and that the inverse Fourier transform may be applied to it. So $\hat{f}(\omega)$ should be a holomorphic hyperfunction at $\omega = 0$.

Example 5.18. Let $a \neq 0$ and real. Take $f(t) = e^{iat}$.

$$\mathcal{H}[e^{iat}](t) = i\mathcal{F}^{-1}[\operatorname{sgn}(\omega)\,2\pi\,\delta(\omega - a)](t) = i\mathcal{F}^{-1}[\operatorname{sgn}(a)\,2\pi\,\delta(\omega - a)](t)$$
$$= i\,\operatorname{sgn}(a)\,\mathcal{F}^{-1}[2\pi\,\delta(\omega - a)](t) = i\,\operatorname{sgn}(a)\,e^{iat}.$$

If we apply again (5.47) to $\mathcal{H}[f](t)$, we obtain

$$\mathcal{H}[\mathcal{H}[f]](t) = -\,\mathcal{F}^{-1}[\operatorname{sgn}(\omega)\,\{\operatorname{sgn}(\omega)\,\hat{f}(\omega)\}](t).$$

Unfortunately we cannot always write for all hyperfunctions

$$\operatorname{sgn}(\omega)\,\{\operatorname{sgn}(\omega)\,\hat{f}(\omega)\} = \hat{f}(\omega).$$

This reflects what we already know that $\mathcal{H}[\mathcal{H}[f]](t) = -f(t)$ does not hold without further restrictions (see Proposition 5.4). Nevertheless, Fourier transformation may be helpful to obtain Hilbert transform pairs.

Example 5.19. Let us compute the Hilbert transform of $f(t) = \exp(-at^2),\ a > 0$. Its Fourier transform is

$$\hat{f}(\omega) = \sqrt{\frac{\pi}{a}}\,e^{-\omega^2/(4a)},$$

and, by using the result (4.92) we get with Dawson's function

$$\mathcal{F}^{-1}[\operatorname{sgn}\omega\,e^{-\omega^2/(4a)}](x) = 2i\,\frac{\sqrt{a}}{\pi}\,\mathrm{Daw}(\sqrt{a}x).$$

We thus obtain

$$\mathcal{H}[e^{-at^2}](t) = i\,\mathcal{F}^{-1}[\operatorname{sgn}(\omega)\,\hat{f}(\omega)](t) = i\,\sqrt{\frac{\pi}{a}}\mathcal{F}^{-1}[\operatorname{sgn}\omega\,e^{-\omega^2/(4a)}](t)$$
$$= i\,\sqrt{\frac{\pi}{a}}\,2i\,\frac{\sqrt{a}}{\pi}\,\mathrm{Daw}(\sqrt{a}t),$$

and finally,

$$e^{-at^2} \overset{\mathcal{H}}{\circ\!-\!\bullet} -\frac{2}{\sqrt{\pi}}\,\mathrm{Daw}(\sqrt{a}\,t) \tag{5.48}$$

in agreement with (5.10).

Example 5.20. Take $\mathrm{fp}(1/t^m)$ with its Fourier transform

$$\hat{f}(\omega) = \frac{\pi}{i^m \, (m-1)!} \, \omega^{m-1} \, \mathrm{sgn}(\omega),$$

thus

$$i \, \mathrm{sgn}(\omega) \, \hat{f}(\omega) = \frac{\pi}{i^m \, (m-1)!} \, \omega^{m-1}.$$

By taking into account that $\delta^{(m-1)}(t) \lhd - \blacktriangleright i^{m-1} \, \omega^{m-1}$, we find again

$$\mathrm{fp}\frac{1}{t^m} \circ\!\!\overset{\mathcal{H}}{\longrightarrow}\!\!\bullet\; \pi \frac{(-1)^{m-1}}{(m-1)!} \, \delta^{(m-1)}(t). \tag{5.49}$$

Example 5.21. For $\alpha \notin \mathbb{Z}$ we have

$$|t|^\alpha \, \mathrm{sgn}\, t \lhd - \blacktriangleright 2i \, \Gamma(\alpha+1) \, \sin(\frac{\pi}{2}(\alpha+1)) \, \frac{\mathrm{sgn}\,\omega}{|\omega|^{\alpha+1}}.$$

Multiplying the Fourier transform on the right-hand side by $i \, \mathrm{sgn}\,\omega$ gives

$$- 2\,\Gamma(\alpha+1) \, \sin(\frac{\pi}{2}(\alpha+1)) \, \frac{1}{|\omega|^{\alpha+1}}$$
$$= - \tan(\frac{\pi}{2}(\alpha+1)) \, 2\,\Gamma(\alpha+1) \, \cos(\frac{\pi}{2}(\alpha+1)) \, \frac{1}{|\omega|^{\alpha+1}}.$$

By using the Fourier transform of $-|t|^\alpha$ (see (4.35)), we find

$$|t|^\alpha \, \mathrm{sgn}\, t \circ\!\!\overset{\mathcal{H}}{\longrightarrow}\!\!\bullet\; - \cot(\frac{\pi}{2}\alpha) \, |t|^\alpha, \tag{5.50}$$

provided $\Re\alpha < 0$ and viewed as a correspondence between ordinary functions. However, if we adopt the point of view that

$$\mathcal{H}[f](t) := i \, \mathcal{F}^{-1}[\mathrm{sgn}(\omega) \, \hat{f}(\omega)](t)$$

every time whenever the right-hand side makes sense, then it suffices that $\alpha \notin \mathbb{Z}$.

Example 5.22. Compute the Hilbert transform of $u(t)/t^m$, $m \in \mathbb{N}$. The Fourier transform is

$$-\frac{(-i\omega)^{m-1}}{(m-1)!} \, \{\log|\omega| + i\frac{\pi}{2} \, \mathrm{sgn}\,\omega + \gamma - \sum_{k=1}^{m-1} \frac{1}{k}\}$$

by (4.32). We have to find the inverse Fourier transform of

$$- i \frac{(-i\omega)^{m-1}}{(m-1)!} \, \{\mathrm{sgn}\,\omega \log|\omega| + i\frac{\pi}{2} + \mathrm{sgn}\,\omega(\gamma - \sum_{k=1}^{m-1} \frac{1}{k})\}$$
$$= \frac{(-i)^m}{(m-1)!} \, \omega^{m-1} \, \mathrm{sgn}\,\omega \log|\omega| + \frac{\pi}{2(m-1)!}(-i\omega)^{m-1}$$
$$- \frac{(-i)^m}{(m-1)!} \, \Psi(m)\omega^{m-1} \, \mathrm{sgn}\,\omega.$$

This gives

$$\frac{1}{\pi}\{\Psi(m)\,\mathrm{fp}\frac{1}{t^m}-\frac{\log|t|}{t^m}\}+\frac{(-1)^{m-1}\pi}{2(m-1)!}\delta^{(m-1)}(t)-\frac{\Psi(m)}{\pi}\,\mathrm{fp}\frac{1}{t^m}.$$

Finally, we obtain

$$\frac{u(t)}{t^m}\ \circ\!\!-\!\!\!-\!\!\bullet\ -\frac{1}{\pi}\frac{\log|t|}{t^m}+\frac{(-1)^{m-1}\pi}{2(m-1)!}\delta^{(m-1)}(t).\tag{5.51}$$

Similarly, the reader may find

$$\frac{u(-t)}{t^m}\ \circ\!\!-\!\!\!-\!\!\bullet\ \frac{1}{\pi}\frac{\log|t|}{t^m}+\frac{(-1)^{m-1}\pi}{2(m-1)!}\delta^{(m-1)}(t).\tag{5.52}$$

Sometimes Laplace transformation can also be used to compute Hilbert transforms.

Example 5.23. Take $f(t)=\exp(-a|x|)$, $a>0$. Because

$$\mathcal{H}[f](t):=\frac{1}{\pi}\,\mathrm{pv}\int_{-\infty}^{\infty}\frac{e^{-a|x|}}{x-t}\,dx=\frac{2t}{\pi}\,\mathrm{pv}\int_0^{\infty}e^{-ax}\frac{dx}{x^2-t^2},$$

for $t>0$, we can write

$$\mathcal{H}[f](t)=\frac{2t}{\pi}\,\mathcal{L}[\frac{u(x)}{x^2-t^2}](a)=-\frac{1}{\pi}\{e^{-at}\,\mathrm{Ei}(at)+e^{at}\,E_1(at)\}$$

where we have used (3.71). The expression on the right-hand side is not defined for $t<0$, but by taking into account the odd parity of the Hilbert transform, we finally obtain

$$\mathcal{H}[e^{-a|t|}](t)=-\,u(t)\{\frac{1}{\pi}\{e^{-at}\,\mathrm{Ei}(at)+e^{at}\,E_1(at)\}\}$$
$$+u(-t)\{\frac{1}{\pi}\{e^{at}\,\mathrm{Ei}(-at)+e^{-at}\,E_1(-at)\}\}.\tag{5.53}$$

Problem 5.11. By using again

$$\mathcal{H}[f(t)](t)=i\,\mathcal{F}^{-1}[\mathrm{sgn}\,\omega\,\mathcal{F}[f(t)](\omega)](t),\tag{5.54}$$

establish

$$|t|^{\alpha}\ \circ\!\!-\!\!\!-\!\!\bullet\ \tan(\frac{\pi}{2}\alpha)\,|t|^{\alpha}\,\mathrm{sgn}\,t,\quad\alpha\notin\mathbb{Z},\tag{5.55}$$

$$\frac{\log|t|}{t^m}\ \circ\!\!-\!\!\!-\!\!\bullet\ \frac{\pi}{2}\frac{\mathrm{sgn}\,t}{t^m},\tag{5.56}$$

$$\frac{\mathrm{sgn}\,t}{t^m}\ \circ\!\!-\!\!\!-\!\!\bullet\ -\frac{2}{\pi}\frac{\log|t|}{t^m}.\tag{5.57}$$

Problem 5.12. By using $u(t)\,t^{\alpha}=\{\mathrm{sgn}(t)|t|^{\alpha}+|t|^{\alpha}\}/2$, $u(-t)\,t^{\alpha}=\{-\mathrm{sgn}(t)|t|^{\alpha}+|t|^{\alpha}\}/2$ establish for $\alpha\notin\mathbb{Z}$,

$$u(t)\,t^{\alpha}\ \circ\!\!-\!\!\!-\!\!\bullet\ \begin{cases}-\cot(\pi\alpha)\,t^{\alpha},&t>0\\-\frac{1}{\sin(\pi\alpha)}\,|t|^{\alpha}&t<0,\end{cases}\tag{5.58}$$

and

$$u(-t)\,|t|^\alpha \overset{\mathcal{H}}{\circ\!\!-\!\!\bullet} \begin{cases} \cot(\pi\alpha)\,|t|^\alpha, & t < 0 \\ \frac{1}{\sin(\pi\alpha)}\,t^\alpha, & t > 0. \end{cases} \tag{5.59}$$

Then, for $\alpha = -n - 1/2$, $n \in \mathbb{N}_0$, obtain

$$\frac{u(t)}{t^{n+1/2}} \overset{\mathcal{H}}{\circ\!\!-\!\!\bullet} u(-t)\,(-1)^{n+1}\,\frac{1}{|t|^{n+1/2}} \tag{5.60}$$

and

$$\frac{u(-t)}{|t|^{n+1/2}} \overset{\mathcal{H}}{\circ\!\!-\!\!\bullet} u(t)\,(-1)^n\,\frac{1}{t^{n+1/2}}. \tag{5.61}$$

Problem 5.13. Show that

$$\mathcal{H}[t\,e^{-a|t|}](t) = \frac{2}{a\pi} - u(t)\{\frac{t}{\pi}\,\{e^{-at}\,\text{Ei}(at) + e^{at}\,E_1(at)\}\}$$
$$+ u(-t)\{\frac{t}{\pi}\,\{e^{at}\,\text{Ei}(-at) + e^{-at}\,E_1(-at)\}\}. \tag{5.62}$$

5.2   Analytic Signals and Conjugate Hyperfunctions

In signal analysis a hyperfunction of the form

$$x_f(t) := f(t) - i\,\mathcal{H}[f](t) \tag{5.63}$$

is said to be the *analytic signal associated with* $f(t)$. Let us consider the analytic signal associated with the Dirac hyperfunction, sometimes called *the complex Dirac impulse*:

$$x_\delta(t) := \delta(t) + i\,\frac{1}{\pi}\,\text{fp}\frac{1}{t}. \tag{5.64}$$

By taking the Fourier transform of the complex Dirac impulse,

$$x_\delta(t) \triangleleft\!-\!\blacktriangleright 1 + i\,\frac{1}{\pi}\,\frac{\pi}{i}\,\text{sgn}\,\omega = 1 + \text{sgn}\,\omega = 2u(\omega),$$

we see that the Fourier transform of the complex Dirac impulse is just the doubled unit-step, or conversely, the inverse Fourier transform of the doubled unit-step gives the complex Dirac impulse.

Because $\delta(t) \star f(t) = \delta(t) \otimes f(t) = f(t)$, and

$$\mathcal{H}[f](t) := -\frac{1}{\pi}\,\text{fp}\frac{1}{t} \otimes f(t),$$

we may write for an analytic signal $x_f(t) = f(t) - i\,\mathcal{H}[f](t)$,

$$x_f(t) = \{\delta(t) + i\,\frac{1}{\pi}\,\text{fp}\frac{1}{t}\} \otimes f(t) = x_\delta(t) \otimes f(t) \tag{5.65}$$

which means that the associated analytic signal is obtained by convolution of the complex Dirac impulse and the signal. Thus, by taking the Fourier transform, and using the convolution property, we obtain

$$x_f(t) \triangleleft\!-\!\blacktriangleright 2u(\omega)\,\hat{f}(\omega). \tag{5.66}$$

This shows that the spectrum of the associated analytic signal is entirely concentrated on the positive real axis, or, more precisely, *if we pass from the signal $f(t)$ to its associated analytic signal $x_f(t)$, the negative part of the spectrum of $f(t)$ is cut away and the positive part is doubled.*

Let $f(t) = f(t + T) = [F(z)]$ be a periodic hyperfunction of period T. In Section 2.6 we learned that a periodic hyperfunction has a strong defining function with the same period T, given by its Fourier series

$$\tilde{F}(z) = c_0 \mathbf{1}(z) + \sum_{k=1}^{\infty} c_k\, e^{ik\omega\, z}\, \mathbf{1}_+(z) + \sum_{k=1}^{\infty} c_{-k}\, e^{-ik\omega\, z}\, \mathbf{1}_-(z).$$

The periodic hyperfunction itself has the Fourier series

$$f(t) = c_0 + \sum_{k=1}^{\infty} c_k\, e^{ik\omega\, t} + \sum_{k=1}^{\infty} c_{-k}\, e^{-ik\omega\, t}$$

$$= \mathrm{pv} \sum_{k=-\infty}^{\infty} c_k\, e^{ik\omega\, t}.$$

Because the Hilbert transform of the hyperfunction $f(t) = f(t + T)$ is given by

$$\mathcal{H}[f](t) = 2\, i\, \left[\tilde{F}(z)\, \mathbf{1}(z) \right],$$

we see that the Hilbert transform is again a periodic hyperfunction with the same period T, and therefore, has a Fourier series, say,

$$\mathcal{H}[f](t) = \hat{c}_0 + \sum_{k=1}^{\infty} \hat{c}_k\, e^{ik\omega\, t} + \sum_{k=1}^{\infty} \hat{c}_{-k}\, e^{-ik\omega\, t}$$

$$= \mathrm{pv} \sum_{k=-\infty}^{\infty} \hat{c}_k\, e^{ik\omega\, t}.$$

Because

$$\mathbf{1}^2(z) = \frac{1}{4} = \text{const.},$$

$$\mathbf{1}_+(z)\, \mathbf{1}(z) = \frac{1}{2}\, \mathbf{1}_+(z),\ \ \mathbf{1}_-(z)\, \mathbf{1}(z) = -\frac{1}{2}\, \mathbf{1}_+(z),$$

we obtain for the standard defining function of the Hilbert transform

$$\tilde{F}(z) = \sum_{k=1}^{\infty} ic_k\, e^{ik\omega\, z}\, \mathbf{1}_+(z) - \sum_{k=1}^{\infty} ic_{-k}\, e^{-ik\omega\, z}\, \mathbf{1}_-(z),$$

i.e.,

$$\hat{c}_0 = 0,\ \ \hat{c}_k = i\, c_k,\ \ \hat{c}_{-k} = -i\, c_{-k}.$$

This yields the Fourier series of the Hilbert transform of $f(t) = f(t + T)$ in the form

$$\mathcal{H}[f](t) = \sum_{k=1}^{\infty} i\, c_k\, e^{ik\omega\, t} - \sum_{k=1}^{\infty} i\, c_{-k}\, e^{-ik\omega\, t}$$

$$= \mathrm{pv} \sum_{k=-\infty, k \neq 0}^{\infty} i\, \mathrm{sgn}(k\omega)\, c_k\, e^{ik\omega\, t}.$$

We would have obtained the same result if we simply had applied the Hilbert transform term-wise.

Definition 5.4. The Fourier series

$$\text{pv} \sum_{k=-\infty, k\neq 0}^{\infty} i \, \text{sgn}(k\omega) \, c_k \, e^{ik\omega t} \tag{5.67}$$

is called the *conjugate Fourier series* of

$$\text{pv} \sum_{k=-\infty}^{\infty} c_k \, e^{ik\omega t}. \tag{5.68}$$

If we consider the analytic signal $x_f(t) = f(t) - i\mathcal{H}[f](t)$ associated with a periodic hyperfunction $f(t) = f(t+T)$, we see (with $\omega > 0$) that

$$x_f(t) = \text{pv} \sum_{k=-\infty}^{\infty} c_k \, e^{ik\omega t} - i \, \text{pv} \sum_{k=-\infty, k\neq 0}^{\infty} i \, \text{sgn}(k) \, c_k \, e^{ik\omega t}$$

$$= c_0 + 2 \sum_{k=1}^{\infty} c_k \, e^{ik\omega t}$$

which shows again *that the associated analytic signal of a periodic hyperfunction has the (discrete) spectrum concentrated on the positive real axis*. Moreover, if the given hyperfunction $f(t) = f(t+T)$ is real-valued, we obtain the real form of the two Fourier series,

$$f(t) = \frac{A_0}{2} + \sum_{k=1}^{\infty} (A_k \cos k\omega t + B_k \sin k\omega t),$$

$$\mathcal{H}[f](t) = \sum_{k=1}^{\infty} (B_k \cos k\omega t - A_k \sin k\omega t).$$

This results from the relations

$$c_k = \frac{A_k}{2} - i\frac{B_k}{2}, \ A_k = 2 \, \Re c_k, \ B_k = -2 \, \Im c_k, \ c_{-k} = \overline{c_k}.$$

Details are left to the reader. If the energy of a signal is defined to be

$$E(f) := \sum_{k=-\infty, \ k\neq 0}^{\infty} |c_k|^2,$$

it is also seen that the given $f(t)$ and the associated analytic signal $x_f(t)$ have the same energy.

We conclude this section by deriving another formula for the computation of the Hilbert transform of a given periodic hyperfunction $f(t) = f(t+T)$. Suppose the periodic hyperfunction is given in the form

$$f(t) = \sum_{k=-\infty}^{\infty} f_0(t - kT)$$

where $f_0(t) = \chi_{(a,a+T)} f(t)$ is the projection of $f(t)$ on the interval of periodicity $(a, a + T)$ (often we take $a = 0$ or $a = -T/2$). We may then write

$$f(t) = \sum_{k=-\infty}^{\infty} \delta(t - kT) \star f_0(t).$$

We now use (5.14), i.e.,

$$\sum_{k=-\infty}^{\infty} \delta(t - kT) \overset{\mathcal{H}}{\circ\!\!-\!\!\bullet} -\frac{1}{T} \operatorname{fp} \cot(\frac{\pi t}{T})$$

and Proposition 5.12 and get

$$\mathcal{H}[f(t)](t) = \mathcal{H}[\sum_{k=-\infty}^{\infty} \delta(t - kT) \star f_0(t)]$$

$$= \mathcal{H}[\sum_{k=-\infty}^{\infty} \delta(t - kT)](t) \star f_0(t) = -\frac{1}{T} \operatorname{fp} \cot(\frac{\pi t}{T}) \star f_0(t).$$

We have established the following result.

Proposition 5.13. *The Hilbert transform of the periodic hyperfunction* $f(t) = f(t + T)$ *can also be obtained by*

$$f(t) = f(t + T) \overset{\mathcal{H}}{\circ\!\!-\!\!\bullet} \left[-\frac{1}{T} \int_a^{a+T} \cot(\frac{\pi(z - x)}{T}) f(x) \, dx \right]$$

$$= -\frac{1}{T} \int_a^{a+T} \cot(\frac{\pi(t - x)}{T}) f(x) \, dx. \tag{5.69}$$

Example 5.24. Let $f(t) = \operatorname{sgn}(\cos t)$, $a = -\pi/2, T = 2\pi$. Then $f_0(t) = 1$ for $-\pi/2 < t < \pi/2$, and $f_0(t) = -1$ for $\pi/2 < t < 3\pi/2$. Thus,

$$\mathcal{H}[\operatorname{sgn}(\cos t)](t) = \frac{1}{2\pi} \{ -\int_{-\pi/2}^{\pi/2} \cot(\frac{t - x}{2}) \, dx + \int_{\pi/2}^{3\pi/2} \cot(\frac{t - x}{2}) \, dx \}.$$

Computing the two elementary integrals yields

$$-\frac{2}{\pi} \log |\tan(\frac{t}{2} + \pi/4)|.$$

By Proposition 5.11 we obtain

$$\operatorname{sgn}(\cos(\omega t + b)) \overset{\mathcal{H}}{\circ\!\!-\!\!\bullet} -\frac{2}{\pi} \log |\tan(\frac{\omega t}{2} + \frac{\pi}{4} + \frac{b}{2})|,$$

and, for $b = 0$ and $b = -\pi/2$, finally,

$$\operatorname{sgn}(\cos \omega t) \overset{\mathcal{H}}{\circ\!\!-\!\!\bullet} -\frac{2}{\pi} \log |\tan(\frac{\omega t}{2} + \frac{\pi}{4})|, \tag{5.70}$$

$$\operatorname{sgn}(\sin \omega t) \overset{\mathcal{H}}{\circ\!\!-\!\!\bullet} -\frac{2}{\pi} \log |\tan(\frac{\omega t}{2})|. \tag{5.71}$$

5.3   Integral Equations IV

Here we consider some integral equations of the type

$$\text{pv} \int_a^b \frac{f(x)}{x - t} \, dx = g(t) \tag{5.72}$$

where $g(t)$ is the given and $f(t)$ the unknown hyperfunction.

The case $a = -\infty$, $b = \infty$

We interpret the singular integral equation

$$\text{pv} \int_{-\infty}^{\infty} \frac{f(x)}{x - t} \, dx = g(t) \tag{5.73}$$

as

$$\mathcal{H}[f(t)](t) = \frac{1}{\pi} g(t). \tag{5.74}$$

According to Proposition 5.4 a solution is

$$f(t) = -\frac{1}{\pi} \mathcal{H}[g(t)](t) + c,$$

provided the second member $g(t)$ is in $\mathcal{B}_0(\mathbb{R})$ (see Definition 2.14). Now, it is possible that the hyperfunction $g(t)$ is not in $\mathcal{B}_0(\mathbb{R})$ but its derivative $D^k g(t)$ for some k is. By using the property $\mathcal{H}[D^k g(t)](t) = D^k \mathcal{H}[g](t)$, we may first solve

$$\mathcal{H}[h(t)](t) = \frac{1}{\pi} D^k g(t) \in \mathcal{B}_0(\mathbb{R}) \tag{5.75}$$

for $h(t)$ and then find $f(t)$ by k-fold integration.

Example 5.25. Solve the integral equation

$$\text{pv} \int_{-\infty}^{\infty} \frac{f(x)}{x - t} \, dx = \chi_{a,b}(t). \tag{5.76}$$

Because $\chi_{a,b}(t) \in \mathcal{B}_0(\mathbb{R})$ the solution is given by

$$f(t) = -\frac{1}{\pi} \mathcal{H}[\chi_{a,b}(t)](t) + c = -\frac{1}{\pi^2} \log\left|\frac{b - t}{a - t}\right| + c. \tag{5.77}$$

Example 5.26. Solve the integral equation

$$\text{pv} \int_{-\infty}^{\infty} \frac{f(x)}{x - t} \, dx = |t|^\alpha, \quad \alpha \notin \mathbb{Z}. \tag{5.78}$$

A) For $\Re \alpha < 0$ we have that $|t|^\alpha \in \mathcal{B}_0(\mathbb{R})$, thus

$$f(t) = -\frac{1}{\pi} \mathcal{H}[|t|^\alpha](t) + c.$$

Because

$$|t|^\alpha \overset{\mathcal{H}}{\circ\!\!-\!\!\bullet} \tan\left(\frac{\pi}{2}\alpha\right) |t|^\alpha \operatorname{sgn} t$$

we obtain

$$f(t) = -\frac{1}{\pi} \tan(\frac{\pi}{2}\alpha) |t|^\alpha \operatorname{sgn} t + c.$$

B) Now let $0 < \Re\alpha < 1$. Then, $|t|^\alpha \notin \mathcal{B}_0(\mathbb{R})$, but $D|t|^\alpha = \alpha |t|^{\alpha-1} \operatorname{sgn} t \in \mathcal{B}_0(\mathbb{R})$. Therefore,

$$h(t) = Df(t) = -\frac{\alpha}{\pi}\mathcal{H}[|t|^{\alpha-1}\operatorname{sgn} t](t) + c.$$

Because

$$|t|^{\alpha-1}\operatorname{sgn} t \circ \overset{\mathcal{H}}{\longrightarrow} \bullet - \cot(\frac{\pi}{2}(\alpha-1))|t|^{\alpha-1}$$

we obtain

$$h(t) = \frac{\alpha}{\pi}\tan(\frac{\pi}{2}\alpha)|t|^{\alpha-1} + c.$$

Integrating the result yields

$$f(t) = -\frac{1}{\pi}\tan(\frac{\pi}{2}\alpha)|t|^\alpha \operatorname{sgn} t + ct + c'. \tag{5.79}$$

Since the Hilbert transform of t does not exist, we must choose $c = 0$ in order to find a solution having a Hilbert transform, and we again obtain the same expression as under the hypothesis $\Re\alpha < 0$. Continuing in this way we may establish similar results for all $\alpha \notin \mathbb{Z}$.

Problem 5.14. Show that a solution of the integral equation

$$\operatorname{pv}\int_{-\infty}^\infty \frac{f(x)}{x - t}\,dx = |t|^\alpha \operatorname{sgn} t, \quad \alpha \notin \mathbb{Z}$$

is given by

$$f(t) = \frac{1}{\pi}\cot(\frac{\pi}{2}\alpha)|t|^\alpha + c.$$

Example 5.27. Solve the integral equation

$$\operatorname{pv}\int_{-\infty}^\infty \frac{f(x)}{x - t}\,dx = \operatorname{fp}\frac{1}{t^m}, \quad m \in \mathbb{N}. \tag{5.80}$$

Because $\operatorname{fp}(1/t^m) \in \mathcal{B}_0(\mathbb{R})$, we immediately have

$$f(t) = -\frac{1}{\pi}\mathcal{H}[\operatorname{fp}\frac{1}{t^m}](t) + c = -\frac{(-1)^{m-1}}{(m-1)!}\delta^{(m-1)}(t) + c. \tag{5.81}$$

Problem 5.15. Show that a solution of the integral equation

$$\operatorname{pv}\int_{-\infty}^\infty \frac{f(x)}{x - t}\,dx = \delta^{(m)}(t), \quad m \in \mathbb{N}_0$$

is given by

$$f(t) = -\frac{(-1)^{m+1}m!}{\pi^2}\operatorname{fp}\frac{1}{t^{m+1}} + c.$$

Example 5.28. The integral equation

$$\text{pv} \int_{-\infty}^{\infty} \frac{f(x)}{x-t} \, dx = t^m, \quad m \in \mathbb{N}$$

has no solution having a Hilbert transform. To show this assume that the hyperfunction $f(t)$ is a solution. This would imply

$$\mathcal{H}[D^{m+1} f(t)] = \frac{1}{\pi} D^{m+1} t^m = 0,$$

thus, $D^{m+1} f(t) = c_0$. Then n-fold integration yields

$$f(t) = c_0 \frac{t^m}{m!} + c_1 \frac{t^{m-1}}{(m-1)!} + \cdots + c_{m+1}.$$

But this hyperfunction only has a Hilbert transform if $c_0 = c_1 = \cdots = c_m = 0$. But in this case the Hilbert transform of $f(t) = c_{m+1}$ becomes zero and is in contradiction with the given t^m.

The case $a = 0, b = \infty$

Consider the singular integral equation

$$\text{pv} \int_0^{\infty} \frac{f(x)}{x-t} \, dx = g(t), \quad 0 < t < \infty. \tag{5.82}$$

By using the projection $h(t) := \chi_{(0,\infty)} f(t)$, the given integral equation reads as

$$\mathcal{H}[h](t) = \frac{1}{\pi} g(t), \tag{5.83}$$

where $h(t) = [H(z)]$ is a right-sided hyperfunction. This means that the defining function $H(z)$ is real analytic on the negative part of the real axis. From (5.60) it is seen that we may add to a particular solution $h(t)$ any linear combination of $u(t) t^{-n-1/2}, n \in \mathbb{N}_0$ to obtain another solution, since the Hilbert transform of such a power hyperfunction vanishes for $t > 0$. So, if we have found a particular solution $h(t)$ of the integral equation, the expression

$$h(t) + u(t) \sum_{n \geq 0} \frac{c_n}{t^{n+1/2}},$$

for arbitrary constants c_k, is still a solution. Let us now find a particular solution. The procedure presented in the book [19, Imai] is rather tricky. First, we observe that the function $(-z)^{-1/2}$ is real analytic on the negative part of the real axis, if the fractional power denotes the principal branch. Thus, the function $K(z) := (-z)^{-1/2} H(z)$ is also real analytic on the negative part of the real axis. We now assume as a working hypothesis that $K(z)$ is a strong defining function of a hyperfunction $k(t)$ which is also a right-sided hyperfunction. This means that $k(t)$ has a Hilbert transform. By Proposition 5.5 we have

$$k(t) = K_+(t+i0) - K_-(t-i0),$$
$$\mathcal{H}[k](t) = i\{K_+(t+i0) - K_-(t-i0)\}.$$

(the reader is advised to draw a figure). The integrand is holomorphic inside D, hence by Cauchy's integral theorem we have

$$\int_{\partial D} \tau^{s-1} e^{-\tau} d\tau = 0,$$

where ∂D denotes the boundary of D. On the small quarter-circle with radius ϵ, the contribution to the contour integral tends to zero as $\epsilon \to 0$, provided $\Re s > 0$. On the large quarter-circle with radius R, the contribution to the contour integral also tends to zero as $R \to \infty$, provided $|\tau^{s-1}|$ tends to zero uniformly with respect to $\arg \tau$. This follows from Jordan's Lemma (see at the end of Sections 3.2.1 and 2.2). Therefore, for $0 < \Re s < 1$, we arrive at

$$\int_0^{-i\infty} \tau^{s-1} e^{-\tau} d\tau = \int_0^{\infty} \tau^{s-1} e^{-\tau} d\tau = \Gamma(s).$$

Thus,

$$\mathcal{M}[e^{ix}](s) = e^{i\frac{\pi}{2} s}\Gamma(s), \quad 0 < \Re s < 1. \tag{6.8}$$

By temporarily restricting s to real values in the interval $(0, 1)$, taking the real and imaginary parts on both sides, and finally invoking analytic continuation into the vertical strip $0 < \Re s < 1$, we obtain

$$\mathcal{M}[\cos x](s) = \cos(\frac{\pi}{2} s)\,\Gamma(s), \tag{6.9}$$

$$\mathcal{M}[\sin x](s) = \sin(\frac{\pi}{2} s)\,\Gamma(s). \tag{6.10}$$

Because the sine function has a zero at $s = 0$, the strip of convergence for the second relation can be enlarged to $-1 < \Re s < 1$ by analytic continuation.

Example 6.3. $f(x) = (1 + x)^{-n}$, $n \in \mathbb{N}$.

$$\mathcal{M}[\frac{1}{(1 + x)^n}](s) = \int_0^{\infty} \frac{x^{s-1}}{(1 + x)^n} dx$$

$$= \int_0^1 t^{s-1}(1 - t)^{n-s-1} dt,$$

where we have made the change of variables $x = t/(1 - t)$ in the Mellin integral. The last integral equals $B(s, n - s) = \Gamma(s)\Gamma(n - s)/\Gamma(n)$. Therefore,

$$\mathcal{M}[\frac{1}{(1 + x)^n}](s) = \frac{\Gamma(s)\Gamma(n - s)}{\Gamma(n)}, \tag{6.11}$$

with $0 < \Re s < n$. For $n = 1$, and by using the reflection formula for the Gamma function, we obtain

$$\mathcal{M}[\frac{1}{1 + x}](s) = \frac{\pi}{\sin(\pi s)}, \tag{6.12}$$

with $0 < \Re s < 1$.

Example 6.4. The function $1/(1 + z)$ of the complex variable z is holomorphic for $|\arg z| < \pi$, the same is true for z^{s-1} for any fixed s, thus, $g(z) = z^{s-1}/(1 + z)$

is holomorphic for $|\arg z| < \pi$, too. Consider now the sector $D(R, \alpha) = \{z \mid 0 \le \arg z \le \alpha,\ 0 \le |z| \le R\}$ where $g(z)$ is holomorphic. The contour integral of $g(z)$ along the boundary $\partial D(R, \alpha)$ vanishes by Cauchy's theorem. If we can show that the contour integral tends to zero on the arc $Re^{i\tau}$, $0 \le \tau \le \alpha$ as $R \to \infty$, we shall have

$$\int_0^{\infty e^{i\alpha}} \frac{z^{s-1}}{1+z}\, dz = \int_0^\infty \frac{x^{s-1}}{1+x}\, dx = \frac{\pi}{\sin(\pi s)}.$$

Let us estimate the contour integral over the arc. We have

$$\left| \int_0^\alpha \frac{R^{s-1}\, e^{i\tau(s-1)}\, Rie^{i\tau}}{1 + Re^{i\tau}}\, d\tau \right| = O(R^{\Re s - 1}) \to 0$$

as $R \to \infty$ as long as $\Re s < 1$. Therefore, for $0 < \Re s < 1$,

$$\frac{\pi}{\sin(\pi s)} = e^{i\alpha(s-1)} \int_0^\infty \frac{t^{s-1}}{1 + te^{i\alpha}}\, e^{i\alpha}\, dt$$

$$= e^{i\alpha s} \int_0^\infty t^{s-1} \frac{1 + t\cos\alpha - it\sin\alpha}{t^2 + 2t\cos\alpha + 1}\, dt.$$

Assuming temporarily s real, separating the expressions into real and imaginary parts, and eventually invoking analytic continuation to the whole vertical strip, yields finally for $0 < \Re s < 1$ and $-\pi < \alpha < \pi$,

$$\mathcal{M}[\frac{1 + x\cos\alpha}{x^2 + 2x\cos\alpha + 1}](s) = \frac{\pi\,\cos(\alpha s)}{\sin(\pi s)}, \tag{6.13}$$

$$\mathcal{M}[\frac{x\sin\alpha}{x^2 + 2x\cos\alpha + 1}](s) = \frac{\pi\,\sin(\alpha s)}{\sin(\pi s)}. \tag{6.14}$$

Problem 6.1. Establish the following Mellin transforms $(a > 0)$:

$$\mathcal{M}[Y(x - a)x^{-\mu}](s) = -\frac{a^{s-\mu}}{s - \mu}, \quad \Re s < \Re\mu,$$

$$\mathcal{M}[Y(a - x)x^{-\mu}](s) = \frac{a^{s-\mu}}{s - \mu}, \quad \Re s > \Re\mu.$$

Compare this with Example 6.7

Example 6.5. Let us compute the Mellin transform of the Bessel function $J_\nu(x)$. For doing so, we use the result

$$\int_0^\infty e^{-zx^2} x^\lambda\, dx = \frac{1}{2} z^{-\frac{\lambda+1}{2}} \Gamma(\frac{\lambda+1}{2}), \tag{6.15}$$

valid for $\Re\lambda > -1$ and $-\pi < \arg z < \pi$, and Schläfli's integral representation of the Bessel function

$$J_\nu(x) = \left(\frac{x}{2}\right)^\nu \frac{1}{2\pi i} \int_{-\infty}^{(0+)} e^{\zeta - x^2/(4\zeta)} \frac{d\zeta}{\zeta^{\nu+1}} \tag{6.16}$$

(see, for example [29, Olver p. 58]). Then,

$$\tilde{f}(s) = \int_0^\infty x^{s-1} J_\nu(x) \, dx$$

$$= \int_0^\infty \left(\frac{x}{2}\right)^\nu \frac{x^{s-1}}{2\pi i} \int_{-\infty}^{(0+)} e^{\zeta - x^2/(4\zeta)} \frac{d\zeta}{\zeta^{\nu+1}} \, dx$$

$$= \frac{1}{2\pi i \, 2^\nu} \int_{-\infty}^{(0+)} e^\zeta \, \zeta^{-\nu-1} \int_0^\infty x^{\nu+s-1} \, e^{-x^2/(4\zeta)} \, dx \, d\zeta,$$

where we have interchanged the order of integration. By using the above integral with $\lambda = \nu + s - 1$ and $z = 1/(4\zeta)$ we obtain

$$\tilde{f}(s) = 2^{s-1} \Gamma(\frac{\nu+s}{2}) \frac{1}{2\pi i} \int_{-\infty}^{(0+)} e^\zeta \, \zeta^{-\frac{\nu-s+2}{2}} \, d\zeta = 2^{s-1} \frac{\Gamma(\frac{\nu+s}{2})}{\Gamma(\frac{\nu-s+2}{2})},$$

where Hankel's loop integral of the Gamma function has been used. The condition $\Re\lambda > -1$ transforms into $\Re s > -\Re\nu$. Furthermore $\Re(\nu - s + 2)$ must be positive. Finally we obtain

$$J_\nu(x) \;\circ\!\!\!-\!\!\!\stackrel{\mathcal{M}}{\bullet}\; 2^{s-1} \frac{\Gamma(\frac{\nu+s}{2})}{\Gamma(\frac{\nu-s+2}{2})}, \qquad -\Re\nu < \Re s < \Re\nu + 2. \tag{6.17}$$

6.2 Mellin Transforms of Hyperfunctions

Having defined the Laplace transform of hyperfunctions, we might use (6.2) for defining their Mellin transforms. This is feasible because if $f(x) = [F_+(z), F_-(z)]$ is a hyperfunction defined on the positive part of the real axis, then

$$f(e^{-t}) := [-F_-(e^{-z}), -F_+(e^{-z})] \tag{6.18}$$

is a well-defined hyperfunction (see Sections 1.3.1 and 1.3.2). Notice that from

$$z = x + iy, \quad e^{-z} = e^{-x} \{\cos x - i \sin y\}$$

we have that

$$0 < \Im z < \pi \Longrightarrow \Im e^{-z} < 0, \quad -\pi < \Im z < 0 \Longrightarrow \Im e^{-z} > 0,$$

which is the reason for the swap of the upper and lower component in (6.18).

Definition 6.1. Let $f(x) = [F_+(z), F_-(z)]$ be a hyperfunction defined on the positive part of the real axis (the behavior of $f(x)$ on the negative part does not matter). The *Mellin transform of $f(x)$* is defined and denoted by

$$\tilde{f}(s) = \mathcal{M}[f(x)](s) := \mathcal{L}[f(e^{-t})](s) \tag{6.19}$$

where the hyperfunction $f(e^{-t})$ is given by (6.18).

Example 6.6. Because $\phi(t) = e^{-t}$ is a decreasing and strictly positive function, we have $\delta(e^{-t}) = 0$, and therefore,

$$\mathcal{M}[\delta(x)](s) = 0. \tag{6.20}$$

Consider now $\delta(x - a)$ with $a > 0$. Then, the real analytic function $\phi(t) = e^{-t} - a$ has exactly one zero $t_0 = -\log a$. By using

$$\delta(\phi(t)) = \frac{1}{|\phi'(t_0)|} \delta(t - t_0),$$

we obtain $\delta(e^{-t} - a) = (1/a)\,\delta(t + \log a)$ whose Laplace transform is $\exp(\log a\, s)/a$, i.e., a^{s-1}. Thus,

$$\mathcal{M}[\delta(x - a)](s) = a^{s-1} \quad (a > 0). \tag{6.21}$$

If $a \leq 0$ we have $\mathcal{M}[\delta(x - a)](s) = 0$. As a consequence we have for $T > 0$,

$$\mathcal{M}\left[\sum_{k=-\infty}^{\infty} \delta(x - kT)\right](s) = \mathcal{M}\left[\sum_{k=1}^{\infty} \delta(x - kT)\right](s) = \sum_{k=1}^{\infty} (kT)^{s-1}$$

$$= T^{s-1} \sum_{k=1}^{\infty} k^{s-1} = T^{s-1} \sum_{k=1}^{\infty} \frac{1}{k^{1-s}} \tag{6.22}$$

$$= T^{s-1}\, \zeta(1 - s),$$

where $\zeta(z)$ denotes Riemann's zeta function which is defined for $\Re s > 1$, thus, s must be restricted to $\Re s < 0$ (see also Section 6.6.4).

Consider now $\delta'(x - a), a > 0$. By using (1.27)

$$\delta'(\phi(t)) = \frac{1}{\phi'(t_0)|\phi'(t_0)|} \left\{ \delta'(t - t_0) + \frac{\phi''(t_0)}{\phi'(t_0)} \delta(t - t_0) \right\}$$

with $\phi(t) = e^{-t} - a$ and $t_0 = -\log a$, we obtain

$$\delta'(e^{-t} - a) = \frac{1}{-a^2} \left\{ \delta'(t + \log a) - \delta(t + \log a) \right\}$$

whose Laplace transform is $(1 - s)\, a^{s-2}$. Thus,

$$\mathcal{M}[\delta'(x - a)](s) = (1 - s)\, a^{s-2} \quad (a > 0), \tag{6.23}$$

and, if $a \leq 0$, $\mathcal{M}[\delta'(x - a)](s) = 0$.

Example 6.7. Let us investigate the Mellin transform of the hyperfunction $f(x) = x^n\, u(x - a)$ with $a > 0$, $n \in \mathbb{N}$. Of course, we can compute this Mellin transform by taking the ordinary function $x \to x^n\, Y(x - a)$ and using (6.1). But it is instructive to do it now by using our definition of the Mellin transform of a hyperfunction. By Example 1.16 we know that

$$u(e^{-t} - a) = u(-(t + \log a)),$$

thus, multiplying both sides with the real analytic function e^{-nt} yields

$$(e^{-t})^n\, u(e^{-t} - a) = e^{-nt}\, u(-(t + \log a)) = \left[\frac{e^{-nz}}{2\pi i} \log(z + \log a) \right].$$

The Laplace transform of this hyperfunction reduces to the classical Laplace integral, i.e.,

$$\mathcal{M}[x^n u(x-a)](s) = \mathcal{L}[(e^{-t})^n\, u(e^{-t}-a)](s)$$

$$= \int_{-\infty}^{-\log a} e^{-st}\, e^{-nt}\, dt = -\frac{a^{n+s}}{n+s},$$

provided that $\Re(n+s) < 0$. Finally, for $a > 0$, we find

$$\mathcal{M}[x^n u(x-a)](s) = -\frac{a^{n+s}}{n+s}, \quad \Re s < -n. \tag{6.24}$$

By using $\chi_{(a,b)} = u(x-a) - u(x-b)$, $0 < a < b$, we obtain

$$\mathcal{M}[\chi_{(a,b)} x^n](s) = \frac{1}{n+s}\{b^{n+s} - a^{n+s}\} \tag{6.25}$$

with $\Re s < -n$, and, for $n = 0$,

$$\mathcal{M}[\chi_{(a,b)}](s) = \frac{1}{s}\{b^s - a^s\} \tag{6.26}$$

with $\Re s < 0$.

Problem 6.2. Show that

$$\mathcal{M}[x^n u(a-x)](s) = \frac{a^{n+s}}{n+s}, \quad \Re s > -n. \tag{6.27}$$

6.3 Operational Properties

6.3.1 Linearity

It is clear that the Mellin transformation behaves as a linear operator, provided that the intersection of the strips of convergence of the involved hyperfunctions is not empty. So if $\mathcal{M}[f_k(x)](s) = \tilde{f}_k(s)$ for $\alpha_k < \Re s < \beta_k$, then

$$\mathcal{M}[\sum_{k=1}^{n} c_k\, f_k(x)](s) = \sum_{k=1}^{n} c_k\, \mathcal{M}[f_k(x)](s), \tag{6.28}$$

provided $\alpha < \Re s < \beta$, with $\alpha := \max\{\alpha_1, \ldots, \alpha_n\}$, and $\beta := \min\{\beta_1, \ldots, \beta_n\}$.

6.3.2 Scale Changes

Let the hyperfunction $f(x)$ have the Mellin transform $\tilde{f}(s)$ for $\alpha < \Re s < \beta$. We are seeking the Mellin transform of $f(ax)$, $a > 0$. By using the linear substitution rule for the Laplace transformation we obtain

$$\mathcal{M}[f(ax)](s) = \mathcal{L}[f(ae^{-t})](s) = \mathcal{L}[f(e^{-(t-\log a)})](s)$$

$$= e^{-s\log a}\, \mathcal{L}[f(e^{-t})](s) = a^{-s}\, \mathcal{M}[f(x)](s).$$

The strip of convergence remains unchanged. Thus, we obtain

Proposition 6.1. *Let the hyperfunction* $f(x) = [F_-(z), F_+(z)]$ *be defined for* $x > 0$ *with* $\mathcal{M}[f(x)](s) = \tilde{f}(s)$ *for* $\alpha < \Re s < \beta$, *then, with* $a > 0$, *we have*

$$\mathcal{M}[f(a\,x)](s) = a^{-s}\,\mathcal{M}[f(x)](s) = a^{-s}\,\tilde{f}(s). \quad (6.29)$$

For example,

$$\mathcal{M}[e^{-ax}](s) = a^{-s}\,\Gamma(s), \quad \Re s > 0. \quad (6.30)$$

What is the Mellin transform of $f(x^a)$ knowing the Mellin transform of $f(x)$ with the strip of convergence $\alpha < \Re s < \beta$? For $a \neq 0$, we have by using again the linear substitution rule for the Laplace transformation,

$$\mathcal{M}[f(x^a)](s) = \mathcal{L}[f(e^{-at})](s) = \frac{1}{|a|}\,\mathcal{L}[f(e^{-t})](\frac{s}{a})$$

$$= \frac{1}{|a|}\,\mathcal{M}[f(x)](\frac{s}{a}),$$

with $a\,\alpha < \Re s < a\,\beta$, and $a\,\beta < \Re s < a\,\alpha$, according to $a > 0$ or $a < 0$, respectively. Therefore,

Proposition 6.2. *Let the hyperfunction* $f(x) = [F_-(z), F_+(z)]$ *be defined for* $x > 0$ *with* $\mathcal{M}[f(x)](s) = \tilde{f}(s)$ *for* $\alpha < \Re s < \beta$, *then, with* $a \neq 0$ *real,*

$$\mathcal{M}[f(x^a)](s) = \frac{1}{|a|}\,\mathcal{M}[f(x)](\frac{s}{a}) = \frac{1}{|a|}\,\tilde{f}(\frac{s}{a}), \quad (6.31)$$

with $a\,\alpha < \Re s < a\,\beta$, *and* $a\,\beta < \Re s < a\,\alpha$, *according to* $a > 0$ *or* $a < 0$, *respectively.*

For example, with $a > 0$ and $b \neq 0$ real,

$$\mathcal{M}[e^{-ax^b}](s) = \frac{a^{-s/b}}{|b|}\,\Gamma(\frac{s}{b}). \quad (6.32)$$

Problem 6.3. By using (6.13), (6.14) establish for $0 < |\alpha| < \beta$ and for $0 < \Re s < \pi/\beta$,

$$\mathcal{M}^{-1}[\frac{\cos(\alpha s)}{\sin(\beta s)}](x) = \frac{1}{\beta}\,\frac{1 + x^{\pi/\beta}\,\cos(\alpha\pi/\beta)}{x^{2\beta/\pi} + 2\,x^{\pi/\beta}\,\cos(\alpha\pi/\beta) + 1}, \quad (6.33)$$

$$\mathcal{M}^{-1}[\frac{\sin(\alpha s)}{\sin(\beta s)}](x) = \frac{1}{\beta}\,\frac{x^{\pi/\beta}\,\sin(\alpha\pi/\beta)}{x^{2\pi/\beta} + 2\,x^{\pi/\beta}\,\cos(\alpha\pi/\beta) + 1}. \quad (6.34)$$

6.3.3   Multiplication by $(\log x)^n$

Because the hyperfunction $f(x) = [F_-(z), F_+(z)]$ is defined for $x > 0$, and only the behavior for $x > 0$ matters, we can multiply it with the function $(\log x)^n$ which is real analytic for $x > 0$ and obtain $(\log x)^n\,f(x) = [(\log z)^n\,F_-(z), (\log z)^n\,F_+(z)]$. If now $\mathcal{M}[f(x)](s) = \tilde{f}(s)$ for $\alpha < \Re s < \beta$, and $n \in \mathbb{N}$, then, by using the Multiplication Rule for Laplace transformation, we obtain

$$\mathcal{M}[(\log x)^n\,f(x)](s) = \mathcal{L}[(\log(e^{-t}))^n\,f(e^{-t})](s) = \mathcal{L}[(-t)^n\,f(e^{-t})](s)$$

$$= \frac{d^n}{ds^n}\mathcal{L}[f(e^{-t})](s) = \frac{d^n}{ds^n}\mathcal{M}[f(x)](s)$$

where the strip of convergence remains unchanged. Thus,

Proposition 6.3. *Let the hyperfunction $f(x) = [F_-(z), F_+(z)]$ be defined for $x > 0$ and suppose that $\mathcal{M}[f(x)](s) = \tilde{f}(s)$ for $\alpha < \Re s < \beta$, then*

$$\mathcal{M}[(\log x)^n f(x)](s) = \frac{d^n}{ds^n}\mathcal{M}[f(x)](s) = \tilde{f}^{(n)}(s). \tag{6.35}$$

We may also call this rule the Image Derivation Rule.

Example 6.8. For $a > 0$ we have,

$$\mathcal{M}[(\log x)^n \delta(x - a)](s) = \frac{d^n}{ds^n}a^{s-1} = (\log a)^n a^{s-1}. \tag{6.36}$$

From Example 6.1 we obtain

$$\mathcal{M}[(\log x)^n e^{-x}](s) = \frac{d^n}{ds^n}\Gamma(s), \quad (0 < \Re s). \tag{6.37}$$

6.3.4 Multiplication by x^μ, $\mu \in \mathbb{C}$

Again, if the hyperfunction $f(x) = [F_+(z), F_-(z)]$ is defined for $x > 0$ and only its behavior for $x > 0$ matters, we can multiply it by the function x^μ which is real analytic for $x > 0$ and obtain $x^\mu f(x) = [z^\mu F_+(z), z^\mu F_-(z)]$. If now $\mathcal{M}[f(x)](s) = \tilde{f}(s)$ for $\alpha < \Re s < \beta$, then, by using the Image Translation Rule for the Laplace transformation, we obtain

$$\mathcal{M}[x^\mu f(x)](s) = \mathcal{L}[e^{-\mu t} f(e^{-t})](s)$$
$$= \mathcal{L}[f(e^{-t})](s + \mu) = \mathcal{M}[f(x)](s + \mu).$$

The strip of convergence is now translated and becomes $\alpha - \Re\mu < \Re s < \beta - \Re\mu$.

Proposition 6.4. *Let the hyperfunction $f(x) = [F_+(z), F_-(z)]$ be defined for $x > 0$ with $\mathcal{M}[f(x)](s) = \tilde{f}(s)$ for $\alpha < \Re s < \beta$, then for $\alpha - \Re\mu < \Re s < \beta - \Re\mu$ we have*

$$\mathcal{M}[x^\mu f(x)](s) = \mathcal{M}[f(x)](s + \mu) = \tilde{f}(s + \mu). \tag{6.38}$$

We may also call this rule the Image Translation Rule.

Example 6.9. For $a > 0$ we have,

$$\mathcal{M}[x^\mu \delta(x - a)](s) = a^{s+\mu-1}. \tag{6.39}$$

From Example 6.1 we obtain

$$\mathcal{M}[x^\mu e^{-x}](s) = \Gamma(s + \mu), \quad (-\Re\mu < \Re s). \tag{6.40}$$

6.3.5 Reflection

Let $\mathcal{M}[f(x)](s) = \tilde{f}(s)$ with $\alpha < \Re s < \beta$. From the Multiplication Rule with $\mu = 1$ we have then

$$\mathcal{M}[x f(x)](s) = \mathcal{M}[f(x)](s + 1), \quad \alpha - 1 < \Re s < \beta - 1,$$

and from Proposition 6.2 with $a = -1$,

$$\mathcal{M}[x^{-1} f(x^{-1})](s) = \frac{1}{|-1|}\mathcal{M}[f(x)](\frac{s}{-1} + 1)$$

with $-(\alpha - 1) > \Re s > -(\beta - 1)$. This gives

Proposition 6.5. *If the hyperfunction $f(x)$ has the Mellin transform $\tilde{f}(s)$ with the strip of convergence $\alpha < \Re s < \beta$, then*

$$\mathcal{M}[\frac{1}{x} f(\frac{1}{x})](s) = \tilde{f}(1 - s) \tag{6.41}$$

with $1 - \beta < \Re s < 1 - \alpha$.

Together with the Multiplication Rule we obtain

$$\mathcal{M}[f(\frac{1}{x})](s) = \tilde{f}(-s) \tag{6.42}$$

with $-\beta < \Re s < -\alpha$.

Example 6.10.

$$\mathcal{M}[\frac{e^{-a/x}}{x}](s) = a^{s-1}\, \Gamma(1 - s), \quad \Re s < 1, \tag{6.43}$$

$$\mathcal{M}[\chi_{(a,b)}\frac{1}{x^n}](s) = \frac{b^{n-s} - a^{n-s}}{n - s} \tag{6.44}$$

with $n < \Re s$, $n \in \mathbb{N}$, $0 < a < b$.

6.3.6 Differentiation Rules

Let $f(x) = [F_+(z), F_-(z)]$ and $g(x) := f(e^{-x}) = [-F_-(e^{-z}), -F_+(e^{-z})]$. Then,

$$Dg(x) = [-F'_-(e^{-z})e^{-z}(-1), -F'_+(e^{-z})e^{-z}(-1)]$$
$$= -e^{-x}[-F'_-(e^{-z}), -F'_+(e^{-z})] = -e^{-x} f'(e^{-x}),$$

thus, by using the Differentiation Rule of the Laplace transformation,

$$\mathcal{M}[f'(x)](s) = \mathcal{L}[f'(e^{-x})](s) = \mathcal{L}[-e^x\, Df(e^{-x})](s)$$
$$= -\,\mathcal{L}[Df(e^{-x})](s)\big|_{s \to s-1} = -s\,\, \mathcal{L}[f(e^{-x})](s)\big|_{s \to s-1}$$
$$= -(s - 1)\,\mathcal{M}[f(x)](s - 1).$$

We have proved

Proposition 6.6. *If $\mathcal{M}[f(x)](s) = \tilde{f}(s)$, for $\alpha < \Re s < \beta$, then*

$$\mathcal{M}[f'(x)](s) = -(s - 1)\,\tilde{f}(s - 1) \tag{6.45}$$

for $\alpha + 1 < \Re s < \beta + 1$.

By iteration on n, we obtain

$$\mathcal{M}[D^n f(x)](s) = (-1)^n (s - 1)(s - 2)\ldots(s - n)\,\tilde{f}(s - n)$$
$$= (1 - s)_n\, \tilde{f}(s - n) \tag{6.46}$$

with $\alpha + n < \Re s < \beta + n$, and where Pochhammer's symbol $(a)_n := a(a + 1)\ldots (a + n - 1)$ has been used.

By using (6.21) we obtain, for $a > 0$,

$$\mathcal{M}[\delta^{(n)}(x - a)](s) = (-1)^n (s - 1)(s - 2)\ldots(s - n)\, a^{s-n-1}. \tag{6.47}$$

By the Multiplication Rule and the Differentiation Rule we obtain

$$\mathcal{M}[x\,f'(x)](s) = -s\,\mathcal{M}[f(x)](s),$$

and the strip of convergence remains the same. By iteration on n we get

$$\mathcal{M}[(x\,Df(x))^n](s) = (-s)^n\,\mathcal{M}[f(x)](s). \tag{6.48}$$

For $n = 2$ this gives

$$\mathcal{M}[x^2\,D^2 f(x) + x Df(x)](s) = s^2\,\mathcal{M}[f(x)](s). \tag{6.49}$$

By using the Multiplication Rule with $\mu = -2$, and assuming $\mathcal{M}[f(x)](s) = \tilde{f}(s)$ with $\alpha < \Re s < \beta$, we obtain

$$\mathcal{M}[D^2 f(x) + \frac{1}{x}\,Df(x)](s) = (s-2)^2\,\mathcal{M}[f(x)](s-2) \tag{6.50}$$

with now $\alpha - 2 < \Re s < \beta - 2$.

Problem 6.4. Show for $a > 0$, $n = 0, 1, 2, \ldots$ that (6.47) can be written as

$$\mathcal{M}[\delta^{(n)}(x - a)](s) = \frac{a^{-(1-s+n)}\,\Gamma(1-s+n)}{\Gamma(1-s)}. \tag{6.51}$$

6.3.7 Integration Rules

Let $f(x) = [F_+(z), F_-(z)]$ be a hyperfunction having the Mellin transform $\tilde{f}(s)$ in the strip of convergence $\alpha < \Re s < \beta$. In Section 3.5.6 we introduced the primitive or anti-derivative of the hyperfunction $f(x)$,

$$_0 D^{-1} f(x) := \left[\int_0^z F_+(\zeta)\,d\zeta, \int_0^z F_-(\zeta)\,d\zeta\right].$$

Let us find its Mellin transform. We have

$$\mathcal{M}[_0 D^{-1} f(x)](s) = \mathcal{L}[_0 D^{-1} f(e^{-t})](s)$$

$$= \mathcal{L}\left[-\int_0^{e^{-z}} F_-(\zeta)\,d\zeta, -\int_0^{e^{-z}} F_+(\zeta)\,d\zeta\right](s).$$

By the change of variables $\zeta = \exp(-\tau)$, the last expression on the right-hand side transforms into

$$\mathcal{L}\left[-\int_\infty^z F_-(e^{-\tau})\,e^{-\tau}\,(-d\tau), -\int_\infty^z F_+(e^{-\tau})\,e^{-\tau}\,(-d\tau)\right](s)$$

$$= \mathcal{L}\left[-\int_z^\infty F_-(e^{-\tau})\,e^{-\tau}\,d\tau, -\int_z^\infty F_+(e^{-\tau})\,e^{-\tau}\,d\tau\right](s)$$

$$= \mathcal{L}[_\infty D^{-1}(e^{-t}\,f(e^{-t})](s).$$

By using the Integration Rule 3.122 of the Laplace transformation, we obtain

$$\mathcal{L}[_\infty D^{-1}(e^{-t}\,f(e^{-t}))](s) = -\frac{1}{s}\,\mathcal{L}[e^{-t}\,f(e^{-t})](s),$$

provided that $\alpha < \Re s < \min[\beta, 0]$ has a positive width, i.e., $\alpha < 0$. For the last expression we obtain

$$-\frac{1}{s}\,\mathcal{M}[x\,f(x)](s) = -\frac{1}{s}\,\mathcal{M}[f(x)](s+1)$$

with $\alpha - 1 < \Re s < \min[\beta, 0] - 1$. We have proved

Proposition 6.7. *Let $f(x) = [F_+(z), F_-(z)]$ be a hyperfunction having the Mellin transform $\tilde{f}(s)$ in the strip of convergence $\alpha < \Re s < \beta$ with $\alpha < 0$, then*

$$\mathcal{M}[_0 D^{-1} f(x)](s) = -\frac{1}{s}\,\tilde{f}(s+1) \tag{6.52}$$

for $\alpha - 1 < \Re s < \min[\beta - 1, -1]$.

Similarly, one might prove (details are left to the reader)

Proposition 6.8. *Let $f(x) = [F_+(z), F_-(z)]$ be a hyperfunction having the Mellin transform $\tilde{f}(s)$ in the strip of convergence $1 < \alpha < \Re s < \beta$, then*

$$\mathcal{M}[_\infty D^{-1} f(x)](s) = \mathcal{M}[-\int_x^\infty f(\xi)\, d\xi](s) = -\frac{1}{s}\,\tilde{f}(s+1) \tag{6.53}$$

for $\alpha - 1 < \Re s < \beta - 1$.

Example 6.11. If $f(x) = \delta(x - a)$, $a > 0$, then $\tilde{f}(s) = a^{s-1}$ with $\alpha = -\infty < \Re s < \infty = \beta$. Then,

$$\mathcal{M}[_0 D^{-1}\delta(x-a)](s) = -\frac{a^s}{s}$$

with $-\infty < \Re s < -1$. By analytic continuation the strip of convergence can be extended to $-\infty < \Re s < 0$.

Example 6.12. We have

$$\mathcal{M}[e^{-x^2}](s) = \frac{1}{2}\Gamma(\frac{s}{2})$$

with $\alpha = 0 < \Re s < \infty = \beta$. Thus,

$$\mathcal{M}[_\infty D^{-1} e^{-x^2}](s) = -\frac{1}{2\,s}\Gamma(\frac{s+1}{2}).$$

The Mellin transform of the complementary Error Function becomes then

$$\mathcal{M}[\text{erfc}(x)](s) = \mathcal{M}[\frac{2}{\sqrt{\pi}}\int_x^\infty e^{-t^2}\, dt](s)$$

$$= -\frac{2}{\sqrt{\pi}}\,\mathcal{M}[_\infty D^{-1} e^{-x^2}](s) = \frac{1}{\sqrt{\pi}\,s}\Gamma(\frac{s+1}{2}), \tag{6.54}$$

with $0 < \Re s < \infty$.

6.4 Inverse Mellin Transformation

We have defined the Mellin transform $\mathcal{M}[f(x)](s)$ of a hyperfunction $f(x) = [F_+(z), F_-(z)]$ on the positive part of the real axis to be the Laplace transform

$$\mathcal{L}[f(e^{-t})](s) = \tilde{f}(s),$$

provided the Laplace integral converges in a vertical strip $\alpha < \Re s < \beta$ of the complex s-plane. If the Laplace integral converges absolutely in the vertical strip, and the ordinary function $g(t) = f(e^{-t})$ is locally of bounded variation and assumed to be normalized, then the complex inversion formula

$$g(t) = f(e^{-t}) = \frac{1}{2\pi i} \, \mathrm{pv} \int_{c-i\infty}^{c+i\infty} e^{st} \, \tilde{f}(s) \, ds$$

holds. "Normalized" means that at points t where $g(t)$ has a discontinuity of the first kind, it is redefined to be $g(t) = \{g(t+0) + g(t-0)\}/2$.

The case where $g(t) = [G_+(z), G_-(z)]$ is a hyperfunction is now covered by putting

$$g(t) = \left[\frac{1}{2\pi i} \int_c^{c+i\infty} e^{sz} \, \tilde{f}(s) \, ds, -\frac{1}{2\pi i} \int_{c-i\infty}^c e^{sz} \, \tilde{f}(s) \, ds \right].$$

In the first integral we have $\Im z > 0$, and in the second one $\Im z < 0$. This ensures that the integrals converge absolutely as long as $\tilde{f}(s)$ does not increase more rapidly than any polynomial in s, possibly multiplied by a term $\exp(\lambda s)$ with a real λ. If now $f(x) = [F_+(z), F_-(z)]$ then

$$g(t) = f(e^{-t}) = \left[-F_-(e^{-z}), -F_+(e^{-z}) \right] = [G_+(z), G_-(z)],$$

where the range of the variable z is restricted to $-\pi < \Im z < \pi$. But $g(t) = [G_+(z), G_-(z)]$ implies then

$$f(x) = [-G_-(-\log z), -G_+(-\log z)] = [F_+(z), F_-(z)],$$

where $\log z$ denotes the principal branch. Therefore,

$$F_+(z) = -G_-(-\log z) = \frac{1}{2\pi i} \int_{c-i\infty}^c e^{s(-\log z)} \, \tilde{f}(s) \, ds$$

$$= \frac{1}{2\pi i} \int_{c-i\infty}^c z^{-s} \, \tilde{f}(s) \, ds,$$

and

$$F_-(z) = -G_+(-\log z) = -\frac{1}{2\pi i} \int_c^{c+i\infty} e^{s(-\log z)} \, \tilde{f}(s) \, ds$$

$$= -\frac{1}{2\pi i} \int_c^{c+i\infty} z^{-s} \, \tilde{f}(s) \, ds.$$

We have established

Proposition 6.9. *If $\tilde{f}(s)$ is holomorphic in the vertical strip $\alpha < \Re s < \beta$ and satisfies within it the estimate $f(s) = O(s^n \exp(\lambda s))$, $n \in \mathbb{N}$, $\lambda \in \mathbb{R}$ as $|\Im s| \to \infty$, the inverse Mellin transform yielding the hyperfunction $f(x) = [F_+(z), F_-(z)]$ is given by*

$$F_+(z) = \frac{1}{2\pi i} \int_{c-i\infty}^{c} z^{-s} \tilde{f}(s)\, ds, \quad \Im z > 0,$$

$$F_-(z) = -\frac{1}{2\pi i} \int_{c}^{c+i\infty} z^{-s} \tilde{f}(s)\, ds, \quad \Im z < 0,$$

(6.55)

with $\alpha < c < \beta$.

Example 6.13. Take $\tilde{f}(s) = a^{s-1}$, $a > 0$ (the mentioned λ in the above proposition is $\log a$). Let $\Im z > 0$, thus $0 < \arg z < \pi$. We can take $c = 0$.

$$F_+(z) = \frac{1}{2\pi i} \int_{-i\infty}^{0} z^{-s} a^{s-1}\, ds = \frac{1}{2\pi a} \int_{-\infty}^{0} \left(\frac{a}{z}\right)^{i\eta} d\eta$$

$$= \frac{1}{2\pi i\, a\, \log(\frac{a}{z})}$$

since $0 < \arg z$. Similarly, for $\Im z < 0$, thus $-\pi < \arg z < 0$, we find

$$F_-(z) = -\frac{1}{2\pi i} \int_{0}^{i\infty} z^{-s} a^{s-1}\, ds = \frac{1}{2\pi i\, a\, \log(\frac{a}{z})}$$

because $\arg z < 0$. Therefore, we obtain

$$f(x) = [F(z)] = \left[\frac{1}{2\pi i\, a\, \log(\frac{a}{z})}\right].$$

For $x < 0$ we have

$$F(x + i0) - F(x - i0) = \frac{1}{a\,(\pi^2 + (\log(\frac{|x|}{a}))^2)}.$$

On the range $(0, a) \cup (a, \infty)$, $F(z)$ is real analytic, thus $F(x+i0) - F(x-i0) = 0$. Near $z = a$, we have the expansion

$$F(z) = -\frac{1}{2\pi i\, a\, \log(\frac{z}{a})} = -\frac{1}{2\pi i\, a\, \log(1 + \frac{z-a}{a})}$$

$$= -\frac{1}{2\pi i\, a} \left\{ \frac{1}{\frac{z-a}{a} - \frac{(z-a)^2}{2a^2} + - \cdots} \right\}$$

$$= -\frac{1}{2\pi i\,(z-a)} \{1 + (z-a)/2 - + \cdots\}.$$

This finally shows that

$$f(x) = \frac{1}{a\,(\pi^2 + (\log(\frac{|x|}{a}))^2)}\, u(-x) + \delta(x - a).$$

Since only the restriction on the positive part of the real axis matters we arrive at

$$\mathcal{M}^{-1}[a^{s-1}](x) = \delta(x - a), \quad a > 0,$$

(6.56)

as is to be expected.

6.5 $\mathcal{M}$-Convolutions

The $\mathcal{M}$-convolution of two ordinary functions $f(x)$ and $g(x)$ is defined and denoted by

$$f(x) * g(x) = \int_0^\infty \frac{1}{t} f\left(\frac{x}{t}\right) g(t)\, dt.$$

We shall now carry over this property to hyperfunctions and assume that $f(x) = [F_+(z), F_-(z)]$ and $g((x) = [G_+(z), G_-(z)]$ are two given hyperfunctions defined on the positive part of the real line. For $\Im z > 0$ fixed and $x > 0$, the function $F_+(z/x)/x$ is real analytic in x. Likewise the same holds for $\Im z < 0$ fixed and $F_-(z/x)/x$. Therefore, we obtain two well-defined hyperfunctions of the variable $x \in \mathbb{R}_+$,

$$\frac{1}{x} F_+\left(\frac{z}{x}\right) g(x) \text{ for } \Im z > 0, \quad \frac{1}{x} F_-\left(\frac{z}{x}\right) g(x) \text{ for } \Im z < 0$$

whose integrals

$$H_+(z) := \int_0^\infty \frac{1}{x} F_+\left(\frac{z}{x}\right) g(x)\, dx, \quad \Im z > 0, \tag{6.57}$$

$$H_-(z) := \int_0^\infty \frac{1}{x} F_-\left(\frac{z}{x}\right) g(x)\, dx, \quad \Im z < 0 \tag{6.58}$$

are defined as integrals over hyperfunctions, provided they are convergent.

Definition 6.2. We call the hyperfunction $h(x) := [H_+(z), H_-(z)]$ the *Mellin *convolution* of $f(x)$ and $g(x)$, denoted by $h(t) = f(x) * g(x)$.

Example 6.14. With $f(x) = \delta(x - a)$, $a > 0$, we obtain for $x > 0$,

$$\delta(x - a) * g(x) = \left[-\frac{1}{2\pi i} \int_0^\infty \frac{g(x)}{x\left(\frac{z}{x} - a\right)}\, dx \right]$$

$$= \left[\frac{1}{2\pi i a} \int_0^\infty \frac{g(x/a)}{x - z}\, dx \right] = \frac{1}{a} \chi_{(0,\infty)} g(x/a).$$

Hence, because only the behavior for $x > 0$ matters, we write

$$\delta(x - a) * g(x) = \frac{1}{a} g\left(\frac{x}{a}\right). \tag{6.59}$$

Especially, for $a = 1$, and $g(x) = \delta(x - b)$, $b > 0$,

$$\delta(x - 1) * g(x) = g(x), \quad \delta(x - a) * \delta(x - b) = \delta(x - ab),$$

respectively.

Problem 6.5. For $a > 0$ and for $x > 0$ establish

$$\delta^{(n)}(x - a) * g(x) = \frac{1}{a^{n+1}} D^n\{x^n g\left(\frac{x}{a}\right)\}. \tag{6.60}$$

Suppose now that the hyperfunction $h(x) = f(x) * g(x)$ exists and that $f(x)$ and $g(x)$ have Mellin transforms $\hat{f}(s)$ and $\tilde{g}(s)$. Let us compute the Mellin transform of $h(x)$. We have

$$\mathcal{M}[h(x)](s) = \mathcal{L}[h(e^{-t}](s) = \mathcal{L}[[-H_-(e^{-z}), -H_+(e^{-z})]](s)$$

$$= \mathcal{L}\left[\left[-\int_0^\infty \frac{1}{x} F_-\left(\frac{e^{-z}}{x}\right) g(x)\, dx, -\int_0^\infty \frac{1}{x} F_+\left(\frac{e^{-z}}{x}\right) g(x)\, dx \right]\right](s).$$

A change of variables $x = e^{-t}$ in the integrals yields

$$\mathcal{M}[h(x)](s) = \mathcal{L}\Big[\Big[-\int_{-\infty}^{\infty} F_-(e^{-(z-t)}g(e^{-t})\,dt, -\int_{-\infty}^{\infty} F_+(e^{-(z-t)}g(e^{-t})\,dt\Big]\Big](s)$$
$$= \mathcal{L}[f(e^{-t}) \star g(e^{-t})](s) = \mathcal{L}[f(e^{-t})](s)\,\mathcal{L}[g(e^{-t})](s)$$
$$= \mathcal{M}[f(x)](s)\,\mathcal{M}[g(x)](s).$$

Here we have used the Convolution Rule (Proposition 3.22) and the assumption that the hyperfunctions $f(e^{-t})$ and $g(e^{-t})$ have a non-void overlap of their strips of convergence. Therefore, we obtain the *Convolution Rule for the Mellin transformation*.

Proposition 6.10. *Assume that the hyperfunctions $f(x)$ and $g(x)$ are both defined on the positive part of the real axis and have Mellin transforms $\tilde{f}(s)$ and $\tilde{g}(s)$ with a non-void intersection $\alpha < \Re s < \beta$ of positive width of their strips of convergence. Then, the following convolution rule,*

$$\mathcal{M}[f(x) * g(x)](s) = \mathcal{M}[f(x)](s)\,\mathcal{M}[g(x)](s), \tag{6.61}$$

holds for $\alpha < \Re s < \beta$.

Example 6.15. From $\mathcal{M}[x^\mu \delta(x-a)](s) = a^{s+\mu-1}$ and $\mathcal{M}[x^\nu \delta(x-b)](s) = b^{s+\nu-1}$ we have, with $a > 0$, $b > 0$,

$$x^\mu \delta(x-a) * x^\nu \delta(x-b) \ \overset{\mathcal{M}}{\circ\!\!-\!\!\bullet}\ b^{\nu-\mu}\, x^\mu\, \delta(x-ab) = a^{\mu-\nu}\, x^\nu\, \delta(x-ab). \tag{6.62}$$

With $\mathcal{M}[f_1(x)](s) = \tilde{f}_1(s)$ and $\mathcal{M}[f_2(x)](s) = \tilde{f}_2(s)$ and the strips of converges $\alpha_1 < \Re s < \beta_1$ and $\alpha_2 < \Re s < \beta_2$, respectively, we obtain, by using the Convolution Rule and the Multiplication Rule,

$$f_1(x) * x^\mu\, f_2(x) \ \overset{\mathcal{M}}{\circ\!\!-\!\!\bullet}\ \tilde{f}_1(s)\tilde{f}_2(s+\mu), \tag{6.63}$$

where $\max[\alpha_1 - \Re\mu, \alpha_2] < \Re s < \min[\beta_1 - \Re\mu, \beta_2]$.

In the case of ordinary functions the expression on the left-hand side reads

$$\int_0^{\infty} t^{\mu-1}\, f_1\Big(\frac{x}{t}\Big)\, f_2(t)\, dt.$$

Replacing $f_2(x)$ by $f_2(1/x)/x$ and using the Reflection Property gives with $\mu = 0$,

$$\int_0^{\infty} t^{-1}\, f_1\Big(\frac{x}{t}\Big)\, f_2\Big(\frac{1}{t}\Big)\frac{1}{t}\, dt \ \overset{\mathcal{M}}{\circ\!\!-\!\!\bullet}\ \tilde{f}_1(s)\, \tilde{f}_2(1-s)$$

with $\max[\alpha_1, 1 - \beta_2] < \Re s < \min[\beta_1, 1 - \alpha_2]$. A change of variables $t = 1/\tau$ in the integral on the left-hand side implies

$$\int_0^{\infty} f_1(\tau x)\, f_2(\tau)\, d\tau \ \overset{\mathcal{M}}{\circ\!\!-\!\!\bullet}\ \tilde{f}_1(s)\, \tilde{f}_2(1-s), \tag{6.64}$$

with $\max[\alpha_1, 1 - \beta_2] < \Re s < \min[\beta_1, 1 - \alpha_2]$. This reads also as

$$\mathcal{M}^{-1}[\tilde{f}_1(s)\, \tilde{f}_2(1-s)](x) = \int_0^{\infty} f_1(\tau x)\, f_2(\tau)\, d\tau. \tag{6.65}$$

Example 6.16. By using (6.6) we get

$$M^{-1}[\Gamma(s)\,\tilde{f}(1-s)](x) = \int_0^\infty e^{-\tau x}\,f(\tau)\,d\tau = \mathcal{L}_+[f(t)](x) \qquad (6.66)$$

which shows that the one-sided Laplace transform of an ordinary function $f(t)$ can be seen as an inverse Mellin transform.

As in the opening of this section, we may define for hyperfunctions $f(x) = [F_+(z), F_-(z)]$ and $g(x) = [G_+(z), G_-(z)]$, both defined on the positive part of the real axis,

$$H_+(z) := \int_0^\infty F_+(zx)\,g(x)\,dx, \quad \Im z > 0, \qquad (6.67)$$

$$H_-(z) := \int_0^\infty F_-(zx)\,g(x)\,dx, \quad \Im z < 0. \qquad (6.68)$$

We call the hyperfunction $h(x) := [H_+(z), H_-(z)]$ the *Mellin* $\circ$*convolution of* $f(x)$ *and* $g(x)$ denoted by $h(x) = f(x) \circ g(x)$. The relation (6.64) becomes then the $\circ$convolution theorem:

$$f(x) \circ g(x) \; \circ\!\!\overset{\mathcal{M}}{-\!\!\bullet}\; \tilde{f}(s)\,\tilde{g}(1-s) \qquad (6.69)$$

with $\max[\alpha_1, 1-\beta_2] < \Re s < \min[\beta_1, 1-\alpha_2]$. Here $\alpha_1 < \Re s < \beta_1$ and $\alpha_2 < \Re s < \beta_2$ are the assumed strips of convergence of $\tilde{f}(s)$ and $\tilde{g}(s)$, respectively.

6.5.1 Reciprocal Integral Transforms

Definition 6.3. A pair of integral transforms with multiplicative kernels $k(xt)$ and $h(xt)$ such that

$$F(x) := \int_0^\infty k(xt)\,f(t)\,dt, \quad f(x) := \int_0^\infty h(xt)\,F(t)\,dt \qquad (6.70)$$

holds are said to be *reciprocal to each other.*

Let us assume that all involved functions have Mellin transforms in a common strip of convergence. Then we have by (6.64)

$$\tilde{F}(s) = \tilde{k}(s)\,\tilde{f}(1-s), \qquad (6.71)$$

and

$$\tilde{f}(s) = \tilde{h}(s)\,\tilde{F}(1-s).$$

Replacing in the last relation s by $1 - s$, gives

$$\tilde{f}(1-s) = \tilde{h}(1-s)\,\tilde{F}(s). \qquad (6.72)$$

From (6.71) and (6.72) we obtain the relation

$$\tilde{k}(s)\,\tilde{h}(1-s) = 1. \qquad (6.73)$$

Thus, we have established

Proposition 6.11. *If all involved functions in (6.70) have Mellin transforms having a common strip of convergence, and if $\tilde{k}(s)$ and $\tilde{h}(s)$ are the Mellin transform of the kernel functions $k(x)$ and $h(x)$, respectively, then (6.73) holds.*

Example 6.17. In Chapter 7 we shall study Hankel transforms. Let us show that the conventional Hankel transform and the Hankel-Doetsch transform are self-reciprocal, i.e.,

$$F(y) = \int_0^\infty x\, J_\nu(yx) f(x)\, dx, \quad f(x) = \int_0^\infty y\, J_\nu(yx) F(y)\, dy,$$

$$G(y) = \int_0^\infty J_\nu(2\sqrt{yx}) g(x)\, dx, \quad g(x) = \int_0^\infty J_\nu(2\sqrt{yx}) G(y)\, dy.$$

We prove the second relation and leave the first to verification by the reader. By using (6.17) we have, by the Scale Change Property,

$$k(x) = h(x) = J_\nu(\sqrt{2x}) \overset{\mathcal{M}}{\circ\!\!-\!\!\bullet} \frac{\Gamma(\frac{\nu}{2}+s)}{\Gamma(\frac{\nu}{2}-s+1)}), \quad -\frac{\Re\nu}{2} < \Re s < \frac{\Re\nu}{2} + 1.$$

Thus,

$$\tilde{k}(s) = \frac{\Gamma(\frac{\nu}{2}+s)}{\Gamma(\frac{\nu}{2}-s+1)}), \quad -\frac{\Re\nu}{2} < \Re s < \frac{\Re\nu}{2} + 1,$$

$$\tilde{h}(1-s) = \tilde{k}(1-s) = \frac{\Gamma(\frac{\nu}{2}+1-s)}{\Gamma(\frac{\nu}{2}+s)}), \quad -\frac{\Re\nu}{2} < \Re s < \frac{\Re\nu}{2} + 1$$

which shows that $\tilde{k}(s)\,\tilde{k}(1-s) = 1$.

Problem 6.6. Show that the Fourier-Cosine transform and its inverse transform form a reciprocal pair

$$F(x) = \int_0^\infty \cos(xt)\, f(t)\, dt, \quad f(x) = \frac{2}{\pi} \int_0^\infty \cos(xt)\, F(t)\, dt.$$

6.5.2 Transform of a Product and Parseval's Formula

Let, on the positive part of the real axis, $f(x)$ be a real analytic function, and $h(x) = [H_+(z), H_-(z)]$ a hyperfunction. Then, $f(x)h(x)$ is a well-defined hyperfunction on $\mathbb{R}_+$. Assume that the Mellin transforms of $f(x)$ and $h(x)$ have strips of convergence $\alpha < \Re s < \beta$ and $\alpha' < \Re s < \beta'$, respectively. The Mellin transform of the product $f(x)h(x)$ is defined by

$$\mathcal{M}[f(x)h(x)](s) = \mathcal{L}[f(e^{-t})h(e^{-t})](s),$$

where $f(e^{-t})$ is real analytic on $\mathbb{R}$, and $\alpha'' < \Re s < \beta''$, say. This hyperfunction is specified by

$$f(e^{-t})h(e^{-t}) = \left[-f(e^{-z})H_-(e^{-z}), -f(e^{-z})H_+(e^{-z}) \right],$$

and its Laplace transform is given by

$$-\int_{\gamma^+_{-\infty,\infty}} e^{-sz} f(e^{-z})H_-(e^{-z})\, dz + \int_{\gamma^-_{-\infty,\infty}} e^{-sz} f(e^{-z})H_+(e^{-z})\, dz.$$

The change of variables $\zeta = e^{-z}$ transforms this expression into

$$-\int_{\gamma_{0,\infty}^-} \zeta^{s-1} f(\zeta) H_-(\zeta)\, d\zeta + \int_{\gamma_{0,\infty}^+} \zeta^{s-1} f(\zeta) H_+(\zeta)\, d\zeta. \tag{6.74}$$

By using Proposition 6.9,

$$H_+(\zeta) = \frac{1}{2\pi i} \int_{c-i\infty}^{c} \zeta^{-p} \tilde{h}(p)\, dp, \quad \Im\zeta > 0,$$

$$H_-(\zeta) = -\frac{1}{2\pi i} \int_{c}^{c+i\infty} \zeta^{-p} \tilde{h}(p)\, dp, \quad \Im\zeta < 0$$

and then inserting these expressions into (6.74) followed by an interchange of the order of integration yields

$$\frac{1}{2\pi i} \int_{c}^{c+i\infty} \tilde{h}(p) \int_{\gamma_{0,\infty}^-} \zeta^{s-p-1} f(\zeta)\, d\zeta\, dp$$

$$+ \frac{1}{2\pi i} \int_{c-i\infty}^{c} \tilde{h}(p) \int_{\gamma_{0,\infty}^+} \zeta^{s-p-1} f(\zeta)\, d\zeta\, dp.$$

Because $f(\zeta)$ is real analytic, the two contours $\gamma_{0,\infty}^-$, $\gamma_{0,\infty}^+$ can be collapsed into the interval $(0, \infty)$. Adding the two integrals implies

$$\frac{1}{2\pi i} \int_{c-i\infty}^{c+i\infty} \tilde{h}(p) \int_0^\infty \zeta^{s-p-1} f(\zeta)\, d\zeta\, dp = \frac{1}{2\pi i} \int_{c-i\infty}^{c+i\infty} \tilde{h}(p) \tilde{f}(s-p)\, dp.$$

Now the strip of convergence of $\tilde{f}(s-p)$ given by $-\beta + \Re s < \Re p < -\alpha + \Re s$ and $\alpha' < \Re s < \beta'$ should overlap, hence we must have $\alpha + \alpha' < \alpha'' < \beta'' < \beta + \beta'$. We have established

Proposition 6.12. *On the positive part of the real axis, let $f(x)$ be a real analytic function and $h(x) = [H_+(z), H_-(z)]$ a hyperfunction. Assume that $\tilde{f}(p) = \mathcal{M}[f(x)](p)$ and $\tilde{h}(p) = \mathcal{M}[h(x)](p)$ exist with strips of convergence $\alpha < \Re p < \beta$ and $\alpha' < \Re p < \beta'$, respectively. Moreover, assume that $\alpha + \alpha' < \beta + \beta'$. Let $\Re p = c$ be the vertical line inside the overlapping strip of convergence. Then we have*

$$\mathcal{M}[f(x)h(x)](s) = \frac{1}{2\pi i} \int_{c-i\infty}^{c+i\infty} \tilde{h}(p) \tilde{f}(s-p)\, dp \tag{6.75}$$

for $\alpha + \alpha' < \Re s < \beta + \beta'$.

For $s = 1$ we obtain Parseval's formula for the Mellin transformation.

Corollary 6.13. *If the hypotheses of the above proposition are satisfied for $s = 1$, then we have*

$$\int_0^\infty f(x)\, h(x)\, dx = \frac{1}{2\pi i} \int_{c-i\infty}^{c+i\infty} \tilde{h}(p) \tilde{f}(1-p)\, dp. \tag{6.76}$$

Example 6.18. Take $h(x) = \delta(x-a)$, $a > 0$. The left-hand side gives

$$\int_0^\infty f(x)\delta(x-a)\, dx = f(a),$$

while the right-hand side yields

$$\frac{1}{2\pi i} \int_{c-i\infty}^{c+i\infty} a^{p-1} \, \tilde{f}(1-p) \, dp = \frac{1}{2\pi i} \int_{1-c-i\infty}^{1-c+i\infty} a^{-s} \, \tilde{f}(s) \, ds,$$

thus we obtain the formula for the inverse Mellin transform

$$f(a) = \frac{1}{2\pi i} \int_{c'-i\infty}^{c'+i\infty} a^{-s} \, \tilde{f}(s) \, ds.$$

6.6   Applications

6.6.1   Dirichlet's Problem in a Wedge-shaped Domain

The Laplace operator in plane polar coordinates (r, θ) is

$$\nabla^2 = \frac{\partial^2}{\partial r^2} + \frac{1}{r} \frac{\partial}{\partial r} + \frac{1}{r^2} \frac{\partial^2}{\partial \theta^2}.$$

Laplace's equation $\nabla^2 f(r, \theta) = 0$, or $r^2 \, \nabla^2 f(r, \theta) = 0$ becomes then

$$(r^2 \frac{\partial^2}{\partial r^2} + r \frac{\partial}{\partial r} + \frac{\partial^2}{\partial \theta^2}) f(r, \theta) = 0. \tag{6.77}$$

By taking the Mellin transform with respect to the variable r, i.e., $\tilde{f}(s, \theta) = \mathcal{M}[f(r, \theta)](s)$, and using (6.49) we obtain the image equation, now being an ordinary differential equation

$$s^2 \, \tilde{f}(s, \theta) + \frac{\partial^2 \tilde{f}(s, \theta)}{\partial \theta^2} = 0.$$

Here it has been supposed that Mellin transform and partial differentiation with respect to θ can be interchanged. Let us now solve Dirichlet's problem for the interior of a wedge-shaped infinite domain

$$D = D(\theta_1, \theta_2)) := \{(r, \theta) \,|\, 0 \le \theta_1 < \theta < \theta_2 \le 2\pi, \; 0 < r\} \tag{6.78}$$

with imposed boundary conditions on the two rays of the sector

$$f(r, \theta_1) = f_1(r), \quad f(r, \theta_2) = f_2(r), \quad 0 < r. \tag{6.79}$$

Assuming that the two boundary functions have Mellin transforms

$$\mathcal{M}[f_1(r)](s) = \tilde{f}_1(s), \quad \mathcal{M}[f_2(r)](s) = \tilde{f}_2(s),$$

we obtain the image system, where s functions as a parameter,

$$\frac{\partial^2 \tilde{f}(s, \theta)}{\partial \theta^2} + s^2 \, \tilde{f}(s, \theta) = 0,$$

$$\lim_{\theta \to \theta_1+} \tilde{f}(s, \theta) = \tilde{f}_1(s), \quad \lim_{\theta \to \theta_2-} \tilde{f}(s, \theta) = \tilde{f}_2(s).$$

The solution of the image system is

$$\tilde{f}(s,\theta) = A(s)\,\cos(s\theta) + B(s)\,\sin(s\theta),$$

$$A(s) = \frac{\sin(s\theta_2)}{\sin(s(\theta_2 - \theta_1))}\,\tilde{f}_1(s) - \frac{\sin(s\theta_1)}{\sin(s(\theta_2 - \theta_1))}\,\tilde{f}_2(s),$$

$$B(s) = \frac{\cos(s\theta_1)}{\sin(s(\theta_2 - \theta_1))}\,\tilde{f}_2(s) - \frac{\cos(s\theta_2)}{\sin(s(\theta_2 - \theta_1))}\,\tilde{f}_1(s).$$

For the case with $\theta_1 = 0$, and $f_2(r) = 0$, we obtain the simplified expression

$$\tilde{f}(s,\theta) = \frac{\sin(s\,(\theta_2 - \theta))}{\sin(s\,\theta_2)}\,\tilde{f}_1(s).$$

A) If the boundary data on the horizontal leg is the Dirac impulse $f_1(r) = \delta(r - a)$, $a > 0$, , i.e., $\tilde{f}_1(s) = a^{s-1}$, the solution takes the form

$$f(r,\theta) = \mathcal{M}^{-1}\Big[\frac{\sin(s\,(\theta_2 - \theta))}{\sin(s\,\theta_2)}\,a^{s-1}\Big](r).$$

By using the Convolution Rule for Mellin transforms, we may write

$$f(r,\theta) = \mathcal{M}^{-1}\Big[\frac{\sin(s\,(\theta_2 - \theta))}{\sin(s\,\theta_2)}\Big](r) * \delta(r - a).$$

By using Problem 6.3, we obtain

$$\mathcal{M}^{-1}\Big[\frac{\sin(s\,(\theta_2 - \theta))}{\sin(s\,\theta_2)}\Big](r) = \frac{1}{\theta_2}\,\frac{r^{\pi/\theta_2}\,\sin(\frac{\theta}{\theta_2}\pi)}{r^{2\pi/\theta_2} - 2\,r^{\pi/\theta_2}\cos(\frac{\theta}{\theta_2}\pi) + 1},$$

and by (6.59) we eventually find

$$f(r,\theta) = \frac{1}{a\,\theta_2}\,\frac{(r/a)^{\pi/\theta_2}\,\sin(\frac{\theta}{\theta_2}\pi)}{(r/a)^{2\pi/\theta_2} - 2\,(r/a)^{\pi/\theta_2}\cos(\frac{\theta}{\theta_2}\pi) + 1}. \tag{6.80}$$

B) If the boundary data on the horizontal leg is $f_1(r) = r^n\,u(a - r)$, then, by Problem 6.2,

$$\tilde{f}_1(s) = \frac{a^{n+s}}{n+s}, \quad \Re s > -n, \tag{6.81}$$

and the solution takes the form

$$f(r,\theta) = \mathcal{M}^{-1}\Big[\frac{\sin(s\,(\theta_2 - \theta))}{\sin(s\,\theta_2)}\,\frac{a^{n+s}}{n+s}\Big](r).$$

For the case $n = 0$ the inversion formula gives

$$f(r,\theta) = \frac{1}{2\pi i}\int_{c-i\infty}^{c+i\infty}\Big(\frac{r}{a}\Big)^{-s}\,\frac{\sin(s\,(\theta_2 - \theta))}{s\,\sin(s\,\theta_2)}\,ds,$$

where $0 < c < \pi/\theta_2$. In order to evaluate the above integral by the calculus of residues, we close the vertical contour $(c-iR, c+iR)$ by a circular arc to the right, if $r > a$, and to the left, if $r < a$. Eventually, we let $R \to \infty$. For $r > a$, we obtain

$$f(r,\theta) = -\sum_{k=1}^{\infty}\mathrm{Res}_{s=k\pi/\theta_2}\Big\{\Big(\frac{r}{a}\Big)^{-s}\,\frac{\sin(s\,(\theta_2 - \theta))}{s\,\sin(s\,\theta_2)}\Big\},$$

which gives

$$f(r,\theta) = \frac{1}{\pi}\sum_{k=1}^{\infty}\left(\frac{r}{a}\right)^{-k\pi/\theta_2}\frac{\sin(k\pi\frac{\theta}{\theta_2})}{k}. \tag{6.82}$$

For $r < a$, we obtain

$$f(r,\theta) = \sum_{k=0}^{-\infty}\operatorname{Res}_{s=k\pi/\theta_2}\left\{\left(\frac{r}{a}\right)^{-s}\frac{\sin(s(\theta_2-\theta))}{s\,\sin(s\theta_2)}\right\},$$

which gives

$$f(r,\theta) = 1 - \frac{\theta}{\theta_2} - \frac{1}{\pi}\sum_{k=1}^{\infty}\left(\frac{r}{a}\right)^{k\pi/\theta_2}\frac{\sin(k\pi\frac{\theta}{\theta_2})}{k}. \tag{6.83}$$

Problem 6.7. Show that for $n \in \mathbb{N}$, we have for $0 < r < a$,

$$f(r,\theta) = r^n\frac{\sin((\theta_2-\theta)n)}{\sin(n\theta_2)} + a^n\sum_{k=0}^{\infty}\left(\frac{r}{a}\right)^{k\pi/\theta_2}\frac{\sin(k\pi\frac{\theta}{\theta_2})}{\theta_2 n - k},$$

and for $r > a$,

$$f(r,\theta) = a^n\sum_{k=1}^{\infty}\left(\frac{r}{a}\right)^{-k\pi/\theta_2}\frac{\sin(k\pi\frac{\theta}{\theta_2})}{\theta_2 n + k}.$$

6.6.2 Euler's Differential Equation

The linear differential equation

$$\{x^n D^n + a_1 x^{n-1}D^{n-1} + \cdots + xD + a_n\}y(x) = f(x) \tag{6.84}$$

with constants a_k is called *Euler's Differential Equation of order n* . If $\tilde{y}(s) = \mathcal{M}[y(x)](s)$ with $\alpha < \Re s < \beta$, then, by the Multiplication Rule and the Differentiation Rule, we have

$$\mathcal{M}[x^n D^n f(x)](s) = (-1)^n s(s+1)\cdots(s+n-1)\tilde{y}(s) = (-1)^n(s)_n\,\tilde{y}(s)$$

where again we have $\alpha < \Re s < \beta$. We continue with $n = 2$. The homogeneous Euler Differential equation of order two,

$$\{x^2 D^2 + a_1 xD + a_2\}y(x) = 0, \tag{6.85}$$

has the general solution

$$y(x) = c_1 x^{\lambda_1} + c_2 x^{\lambda_2},$$

provided the characteristic equation

$$\lambda^2 + (a_1 - 1)\lambda + a_2 = 0$$

has two different roots λ_1 and λ_2. If the characteristic equation has a double root λ_1, i.e., if $(a_1 - 1)^2 = 4a_2$, the general solution is given by

$$y(x) = c_1 x^{\lambda_1} + c_2 x^{\lambda_1}\log x.$$

Observe that in both cases these solutions of the homogeneous equation have no Mellin transforms. However, a solution of the inhomogeneous Euler equation may sometimes be found by Mellin transformation (see the next example). With $\mathcal{M}[f(x)](s) = \tilde{f}(s)$, the image equation becomes

$$(s^2 - (a_1 - 1)s + a_2)\,\tilde{y}(s) = \tilde{f}(s). \tag{6.86}$$

Example 6.19. For

$$\{x^2 D^2 + 2xD - 2\}y(x) = u(a - x)\,x^{-3}, \quad a > 0$$

the Mellin transform of a particular solution is found to be

$$\tilde{y}(s) = \frac{a^{s-3}}{(s+1)(s-2)(s-3)}, \quad \Re s > 3.$$

Partial fraction decomposition yields

$$\tilde{y}(s) = \frac{1}{4}\frac{a^{s-3}}{s-3} - \frac{1}{3}\frac{a^{s-3}}{s-2} + \frac{1}{12}\frac{a^{s-3}}{s+1}$$

$$= \frac{1}{4}\frac{a^{s-3}}{s-3} - \frac{1}{3a}\frac{a^{s-2}}{s-2} + \frac{1}{12a^4}\frac{a^{s+1}}{s+1}.$$

Using Problem 7.1 we obtain the particular solution

$$y(x) = \frac{1}{4}\,u(a - x)\{x^{-3} - \frac{1}{3a}x^{-2} + \frac{1}{12a^4}x\}. \tag{6.87}$$

Example 6.20. Solve

$$\{x^2 D^2 + 2xD - 2\}y(x) = \delta(x - a), \quad 0 < a.$$

The same technique as in the preceding example yields

$$\tilde{y}(s) = \frac{a^{s-1}}{(s+1)(s-2)} = -\frac{1}{3}\frac{a^{s-1}}{s+1} + \frac{1}{3}\frac{a^{s-1}}{s-2}$$

$$= -\frac{1}{3a^2}\frac{a^{s+1}}{s+1} + \frac{a}{3}\frac{a^{s-2}}{s-2}.$$

We have three strips of convergence $\Re s < -1$, $-1 < \Re s < 2$, and $2 < \Re s$, corresponding to three different particular solutions:
A) Corresponding to $\Re s < -1$, we find

$$y(x) = u(x - a)\{\frac{1}{3a^2}x - \frac{a}{3}\frac{1}{x^2}\}.$$

B) Corresponding to $-1 < \Re s < 2$, we have

$$y(x) = -\frac{1}{3a^2}\,x\,u(a - x) - \frac{a}{3}\frac{1}{x^2}\,u(x - a).$$

C) Corresponding to $2 < \Re s$ gives the solution

$$y(x) = u(a - x)\{-\frac{1}{3a^2}x + \frac{a}{3}\frac{1}{x^2}\}.$$

The general solution of the homogeneous equation is $y_0(x) = c_1 x + c_2/x^2$.

6.6.3 Integral Equations V

Consider the integral equation

$$\int_0^\infty k(xt)\, y(t)\, dt = f(x), \quad x > 0 \tag{6.88}$$

with a given kernel $k(x)$ and a given right-hand side $f(x)$. The integral equation is a Mellin $\circ$convolution equation $k(x) \circ y(x) = f(x)$. We assume that all involved functions or hyperfunctions have Mellin transforms. The $\circ$convolution theorem yields then $\tilde{k}(s)\tilde{y}(1 - s) = \tilde{f}(s)$. It is further assumed that the left-hand side and the right-hand side have a non-void overlapping strip of convergence. With $\tilde{h}(s) := 1/\tilde{k}(1 - s)$ we find, at least formally,

$$y(x) = \mathcal{M}^{-1}[\tilde{h}(s)\, \tilde{f}(1 - s)](x) = h(x) \circ f(x), \tag{6.89}$$

or, in the case of ordinary functions,

$$y(x) = \int_0^\infty h(xt)\, f(t)\, dt, \tag{6.90}$$

provided $\tilde{h}(s) := 1/\tilde{k}(1 - s)$ has an inverse Mellin transform.

Note that for the kernel $k(x) = e^{-x}$ any known correspondence of a one-sided Laplace transform yields a solution of an integral equation of type (6.88), for example,

$$\int_0^\infty e^{-xt}\, y(t)\, dt = \frac{x}{x^2 + a^2}, \quad x > 0$$

has the solution $y(x) = \cos(ax)$.

Example 6.21. Solve the Laplace integral equation

$$\int_0^\infty e^{-xt}\, y(t)\, dt = x^n\, e^{-ax}, \quad a > 0. \tag{6.91}$$

We have

$$\tilde{f}(s) = \mathcal{M}[x^n\, e^{-ax}](s) = a^{-(s+n)}\, \Gamma(s + n),$$

and, $\tilde{h}(s) = 1/\Gamma(1 - s)$. Therefore,

$$y(x) = \mathcal{M}^{-1}\Big[\frac{a^{-(1-s+n)}\, \Gamma(1 - s + n)}{\Gamma(1 - s)}\Big](x)$$

$$= \mathcal{M}^{-1}[(-1)^n (s - 1)(s - 2)\cdots(s - n)\, a^{s-n-1}](x) = \delta^{(n)}(x - a)$$

as it is to be expected.

Let us now study the integral equation

$$\int_0^\infty k(xt)\, y(t)\, dt + f(x) = y(x), \quad x > 0, \tag{6.92}$$

or $k(x) \circ y(x) + f(x) = y(x)$. By taking the Mellin transform of all members we obtain the image equations

$$\tilde{k}(s)\, \tilde{y}(1 - s) + \tilde{f}(s) = \tilde{y}(s),$$
$$\tilde{k}(1 - s)\, \tilde{y}(s) + \tilde{f}(1 - s) = \tilde{y}(1 - s),$$

where the second equation follows from the first one by replacing s by $1 - s$. Eliminating $y(1 - s)$ from the two equations yields formally

$$y(x) = \mathcal{M}^{-1}\left[\frac{\tilde{f}(s) + \tilde{k}(s)\,\tilde{f}(1 - s)}{1 - \tilde{k}(s)\,\tilde{k}(1 - s)}\right](x). \tag{6.93}$$

Clearly, the usual assumptions concerning the existence of the various Mellin transforms and the overlapping of the strips of convergence have to be made.

6.6.4 Summation of Series

The Mellin transformation is often useful for finding the sum of an infinite or finite series. The method is based on the formula of the inverse Mellin transformation and on interchanging integration and summation. The Hurwitz function and the Riemann zeta function are usually involved. The *Hurwitz zeta function* is defined by

$$\zeta(s, a) := \sum_{k=0}^{\infty} \frac{1}{(k + a)^s}, \tag{6.94}$$

for $\Re s > 1$ and $0 < \Re a \leq 1$, and by analytic continuation elsewhere. For $a = 1$, we obtain the *Riemann zeta function*

$$\zeta(s) := \zeta(s, 1) = \sum_{k=1}^{\infty} \frac{1}{k^s}. \tag{6.95}$$

Because the series converge absolutely and uniformly in any compact domain within the right half-plane $\Re s > 1$, both functions are holomorphic in this half-plane. The only singularity of the Riemann zeta function is a simple pole of residue 1 at $s = 1$, furthermore we have $\zeta(0) = -1/2$. We yet cite *Riemann's reflection formula* of the zeta function:

$$\zeta(1 - s) = 2^{1-s}\,\pi^{-s}\,\cos\left(\frac{\pi}{2}s\right)\Gamma(s)\,\zeta(s). \tag{6.96}$$

The Riemann zeta function has no zeros in the half-plane $\Re s > 1$, and the only zeros in the negative half-plane $\Re s < 0$ are located at $-2, -4, -6, \ldots$. It remains an unproved conjecture of Riemann that all possible zeros within the strip $0 \leq \Re s \leq 1$ are located on the midline $\Re z = 1/2$. For more details see, for example, [29, Olver p.61 - 64]. The Riemann zeta function often comes up in Mellin transforms.

Example 6.22. Let $f(x) = 1/(\exp(x) - 1)$. Its Mellin transform becomes

$$\tilde{f}(s) = \int_0^\infty \frac{x^{s-1}}{e^x - 1}\,dx = \int_0^\infty \frac{x^{s-1}\,e^{-x}}{1 - e^{-x}}\,dx,$$

and, with $x > 0$, we may write

$$\frac{e^{-x}}{1 - e^{-x}} = \sum_{k=1}^{\infty} e^{-kx}.$$

A term by term integration yields

$$\tilde{f}(s) = \int_0^\infty x^{s-1} \sum_{k=1}^\infty e^{-kx}\, dx = \sum_{k=1}^\infty \int_0^\infty e^{-kx}\, x^{s-1}\, dx$$

$$= \sum_{k=1}^\infty \frac{\Gamma(s)}{s^k} = \Gamma(s)\,\zeta(s).$$

Thus,

$$\mathcal{M}[\frac{1}{e^x - 1}](s) = \Gamma(s)\,\zeta(s), \quad \Re s > 1. \tag{6.97}$$

By the Scale Change property, we get

$$\mathcal{M}[\frac{2}{e^{2x} - 1}](s) = 2\, 2^{-s}\, \Gamma(s)\,\zeta(s),$$

and, by using the identity,

$$\frac{1}{e^x - 1} - \frac{1}{e^x + 1} = \frac{2}{e^{2x} - 1},$$

we obtain, still for $\Re s > 1$,

$$\mathcal{M}[\frac{1}{e^x + 1}](s) = \Gamma(s)\,\zeta(s) - 2^{-s+1}\,\Gamma(s)\,\zeta(s) = (1 - 2^{1-s})\,\Gamma(s)\,\zeta(s). \tag{6.98}$$

Problem 6.8. Establish that for $\Re s > 1$ we have

$$\sum_{k=1}^\infty \frac{(-1)^{k-1}}{k^s} = (1 - 2^{1-s})\,\zeta(s). \tag{6.99}$$

Let now $f(x) = [F_+(z), F_-(z)]$ be a given hyperfunction defined on the positive part of the real axis. Assume that the series

$$\sum_{k=0}^\infty f(k + x) = g(x)$$

converges in the sense of hyperfunctions to $g(x)$ (see Section 2.1). Assume further that $f(x)$ has the Mellin transform $\tilde{f}(s)$ with the strip of convergence $\alpha < \Re s < \beta$. Then by Proposition 6.9 we have for $f(k + x) = [F_+(k + z), F_-(k + z)]$,

$$F_+(k + z) = \frac{1}{2\pi i} \int_{c-i\infty}^c (k + z)^{-s}\, \tilde{f}(s)\, ds, \quad \Im z > 0,$$

$$F_-(k + z) = -\frac{1}{2\pi i} \int_c^{c+i\infty} (k + z)^{-s}\, \tilde{f}(s)\, ds, \quad \Im z < 0$$

with $\alpha < c < \beta$. This implies

$$G_+(z) := \sum_{k=0}^\infty F_+(k + z) = \frac{1}{2\pi i} \int_{c-i\infty}^c \zeta(s, z)\, \tilde{f}(s)\, ds, \quad \Im z > 0,$$

$$G_-(z) := \sum_{k=0}^\infty F_-(k + z) = -\frac{1}{2\pi i} \int_{c-i\infty}^c \zeta(s, z)\, \tilde{f}(s)\, ds, \quad \Im z < 0,$$

where we have interchanged the order of summation and integration. If the hyperfunction $g(x) = [G_+(z), G_-(z)]$ has a value at $x = a$, we obtain the formula

$$g(a) = \sum_{k=0}^{\infty} f(k+a) = G_+(a+i0) - G_-(a-i0)$$

$$= \frac{1}{2\pi i} \int_{c-i\infty}^{c+i\infty} \zeta(s,a) \tilde{f}(s) \, ds. \tag{6.100}$$

For $a = 1$, we get

$$\sum_{k=1}^{\infty} f(k) = \frac{1}{2\pi i} \int_{c-i\infty}^{c+i\infty} \zeta(s) \tilde{f}(s) \, ds. \tag{6.101}$$

Also, by the Scale Change Property we have $\mathcal{M}[f(kx)] = k^{-s} \tilde{f}(s)$, thus, for $f(kx) = [F_+(kz), F_-(kz]$ we obtain

$$F_+(kz) = \frac{1}{2\pi i} \int_{c-i\infty}^{c} (zk)^{-s} \tilde{f}(s) \, ds, \quad \Im z > 0,$$

$$F_-(kz) = -\frac{1}{2\pi i} \int_{c}^{c+i\infty} (zk)^{-s} \tilde{f}(s) \, ds, \quad \Im z < 0.$$

Summing and interchanging sum and integral gives

$$G_+(z) := \sum_{k=1}^{\infty} F_+(kz) = \frac{1}{2\pi i} \int_{c-i\infty}^{c} z^{-s} \zeta(s) \tilde{f}(s) \, ds, \quad \Im z > 0,$$

$$G_-(z) := \sum_{k=1}^{\infty} F_-(kz) = -\frac{1}{2\pi i} \int_{c-i\infty}^{c} z^{-s} \zeta(s) \tilde{f}(s) \, ds, \quad \Im z < 0,$$

and we obtain the hyperfunction

$$g(x) = \sum_{k=1}^{\infty} f(kx) = [G_+(z), G_-(z)]. \tag{6.102}$$

Again, if the hyperfunction $g(x)$ has a value at x, we obtain the formula

$$\sum_{k=1}^{\infty} f(kx) = G_+(x+i0) - G_-(x-i0)$$

$$= \frac{1}{2\pi i} \int_{c-i\infty}^{c+i\infty} x^{-s} \zeta(s) \tilde{f}(s) \, ds. \tag{6.103}$$

It is this formula which applies to most applications.

Example 6.23. Let us find the sum of the series

$$\sum_{k=1}^{\infty} \frac{\cos(kx)}{k^2}. \tag{6.104}$$

With $f(x) = \cos x / x^2$ we have

$$\sum_{k=1}^{\infty} f(kx) = \frac{1}{x^2} \sum_{k=1}^{\infty} \frac{\cos(kx)}{k^2},$$

and, $\tilde{f}(s) = \Gamma(s-2)\cos(\frac{\pi}{2}(s-2))$ with $2 < \Re s < 3$. Therefore,

$$\sum_{k=1}^{\infty} f(kx) = \frac{1}{2\pi i} \int_{c-i\infty}^{c+i\infty} x^{-s}\,\zeta(s)\,\Gamma(s-2)\,\cos(\frac{\pi}{2}(s-2))\,ds,$$

where $2 < c < 3$. By using

$$\Gamma(s-2) = \frac{\Gamma(s)}{(s-2)(s-1)},$$

we obtain

$$\sum_{k=1}^{\infty} f(kx) = -\frac{1}{2\pi i} \int_{c-i\infty}^{c+i\infty} x^{-s}\,\frac{\zeta(s)\,\Gamma(s)\,\cos(\frac{\pi}{2}s)}{(s-2)(s-1)}\,ds$$

$$= -\frac{1}{4\pi i} \int_{c-i\infty}^{c+i\infty} (\frac{2\pi}{x})^{s}\,\frac{\zeta(1-s)}{(s-2)(s-1)}\,ds,$$

where we have used Riemann's reflection formula (6.96). We close the vertical integration contour $(c-iR, c+iR)$ by the horizontal segment $(c+iR, iR)$ followed by the semi-circle with radius R in the left-half plane, followed by the horizontal segment $(-iR, c-iR)$. If we restrict x to the interval $(0, 2\pi)$, it can be seen that the contribution to the contour integral from the two horizontal segments and the left semicircle tends to zero as $R \to \infty$. This implies that we finally have

$$\sum_{k=1}^{\infty} \frac{\cos(kx)}{k^2} = -\frac{x^2}{2}\,\mathrm{Res}_{\{0,1,2\}}\{(\frac{2\pi}{x})^{s}\,\frac{\zeta(1-s)}{(s-2)(s-1)}\},$$

where $0, 1, 2$ are simple poles. The sum of the residues yields

$$\frac{-1}{(0-2)(0-1)} + \frac{2\pi}{x}\,\frac{\zeta(0)}{1-2} + (\frac{2\pi}{x})^2\,\frac{\zeta(-1)}{2-1}$$

$$= -\frac{1}{2} + \frac{\pi}{x} - \frac{\pi^2}{3x^2}.$$

Finally, we obtain for $0 < x < 2\pi$,

$$\sum_{k=1}^{\infty} \frac{\cos(kx)}{k^2} = \frac{\pi^2}{6} - \frac{\pi}{2}x + \frac{x^2}{4}. \tag{6.105}$$

Chapter 7

Hankel Transforms

First we show that the conventional Hankel transform pair arises in a natural way when, in the two-dimensional Fourier transformation, polar coordinates are introduced. Unfortunately, no firm convention about the definition of the Hankel transform pair is established. We shall use the most widespread one. In order to lay the groundwork for the theory of Hankel transformation of hyperfunctions, we present a concise exposition of the various cylinder functions, the integrals of Lommel and MacRobert's proof of the inversion formula. The Hankel transform of a hyperfunction defined on the positive part of the real line is then defined by using the Hankel functions for the kernel. Along the line of MacRobert's proof and using the integrals of Lommel, we then prove that the defined Hankel transform of a hyperfunction is a self-reciprocal transformation when restricted to the strictly positive part of the real axis. The operational rules known for the Hankel transformation of ordinary functions are then carried over to the Hankel transformation of hyperfunctions. The chapter closes with a few applications about problems of mathematical physics.

7.1   Hankel Transforms of Ordinary Functions

7.1.1   Genesis of the Hankel Transform

Let us start with the two-dimensional conventional Fourier transform pair in its symmetrical form

$$F(k_1, k_2) = \frac{1}{2\pi} \int_{-\infty}^{\infty} \int_{-\infty}^{\infty} e^{-i\,(k_1 x_1 + k_2 x_2)}\, f(x_1, x_2)\, dx_1 dx_2, \qquad (7.1)$$

$$f(x_1, x_2) = \frac{1}{2\pi} \int_{-\infty}^{\infty} \int_{-\infty}^{\infty} e^{i\,(k_1 x_1 + k_2 x_2)}\, F(k_1, k_2)\, dk_1 dk_2. \qquad (7.2)$$

We introduce polar coordinates in the (x_1, x_2)-plane and in the (k_1, k_2)-plane;

$$x_1 = r \cos\theta, \qquad\qquad x_2 = r \sin\theta, \qquad\qquad r = \sqrt{x_1^2 + x_2^2},$$

$$k_1 = -\rho \sin\phi, \qquad\qquad k_2 = \rho \cos\phi, \qquad\qquad \rho = \sqrt{k_1^2 + k_2^2}.$$

The above Fourier transform pair then becomes

$$G(\rho, \phi) = \frac{1}{2\pi} \int_0^{2\pi} \int_0^\infty e^{-i r \rho \sin(\theta - \phi)} g(r, \theta) \, r \, dr \, d\theta, \tag{7.3}$$

$$g(r, \theta) = \frac{1}{2\pi} \int_0^{2\pi} \int_0^\infty e^{i r \rho \sin(\theta - \phi)} G(\rho, \phi) \, \rho \, d\rho \, d\phi, \tag{7.4}$$

where we have set

$$g(r, \theta) := f(r \cos\theta, r \sin\theta), \quad G(\rho, \phi) := F(-\rho \sin\phi, \rho \cos\phi).$$

The next step consists in expanding the 2π-periodic functions into their Fourier series

$$G(\rho, \phi) = \sum_{n=-\infty}^\infty G_n(\rho) \, e^{in\phi}, \quad g(r, \theta) = \sum_{m=-\infty}^\infty g_m(r) \, e^{im\theta}, \tag{7.5}$$

with their Fourier coefficients

$$G_n(\rho) = \frac{1}{2\pi} \int_0^{2\pi} G(\rho, \phi) \, e^{-in\phi} \, d\phi, \tag{7.6}$$

$$g_m(r) = \frac{1}{2\pi} \int_0^{2\pi} g(r, \theta) \, e^{-im\theta} \, d\theta. \tag{7.7}$$

We want to express $G_n(\rho)$ in terms of $g_m(r)$. Thus, in (7.6) we substitute (7.3) in which (7.5) has previously been substituted.

$$G_n(\rho) = \frac{1}{2\pi} \int_0^{2\pi} \frac{1}{2\pi} \int_0^{2\pi} \int_0^\infty e^{-i r \rho \sin(\theta - \phi)}$$

$$\times \sum_{m=-\infty}^\infty g_m(r) \, e^{im\theta} \, r \, dr \, d\theta \, e^{-in\phi} \, d\phi.$$

Interchanging the order of integration and summation yields

$$G_n(\rho) = \frac{1}{2\pi} \int_0^\infty r \sum_{m=-\infty}^\infty g_m(r) \frac{1}{2\pi}$$

$$\times \int_0^{2\pi} \int_0^{2\pi} e^{-i r \rho \sin(\theta - \phi)} e^{im\theta} \, d\theta \, e^{-in\phi} \, d\phi \, dr.$$

In the innermost integral on θ we make the change of variables $\theta - \phi = -\alpha$. This yields

$$G_n(\rho) = \frac{1}{2\pi} \int_0^\infty r \sum_{m=-\infty}^\infty g_m(r) \frac{1}{2\pi}$$

$$\times \int_0^{2\pi} \int_{\phi-2\pi}^\phi e^{i r \rho \sin\alpha} e^{-im\alpha} \, d\alpha \, e^{-i(n-m)\phi} \, d\phi \, dr.$$

The integral over α can be replaced by

$$\int_{\phi-2\pi}^\phi e^{i r \rho \sin\alpha} e^{-im\alpha} \, d\alpha = \int_{-\pi}^\pi e^{i r \rho \sin\alpha - im\alpha} \, d\alpha$$

due to the 2π-periodicity of the integrand. We finally obtain

$$G_n(\rho) = \int_0^\infty r \sum_{m=-\infty}^\infty g_m(r) \, \frac{1}{2\pi}$$

$$\times \int_0^{2\pi} e^{-i(n-m)\phi} \, d\phi \, \frac{1}{2\pi} \int_{-\pi}^\pi e^{i r \rho \sin\alpha - i m\alpha} \, d\alpha \, dr.$$

Now

$$\frac{1}{2\pi} \int_0^{2\pi} e^{-i(n-m)\phi} \, d\phi = \delta_n^m,$$

$$J_m(r\rho) := \frac{1}{2\pi} \int_{-\pi}^\pi e^{i(r\rho\sin\alpha - m\alpha)} \, d\alpha \tag{7.8}$$

where δ_n^m denotes the Kronecker delta and $J_m(z)$ is the Bessel function of integer order m (see [29, Olver p.56] or the following section). Therefore, we have obtained

$$G_n(\rho) = \int_0^\infty r \, J_n(\rho r) \, g_n(r) \, dr. \tag{7.9}$$

In exactly the same way, but starting with (7.7), we can establish the converse pair

$$g_n(r) = \int_0^\infty \rho \, J_n(\rho r) \, G_n(\rho) \, d\rho. \tag{7.10}$$

Definition 7.1. The integral transform

$$G_n(\rho) := H_n[g(r)](\rho) := \int_0^\infty r \, J_n(\rho r) \, g(r) \, dr \tag{7.11}$$

with its inverse transform

$$g(r) := H_n^{-1}[G_n(\rho)](r) := \int_0^\infty \rho \, J_n(\rho r) \, G_n(\rho) \, d\rho \tag{7.12}$$

is called *the Hankel transform of order n.*

The pairs (7.9), (7.10) may be modified in order to obtain a still more symmetrical form,

$$\sqrt{\rho} \, G_n(\rho) = \int_0^\infty \sqrt{\rho r} \, J_n(\rho r) \, \sqrt{r} \, g_n(r) \, dr, \tag{7.13}$$

$$\sqrt{r} \, g_n(r) = \int_0^\infty \sqrt{\rho r} \, J_n(\rho r) \sqrt{\rho} \, G_n(\rho) \, d\rho, \tag{7.14}$$

and some authors prefer to deal with this modified Hankel transform pair $\tilde{G}(\rho) := \sqrt{\rho} \, G_n(\rho)$ and $\tilde{g}(r) := \sqrt{r} \, g_n(r)$. There is still another form. In order to find it, we make the change of variables in the Hankel transform pair: $r = \sqrt{2x}$, $\rho = \sqrt{2y}$. Then, with $f(x) := g(\sqrt{2x})$, and $F(y) := G(\sqrt{2y})$ we obtain the pair

$$F_n(y) = \mathbf{H}_n[f(x)] := \int_0^\infty J_n(2\sqrt{xy}) \, f(x) \, dx, \tag{7.15}$$

$$f(x) = \mathbf{H}_n^{-1}[F(y)](x) := \int_0^\infty J_n(2\sqrt{xy}) \, F_n(y) \, dy. \tag{7.16}$$

Gustav Doetsch uses this form in his books on Laplace transforms so let us call it the *Hankel-Doetsch transform* or the Hankel transform in the Doetsch form. The correspondences of a Hankel transform pair and of a Hankel-Doetsch transform pair are denoted by

$$g(r) \circ \overset{H_n}{-\!\!\!-} \bullet G(\rho), \quad f(x) \circ \overset{\mathbf{H}_n}{-\!\!\!-} \bullet F(y),$$

respectively. A given Hankel transform pair implies a Hankel-Doetsch transform pair by the rule:

$$g(r) \circ \overset{H_n}{-\!\!\!-} \bullet G_n(\rho) \Longrightarrow g(\sqrt{2r}) \circ \overset{\mathbf{H}_n}{-\!\!\!-} \bullet G_n(\sqrt{2\rho}), \tag{7.17}$$

while a given Hankel-Doetsch pair gives the Hankel pair by

$$f(x) \circ \overset{\mathbf{H}_n}{-\!\!\!-} \bullet F_n(y) \Longrightarrow f(\frac{x^2}{2}) \circ \overset{H_n}{-\!\!\!-} \bullet F_n(\frac{y^2}{2}). \tag{7.18}$$

The operational laws of the Hankel and the Hankel-Doetsch transformations are not the same. We shall primarily work with the Hankel transform, however, it should be pointed out, that the Hankel-Doetsch form may offer some advantage by exploiting a connection between the Laplace and Hankel transformation. Also, for the theoretical groundwork, we shall henceforth rather use the variables x, y instead of r, ρ.

7.1.2 Cylinder Functions

In the sequel we shall need some facts about cylinder functions. Any solution of Bessel's differential equation

$$z^2 \frac{d^2 u}{dz^2} + z \frac{du}{dz} + (z^2 - n^2) u(z) = 0 \tag{7.19}$$

is called a cylinder function. The most familiar cylinder functions are the *Bessel functions of the first kind of order* n denoted by $J_n(z)$. For $n \in \mathbb{Z}$ the functions $J_n(z)$ can be defined as the coefficient in the Laurent expansion

$$e^{\frac{z}{2}(t - \frac{1}{t})} = \sum_{n=-\infty}^{\infty} J_n(z) t^n.$$

The formula for the coefficients immediately gives

$$J_n(z) = \frac{1}{2\pi i} \int_C e^{\frac{z}{2}(t - \frac{1}{t})} t^{-n-1} \, dt \tag{7.20}$$

where the contour C encircles $t = 0$ once counter-clock wise. If we represent the contour by $C : t = \exp(i\alpha)$, $-\pi < \alpha < \pi$, we obtain (7.8). The change of variables $t = 2\tau/z$ transforms (7.20) into

$$J_n(z) = \frac{1}{2\pi i} \left(\frac{z}{2}\right)^n \int_C e^{\left(\tau - \frac{z^2}{4\tau}\right)} \tau^{-n-1} \, d\tau \tag{7.21}$$

with a contour C of the same type as before. Expanding the integrand into a series of powers of z and interchanging integration and summation yields

$$J_n(z) = \left(\frac{z}{2}\right)^n \sum_{k=0}^{\infty} \frac{(-1)^k}{k!\,(n+k)!} \left(\frac{z}{2}\right)^{2k}. \tag{7.22}$$

From this we can establish the property $J_{-n}(z) = (-1)^n J_n(z)$ by first replacing $(n+k)!$ by $\Gamma(n+k+1)$ and using the fact that $1/\Gamma(n)$ is vanishing for a non-positive integer n. So far we have assumed $n \in \mathbb{Z}$. The series (7.22) can be used to define the Bessel functions of order ν, where ν is any real or complex number:

$$J_\nu(z) = \left(\frac{z}{2}\right)^\nu \sum_{k=0}^{\infty} \frac{(-1)^k}{k!\,\Gamma(\nu+k+1)} \left(\frac{z}{2}\right)^{2k}. \tag{7.23}$$

The Bessel functions of the first kind of integer order are entire functions in z, while now those of a non-integral order ν become multi-valued functions. The principal branch is selected by making an incision in the complex plane along the negative part of the real axis and choosing the principal branch of the power z^ν. Let us now use

$$f_\nu(z) := \frac{1}{2\pi i} \left(\frac{z}{2}\right)^\nu \int_\gamma e^{\left(\tau - \frac{z^2}{4\tau}\right)} \tau^{-\nu-1}\, d\tau$$

with a contour γ not yet specified. By inserting $f_\nu(z)$ into Bessel's equation and performing the differentiation under the integral sign, we obtain

$$\frac{d^2 f_\nu}{dz^2} + \frac{1}{z}\frac{df_\nu}{dz} + \left(1 - \frac{n^2}{z^2}\right) f_\nu(z)$$

$$= \frac{1}{2\pi i} \left(\frac{z}{2}\right)^\nu \int_\gamma \left(1 - \frac{\nu+1}{\tau} + \frac{z^2}{4\tau^2}\right) e^{\left(\tau - \frac{z^2}{4\tau}\right)} \tau^{-\nu-1}\, d\tau$$

$$= -\frac{1}{2\pi i} \left(\frac{z}{2}\right)^\nu \int_\gamma \frac{d}{d\tau} \left(e^{\left(\tau - \frac{z^2}{4\tau}\right)} \tau^{-\nu-1}\right) d\tau.$$

Now we notice that if $\nu = n$ is an integer and the closed contour γ is as C in (7.21), the integral in the last line becomes zero because

$$e^{\left(\tau - \frac{z^2}{4\tau}\right)} \tau^{-\nu-1}$$

is a one-valued function. This shows that $J_n(z)$ of integer order is a cylinder function. If ν is any number, the above function is still holomorphic in the complex plane with a cut along the negative part of the real axis, moreover it is exponentially damped as $\Re\tau \to -\infty$. Therefore, if we take for γ the familiar Hankel loop (see Figure 1.13), then the integral in the last line becomes again zero, which shows that

$$J_\nu(z) = \frac{1}{2\pi i} \left(\frac{z}{2}\right)^\nu \int_{-\infty}^{(0+)} e^{\left(\tau - \frac{z^2}{4\tau}\right)} \tau^{-\nu-1}\, d\tau \tag{7.24}$$

is also a cylinder function. By the way, from this expression we can again establish the series (7.23) by using (3.1).

If ν is not an integer, then $J_\nu(z)$ and $J_{-\nu}(z)$ form a pair of two independent solutions of Bessel's equation. Thus, $AJ_\nu(z) + BJ_{-\nu}(z)$ with two arbitrary complex coefficients A, B constitutes the general solution of Bessel's equation. If $\nu = n$ is an integer, $J_n(z)$ and $J_{-n}(z)$ are not independent as we have seen above, and $AJ_n(z) + BJ_{-n}(z)$ cannot constitute a general solution.

If we temporarily restrict z to positive values and make in (7.24) the change of variables $\tau = zh/2$, we obtain (the Hankel loop contour is transformed into a similar Hankel loop)

$$J_\nu(z) = \frac{1}{2\pi i} \int_{-\infty}^{(0+)} e^{\frac{z}{2}(h-1/h)} \, h^{-\nu-1} \, dh. \tag{7.25}$$

Analytic continuation extends the validity of this formula to the right half-plane $\Re z > 0$. Now we set $h = \exp(t)$ and obtain Schläfli's representation

$$J_\nu(z) = \frac{1}{2\pi i} \int_S e^{z \sinh t - \nu t} \, dt. \tag{7.26}$$

If the Hankel loop is deformed into a contour comprising the lower side of the negative real axis from $-\infty$ to -1, the unit-circle, and the upper side of the negative real axis from -1 to $-\infty$, the corresponding Schläfli contour S has the shape indicated by Figure 7.1.

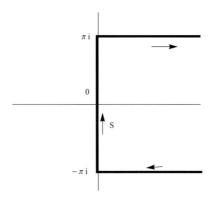

Figure 7.1: Schläfli's contour S

We consider now the integral expression (7.26) but with an arbitrary contour γ, i.e.,

$$g_\nu(z) = \frac{1}{2\pi i} \int_\gamma e^{z \sinh t - \nu t} \, dt. \tag{7.27}$$

We apply Bessel's differential operator to $g_\nu(z)$ assuming that differentiation under the integral is permitted. An elementary calculation yields

$$\left(z^2 \frac{d^2}{dz^2} + z \frac{d}{dz} + (z^2 - \nu^2) \right) g_\nu(z)$$

$$= \frac{1}{2\pi i} \int_\gamma \frac{\partial^2}{\partial t^2} \left(e^{z \sinh t - \nu t} \right) + 2\nu \frac{\partial}{\partial t} \left(e^{z \sinh t - \nu t} \right) \, dt.$$

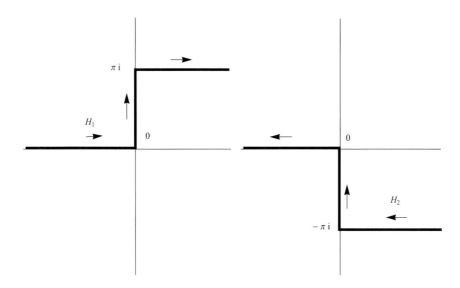

Figure 7.2: Hankel contours H_1 and H_2

If at the endpoints of the contour γ the expression

$$h(t, z) := \frac{\partial}{\partial t} \left(e^{z \, \sinh t - \nu t} \right) + 2\nu \left(e^{z \, \sinh t - \nu t} \right)$$
$$= \left(e^{z \, \sinh t - \nu t} \right) (\nu + z \cosh t)$$

vanishes, then $g_\nu(z)$ is a cylinder function. This is the case for $\gamma = S$ as the reader may check by observing that for $\Re z > 0$ we have $h(\infty \pm i\pi, z) = 0$. This confirms again that $J_\nu(z)$ is a cylinder function. However, the crucial observation to be made here is that we have also $h(-\infty, z) = 0$. This implies that

$$H_\nu^{(1)}(z) = \frac{1}{\pi i} \int_{H_1} e^{z \, \sinh t - \nu t} \, dt \qquad (7.28)$$

and

$$H_\nu^{(2)}(z) = \frac{1}{\pi i} \int_{H_2} e^{z \, \sinh t - \nu t} \, dt, \qquad (7.29)$$

with the contours shown in Figure 7.2 being cylinder functions, too.

Definition 7.2. The two cylinder functions defined by (7.28) and (7.29) are called *Hankel functions or Bessel functions of the third kind of order ν.*

From the shape of the contours in Figure 7.1 and 7.2, we immediately conclude that

$$J_\nu(z) = \frac{1}{2} \left\{ H_\nu^{(1)}(z) + H_\nu^{(2)}(z) \right\}. \qquad (7.30)$$

The asymptotic expansions for the Hankel functions as $z \to \infty$ are

$$H_\nu^{(1)}(z) \approx \sqrt{\frac{2}{\pi z}} \exp\{i(z - \frac{\nu \pi}{2} - \frac{\pi}{4})\} \sum_{k=0}^{\infty} i^k \frac{A_k(\nu)}{z^k}, \qquad (7.31)$$

valid inside the sector $-\pi < \arg(z) < 2\pi$, and

$$H_\nu^{(2)}(z) \approx \sqrt{\frac{2}{\pi z}} \exp\{-i(z - \frac{\nu\pi}{2} - \frac{\pi}{4})\} \sum_{k=0}^{\infty} i^k \frac{A_k(\nu)}{z^k}, \tag{7.32}$$

valid inside the sector $-2\pi < \arg(z) < \pi$. The coefficients are

$$A_0(\nu) = 1, \quad A_k(\nu) = \frac{(4\nu^2 - 1^2)(4\nu^2 - 3^2)\dots(4\nu^2 - (2k-1)^2)}{k!\, 8^k}. \tag{7.33}$$

See [29, Olver p.238]. The leading terms are

$$H_\nu^{(1)}(z) \approx \sqrt{\frac{2}{\pi z}} \exp\{i(z - \frac{\nu\pi}{2} - \frac{\pi}{4})\}, \tag{7.34}$$

and

$$H_\nu^{(2)}(z) \approx \sqrt{\frac{2}{\pi z}} \exp\{-i(z - \frac{\nu\pi}{2} - \frac{\pi}{4})\}. \tag{7.35}$$

This implies that the leading term of the asymptotic expansion for Bessel's function becomes

$$J_\nu(z) \approx \sqrt{\frac{2}{\pi z}} \cos(z - \frac{\nu\pi}{2} - \frac{\pi}{4}), \quad z \to \infty, \tag{7.36}$$

valid in the common sector $-\pi < \arg z < \pi$.

The integrals in (7.28) and (7.29) converge uniformly with respect to z varying in any compact sub-domain not containing the origin. Thus, in any point $z \neq 0$ the two Hankel functions are holomorphic. However, for $z = 0$ the integrals diverge. Thus, we expect that the Hankel functions are singular and become infinite at $z = 0$.

With the Bessel functions of the third kind we now define the Bessel functions of the second kind.

Definition 7.3. The *Bessel function of the second kind of order ν* are defined and denoted by

$$Y_\nu(z) := \frac{1}{2i} \{H_\nu^{(1)}(z) - H_\nu^{(2)}(z)\}. \tag{7.37}$$

By using (7.30) and (7.37) we obtain immediately

$$H_\nu^{(1)}(z) = J_\nu(z) + iY_\nu(z), \tag{7.38}$$
$$H_\nu^{(2)}(z) = J_\nu(z) - iY_\nu(z). \tag{7.39}$$

For integral order $\nu = n = 0, 1, 2, \dots$ the Bessel functions of the first kind are entire functions, while the two Hankel functions are singular at $z = 0$. This implies that the Bessel functions of the second kind must be singular at $z = 0$. Indeed the Bessel functions of the second kind have the following expansion about $z = 0$, for $n \in \mathbb{N}_0$:

$$Y_0(z) = \frac{2}{\pi} \log(\frac{z}{2}) J_0(z) + \phi_0(z), \tag{7.40}$$

$$Y_n(z) = -\frac{2^n}{\pi z^n} \sum_{k=0}^{n-1} \frac{(n-k-1)!}{4^k\, k!} z^{2k} + \frac{2}{\pi} \log(\frac{z}{2}) J_n(z) + \phi_n(z), \tag{7.41}$$

with the entire function

$$\phi_n(z) = -\frac{z^n}{2^n \pi} \sum_{k=0}^{\infty} \frac{\Psi(k+1) + \Psi(n+k+1)}{k! \, (n+k)! \, 4^k} (-z)^{2k}. \tag{7.42}$$

(See [1, Abramowitz-Stegun p. 104].)

The integral representations (7.26), (7.28), (7.29) imply that the Bessel functions of the first and third kind, and consequently the Bessel functions of the second kind, satisfy the following *recurrence relations for cylinder functions.*

Proposition 7.1. *Let $f_\nu(z)$ denote any one of the cylinder functions of the first, second or third kind, then*

$$f_{\nu-1}(z) + f_{\nu+1}(z) = \frac{2\nu}{z} f_\nu(z), \tag{7.43}$$

$$f_{\nu-1}(z) - f_{\nu+1}(z) = 2 f_\nu'(z) \tag{7.44}$$

hold, and consequently

$$f_{\nu+1}(z) = \frac{\nu}{z} f_\nu(z) - f_\nu'(z), \tag{7.45}$$

$$f_{\nu-1}(z) = \frac{\nu}{z} f_\nu(z) + f_\nu'(z). \tag{7.46}$$

(See, for example the book of Olver or [1, Abramowitz-Stegun p. 105].)

Problem 7.1. By using the above recurrence relations, show that

$$\frac{d}{dz} \left(z \, f_\nu(az) \right) = a \left(\frac{\nu+1}{2\nu} \right) z f_{\nu-1}(az) - a \left(\frac{\nu-1}{2\nu} \right) z f_{\nu+1}(az), \tag{7.47}$$

and

$$\frac{d}{dx} \left(\frac{f_\nu(y x)}{x^\nu} \right) = -\frac{y}{x^\nu} f_{\nu+1}(y x), \quad \frac{d}{dx} \left(x^\nu f_\nu(y x) \right) = y x^\nu f_{\nu-1}(y x). \tag{7.48}$$

Problem 7.2. Show that

$$az \, \{f_{\nu-1}^{(m)}(az) + f_{\nu+1}^{(m)}(az)\} + m\{f_{\nu-1}^{(m-1)}(az) + f_{\nu+1}^{(m-1)}(az)\} = 2\nu f^{(m)}(az).$$

Let us finish this section by computing the one-sided Laplace transform of $J_\nu(t)$. In (7.25) we make the change of variable $h = \sqrt{s^2 + 1} + s$. Then,

$$\frac{1}{2} \left(h - \frac{1}{h} \right) = s, \quad dh = \frac{\sqrt{s^2 + 1} + s}{\sqrt{s^2 + 1}} ds,$$

whereas the Hankel loop in the h-plane is transformed into a similar Hankel loop in the s-plane. Therefore, for $t > 0$, we have

$$J_\nu(t) = \frac{1}{2\pi i} \int_{-\infty}^{(0+)} e^{ts} \frac{(\sqrt{s^2+1} + s)^{-\nu}}{\sqrt{s^2+1}} ds. \tag{7.49}$$

By using Jordan's lemma, and assuming $\Re \nu > -1$, the upper horizontal ray of the Hankel loop can be bent up, and the lower ray bent down, thereby becoming the vertical straight line $\Re s = c > 0$ without changing the value of the integral. Thus,

$$J_\nu(t) = \frac{1}{2\pi i} \int_{c-i\infty}^{c+i\infty} e^{ts} \frac{(\sqrt{s^2+1}+s)^{-\nu}}{\sqrt{s^2+1}} \, ds$$

$$= \mathcal{L}^{-1}[\frac{1}{(\sqrt{s^2+1}+s)^\nu \sqrt{s^2+1}}](t)$$

due to the complex inversion formula. Thus, for $\Re\nu > -1$,

$$\mathcal{L}_+[J_\nu(t)](s) = \frac{1}{(\sqrt{s^2+1}+s)^\nu \sqrt{s^2+1}} = \frac{\sqrt{s^2+1}-s)^\nu}{\sqrt{s^2+1}}.$$

By using the Similarity Rule, we obtain for $y > 0$,

$$\mathcal{L}_+[J_\nu(y\,t)](s) == \frac{\sqrt{s^2+y^2}-s)^\nu}{y^\nu \sqrt{s^2+y^2}}. \tag{7.50}$$

Problem 7.3. Use (7.40), (7.41), (7.42) to establish the following hyperfunctions.

$$u(-x)J_0(x) = \left[-\frac{i}{4}Y_0(z)\right], \quad u(x)J_0(x) = \left[\frac{i}{4}Y_0(-z)\right],$$

$$u(-x)J_1(x) + \delta(x) = \left[-\frac{i}{4}Y_1(z)\right], \quad -u(x)J_1(x) + \delta(x) = \left[\frac{i}{4}Y_1(-z)\right].$$

7.1.3   Lommel's Integral

For MacRobert's proof of the Hankel inversion formula in the next section and subsequently for Hankel transforms of hyperfunctions, we shall need relations known as Lommel's integral.

Let $U_\nu(x)$ and $V_\nu(x)$ denote any of the cylinder functions of the first, second or third kind. From Problem 7.1 we have the relations

$$\frac{d}{dx}(x^\nu U_\nu(a\,x)) = a\,x^\nu U_{\nu-1}(a\,x), \quad \frac{d}{dx}(x^{-\nu} V_\nu(b\,x)) = -b\,x^{-\nu} V_{\nu+1}(b\,x).$$

Let us consider the integral

$$I_\nu(\epsilon,\lambda) := \int_\epsilon^\lambda x\, U_\nu(ax)\, V_\nu(bx)\, dx \tag{7.51}$$

for $0 < \epsilon < \lambda < \infty$. This integral can be written in the form

$$I_\nu(\epsilon,\lambda) := \int_\epsilon^\lambda x^{\nu+1} U_\nu(ax)\, x^{-\nu}\, V_\nu(bx)\, dx.$$

Integration by parts and using the above relations yields

$$I_\nu(\epsilon,\lambda) = \left[\frac{1}{a}x^{\nu+1}U_{\nu+1}(ax)\, x^{-\nu}\, V_\nu(bx)\right]_\epsilon^\lambda$$

$$+ \frac{b}{a}\int_\epsilon^\lambda x^{\nu+1} U_{\nu+1}(ax)\, x^{-\nu}\, V_{\nu+1}(bx)\, dx$$

$$= \left[\frac{1}{a}x^{\nu+1}U_{\nu+1}(ax)\, x^{-\nu}\, V_\nu(bx)\right]_\epsilon^\lambda + \frac{b}{a}\, I_{\nu+1}(\epsilon,\lambda).$$

By exchanging a with b and U_ν with V_ν, the expression $I_\nu(\epsilon, \lambda)$ remains unchanged, and we find

$$I_\nu(\epsilon, \lambda) = \left[\frac{1}{b} x^{\nu+1} V_{\nu+1}(bx)\, x^{-\nu}\, U_\nu(ax)\right]_\epsilon^\lambda + \frac{a}{b} I_{\nu+1}(\epsilon, \lambda).$$

From these two equations we eliminate $I_{\nu+1}(\epsilon, \lambda)$ and arrive at

$$\int_\epsilon^\lambda x\, U_\nu(ax)\, V_\nu(bx)\, dx$$

$$= \left[\frac{x}{a^2 - b^2} \{a\, U_{\nu+1}(ax)\, V_\nu(bx) - b\, U_\nu(ax)\, V_{\nu+1}(bx)\}\right]_\epsilon^\lambda \tag{7.52}$$

which is Lommel's integral. By using (7.45), we obtain yet another form

$$\int_\epsilon^\lambda x\, U_\nu(ax)\, V_\nu(bx)\, dx$$

$$= \left[\frac{x}{a^2 - b^2} \{b\, U_\nu(ax)\, V_\nu'(bx) - a\, U_\nu'(ax)\, V_\nu(bx)\}\right]_\epsilon^\lambda.$$

There are two special cases of (7.52) which are important for us. The first one is with $U_\nu(x) = V_\nu(x) = J_\nu(x)$. From

$$J_\nu(x) \approx \frac{x^\nu}{2^\nu\, \Gamma(\nu + 1)}, \qquad x \to 0+ \tag{7.53}$$

it can be seen that the bracket term on the right-hand side of (7.52) tends to zero, as $\epsilon \to 0+$, provided that $\Re\nu > -1$. This yields the relation

$$\int_0^\lambda x\, J_\nu(ax)\, J_\nu(bx)\, dx$$

$$= \frac{\lambda}{a^2 - b^2} \{a\, J_{\nu+1}(a\lambda)\, J_\nu(b\lambda) - b\, J_\nu(a\lambda)\, J_{\nu+1}(b\lambda)\}. \tag{7.54}$$

In the second case, we take $U_\nu(ax) = H_\nu^{(1)}(ax)$ with $\Im a > 0$ and $V_\nu(bx) = J_\nu(bx)$ or $U_\nu(ax) = H_\nu^{(2)}(ax)$ with $\Im a < 0$ and $V_\nu(bx) = J_\nu(bx)$, with $b > 0$, and where we let $\epsilon \to 0+$ and $\lambda \to \infty$. From (7.34), (7.35), and (7.36) we conclude that the bracket term on the right-hand side of (7.52) tends to zero as $\lambda \to \infty$. Because (see [1, Abramowitz-Stegun p. 104])

$$H_0^{(1)}(x) \approx \frac{2i}{\pi} \log x, \quad H_0^{(2)}(x) \approx -\frac{2i}{\pi} \log x, \tag{7.55}$$

$$H_\nu^{(1)}(x) \approx -\frac{i}{\pi} \Gamma(\nu)\, (\frac{x}{2})^{-\nu}, \quad H_\nu^{(2)}(x) \approx \frac{i}{\pi} \Gamma(\nu)\, (\frac{x}{2})^{-\nu} \tag{7.56}$$

as $x \to 0+$ and $\Re\nu > 0$, we find that the bracket term on the right-hand side of (7.52) tends to a finite limit as $\epsilon \to 0+$. This results, for $\Re\nu > 0, b > 0$, in

$$\int_0^\infty x\, H_\nu^{(1)}(ax)\, J_\nu(bx)\, dx = \frac{1}{a^2 - b^2} \frac{2i}{\pi} (\frac{b}{a})^\nu, \quad \Im a > 0,$$

$$\int_0^\infty x\, H_\nu^{(2)}(ax)\, J_\nu(bx)\, dx = -\frac{1}{a^2 - b^2} \frac{2i}{\pi} (\frac{b}{a})^\nu, \quad \Im a < 0. \tag{7.57}$$

The relations (7.57) remain valid for a complex b as long as $\Im((a - b)\lambda) > 0$ in the first formula, and $\Im((a - b)\lambda) < 0$ in the second one as $\Re\lambda$ tends to infinity.

7.1.4 MacRobert's Proof

MacRobert has shown that for holomorphic functions the inversion formula for the Hankel transform can be established by the method of contour integration. More precisely, we have

Proposition 7.2. *Let $0 \le \epsilon < \lambda \le \infty$, and assume that the function $f(x)$ is holomorphic in a neighborhood containing the interval (ϵ, λ). Then, for $\Re \nu > -1$,*

$$g_\nu(y) := \int_\epsilon^\lambda x \, J_\nu(yx) \, f(x) \, dx \tag{7.58}$$

implies

$$\chi_{(\epsilon, \lambda)} f(x) = \int_0^\infty y \, J_\nu(yx) \, g_\nu(y) \, dy. \tag{7.59}$$

See also [34, Sneddon p.53]. If we let $\epsilon \to 0+$ and $\lambda \to \infty$ we obtain the transform pair of Definition 7.1. Note that the transform pair in those definitions was established without the hypothesis that $f(x)$ is real analytic on the interval (ϵ, λ), however, the order of the Hankel transform was supposed to be an integer.

Proof. For $x > 0$ and temporarily $a > 0$, let

$$I(x) := \int_a^\infty y \, J_\nu(yx) \, g_\nu(y) \, dy = \int_a^\infty y \, J_\nu(yx) \int_\epsilon^\lambda t \, J_\nu(yt) \, f(t) \, dt \, dy.$$

Using (7.30) we express $J_\nu(ty)$ by the Hankel functions and obtain

$$I(x) = \int_a^\infty y \, J_\nu(yx) \frac{1}{2} \int_\epsilon^\lambda t \, \{H_\nu^{(1)}(yt) + H_\nu^{(2)}(yt)\} \, f(t) \, dt \, dy$$

$$= \frac{1}{2} \int_a^\infty y \, J_\nu(yx) \int_\epsilon^\lambda t \, H_\nu^{(1)}(yt) \, f(t) \, dt \, dy$$

$$+ \frac{1}{2} \int_a^\infty y \, J_\nu(yx) \int_\epsilon^\lambda t \, H_\nu^{(2)}(yt) \, f(t) \, dt \, dy.$$

The integrands in the inner integrals on the right-hand side are holomorphic functions of t which enables us to replace the contour $(\epsilon, \lambda) \subset \mathbb{R}$ by the contours $\gamma_{\epsilon, \lambda}^+$ and $\gamma_{\epsilon, \lambda}^-$, lying in the upper and lower half-plane, respectively. Thus,

$$I(x) = \frac{1}{2} \int_a^\infty y \, J_\nu(yx) \int_{\gamma_{\epsilon, \lambda}^+} z \, H_\nu^{(1)}(yz) \, f(z) \, dz \, dy$$

$$+ \frac{1}{2} \int_a^\infty y \, J_\nu(yx) \int_{\gamma_{\epsilon, \lambda}^-} z \, H_\nu^{(2)}(yz) \, f(z) \, dz \, dy.$$

$H_\nu^{(1)}(yz)$ and $H_\nu^{(2)}(yz)$ decay exponentially as $y \to \infty$ for $\Im z > 0$ and $\Im z < 0$, respectively, which allows us to reverse the order of integration

$$I(x) = \frac{1}{2} \int_{\gamma_{\epsilon, \lambda}^+} z \, f(z) \int_a^\infty y \, J_\nu(yx) \, H_\nu^{(1)}(yz) \, dy \, dz$$

$$+ \frac{1}{2} \int_{\gamma_{\epsilon, \lambda}^-} z \, f(z) \int_a^\infty y \, J_\nu(yx) \, H_\nu^{(2)}(yz) \, dy \, dz.$$

There is a subtlety in this step since the mentioned limit behavior of the Hankel function fails to be true at the real endpoint λ. However, this difficulty can be overcome by using the comment at the end of Section 7.1.3. Now we let a tend to zero and apply Lommel's formulas (7.57) to obtain

$$I(x) = \frac{1}{2} \int_{\gamma_{\epsilon,\lambda}^+} z\, f(z)\, \frac{1}{z^2 - x^2}\, \frac{2i}{\pi} \left(\frac{x}{z}\right)^\nu dz - \frac{1}{2} \int_{\gamma_{\epsilon,\lambda}^-} z\, f(z)\, \frac{1}{z^2 - x^2} \left(\frac{x}{z}\right)^\nu dz$$

$$= -\frac{i}{\pi} \oint_{(\epsilon,\lambda)} z\, f(z)\, \frac{1}{z^2 - x^2} \left(\frac{x}{z}\right)^\nu dz = \frac{1}{2\pi i} \oint_{(\epsilon,\lambda)} \frac{2z f(z)}{z+x} \left(\frac{x}{z}\right)^\nu \frac{dz}{z-x}.$$

Inside the closed contour (ϵ, λ) the integrand has only the pole $z = x$, the other pole being outside the closed contour. Therefore, for $\epsilon < x < \lambda$ we have

$$I(x) = \mathrm{Res}_{z=x} \left\{ \frac{2z f(z)}{z+x} \left(\frac{x}{z}\right)^\nu \frac{1}{z-x} \right\} = \frac{2x f(x)}{x+x} \left(\frac{x}{x}\right)^\nu = f(x). \tag{7.60}$$

For $0 < x < \epsilon$ or $\lambda < x$ we get $I(x) = 0$. $\qquad\square$

7.1.5 Some Hankel Transforms of Ordinary Functions

Some Hankel transforms may be computed by using known Laplace transforms. Indeed, we have for example

Proposition 7.3. *If the one-sided Laplace transform of $t\, f(t)\, J_\nu(t)$ exists, then we have*

$$\mathcal{H}_\nu[e^{-sx} f(x)](y) = \mathcal{L}_+[f(t)t J_\nu(t\, y)](s). \tag{7.61}$$

Proof.

$$\mathcal{H}_\nu[e^{-sx} f(x)](y) = \int_0^\infty x\, e^{-sx} f(x) J_\nu(yx)\, dx = \mathcal{L}_+[f(t)t J_\nu(t\, y)](s). \qquad\square$$

With $f(x) = 1/x$ and (7.50) we find for $\Re a > 0$ and $\Re \nu > -1$,

$$\frac{e^{-ax}}{x} \overset{H_\nu}{\circ\!-\!\bullet} \frac{(\sqrt{a^2 + y^2} - a)^\nu}{y^\nu \sqrt{a^2 + y^2}}. \tag{7.62}$$

By the self-reciprocity of the Hankel transform, we obtain, for $\nu = 0$,

$$\frac{1}{\sqrt{a^2 + x^2}} \overset{H_0}{\circ\!-\!\bullet} \frac{e^{-ay}}{y}. \tag{7.63}$$

Also, for $m = 0, 1, 2, \ldots$,

$$\mathcal{H}_\nu[x^m e^{-sx}](y) = \mathcal{L}_+[t^{m+1} J_\nu(t\, y)](s)$$

$$= (-1)^{m+1} \frac{d^{m+1}}{ds^{m+1}} \frac{(\sqrt{s^2 + y^2} - s)^\nu}{y^\nu \sqrt{s^2 + y^2}};,$$

for example with $\nu = 0$, $m = 0$,

$$e^{-ax} \overset{H_0}{\circ\!-\!\bullet} \frac{a}{(a^2 + y^2)^{3/2}}. \tag{7.64}$$

The above result is a special case of the following situation. Using the known Laplace correspondence

$$\mathcal{L}_+[t^{\nu+1}\, J_\nu(yt)](s) = \frac{2^{\nu+1}\, \Gamma(\nu+3/2)\, s\, y^\nu}{\sqrt{\pi}\, (s^2+y^2)^{\nu+3/2}},$$

for $\Re s > 0$, we get

$$x^\nu e^{-ax} \circ\!\!\!-\!\!\!\overset{H_\nu}{\bullet}\, \frac{2^{\nu+1}\, \Gamma(\nu+3/2)\, a\, y^\nu}{\sqrt{\pi}\, (a^2+y^2)^{\nu+3/2}} \tag{7.65}$$

for $\Re \nu > -1$, $\Re a > 0$. By using the self-reciprocity of the Hankel transform we obtain

$$\frac{x^\nu}{(a^2+x^2)^{\nu+3/2}} \circ\!\!\!-\!\!\!\overset{H_\nu}{\bullet}\, \frac{\sqrt{\pi}}{2^{\nu+1}\, \Gamma(\nu+3/2)\, a}\, y^\nu e^{-ay}.$$

Example 7.1. By using the substitution $x = \sqrt{t}$, we obtain

$$H_\nu[x^\mu\, e^{-ax^2}](y) = \int_0^\infty x^{\mu+1}\, J_\nu(yx) e^{-ax^2}\, dx = \int_0^\infty t^{(\mu+1)/2}\, J_\nu(y\sqrt{t}) e^{-at}\, \frac{dt}{2\sqrt{t}}$$

$$= \frac{1}{2} \int_0^\infty e^{-at} t^{\mu/2}\, J_\nu(y\sqrt{t})\, dt = \frac{1}{2}\, \mathcal{L}_+[t^{\mu/2}\, J_\nu(y\sqrt{t})](a).$$

There is now a whole family of Laplace correspondences that can be used, for example,

$$\mathcal{L}_+[t^{\frac{\nu}{2}+n}\, J_\nu(a\sqrt{t})](s) = n!\, \left(\frac{a}{2}\right)^\nu \frac{\exp(-a^2/(4s))}{s^{\nu+n+1}}\, L_n^\nu\left(\frac{a^2}{4s}\right)$$

valid for $\Re s > 0$, $\Re(\nu+n) > -1$, $n \in \mathbb{N}$, and where

$$L_n^\nu(x) := \frac{e^x}{n!\, x^\nu}\, \frac{d^n}{dx^n}\{e^{-x} x^{n+\nu}\} \tag{7.66}$$

is Laguerre's polynomial. Thus, for $\Re(\nu+n) > -1$, $\Re a > 0$, we obtain

$$x^{\nu+2n}\, e^{-ax^2} \circ\!\!\!-\!\!\!\overset{H_\nu}{\bullet}\, n!\, \left(\frac{y}{2}\right)^\nu \frac{\exp(-y^2/(4a))}{2a^{\nu+n+1}}\, L_n^\nu\left(\frac{y^2}{4a}\right). \tag{7.67}$$

Especially,

$$e^{-ax^2} \circ\!\!\!-\!\!\!\overset{H_0}{\bullet}\, \frac{\exp(-y^2/(4a))}{2a}, \qquad x e^{-ax^2} \circ\!\!\!-\!\!\!\overset{H_1}{\bullet}\, \frac{y}{4}\, \frac{\exp(-y^2/(4a))}{a^2}. \tag{7.68}$$

Problem 7.4. In the first correspondence of (7.68) put $a = \kappa + i\omega$, $\kappa > 0$ and show, by equating real part and imaginary part on both sides, that we have

$$e^{-\kappa x^2}\cos(\omega x^2) \circ\!\!\!-\!\!\!\overset{H_0}{\bullet}\, 2\Omega\, e^{-\kappa \Omega y^2}\{\kappa\, \cos(\omega\Omega y^2) + \omega\, \sin(\omega\Omega y^2)\}, \tag{7.69}$$

$$e^{-\kappa x^2}\sin(\omega x^2) \circ\!\!\!-\!\!\!\overset{H_0}{\bullet}\, 2\Omega\, e^{-\kappa \Omega y^2}\{-\omega\, \cos(\omega\Omega y^2) + \kappa\, \sin(\omega\Omega y^2)\}, \tag{7.70}$$

where

$$\Omega = \frac{1}{4(\kappa^2+\omega^2)}.$$

Problem 7.5. Use the Laplace correspondence

$$Y(t)\, J_\nu(a\sqrt{t})\, J_\nu(b\sqrt{t}) \circ\!-\!\bullet\; \frac{1}{s}\exp(-\frac{a^2+b^2}{4s})\, I_\nu(\frac{ab}{2s})$$

to establish for $a > 0$, $b > 0$,

$$e^{-ax^2}\, J_\nu(bx) \circ\!\!\underset{}{\overset{H_\nu}{-}}\!\!\bullet\; \frac{1}{2a}\, e^{-(y^2+b^2)/(4a)}\, I_\nu(\frac{by}{2a}). \tag{7.71}$$

Example 7.2. Results on Mellin transforms can also be used, for example,

$$\mathcal{H}_\nu[x^{s-2}](y) = \int_0^\infty x^{s-1}\, J_\nu(yx)\, dx = \mathcal{M}[J_\nu(yx)](s)$$

$$= y^{-s}\, 2^{s-1}\, \frac{\Gamma(\frac{\nu+s}{2})}{\Gamma(\frac{\nu-s+2}{2})}$$

with $-\Re\nu < \Re s < \Re\nu + 2$, where (6.17) was used together with the Scale Change Rule for Mellin transforms. Thus,

$$x^\mu \circ\!\!\underset{}{\overset{H_\nu}{-}}\!\!\bullet\; 2^{\mu+1}\frac{\Gamma(\frac{\nu+\mu+2}{2})}{\Gamma(\frac{\nu-\mu}{2})}\, y^{-\mu-2} \tag{7.72}$$

for $-\Re\nu - 2 < \Re\mu < \Re\nu$. However, by (7.36) it is seen that the integral

$$\int_0^\infty x^{\mu+1}\, J_\nu(x)\, dx$$

only converges if $\Re\mu < -1/2$ (see also Example 7.8). Hence (7.72) is valid in the classical sense only if $-\Re\nu - 2 < \Re\mu < -1/2$. Particularly, for $\nu = 0$, we get

$$\frac{1}{x} \circ\!\!\underset{}{\overset{H_0}{-}}\!\!\bullet\; \frac{1}{y}. \tag{7.73}$$

Example 7.3. Find the Hankel transform of $f(x) = x^\nu\, (a^2 - x^2)^\mu\, Y(a - x)$.

$$g(y) = \mathcal{H}_\nu[f(x)](y) = \int_0^a x^{\nu+1}\, (a^2 - x^2)^\mu\, J_\nu(yx)\, dx.$$

Here the Taylor expansion (7.23) of the Bessel function is used;

$$J_\nu(yx) = \left(\frac{yx}{2}\right)^\nu \sum_{k=0}^\infty \frac{(-1)^k}{k!\,\Gamma(\nu+k+1)}\left(\frac{yx}{2}\right)^{2k}.$$

Interchanging the series and integration yields

$$g(y) = \left(\frac{y}{2}\right)^\nu \sum_{k=0}^\infty \frac{(-1)^k\, y^{2k}}{k!\,\Gamma(\nu+k+1)\, 4^k} \int_0^a x^{2(\nu+k)+1}(a^2 - x^2)^\mu\, dx.$$

The change of variables $x = a\sqrt{t}$ transforms the integral into

$$\frac{1}{2}\, a^{2(k+\nu+\mu+1)}\, B(k+\nu+1, \mu+1) = \frac{1}{2}\, a^{2(k+\nu+\mu+1)}\, \frac{\Gamma(k+\nu+1)\Gamma(\mu+1)}{\Gamma(k+\nu+\mu+2)}.$$

Then,

$$g(y) = \left(\frac{y}{2}\right)^\nu \frac{a^{2(\nu+\mu+1)}\,\Gamma(\mu+1)}{2} \sum_{k=0}^{\infty} \frac{(-1)^k\,(ay)^{2k}}{k!\,\Gamma(\nu+\mu+k+2)\,4^k}.$$

Rearranging the terms and using again the series expansion of the Bessel function eventually yields, for $\nu > -1, \mu > -1$,

$$x^\nu\,(a^2 - x^2)^\mu\,Y(a-x) \overset{H_\nu}{\circ\!\!-\!\!\bullet} 2^\mu\,a^{\nu+\mu+1}\,\frac{\Gamma(\mu+1)}{y^{\mu+1}}\,J_{\nu+\mu+1}(ay). \tag{7.74}$$

Putting $a = 1$ and using the change of variables $x = \sin\theta$ in the Hankel integral yields *Sonine's First Integral,*

$$J_{\nu+\mu+1}(y) = \frac{y^{\mu+1}}{2^\mu\,\Gamma(\mu+1)} \int_0^{\pi/2} \sin^{\nu+1}(\theta)\,\cos^{2\mu+1}(\theta)\,J_\nu(y\sin\theta)\,d\theta, \tag{7.75}$$

as the reader may verify.

7.1.6 Operational Properties

Parseval's Relation

The Hankel transformation does not satisfy any simple convolution property, however, a relation of Parseval's type is readily obtainable.

Proposition 7.4. *Let* $\mathcal{H}_\nu[f(x)](y) = F_\nu(y)$ *and* $\mathcal{H}_\nu[g(x)](y) = G_\nu(y)$, *then*

$$\int_0^\infty y\,F_\nu(y)\,G_\nu(y)\,dy = \int_0^\infty x\,f(x)\,g(x)\,dx. \tag{7.76}$$

Proof.

$$\int_0^\infty y\,F_\nu(y)\,G_\nu(y)\,dy = \int_0^\infty y\,F_\nu(y) \int_0^\infty x\,g(x)J_\nu(yx)\,dx\,dy$$

$$= \int_0^\infty x\,g(x) \int_0^\infty y\,F_\nu(y)\,J_\nu(yx)\,dy\,dx$$

$$= \int_0^\infty x\,g(x)f(x)\,dx,$$

where we have interchanged the order of integration, and then used the self-reciprocity of the Hankel transformation. □

The Similarity and Division Rule

If $\mathcal{H}_\nu[f(x)](y) = F_\nu(y)$, we want to find the Hankel transform of $f(ax)$, where $a > 0$. By a simple change of variables, we obtain immediately

$$\mathcal{H}_\nu[f(ax)](y) = \int_0^\infty x\,J_\nu(yx)\,f(ax)\,dx$$

$$= \frac{1}{a^2} \int_0^\infty t\,J_\nu(\tfrac{y}{a}\,t)\,f(t)\,dt$$

which gives

Proposition 7.5. *If* $\mathcal{H}_\nu[f(x)](y) = F_\nu(y)$, *then for* $a > 0$,

$$\mathcal{H}_\nu[f(a\,x)](y) = \frac{1}{a^2}\,F_\nu(\tfrac{y}{a}). \tag{7.77}$$

Division by x

If $\mathcal{H}_\nu[f(x)](y) = F_\nu(y)$ we want to find the Hankel transform of $f(x)/x$. We shall use (7.43) and obtain,

$$
\begin{aligned}
\mathcal{H}_\nu[\frac{f(x)}{x}](y) &= \int_0^\infty J_\nu(yx)\, f(x)\, dx \\
&= \frac{y}{2\nu} \int_0^\infty x\{J_{\nu-1}(yx) + J_{\nu+1}(yx)\}\, f(x)\, dx \\
&= \frac{y}{2\nu} \{F_{\nu-1}(y) + F_{\nu+1}(y)\}.
\end{aligned}
$$

We have proved

Proposition 7.6. *If* $\mathcal{H}_\nu[f(x)](y) = F_\nu(y)$, *then for* $\nu \neq 0$,

$$\mathcal{H}_\nu[\frac{f(x)}{x}](y) = \frac{y}{2\nu}\{F_{\nu-1}(y) + F_{\nu+1}(y)\}. \tag{7.78}$$

By applying twice this rule, we get for $\nu \neq -1, 0, 1$,

$$\mathcal{H}_\nu[\frac{f(x)}{x^2}](y) = \frac{y^2}{4\nu}\{\frac{1}{\nu-1}F_{\nu-2}(y) + \frac{2\nu}{\nu^2-1}F_\nu(y) + \frac{1}{\nu+1}F_{\nu+2}(y)\}. \tag{7.79}$$

Differentiation Rules

If $\mathcal{H}_\nu[f(x)](y) = F_\nu(y)$, we want to find the Hankel transform of the derivative.

$$
\begin{aligned}
\mathcal{H}_\nu[\frac{df(x)}{dx}](y) &= \int_0^\infty x\frac{df(x)}{dx}\, J_\nu(yx)\, dx \\
&= x\, f(x)\, J_\nu(yx)|_0^\infty - \int_0^\infty f(x)\, \frac{d}{dx}(x\, J_\nu(yx))\, dx.
\end{aligned}
$$

By using Problem 7.1 the integral on the right-hand side becomes

$$
\int_0^\infty f(x)\, y\, \{\left(\frac{\nu+1}{2\nu}\right) x J_{\nu-1}(yx) - \left(\frac{\nu-1}{2\nu}\right) x J_{\nu+1}(yx)\}\, dx
$$

$$
= y\, \{\left(\frac{\nu+1}{2\nu}\right) F_{\nu-1}(y) - \left(\frac{\nu-1}{2\nu}\right) F_{\nu+1}(y)\}.
$$

Thus,

Proposition 7.7. *Let* $\mathcal{H}_\nu[f(x)](y) = F_\nu(y)$, *then*

$$
\mathcal{H}_\nu[\frac{df(x)}{dx}](y) = y\, \{\left(\frac{\nu-1}{2\nu}\right) F_{\nu+1}(y) - \left(\frac{\nu+1}{2\nu}\right) F_{\nu-1}(y)\}
$$
$$
+ x\, f(x)\, J_\nu(yx)|_{x\to 0}^{x\to\infty}. \tag{7.80}
$$

Problem 7.6. By using the Division and the Differentiation Rule, establish

$$\mathcal{H}_\nu[\frac{1}{x}\frac{df(x)}{dx}](y)$$
$$= -\frac{y^2}{4}\left\{\frac{1}{\nu-1}F_{\nu-2}(y) + \frac{2}{\nu^2-1}F_\nu(y) - \frac{1}{\nu+1}F_{\nu+2}(y)\right\} \tag{7.81}$$
$$+ f(x)\,J_\nu(yx)|_{x\to0}^{x\to\infty}.$$

Problem 7.7. By applying twice the above proposition, establish

$$\mathcal{H}_\nu[\frac{d^2f(x)}{dx^2}](y)$$
$$= \frac{y^2}{4}\left\{\frac{\nu+1}{\nu-1}F_{\nu-2}(y) - \frac{2(\nu^2-3)}{\nu^2-1}F_\nu(y) + \frac{\nu-1}{\nu+1}F_{\nu+2}(y)\right\} \tag{7.82}$$
$$+ x\,f'(x)\,J_\nu(yx)|_{x\to0}^{x\to\infty}$$
$$+ x\,y\,f(x)\left\{\frac{\nu-1}{2\nu}J_{\nu+1}(yx) - \frac{\nu+1}{2\nu}J_{\nu-1}(yx)\right\}\Bigg|_{x\to0}^{x\to\infty}.$$

By using (7.81), (7.82) and (7.79) multiplied by ν^2, we obtain the important

Proposition 7.8. Let $\mathcal{H}_\nu[f(x)](y) = F_\nu(y)$, then

$$\mathcal{H}_\nu[\frac{d^2f(x)}{dx^2} + \frac{1}{x}\frac{df(x)}{dx} - \nu^2\frac{f(x)}{x^2}](y) = -y^2\,F_\nu(y)$$
$$+ (f(x) + x\,f'(x))\,J_\nu(yx)|_{x\to0}^{x\to\infty} \tag{7.83}$$
$$+ x\,y\,f(x)\left\{\frac{\nu-1}{2\nu}J_{\nu+1}(yx) - \frac{\nu+1}{2\nu}J_{\nu-1}(yx)\right\}\Bigg|_{x\to0}^{x\to\infty}.$$

By using (7.43) and (7.44) the expression

$$x\,y\,f(x)\left\{\frac{\nu-1}{2\nu}J_{\nu+1}(yx) - \frac{\nu+1}{2\nu}J_{\nu-1}(yx)\right\}$$

can also be written as

$$f(x)\{yx\,J_\nu'(yx) - J_\nu(yx)\}$$

such that the sum of the two boundary terms in (7.83) becomes

$$R_f(y) := (f(x) + x\,f'(x))\,J_\nu(yx)|_{x\to0}^{x\to\infty}$$
$$+ x\,y\,f(x)\left\{\frac{\nu-1}{2\nu}J_{\nu+1}(yx) - \frac{\nu+1}{2\nu}J_{\nu-1}(yx)\right\}\Bigg|_{x\to0}^{x\to\infty}$$
$$= x\,f'(x)\,J_\nu(yx) + yx\,f(x)J_\nu'(yx)|_{x\to0}^{x\to\infty}.$$

If the function $f(x)$ is such that the boundary term $R_f(y)$ becomes zero, which is generally the case in many applications, we have for $\nu = 0$ the operational property

$$\mathcal{H}_0[\frac{d^2f(x)}{dx^2} + \frac{1}{x}\frac{df(x)}{dx}](y) = -y^2\,F_0(y). \tag{7.84}$$

This property makes of the Hankel transformation an important tool for solving boundary value problems having cylindrical symmetry. For by using cylindrical polar coordinates (r, θ, z) with $x = r \cos \theta$, $y = r \sin \theta$, $z = z$, the Laplace operator is expressed as

$$\nabla^2 = \frac{\partial^2}{\partial r^2} + \frac{1}{r}\frac{\partial}{\partial r} + \frac{1}{r^2}\frac{\partial^2}{\partial \theta^2} + \frac{\partial^2}{\partial z^2}.$$

In a problem of axial symmetry the term with the second derivative with respect to θ vanishes, and we have

$$\nabla^2 = \frac{\partial^2}{\partial r^2} + \frac{1}{r}\frac{\partial}{\partial r} + \frac{\partial^2}{\partial z^2}.$$

Then, with $g = g(r, z) \circ \overset{H_0}{-\!\!\!-\!\!\!\bullet} G(\rho, z)$, and taking the zero order Hankel transform with respect to r, we obtain,

$$\nabla^2 g = \frac{\partial^2 g}{\partial r^2} + \frac{1}{r}\frac{\partial g}{\partial r} + \frac{\partial^2 g}{\partial z^2} \circ \overset{H_0}{-\!\!\!-\!\!\!\bullet} -\rho^2 \, G(\rho, z) + \frac{d^2 G}{dz^2}.$$

Example 7.4. The diffusion equation of a boundary value problem with axial symmetry and a constant behavior along the z-axis is modeled by

$$\frac{\partial f}{\partial t} = \kappa \nabla^2 f, \quad (r > 0, t > 0), \tag{7.85}$$

where $u = u(x, y, z, t) =: f(r, t)$, and $\kappa > 0$ is the diffusivity constant, and $h(r) = f(r, 0)$ is the specified initial condition. Taking the zero order Hankel transform with respect to r on both sides of the equation yields the ordinary differential equation

$$\frac{\partial F(\rho, t)}{\partial t} = -\kappa \rho^2 \, F(\rho, t) \quad F(\rho, 0) = \mathcal{H}_0[h(r)](y) = H(\rho),$$

where we have assumed $\mathcal{H}_0[f(r, t)](\rho) = F(\rho, t)$. Its solution is

$$F(y, t) = H(\rho) \, e^{-\kappa \rho^2 t}.$$

Application of the inverse Hankel transform yields

$$f(r, t) = \int_0^\infty \rho \, J_0(\rho r) \, H(\rho) \, e^{-\kappa \rho^2 t} \, d\rho. \tag{7.86}$$

Because we have

$$H(\rho) = \int_0^\infty r' \, J_0(\rho r') \, h(r') dr',$$

we can write

$$f(r, t) = \int_0^\infty \int_0^\infty \rho \, e^{-\kappa \rho^2 t} \, J_0(\rho r) \, r' \, J_0(\rho r') \, h(r') \, dr' \, d\rho.$$

Interchanging the order of integration yields

$$f(r, t) = \int_0^\infty h(r') \, r' \int_0^\infty e^{-\kappa \rho^2 t} \, J_0(\rho r) \, J_0(\rho r') \, \rho \, d\rho \, dr'.$$

The inner integral is transformed into a Laplace integral by the change of variables $x = \kappa \rho^2$:

$$\int_0^\infty e^{-\kappa \rho^2 t} J_0(\rho r) J_0(\rho r') \rho \, d\rho = \frac{1}{2\kappa} \int_0^\infty e^{-xt} J_0(\frac{r}{\sqrt{\kappa}} \sqrt{x}) J_0(\frac{r'}{\sqrt{\kappa}} \sqrt{x}) \, dx.$$

By using the Laplace correspondence

$$Y(x) J_n(a\sqrt{x}) J_n(b\sqrt{x}) \circ\!\!-\!\!\bullet \frac{e^{(a^2+b^2)/(4s)}}{s} I_n(\frac{ab}{2s}),$$

we finally obtain

$$f(r,t) = \frac{1}{2\kappa t} \int_0^\infty e^{-(r^2+r'^2)/(4\kappa t)} I_0(\frac{r\,r'}{2\kappa t}) h(r') r' \, dr'$$

$$= \frac{e^{-r^2/(4\kappa t)}}{2\kappa t} \int_0^\infty e^{-r'^2/(4\kappa t)} I_0(\frac{r\,r'}{2\kappa t}) h(r') r' \, dr'.$$

$$(7.87)$$

If we have an initial condition of the form $h(r) = \delta(r-a)/(2\pi a)$ which corresponds to a concentric source of unit-strength at $r = a > 0$, we obtain

$$f(r,t) = \frac{1}{4\pi\kappa t} e^{-(r^2+a^2)/(4\kappa t)} I_0(\frac{r\,a}{2\kappa t}). \qquad (7.88)$$

7.2 Hankel Transforms of Hyperfunctions

7.2.1 Basic Definitions

We shall denote the complex variables z, ζ by $z = x + iy$ and $\zeta = \xi + i\eta$. Then we have $\zeta z = (\xi x - \eta y) + i(\xi y + \eta x)$. For a fixed and non-real ζ, we note that

(i) if $\Im\zeta = \eta > 0$, then $\Im(\zeta z)$ tends to ∞ as $x = \Re z \to \infty$,

(ii) if $\Im\zeta = \eta < 0$, then $\Im(\zeta z)$ tends to $-\infty$ as $x = \Re z \to \infty$,

(iii) if $\Re\zeta = \xi > 0$, then $\Re(\zeta z)$ tends to ∞ as $x = \Re z \to \infty$.

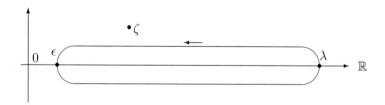

Figure 7.3: Contour (ϵ, λ)

In the complex z-plane consider now the finite closed contour (ϵ, λ) with $0 < \epsilon < \lambda < \infty$ shown in Figure 7.3. The fixed point ζ lies outside the contour. According to (7.34) and (7.35), the expressions $zH_\nu^{(1)}(\zeta z)$ for $\Im\zeta = \eta > 0$, and $zH_\nu^{(2)}(\zeta z)$ for $\Im\zeta = \eta < 0$, with ζ fixed, have the asymptotic behavior

$$z\, H_\nu^{(1)}(\zeta z) \approx \sqrt{\frac{2z}{\pi}} \exp\{i(\zeta z - \frac{\nu\pi}{2} - \frac{\pi}{4})\},$$

and

$$z H_\nu^{(2)}(\zeta z) \approx \sqrt{\frac{2z}{\pi}} \, \exp\{-i(\zeta z - \frac{\nu\pi}{2} - \frac{\pi}{4})\}$$

as $x = \Re z \to \infty$, respectively. Thus, for large positive x,

$$\left| z H_\nu^{(1)}(\zeta z) \right| < A_1 \, e^{-\xi y} \, x e^{-\eta x}, \quad (\eta > 0),$$

and

$$\left| z H_\nu^{(2)}(\zeta z) \right| < A_2 \, e^{\xi y} \, x e^{\eta x}, \quad (\eta < 0).$$

By taking into account that ξ is fixed and $|y|$ remains bounded, we conclude that the two contour integrals

$$\oint_{(\epsilon,\infty)} z H_\nu^{(1)}(\zeta z) \, F(z) \, dz, \quad \Im\zeta > 0,$$

$$\oint_{(\epsilon,\infty)} z H_\nu^{(2)}(\zeta z) \, F(z) \, dz, \quad \Im\zeta < 0$$

converge (and can be differentiated under the integral sign) as long as $|F(z)|$ is bounded by any polynomial in $|z|$.

Let $f(x) = [F(z)]$ be a hyperfunction defined for $x > 0$. For $0 < \epsilon < \lambda$ we define in the upper ζ-half-plane $\Im\zeta > 0$ the holomorphic function

$$G_+(\zeta; \epsilon, \lambda) := -\frac{1}{2} \oint_{(\epsilon,\lambda)} z H_\nu^{(1)}(\zeta z) \, F(z) \, dz, \tag{7.89}$$

and in the lower ζ-half-plane $\Im\zeta < 0$, the holomorphic function

$$G_-(\zeta; \epsilon, \lambda) := \frac{1}{2} \oint_{(\epsilon,\lambda)} z H_\nu^{(2)}(\zeta z) \, F(z) \, dz. \tag{7.90}$$

Definition 7.4. The Hankel transform of the hyperfunction $f(x) = [F(z)]$ defined for $x > 0$ is the hyperfunction

$$g(y) := \lim_{\epsilon \to 0+} \lim_{\lambda \to \infty} [G_+(\zeta; \epsilon, \lambda), G_-(\zeta; \epsilon, \lambda)] = [G_+(\zeta), G_-(\zeta)], \tag{7.91}$$

provided the two limits exist (remember that the two Hankel functions have a singularity at the origin).

We see that our generalized Hankel transform exists for hyperfunctions of slow growth as $x \to \infty$. Thus, we may write

$$g(y) = \left[-\frac{1}{2} \oint_{(0+,\infty)} z H_\nu^{(1)}(\zeta z) \, F(z) \, dz, \frac{1}{2} \oint_{(0+,\infty)} z H_\nu^{(2)}(\zeta z) \, F(z) \, dz \right].$$

We show now that the defined Hankel transform of a hyperfunction reduces to the familiar conventional Hankel transform when $f(x)$ and $g(y)$ are ordinary functions. If the hyperfunction $f(x)$ is an ordinary function, it has a value at any point x, and the function value is $f(x) = \lim_{\delta \to 0+} \{F_+(x + i\delta) - F_-(x - i\delta)\}$. Similarly, if the

Hankel transform $g(y)$ is an ordinary function, we have $g(y) = \lim_{\delta \to 0+} \{G_+(y + i\delta) - G_-(y - i\epsilon)\}$. Now, by using the compressed notation $G_+(y + i0; \epsilon, \lambda)$ for $\lim_{\delta \to 0+} G_+(y + i\delta; \epsilon, \lambda)$ etc. we may write

$$G_+(y + i0; \epsilon, \lambda) - G_-(y - i0, \epsilon, \lambda)$$

$$= -\frac{1}{2} \oint_{(\epsilon, \lambda)} z \, H_\nu^{(1)}((y + i0)z) \, F(z) \, dz - \frac{1}{2} \oint_{(\epsilon, \lambda)} z \, H_\nu^{(2)}((y - i0)z) \, F(z) \, dz$$

$$= -\oint_{(\epsilon, \lambda)} \frac{z}{2} \{H_\nu^{(1)}((y + i0)z) + H_\nu^{(2)}((y - i0)z)\} \, F(z) \, dz.$$

By collapsing the contour to the upper and the lower side of the interval, we can write for the last expression

$$\int_\epsilon^\lambda \frac{x}{2} \{H_\nu^{(1)}(y(x + i0)) + H_\nu^{(2)}(y(x + i0))\} \, F(x + i0) \, dx$$

$$- \int_\epsilon^\lambda \frac{x}{2} \{H_\nu^{(1)}(y(x - i0)) + H_\nu^{(2)}(y(x - i0))\} \, F(x - i0) \, dx$$

$$= \int_\epsilon^\lambda \frac{x}{2} \{H_\nu^{(1)}(yx) + H_\nu^{(2)}(yx)\} \{F(x + i0) - F(x - i0)\} \, dx$$

$$= \int_\epsilon^\lambda x \, J_\nu(yx) \, f(x) \, dx,$$

where (7.30) has been used in the last step. By letting $\epsilon \to 0$, and $\lambda \to \infty$, we finally obtain

$$g(y) = \int_0^\infty x \, J_\nu(yx) \, f(x) \, dx.$$

7.2.2   Transforms of some Familiar Hyperfunctions

Example 7.5. Let $f(x) = \delta(x - a)$, $a > 0$. Choose ϵ such that $0 < \epsilon < a < \lambda$. Then,

$$G_+(\zeta; \epsilon, \lambda) := \frac{1}{2} \frac{1}{2\pi i} \oint_{(\epsilon, \lambda)} \frac{z}{z - a} H_\nu^{(1)}(\zeta z) \, dz$$

$$= \frac{1}{2} \operatorname{Res}_{z=a} \{\frac{z}{z - a} H_\nu^{(1)}(\zeta z)\} = \frac{a}{2} H_\nu^{(1)}(\zeta a)$$

because $H_\nu^{(1)}(\zeta z)$ is holomorphic inside the contour. Similarly,

$$G_-(\zeta; \epsilon, \lambda) := -\frac{1}{2} \frac{1}{2\pi i} \oint_{(\epsilon, \lambda)} \frac{z}{z - a} H_\nu^{(2)}(\zeta z) \, dz$$

$$= -\frac{1}{2} \operatorname{Res}_{z=a} \{\frac{z}{z - a} H_\nu^{(2)}(\zeta z)\} = -\frac{a}{2} H_\nu^{(2)}(\zeta a).$$

By letting $\epsilon \to 0$ and $\lambda \to \infty$, we obtain the hyperfunction (having a value at $y = a$)

$$g(y) = \left[\frac{a}{2} H_\nu^{(1)}(\zeta a), -\frac{a}{2} H_\nu^{(2)}(\zeta a)\right] = \frac{a}{2} \{H_\nu^{(1)}(ay) + H_\nu^{(2)}(ay)\} = a \, J_\nu(ay),$$

thus,

$$\delta(x - a) \overset{H_\nu}{\circ\!\!-\!\!\bullet} a\, J_\nu(ay), \quad a > 0. \tag{7.92}$$

Example 7.6. Let now

$$f(x) = \delta^{(m)}(x - a) = \left[-\frac{(-1)^m\, m!}{2\pi i\, (z - a)^{m+1}} \right]$$

with $a > 0$. Choose again $0 < \epsilon < a < \lambda$. Then,

$$
\begin{aligned}
G_+(\zeta; \epsilon, \lambda) &:= \frac{1}{2} \frac{(-1)^m\, m!}{2\pi i} \oint_{(\epsilon, \lambda)} \frac{z}{(z - a)^{m+1}} H_\nu^{(1)}(\zeta z)\, dz \\
&= \frac{(-1)^m\, m!}{2} \operatorname{Res}_{z=a} \left\{ \frac{z}{(z - a)^{m+1}} H_\nu^{(1)}(\zeta z) \right\} \\
&= \frac{(-1)^m}{2} \zeta^{m-1} \left\{ a\, \zeta H_\nu^{(1)\,(m)}(\zeta a) + m\, H_\nu^{(1)\,(m-1)}(\zeta a) \right\},
\end{aligned}
$$

where for the differentiation Leibniz's formula has been used. Similarly, we find

$$G_-(\zeta; \epsilon, \lambda) := -\frac{(-1)^m}{2} \zeta^{m-1} \left\{ a\, \zeta H_\nu^{(2)\,(m)}(\zeta a) + m\, H_\nu^{(2)\,(m-1)}(\zeta a) \right\}.$$

As above, this yields for $m \in \mathbb{N}$, $a > 0$,

$$\delta^{(m)}(x - a) \overset{H_\nu}{\circ\!\!-\!\!\bullet} (-1)^m\, y^{m-1} \{ a\, y\, J_\nu^{(m)}(ay) + m\, J_\nu^{(m-1)}(ay) \}. \tag{7.93}$$

Example 7.7. Consider now the hyperfunction $f(x) = J_\nu(ax)$, $a > 0$. Restricted to the positive part of the real axis, it is a real analytic function, thus for $x > 0$, we write

$$J_\nu(x) = [J_\nu(az)/2, -J_\nu(az)/2].$$

With $0 < \epsilon < a < \lambda$, we have then

$$G_+(\zeta; \epsilon, \lambda) := -\frac{1}{2} \oint_{(\epsilon, \lambda)} \frac{J_\nu(az)}{2} z\, H_\nu^{(1)}(\zeta z)\, dz = \frac{1}{2} \int_\epsilon^\lambda J_\nu(ax)\, x\, H_\nu^{(1)}(\zeta x)\, dx,$$

$$G_-(\zeta; \epsilon, \lambda) := \frac{1}{2} \oint_{(\epsilon, \lambda)} \frac{J_\nu(az)}{2} z\, H_\nu^{(2)}(\zeta z)\, dz = -\frac{1}{2} \int_\epsilon^\lambda J_\nu(ax)\, x\, H_\nu^{(2)}(\zeta x)\, dx.$$

By using Lommel's formula (7.57) we obtain

$$\lim_{\epsilon \to 0+,\, \lambda \to \infty} G_+(\zeta; \epsilon, \lambda) = \frac{1}{\zeta^2 - a^2} \frac{i}{\pi} \left(\frac{a}{\zeta}\right)^\nu,$$

$$\lim_{\epsilon \to 0+,\, \lambda \to \infty} G_-(\zeta; \epsilon, \lambda) = (-)(-)\frac{1}{\zeta^2 - a^2} \frac{i}{\pi} \left(\frac{a}{\zeta}\right)^\nu.$$

So, for $\Im\zeta > 0$ and $\Im\zeta < 0$, we obtain the same expression which can be written as

$$
\begin{aligned}
&-\frac{1}{a} \frac{1}{2\pi i} \frac{1}{\zeta - a} \left\{ 1 - \frac{\zeta - a}{a} + \left(\frac{\zeta - a}{a}\right)^2 \mp \cdots \right\} \\
&+ \frac{1}{a} \frac{1}{2\pi i} \frac{e^{i\nu\pi}}{\zeta + a} \left\{ 1 + \frac{\zeta + a}{a} + \left(\frac{\zeta - a}{a}\right)^2 + \cdots \right\}.
\end{aligned}
$$

Now we restrict y to $y > 0$ and drop the real analytic parts, i.e., we pass to an equivalent defining function, and obtain

$$J_\nu(ax) \circ\!\!-\!\!\overset{H_\nu}{\bullet} \frac{1}{a}\delta(y - a). \tag{7.94}$$

This is a hint, when compared to (7.92) that our definition of the Hankel transform of a hyperfunction is symmetric or self-reciprocating, i.e., the inverse transform is the same.

Result (7.94) can be used to show that the inverse of the transform of an ordinary function $f(x)$ is the same as the direct transform, for

$$\mathcal{H}_\nu[\mathcal{H}_\nu[f(t)](x)](y) = \mathcal{H}_\nu[\int_0^\infty tJ_\nu(xt)\, f(t)\, dt](y) = \int_0^\infty t\, \mathcal{H}_\nu[J_\nu(xt)](x)\, f(t)\, dt$$

$$= \int_0^\infty t\frac{1}{t}\,\delta(y - t)\, f(t)\, dt = \int_0^\infty \delta(y - t)\, f(t)dt = f(y),$$

hence $\mathcal{H}_\nu[f(t)](x) = \mathcal{H}_\nu^{-1}[f(y)](x)$.

By Example (2.32) the correspondence (7.94) can also be written as

$$\mathcal{H}_\nu[J_\nu(ax)](y) = \frac{1}{y}\delta(y - a).$$

If we revert to the classical Hankel transform notation and writing a' for y, this reads as

$$\int_0^\infty x\, J_\nu(a'x)\, J_\nu(ax)\, dx = \frac{1}{a'}\delta(a' - a) = \frac{1}{a}\delta(a' - a) \tag{7.95}$$

which shows that the system $\{J_\nu(ax)\}$, $a \in \mathbb{R}_+$ forms an orthogonal (continuous) function system with the weight x on the positive part of the real axis.

Example 7.8. Let us now compute the Hankel transform of the hyperfunction $f(x) = x^\mu$ restricted to the positive part of the positive real axis. By (7.72) the Hankel transform exists in the classical sense only if $-2 < \Re(\mu + \nu)$, $\Re\mu < -1/2$. We then have

$$x^\mu \circ\!\!-\!\!\overset{H_\nu}{\bullet} 2^{\mu+1}\frac{\Gamma(\frac{\nu+\mu+2}{2})}{\Gamma(\frac{\nu-\mu}{2})}\, y^{-\mu-2}.$$

Notice now, for a fixed $y > 0$, that the expression on the right-hand side is a holomorphic function of the complex variable μ different from $\mu_m = -2m - 2 - \nu$. It has zeros at $\mu_m = 2m + \nu$, $m = 0, 1, 2, \ldots$. Having shown that the generalized Hankel transform reduces to the classical transform, if the latter exists, we see that the above expression which represents the analytic continuation of the classical Hankel transform for $-2 < \Re(\mu + \nu)$, $\Re\mu < -1/2$ must be the generalized Hankel transform of the hyperfunction x^μ. Therefore, we obtain, for $x > 0$,

$$\mathcal{H}_\nu[x^\mu](y) = \begin{cases} 0, & \text{if } \mu = \nu + 2m, \\ 2^{\mu+1}\frac{\Gamma(\frac{\nu+\mu+2}{2})}{\Gamma(\frac{\nu-\mu}{2})}\, y^{-\mu-2}, & \text{if } \mu \neq -2(m+1) - \nu \end{cases}$$

$$m = 0, 1, 2, \ldots$$

Especially, for $\nu = 0$ and $\mu = n \in \mathbb{N}_0$, we obtain, after some simplifications,

$$\mathcal{H}_0[x^n](y) = \begin{cases} 0, & \text{if } n \text{ even,} \\ (-1)^\ell \, \frac{(1\cdot3\cdot5\cdots(2\ell-1))^2}{y^{2\ell+1}}, & \text{if } n = 2\ell - 1, \end{cases} \tag{7.96}$$

$$1 \; \circ\!\!\xrightarrow{H_0}\!\!\bullet \; 0, \tag{7.97}$$

$$x \; \circ\!\!\xrightarrow{H_0}\!\!\bullet \; -\frac{1}{y^3}, \tag{7.98}$$

$$x^2 \; \circ\!\!\xrightarrow{H_0}\!\!\bullet \; 0, \tag{7.99}$$

$$x^3 \; \circ\!\!\xrightarrow{H_0}\!\!\bullet \; \frac{9}{y^5}. \tag{7.100}$$

Notice that these formulas cannot be read in the opposite sense, i.e., to find, for example, the zero order Hankel transform of $1/x^3$ to be $-y$ which would be erroneous. In fact this Hankel transform does not exist and the hypothesis of the next proposition below is not fulfilled.

Problem 7.8. Use Example 2.33 to establish, for $a > 0$,

$$x^\alpha \, \delta(x - a) \; \circ\!\!\xrightarrow{H_\nu}\!\!\bullet \; a^{\alpha+1} \, J_\nu(ay),$$

$$x^\alpha \, \delta^{(m)}(x - a) \; \circ\!\!\xrightarrow{H_\nu}\!\!\bullet \; (-1)^m \sum_{k=0}^{m} \binom{\alpha}{k} \frac{m!}{(m-k)!} a^{\alpha-k} \, y^{m-k-1}$$

$$\times \left\{ ay \, J_\nu^{(m-k)}(ay) + (m-k) \, J_n u^{(m-k-1)}(ay) \right\}.$$

We shall now establish that the inverse Hankel transform of a hyperfunction is in essence the same as the direct Hankel transform, i.e., that the Hankel transformation of a hyperfunction is a self-reciprocal transformation. We can prove the following form of the inversion theorem.

Proposition 7.9. *Let $f(x) = [F(z)]$ be a hyperfunction defined on the positive part of the real axis. If the twice iterated Hankel transform $\mathcal{H}_\nu[\mathcal{H}_\nu[f]]$ exists, then we have*

$$\mathcal{H}_\nu[\mathcal{H}_\nu[f]](x) = \frac{1}{x^{\nu+1}} \, \chi_{(0,\infty)} \left(x^{\nu+1} f(x) \right). \tag{7.101}$$

So if for $x > 0$ the right-hand side is identified with $f(x)$, we see that we may write $\mathcal{H}_\nu[f] = \mathcal{H}_\nu^{-1}[f]$.

Proof. Let $0 < \epsilon < t < \lambda < \infty$ and $f(t) = [F(\tau)]$ be a given hyperfunction. Let $g(x) = \mathcal{H}_\nu[f(t)](x) = [G_+(z), G_-(z)]$ be its Hankel transform such that

$$G_+(z; \epsilon, \lambda) = -\frac{1}{2} \oint_{(\epsilon,\lambda)} \tau \, H_\nu^{(1)}(z\tau) \, F(\tau) \, d\tau \; \rightarrow G_+(z),$$

$$G_-(z; \epsilon, \lambda) = \frac{1}{2} \oint_{(\epsilon,\lambda)} \tau \, H_\nu^{(2)}(z\tau) \, F(\tau) \, d\tau \; \rightarrow G_-(\tau)$$

as $\epsilon \to 0+$ and $\lambda \to \infty$. By assumption the hyperfunction $g(x)$ has again a Hankel transform $h(y) = \mathcal{H}_\nu[g(x)](y) = [H_+(\zeta), H_-(\zeta)]$ with

$$H_+(\zeta; \epsilon, \lambda) = -\frac{1}{2} \oint_{(\epsilon,\lambda)} z\, H_\nu^{(1)}(\zeta z)\, G(z)\, dz \;\to\; H_+(\zeta),$$

$$H_-(\zeta; \epsilon, \lambda) = \frac{1}{2} \oint_{(\epsilon,\lambda)} z\, H_\nu^{(2)}(\zeta z)\, G(z)\, dz \;\to\; H_-(\zeta)$$

as $\epsilon \to 0+$ and $\lambda \to \infty$.

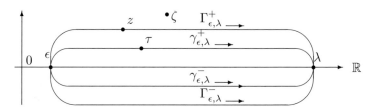

Figure 7.4: Contours in the proof of Proposition 7.9

We now consider in detail $H_+(\zeta; \epsilon, \lambda)$ where ζ is fixed with $\Im\zeta > 0$ and lies outside the closed contour $-\Gamma_{\epsilon,\lambda}^+ + \Gamma_{\epsilon,\lambda}^-$ on which the variable z is running. This contour contains the closed contour $-\gamma_{\epsilon,\lambda}^+ + \gamma_{\epsilon,\lambda}^-$ of the variable τ. See Figure 7.4. By inserting the contour integral over $F(\tau)$ into the contour integral over $G(z)$, we obtain, for $4\,H_+(\zeta; \epsilon, \lambda)$,

$$\int_{\Gamma_{\epsilon,\lambda}^+} z\, H_\nu^{(1)}(\zeta z)\, \Big\{ \int_{\gamma_{\epsilon,\lambda}^+} \tau H_\nu^{(1)}(z\tau)\, F_+(\tau)\, d\tau - \int_{\gamma_{\epsilon,\lambda}^-} \tau H_\nu^{(1)}(z\tau)\, F_-(\tau)\, d\tau \Big\}\, dz$$

$$- \int_{\Gamma_{\epsilon,\lambda}^-} z H_\nu^{(1)}(\zeta z)\, \Big\{ -\int_{\gamma_{\epsilon,\lambda}^+} \tau\, H_\nu^{(2)}(z\tau)\, F_+(\tau)\, d\tau + \int_{\gamma_{\epsilon,\lambda}^-} \tau H_\nu^{(2)}(z\tau)\, F_-(\tau)\, d\tau \Big\}\, dz.$$

By interchanging the order of integration, we get

$$\int_{\gamma_{\epsilon,\lambda}^+} \tau F_+(\tau)\, \Big\{ \int_{\Gamma_{\epsilon,\lambda}^+} z\, H_\nu^{(1)}(\zeta z) H_\nu^{(1)}(z\tau)\, dz + \int_{\Gamma_{\epsilon,\lambda}^-} z H_\nu^{(1)}(\zeta z) H_\nu^{(2)}(z\tau)\, dz \Big\}\, d\tau$$

$$- \int_{\gamma_{\epsilon,\lambda}^-} \tau F_-(\tau)\, \Big\{ \int_{\Gamma_{\epsilon,\lambda}^+} z\, H_\nu^{(1)}(\zeta z) H_\nu^{(1)}(z\tau)\, dz + \int_{\Gamma_{\epsilon,\lambda}^-} z H_\nu^{(1)}(\zeta z) H_\nu^{(2)}(z\tau)\, dz \Big\}\, d\tau.$$

The two integrands of the integrals inside the curly brackets are holomorphic functions of z, thus the contours can be collapsed to the real interval (ϵ, λ). This gives, when we still use (7.30),

$$H_+(\zeta; \epsilon, \lambda) = \frac{1}{2} \int_{\gamma_{\epsilon,\lambda}^+} \tau F_+(\tau) \int_\epsilon^\lambda x\, H_\nu^{(1)}(\zeta x)\, J_\nu(x\tau)\, dx\, d\tau$$

$$- \frac{1}{2} \int_{\gamma_{\epsilon,\lambda}^-} \tau F_-(\tau) \int_\epsilon^\lambda x\, H_\nu^{(1)}(\zeta x)\, J_\nu(x\tau)\, dx\, d\tau.$$

We now let $\epsilon \to 0+$, $\lambda \to \infty$, use Lommel's formulas 7.57), and obtain

$$H_+(\zeta) = \frac{1}{2} \int_{\gamma_{0+,\infty}^+} \tau F_+(\tau) \frac{1}{\zeta^2 - \tau^2} \frac{2i}{\pi} \left(\frac{\tau}{\zeta}\right)^\nu d\tau$$
$$- \frac{1}{2} \int_{\gamma_{0+,\infty}^-} \tau F_-(\tau) \frac{1}{\zeta^2 - \tau^2} \frac{2i}{\pi} \left(\frac{\tau}{\zeta}\right)^\nu d\tau.$$

The reader may now verify that for $H_-(\zeta; \epsilon, \lambda)$, with $\Im\zeta < 0$, we obtain the same expression! Therefore, we can write

$$h(y) = [H(\zeta)] = \left[-\frac{1}{2\pi i} \oint_{(0+,\infty)} \frac{2\tau F(\tau)}{\tau^2 - \zeta^2} \left(\frac{\tau}{\zeta}\right)^\nu d\tau \right].$$

Partial fraction decomposition yields finally

$$h(y) = \left[-\frac{1}{2\pi i} \oint_{(0+,\infty)} \frac{F(\tau)}{\tau - \zeta} \left(\frac{\tau}{\zeta}\right)^{\nu+1} d\tau + \frac{1}{2\pi i} \oint_{(0+,\infty)} \frac{F(\tau)}{\tau + \zeta} \left(\frac{\tau}{\zeta}\right)^{\nu+1} d\tau \right].$$

Consider the first term in the above defining function for $h(y)$. The functions $x^{\nu+1}$ and $1/x^{\nu+1}$ are real analytic functions on $\mathbb{R}_+$. So, if x is restricted to positive values, the hyperfunction $x^{\nu+1} f(x)$ has the defining function $z^{\nu+1} F(z)$. But then, see Section 2.4, we have

$$\frac{1}{x^{\nu+1}} \chi_{(0,\infty)} \left(x^{\nu+1} f(x)\right) = \left[\frac{1}{z^{\nu+1}} \left(-\frac{1}{2\pi i}\right) \oint_{(0+,\infty)} \frac{z^{\nu+1} F(\tau)}{\tau - z} d\tau \right].$$

By a change of variable from τ to $-\tau$, the second term in the defining function for $h(y)$ can be written as

$$\frac{1}{\zeta^{\nu+1}} \left(-\frac{1}{2\pi i}\right) \oint_{(-\infty,0-)} \frac{-(-\tau)^{\nu+1} F(-\tau)}{\tau - \zeta} d\tau = x^{-\nu-1} \chi_{(-\infty,0)} \left((-x)^{\nu+1} f(-x)\right)$$

giving no contribution on $\mathbb{R}_+$, i.e., it does not matter for the Hankel transform, an affair for strictly positive values of the independent variable. Because for strictly positive values of x, we may identify the hyperfunction $\frac{1}{x^{\nu+1}} \chi_{(0,\infty)} \left(x^{\nu+1} f(x)\right)$ with $f(x)$, we have established that

$$h = \mathcal{H}_\nu[\mathcal{H}_\nu[f]] = f \text{ or } \mathcal{H}_\nu^{-1}[f] = \mathcal{H}_\nu[f]. \tag{7.102}$$

$\square$

7.2.3 Operational Properties

We shall now show that most of the operational properties of the Hankel transforms of ordinary functions carry over to the transforms of hyperfunctions.

The Similarity Rule

The Similarity Rule carries over to hyperfunctions. Indeed let $f(x) = [F(z)]$ be a given hyperfunction having the Hankel transform $g(y) = \mathcal{H}_\nu[f(x)](y)$. Let $a > 0$, then $f(ax) = [F(az)]$, and $\mathcal{H}_\nu[f(ax)](y)$ becomes the hyperfunction $[G_{a-}(\zeta; \epsilon, \lambda),$ $G_{a+}(\zeta; \epsilon, \lambda)]$, where

$$G_{a+}(\zeta; \epsilon, \lambda) = -\frac{1}{2} \oint_{(\epsilon, \lambda)} z\, H_\nu^{(1)}(\zeta z)\, F(az)\, dz,$$

$$G_{a-}(\zeta; \epsilon, \lambda) = \frac{1}{2} \oint_{(\epsilon, \lambda)} z\, H_\nu^{(2)}(\zeta z)\, F(az)\, dz.$$

The change of variables $z' = az$ and letting $\epsilon \to 0+,\, \lambda \to \infty$ yields

$$G_{a+}(\zeta; \epsilon, \lambda) = -\frac{1}{a^2}\frac{1}{2} \oint_{(a\epsilon, a\lambda)} z'\, H_\nu^{(1)}(\tfrac{\zeta}{a}z')\, F(z')\, dz' \to \frac{1}{a^2} G_+(\tfrac{\zeta}{a}),$$

$$G_{a-}(\zeta; \epsilon, \lambda) = \frac{1}{a^2}\frac{1}{2} \oint_{(a\epsilon, a\lambda)} z'\, H_\nu^{(2)}(\tfrac{\zeta}{a}\zeta z')\, F(z')\, dz' \to \frac{1}{a^2} G_-(\tfrac{\zeta}{a}).$$

Thus,

$$\mathcal{H}_\nu[f(ax)](y) = \frac{1}{a^2} \mathcal{H}_\nu[f(x)](\tfrac{y}{a}), \tag{7.103}$$

and we have proved

Proposition 7.10. *The Similarity Rule of Proposition 7.5 carries over to hyperfunctions, i.e., (7.103) holds.*

Example 7.9. From (7.92), (7.93) we obtain for $a > 0,\ b > 0$,

$$\delta(ax - b) \ \overset{H_\nu}{\circ\!\!-\!\!\bullet} \ \frac{b}{a^2}\, J_\nu(\tfrac{b}{a}y), \tag{7.104}$$

$$\delta^{(m)}(ax - b) \ \overset{H_\nu}{\circ\!\!-\!\!\bullet} \ \frac{(-1)^m\, y^{m-1}}{a^2}\{\tfrac{b}{a}\, y\, J_\nu^{(m)}(\tfrac{b}{a}y) + m\, J_\nu^{(m-1)}(\tfrac{b}{a}y)\}. \tag{7.105}$$

Bear in mind for the following rules that the Hankel functions are cylinder functions, as are the Bessel functions J_ν, and therefore satisfy the same recurrence relations (see Proposition 7.1). This is the reason why most of the operational properties valid for ordinary functions carry over to hyperfunctions.

The Division Rule

If $f(x) = [F(z)]$ is a given hyperfunction, then for $x > 0$, $f(x)/x = [F(z)/z]$ is also a hyperfunction since $1/x$ is real analytic for $x > 0$ (our approach to Hankel transformation of hyperfunctions is a theory that matters only for positive values of the independent variable). Thus, for the two components of the defining function of $\mathcal{H}_\nu[f(x)/x](y)$ we have

$$G_{1+}(\zeta; \epsilon, \lambda) = -\frac{1}{2} \oint_{(\epsilon, \lambda)} z\, H_\nu^{(1)}(\zeta z)\, \frac{F(z)}{z}\, dz,$$

$$G_{1-}(\zeta; \epsilon, \lambda) = \frac{1}{2} \oint_{(\epsilon, \lambda)} z\, H_\nu^{(2)}(\zeta z)\, \frac{F(z)}{z}\, dz.$$

Let $g_\nu(y) = [G_{\nu,+}(\zeta), G_{\nu,-}(\zeta)] = \mathcal{H}_\nu[f(x)](y)$. We then have

$$-2\,G_{1+}(\zeta;\epsilon,\lambda) = \oint_{(\epsilon,\lambda)} H_\nu^{(1)}(\zeta z)\,F(z)\,dz$$

$$= \frac{\zeta}{2\nu}\int_{(\epsilon,\lambda)} z\{H_{\nu-1}^{(1)}(\zeta z) + H_{\nu+1}^{(1)}(\zeta z)\}\,F(z)\,dz$$

$$= \frac{\zeta}{2\nu}\{G_{\nu-1,+}(\zeta;\epsilon,\lambda) + G_{\nu+1,+}(\zeta;\epsilon,\lambda)\},$$

and similarly,

$$2\,G_{1-}(\zeta;\epsilon,\lambda) = \frac{\zeta}{2\nu}\{G_{\nu-1,-}(\zeta;\epsilon,\lambda) + G_{\nu+1,-}(\zeta;\epsilon,\lambda)\}.$$

Passing to the limits $\epsilon \to 0+, \lambda \to \infty$, we obtain

$$\mathcal{H}_\nu[\frac{f(x)}{x}](y) = \frac{y}{2\nu}\{\mathcal{H}_{\nu-1}[f(x)](y) + \mathcal{H}_{\nu+1}[f(x)](y)\} \qquad (7.106)$$

and we have proved

Proposition 7.11. *Proposition 7.6 carries over to hyperfunctions, i.e., (7.106) holds.*

Problem 7.9. Verify this rule with the correspondences of Problem 7.8.

The Differentiation Rule

Let $f(x) = [F(z)]$ be a hyperfunction, and $g(y) = \mathcal{H}_\nu[f(x)](y)$ its Hankel transform. We are seeking the Hankel transform of $f'(x) = [dF(z)/dz]$. Consider

$$G_+(\zeta;\epsilon,\lambda) = -\frac{1}{2}\oint_{(\epsilon,\lambda)} z\,H_\nu^{(1)}(\zeta z)\,\frac{dF(z)}{dz}\,dz,$$

$$G_-(\zeta;\epsilon,\lambda) = \frac{1}{2}\oint_{(\epsilon,\lambda)} z\,H_\nu^{(2)}(\zeta z)\,\frac{dF(z)}{dz}\,dz.$$

Integration by parts yields

$$-2\,G_+(\zeta;\epsilon,\lambda) = z\,F(z)\,H_\nu^{(1)}(\zeta z)\Big|_{\lambda+i0}^{\lambda-i0} - \oint_{(\epsilon,\lambda)} \frac{d}{dz}\Big(z\,H_\nu^{(1)}(\zeta z)\Big)\,F(z)\,dz.$$

If $f(x)$ is a hyperfunction of slow growth the first term tends to zero as $\lambda \to \infty$ due to the fact that $\Im\zeta > 0$. For the latter term we use Problem 7.1 which gives

$$\frac{d}{dz}\Big(z\,H_\nu^{(1)}(\zeta z)\Big) = \zeta\left(\frac{\nu+1}{2\nu}\right)z H_{\nu-1}^{(1)}(\zeta z) - \zeta\left(\frac{\nu-1}{2\nu}\right)z H_{\nu+1}^{(1)}(\zeta z).$$

Therefore

$$-2\,G_+(\zeta;\epsilon,\infty) = -\zeta\left(\frac{\nu+1}{2\nu}\right)\oint_{(\epsilon,\infty)} z\,H_{\nu-1}^{(1)}(\zeta z)\,F(z)\,dz$$

$$+\zeta\left(\frac{\nu-1}{2\nu}\right)\oint_{(\epsilon,\infty)} z\,H_{\nu+1}^{(1)}(\zeta z)\,F(z)\,dz,$$

and similarly, with $\Im \zeta < 0$,

$$2\, G_-(\zeta; \epsilon, \infty) = -\zeta \left(\frac{\nu+1}{2\nu}\right) \oint_{(\epsilon,\infty)} z\, H^{(2)}_{\nu-1}(\zeta z)\, F(z)\, dz$$
$$+ \zeta \left(\frac{\nu-1}{2\nu}\right) \oint_{(\epsilon,\infty)} z\, H^{(2)}_{\nu+1}(\zeta z)\, F(z)\, dz.$$

Letting $\epsilon \to 0+$ we obtain

$$\mathcal{H}_\nu[Df(x)](y)$$
$$= \left(\frac{\nu-1}{2\nu}\right) y\, \mathcal{H}_{\nu+1}[f(x)](y) - \left(\frac{\nu+1}{2\nu}\right) y\, \mathcal{H}_{\nu-1}[f(x)](y). \tag{7.107}$$

We have proved

Proposition 7.12. *If the hyperfunction $f(x)$ is of slow growth as $x \to \infty$ the Hankel transform of its generalized derivative is given by (7.107).*

Example 7.10. By using the rule and (7.93) we obtain,

$$\mathcal{H}_\nu[D\delta^{(m)}(x-a)](y)$$
$$= \left(\frac{\nu-1}{2\nu}\right) y\, \left\{(-1)^m\, y^{m-1}\{a\, y\, J^{(m)}_{\nu+1}(ay) + m\, J^{(m-1)}_{\nu+1}(ay)\}\right\}$$
$$- \left(\frac{\nu+1}{2\nu}\right) y\, \left\{\{(-1)^m\, y^{m-1}\{a\, y\, J^{(m)}_{\nu-1}(ay) + m\, J^{(m-1)}_{\nu-1}(ay)\}\right\}$$
$$= \frac{1}{2}\, y^{m+1}(-1)^m\, \{a\, J^{(m)}_{\nu+1}(ay) - a\, J^{(m)}_{\nu-1}(ay)\}$$
$$- \frac{1}{2\nu}\, y^m(-1)^m\, ay\, \{J^{(m)}_{\nu+1}(ay) + J^{(m)}_{\nu-1}(ay)\}$$
$$+ \frac{1}{2}\, y^m(-1)^m\, m\, \{J^{(m-1)}_{\nu+1}(ay) - J^{(m-1)}_{\nu-1}(ay)\}$$
$$- \frac{1}{2\nu}\, y^m(-1)^m\, m\, \{J^{(m-1)}_{\nu+1}(ay) + J^{(m-1)}_{\nu-1}(ay)\}.$$

By using Problem 7.2, the fifth and the seventh lines equal

$$-\frac{1}{2\nu}\, y^{m+1}(-1)^m\, 2\nu\, J^{(m)}_\nu(ay)$$

and, by Proposition 7.1, the forth and the sixth lines become

$$-\frac{1}{2}\, y^m(-1)^m\{2a J^{(m+1)}_\nu(ay) + 2m J^{(m)}_\nu(ay).$$

Finally, we obtain

$$\mathcal{H}_\nu[D\delta^{(m)}(x-a)](y) = (-1)^{m+1}\, y^m\{ay J^{(m+1)}_\nu(ay) + (m+1) J^{(m)}_\nu(ay)\}$$

in agreement with (7.93) for $m+1$.

As in the case of ordinary functions the following properties can be established. Details are left to the reader.

Proposition 7.13. *If the hyperfunction* $f(x)$ *is of slow growth as* $x \to \infty$, *and* $g_\nu(y) = \mathcal{H}_\nu[f(x)](y)$, *then*

$$\mathcal{H}_\nu[\frac{1}{x}Df(x)](y)$$
$$= -\frac{y^2}{4}\left\{\frac{1}{\nu-1}g_{\nu-2}(y) + \frac{2}{\nu^2-1}g_\nu(y) - \frac{1}{\nu+1}g_{\nu+2}(y)\right\}, \tag{7.108}$$

$$\mathcal{H}_\nu[D^2f(x)](y)$$
$$= \frac{y^2}{4}\left\{\frac{\nu+1}{\nu-1}g_{\nu-2}(y) - \frac{2(\nu^2-3)}{\nu^2-1}g_\nu(y) + \frac{\nu-1}{\nu+1}g_{\nu+2}(y)\right\}, \tag{7.109}$$

$$\mathcal{H}_\nu[D^2f(x) + \frac{1}{x}Df(x) - \nu^2\frac{f(x)}{x^2}](y) = -y^2\,g_\nu(y). \tag{7.110}$$

7.3 Applications

Example 7.11. A) Heat enters at a constant rate Q a semi-infinite body ($z \geq 0$) of thermal conductivity κ through a disk of radius a (defined by $x^2+y^2 \leq a^2$, $z = 0$). The bottom outside of the disk $x^2 + y^2 > a^2$, $z = 0$ is isolated. Find the steady-state temperature distribution inside the body.

The problem has an axial symmetry with respect to the z-axis. We use planar polar coordinates $x = r\cos\phi$, $y = r\sin\phi$. With the temperature $u(r,z)$ the model becomes

$$\nabla^2 u = 0, \quad \kappa\frac{\partial u(r,0)}{\partial z} = -Y(a-r)\frac{Q}{\pi a^2}. \tag{7.111}$$

The Laplace equation in cylindrical coordinates with axial symmetry is

$$\frac{\partial^2 u}{\partial r^2} + \frac{1}{r}\frac{\partial u}{\partial r} + \frac{\partial^2 u}{\partial z^2} = 0.$$

We take the Hankel transform of order 0 with respect to r. The image variable is denoted by ρ. By assuming that $U(\rho, z) = \mathcal{H}_0[u(r,z)](\rho)$ exists, we have in the image domain

$$-\rho^2\,U(\rho,z) + \frac{\partial^2 U(\rho,z)}{\partial z^2} = 0, \quad \frac{\partial U(\rho,0)}{\partial z} = -\frac{Q}{\pi a^2\kappa}\frac{a}{\rho}J_1(a\rho),$$

where (7.74) is used with $\mu = \nu = 0$. The solution of the ordinary differential equation on the left is $A\exp(-\rho z) + B\exp(\rho z)$. Since the temperature must remain bounded, B must be zero. By using the boundary condition on the right, we obtain

$$U(\rho,z) = \frac{Q}{\pi a\kappa}\frac{J_1(a\rho)}{\rho^2}e^{-\rho z}. \tag{7.112}$$

By the self-reciprocity of the Hankel transform, we find for the inverse transform

$$u(r,z) = \frac{Q}{\pi a\kappa}\mathcal{H}_0[\frac{J_1(a\rho)}{\rho^2}e^{-\rho z}](r)$$
$$= \frac{Q}{\pi a\kappa}\int_0^\infty e^{-\rho z}\frac{J_0(\rho r)J_1(a\rho)}{\rho}\,d\rho. \tag{7.113}$$

For $r = 0$ we obtain

$$u(0, z) = \frac{Q}{\pi a \kappa} \mathcal{L}[u(\rho) \frac{J_1(a\rho)}{\rho}](z) = \frac{Q}{\pi \kappa} \frac{1}{z + \sqrt{z^2 + a^2}}. \tag{7.114}$$

We may use Watson's lemma in order to find an asymptotic expansion for large values of z of the Laplace integral (see for example [29, Olver p. 71]). As $\rho \to 0+$ we have

$$\frac{J_0(\rho r) J_1(a\rho)}{\rho} = \frac{a}{2} - \frac{a}{8} \left(\frac{a^2}{2} + r^2\right) \rho^2 + \ldots$$

which generates the asymptotic expansion

$$u(r, z) = \frac{Q}{\pi \kappa} \left\{ \frac{1}{2z} - \frac{1}{4} (\frac{a^2}{2} + r^2) \frac{1}{z^3} \ldots \right\}, \quad z \to \infty. \tag{7.115}$$

From (7.113) and the fact that $J_1(a\rho)/(a\rho) \to 1/2$ as $a \to 0$ we have in this limit case

$$u(r, z) = \frac{Q}{2\pi\kappa} \int_0^\infty e^{-\rho z} J_0(\rho r) \, d\rho = \frac{Q}{2\pi\kappa} \frac{1}{\sqrt{z^2 + r^2}}. \tag{7.116}$$

B) Consider now the problem where heat enters the body with constant rate Q along a very thin annulus of radius a. The model is then

$$\nabla^2 u = 0, \quad \kappa \frac{\partial u(r, 0)}{\partial z} = -\delta(r - a) \frac{Q}{2\pi a}. \tag{7.117}$$

In the image domain we now obtain

$$-\rho^2 U(\rho, z) + \frac{\partial^2 U(\rho, z)}{\partial z^2} = 0, \quad \frac{\partial U(\rho, 0)}{\partial z} = -\frac{Q}{\pi a^2 \kappa} a J_0(a\rho),$$

where (7.92) has been used. Its solution becomes

$$U(\rho, z) = \frac{Q}{\pi a \kappa} \frac{J_0(a\rho)}{\rho} e^{-\rho z}. \tag{7.118}$$

Again by the self-reciprocity of the Hankel transform, we obtain

$$u(r, z) = \frac{Q}{\pi a \kappa} \mathcal{H}_0[\frac{J_0(a\rho)}{\rho} e^{-\rho z}](r)$$

$$= \frac{Q}{\pi a \kappa} \int_0^\infty e^{-\rho z} J_0(\rho r) J_0(a\rho) \, d\rho.$$

Let us use the Laplace correspondence

$$u(t) J_\nu(at) J_\nu(bt) \circ\!\!-\!\!\bullet \frac{1}{\pi\sqrt{ab}} q_{\nu-1/2} \left(\frac{s^2 + a^2 + b^2}{2ab}\right),$$

valid for $\Re\nu > -1/2$, where the involved Legendre function $q_\nu(x)$ can be expressed by Gauss' hypergeometric function $F(a, b; c; z)$,

$$q_\nu(x) = \frac{\sqrt{\pi} \, \Gamma(1 + \nu)}{2^{\nu+1} \Gamma(\nu + 3/2)} \frac{1}{x^{\nu+1}} F(\nu/2 + 1/2), \nu/2 + 1; \nu + 3/2; 1/x^2).$$

This gives finally

$$u(r,z) = \frac{Q}{\pi a \kappa} \frac{1}{\sqrt{z^2 + a^2 + r^2}} \, F\left(1/4, 3/4; 1; \left(\frac{2ar}{z^2 + a^2 + r^2}\right)^2\right). \qquad (7.119)$$

Note that

$$u(0,z) = \frac{Q}{\pi a \kappa} \frac{1}{\sqrt{z^2 + a^2}}. \qquad (7.120)$$

Example 7.12. Find the electrostatic potential between two infinite grounded plane plates at $z = a$ and $z = -a$ if a point charge q is placed at the origin. If $\Phi(r,z)$ denotes the potential, it must satisfy $\nabla^2 \Phi = 0$ for $r > 0, z \neq 0$. In free space, the potential of a point charge q placed at the origin is

$$\Phi_0(r,z) = \frac{q}{\sqrt{r^2 + z^2}}.$$

It can be verified by direct calculation that for $r > 0, z \neq 0$, we have

$$\nabla^2 \Phi_0 = \left(\frac{\partial^2}{\partial r^2} + \frac{1}{r}\frac{\partial}{\partial r} + \frac{\partial^2}{\partial z^2}\right)\left(\frac{q}{\sqrt{r^2 + z^2}}\right) = 0,$$

where the Laplacian has been expressed in cylindrical coordinates. Therefore, we seek the potential in the form

$$\Phi(r,z) = \Phi_0(r,z) + \Psi(r,z),$$

where $\Psi(r,z)$ must satisfy Laplace's equation for $z \neq 0$. This amounts to

$$\frac{\partial^2 \Psi}{\partial r^2} + \frac{1}{r}\frac{\partial \Psi}{\partial r} + \frac{\partial^2 \Psi}{\partial z^2} = 0, \qquad \frac{q}{\sqrt{r^2 + a^2}} + \Psi(r, \pm a) = 0.$$

By taking the zero order Hankel transform with respect to r of the two equations, we find with $\tilde{\Psi}(\rho,z) = \mathcal{H}_0[\Psi(r,z)](\rho)$,

$$-\rho^2 \tilde{\Psi}(\rho,z) + \frac{\partial^2 \tilde{\Psi}}{\partial z^2} = 0,$$

$$q\frac{e^{-a\rho}}{\rho} + \tilde{\Psi}(\rho, \pm a) = 0,$$

where we have used (7.63). The general solution of the ordinary differential equation is of the form $A\cosh(\rho z) + B\sinh(\rho z)$. On physical grounds the solution is expected to be an even function with respect to z which implies that we must choose $\tilde{\Psi}(\rho,z) = A\cosh(\rho z)$, i.e., $B = 0$. Then, the second equation requires $A = -q\exp(-a\rho)/(\rho\cosh(a\rho))$, thus

$$\tilde{\Psi}(\rho,z) = -q\frac{e^{-a\rho}\cosh(\rho z)}{\rho\cosh(a\rho)}.$$

Eventually, we obtain the solution of the problem in the form

$$\Phi(r,z) = \frac{q}{\sqrt{r^2 + z^2}} - q\int_0^\infty e^{-a\rho}\frac{\cosh(\rho z)}{\cosh(\rho a)} J_0(\rho r)\, d\rho. \qquad (7.121)$$

The integral can be evaluated numerically.

Example 7.13. Consider the problem of plate vibrations of a thin plate of thickness h and of infinite extent, where the undeflected surface is the x, y-plane. The governing equation of motion is (see [15, Graff p. 233])

$$\delta_0 \, \nabla^4 w + \rho_0 h \frac{\partial^2 w}{\partial t^2} = q,$$

where w denotes the elongation of the plate in the z-direction at the point (x, y) at time t, and q denotes the external load at that point. The constant $\delta_0 = Eh^3/(12(1 - \nu^2))$ and the density ρ_0 specifies the material of the plate. If we assume a symmetric situation with respect to the z-axis, and use polar coordinates r, θ, the governing equation, with no load, becomes

$$\delta_0 \left(\frac{\partial^2}{\partial r^2} + \frac{1}{r} \frac{\partial}{\partial r} \right)^2 w(r, t) + \rho_0 h \frac{\partial^2 w}{\partial t^2} = 0. \tag{7.122}$$

Assume the initial conditions

$$w(r, 0) = f(r), \qquad \frac{\partial w(r, 0)}{\partial t} = 0 \tag{7.123}$$

where $f(r)$ denotes the initial elongation. The initial velocity is zero. Let $W(\rho, t)$ denote the zero-order Hankel transform with respect to r of $w(r, t)$. The biharmonic differential operator transforms according to

$$\left(\frac{\partial^2}{\partial r^2} + \frac{1}{r} \frac{\partial}{\partial r} \right)^2 w(r, t) \circ \!\!\!- \!\!\!\overset{H_0}{\bullet} (-\rho^2)^2 W(\rho, t).$$

If $F(\rho) = \mathcal{H}_0[f(r)](\rho)$ and $b^2 = \delta_0/(\rho_0 h)$, the image equation is

$$\frac{\partial^2 W}{\partial t^2} + b^2 \rho^4 W(\rho, t) = 0$$

whose solution is, by taking into account the initial conditions, given by

$$W(\rho, t) = F(\rho) \, \cos(b\rho^2 t). \tag{7.124}$$

We now assume an initial elongation in the form of a symmetric Gaussian shape with respect to the origin $r = 0$, i.e.,

$$f(r) = f_0 \, e^{-r^2/a^2} \circ \!\!\!- \!\!\!\overset{H_0}{\bullet} F(\rho) = \frac{f_0 a^2}{2} \exp(-a^2 \rho^2/4), \tag{7.125}$$

where we have used (7.68). By the self-reciprocity of the Hankel transform we obtain

$$w(r, t) = \frac{f_0 \, a^2}{2} \, \mathcal{H}_0[e^{-a^2 \rho^2/4} \cos(b\rho^2 t)](r). \tag{7.126}$$

We use now the result of Problem 7.4 and eventually get

$$\begin{aligned} w(r, t) = f_0 \, a^2 \Omega(t) \, e^{-a^2 \, \Omega(t) \, r^2} \\ \times \, \{a^2 \, \cos(4 \, b \, t \, \Omega(t) r^2) + 4 \, b \, t \, \sin(4 \, b \, t \, \Omega(t) r^2)\} \end{aligned} \tag{7.127}$$

with $\Omega(t) = 1/(a^4 + 16\,b^2\,t^2)$.

Example 7.14. Consider a semi-infinite media $z > 0$ supposed to be symmetric with respect to the z-axis. Wave propagation then is governed by

$$\frac{\partial^2 w}{\partial r^2} + \frac{1}{r}\frac{\partial w}{\partial r} + \frac{\partial^2 w}{\partial z^2} = \frac{1}{c^2}\frac{\partial^2 w}{\partial t^2}, \quad 0 < r < \infty,\ 0 < z < \infty,\ t > 0,$$

when cylindrical polar coordinates (r, θ, z) are used, and $w = w(r, z, t)$ (no dependence on θ) denotes the propagating disturbance. We assume zero initial conditions

$$w(r, z, 0) = \frac{\partial w}{\partial t}(r, z, 0) = 0,$$

and the boundary conditions

$$\frac{\partial w}{\partial t}(r, 0, t) = u(a - r)\,\delta(t),$$

$$w(r, z, t) \to 0 \text{ as } r \to \infty, \text{ and as } z \to \infty.$$

We shall use a compound Laplace and Hankel transformation. By assuming the existence of $\hat{w}(r, z, s) = \mathcal{L}_+[w(r, z, t)](s)$, we obtain the first image equation

$$\frac{\partial^2 \hat{w}}{\partial r^2} + \frac{1}{r}\frac{\partial \hat{w}}{\partial r} + \frac{\partial^2 \hat{w}}{\partial z^2} = \frac{s^2}{c^2}\,\hat{w}(r, z, s),$$

where the zeros' initial conditions were used. By assuming the existence of $W(\rho, z, s) = \mathcal{H}_0[\hat{w}(r, z, s)](\rho)$ and using Proposition 7.13 we get the second image equation

$$-\rho^2\,W(\rho, z, s) + \frac{\partial^2 W}{\partial z^2} = \frac{s^2}{c^2}\,W(\rho, z, s), \text{ i.e.,}$$

$$\frac{\partial^2 W}{\partial z^2} - \left(\frac{s^2}{c^2} + \rho^2\right) W(\rho, z, s) = 0.$$

The general solution of this ordinary differential equation is

$$W(\rho, z, s) = A(\rho, s)\,\exp(-\frac{1}{c}\,\sqrt{s^2 + c^2\,\rho^2}\,z) + B(\rho, s)\,\exp(\frac{1}{c}\,\sqrt{s^2 + c^2\,\rho^2}\,z),$$

where the $B(\rho, s)$ must be zero in order to satisfy the boundary condition at $z = \infty$. From

$$W(\rho, z, s) = A(\rho, s)\,\exp(-\frac{1}{c}\,\sqrt{s^2 + c^2\,\rho^2}\,z)$$

we obtain

$$\frac{\partial W(\rho, 0, s)}{\partial z} = -A(\rho, s)\,\frac{1}{c}\,\sqrt{s^2 + c^2\,\rho^2}.$$

Taking the Laplace and Hankel transform of the first boundary condition yields

$$\mathcal{H}_0[\mathcal{L}_+[\frac{\partial w}{\partial t}(r, 0, t)](s)](\rho) = \frac{\partial W(\rho, 0, s)}{\partial z} = \frac{a}{\rho}\,J_1(a\rho)$$

where (7.74) was used. Comparing the last two results gives

$$A(\rho, s) = -\frac{a\,c}{\rho}\,\frac{J_1(a\rho)}{\sqrt{s^2 + c^2\,\rho^2}},$$

and finally,

$$W(\rho, z, s) = -\frac{a\,c}{\rho} \frac{J_1(a\rho)}{\sqrt{s^2 + c^2 \rho^2}} \exp(-\frac{1}{c} \sqrt{s^2 + c^2 \rho^2}\, z).$$

By using the Laplace correspondence

$$u(t - \zeta)\, J_0(\xi\, \sqrt{t^2 - \zeta^2})\, \circ\!-\!\bullet\ \frac{1}{\sqrt{s^2 + \xi^2}}\, \exp(-\zeta\, \sqrt{s^2 + \xi^2}),$$

we first revert to

$$-\frac{a\,c}{\rho}\, J_1(a\rho)\, u(t - \frac{z}{c})\, J_0\left(c\rho\, \sqrt{t^2 - \left(\frac{z}{c}\right)^2}\right),$$

and then, using the inverse Hankel transform formula, to

$$w(r, z, t) = -a\,c\,u(t - \frac{z}{c}) \int_0^\infty J_1(a\rho)\, J_0(\rho\, \sqrt{c^2\, t^2 - z^2})\, J_0(\rho r)\, d\rho. \qquad (7.128)$$

Appendix A

Complements

A.1 Physical Interpretation of Hyperfunctions

A.1.1 Flow Fields and Holomorphic Functions

(For the following discussion compare also [26, Needham Chapter 11]). Let us consider a two-dimensional flow field in a domain $\mathcal{D}$ of the real (x, y)-plane defined by its velocity field

$$\mathbf{V} = (v_1(x, y), v_2(x, y), 0)^t, \quad (x, y) \in \mathcal{D}.$$

The curl of the velocity $\mathbf{V}$,

$$\vec{\omega} := \mathrm{curl}\mathbf{V} = (0, 0, \frac{\partial v_2}{\partial x} - \frac{\partial v_1}{\partial y})$$

is called the vortex vector and

$$\omega := \frac{\partial v_2}{\partial x} - \frac{\partial v_1}{\partial y}$$

the *vortex* of the plane flow. The divergence of the velocity $\mathbf{V}$,

$$\theta := \mathrm{div}\mathbf{V} = \frac{\partial v_1}{\partial x} + \frac{\partial v_2}{\partial y},$$

is called the *source* of the flow. With the vortex and the source of the flow we now form the *complex vortex-source quantity*

$$\begin{aligned}
\Omega := \omega + i\theta &= \frac{\partial v_2}{\partial x} - \frac{\partial v_1}{\partial y} + i(\frac{\partial v_1}{\partial x} + \frac{\partial v_2}{\partial y}) \\
&= i\,(\frac{\partial}{\partial x} + i\frac{\partial}{\partial y})(v_1 - iv_2).
\end{aligned} \tag{A.1}$$

We now associate with the physical flow field $\mathbf{V}$ the *complex flow quantity*

$$w := v_1(z, \bar{z}) - i\,v_2(z, \bar{z}) \tag{A.2}$$

which is the *Pólya field* of the complex flow $v_1 + iv_2$. From the real physical plane we have passed to the complex model plane by putting

$$z := x + iy, \quad \bar{z} = x - iy,$$

$$x = x(z, \bar{z}) = \frac{z + \bar{z}}{2}, \quad y = y(z, \bar{z}) = \frac{z - \bar{z}}{2i}.$$

We then have

$$\frac{\partial}{\partial z} = \frac{\partial}{\partial x}\frac{\partial x}{\partial z} + \frac{\partial}{\partial y}\frac{\partial y}{\partial z} = \frac{1}{2}\left(\frac{\partial}{\partial x} - i\frac{\partial}{\partial y}\right),$$

$$\frac{\partial}{\partial \bar{z}} = \frac{\partial}{\partial x}\frac{\partial x}{\partial \bar{z}} + \frac{\partial}{\partial y}\frac{\partial y}{\partial \bar{z}} = \frac{1}{2}\left(\frac{\partial}{\partial x} + i\frac{\partial}{\partial y}\right).$$

From (A.10), (A.2), we obtain for the Pólya field $w = w(z, \bar{z})$ the relation

$$\Omega := \omega + i\theta = 2i\,\frac{\partial w}{\partial \bar{z}}. \tag{A.3}$$

Consider now

$$\iint_{\mathcal{D}} \Omega\, dx dy = \iint_{\mathcal{D}} i\left(\frac{\partial}{\partial x} + i\frac{\partial}{\partial y}\right)(v_1 - iv_2)\, dx dy$$

$$= \iint_{\mathcal{D}} \frac{\partial}{\partial x} i\,(v_1 - iv_2) - \frac{\partial}{\partial y}\,(v_1 - iv_2)\, dx dy$$

$$=: \iint_{\mathcal{D}} \left(\frac{\partial Q}{\partial x} - \frac{\partial P}{\partial y}\right) dx dy.$$

If the domain $\mathcal{D}$ is simply connected and has the boundary or border C, and if v_1, v_2 have continuous derivatives in $\mathcal{D}$, Green's theorem in the plane ensures that

$$\iint_{\mathcal{D}} \Omega\, dx dy = \oint_C P\, dx + Q\, dy = \oint_C (v_1 - iv_2)\, dx + i(v_1 - iv_2)\, dy$$

$$= \oint_C (v_1 - iv_2)\,(dx + idy) = \oint_C w(z, \bar{z})\, dz.$$

Therefore,

$$2i \iint_{\mathcal{D}} \frac{\partial w}{\partial \bar{z}} = \oint_C w(z, \bar{z})\, dz. \tag{A.4}$$

Let us now examine the physical signification of the contour integral over the Pólya field of the flow. We may write

$$dz = dx + idy = |dz|\, e^{i\arg dz} = |dz|\, e^{i\phi} = \sqrt{dx^2 + dy^2}\, e^{i\phi} = ds\, e^{i\phi},$$

where ds is the infinitesimal arc-length element of the contour C, and ϕ its angle with the positive x-axis (the reader is advised to draw a picture). If the real physical velocity $\mathbf{V}$ forms an angle ψ with the positive x-axis, then

$$w = \overline{v_1 + iv_2} = \overline{|\mathbf{V}|e^{i\psi}} = |\mathbf{V}|e^{-i\psi}.$$

If $\alpha := \phi - \psi$ denotes the angle between the tangent vector to C and $\mathbf{V}$, we have

$$w\,dz = |\mathbf{V}|e^{i(\phi-\psi)}\,ds = \{|\mathbf{V}|\cos\alpha + i|\mathbf{V}|\sin\alpha\}\,ds$$
$$=: (v_t + iv_n)\,ds.$$

Here v_t denotes the tangential component and v_n the normal component of physical flow $\mathbf{V}$ with respect to the contour C. The quantities

$$\Gamma(C) := \oint_C v_t\,ds = \Re[\oint_C w\,dz], \tag{A.5}$$

$$\Phi(C) := \oint_C v_n\,ds = \Im[\oint_C w\,dz] \tag{A.6}$$

are called the *circulation* and the *flux* of $\mathbf{V}$ along C, respectively. Now we have

$$\oint_C w(z,\overline{z})\,dz = \Gamma(C) + i\,\Phi(C)$$
$$= 2i \iint_{\mathcal{D}} \frac{\partial w}{\partial \overline{z}} = \iint_{\mathcal{D}} \Omega\,dxdy = \iint_{\mathcal{D}} \omega + i\theta\,dxdy \tag{A.7}$$

which shows that *the circulation along C is the sum of all vortices inside $\mathcal{D}$*:

$$\Gamma(C) = \iint_{\mathcal{D}} \omega\,dxdy, \tag{A.8}$$

and, *the flux through C is the sum of all sources inside $\mathcal{D}$*:

$$\Phi(C) = \iint_{\mathcal{D}} \theta\,dxdy. \tag{A.9}$$

A flow in a domain $\mathcal{D}$ is said to be *irrotational*, if $\omega = 0$ in $\mathcal{D}$; it is called *sourceless*, if $\theta = 0$ in $\mathcal{D}$. The flow is said to be *irrotational and sourceless* in $\mathcal{D}$, if Ω, and thus $\partial w/\partial \overline{z}$ vanishes in $\mathcal{D}$. An irrotational and sourceless flow in a simply connected domain $\mathcal{D}$ generates zero circulation and zero flux along the boundary C of $\mathcal{D}$. Here it is important that the domain is simply connected, i.e., has no holes in it.

A.1.2 Pólya fields and Defining Functions

Example A.1. Consider the physical flow

$$\mathbf{V} = -\frac{1}{2\pi}\frac{1}{x^2+y^2}(-y, x)^t,$$

called a vortex filament, defined on the punctuated unit-disk $\mathcal{D} = \{z \in \mathbb{C} \mid |z| \le 1, z \ne 0\}$ which is not simply connected. The Pólya field becomes

$$w(z) = v_1 - iv_2 = \frac{y}{2\pi(x^2+y^2)} + i\frac{x}{2\pi(x^2+y^2)}$$
$$= \frac{i}{2\pi}\frac{1}{x+iy} = -\frac{1}{2\pi i}\frac{1}{z},$$

and

$$\oint_C w(z)\, dz = -\frac{1}{2\pi i} \oint_C \frac{1}{z}\, dz = -1$$

producing a non-zero circulation along the unit circle in spite of a vanishing curl and divergence throughout $\mathcal{D}$:

$$\omega := \frac{\partial v_2}{\partial x} - \frac{\partial v_1}{\partial y} = \frac{x^2 - y^2}{2\pi(x^2 + Y^2)^2} + \frac{-x^2 + y^2}{2\pi(x^2 + y^2)^2} = 0,$$

$$\theta := \mathrm{div}\,\mathbf{V} = \frac{\partial v_1}{\partial x} + \frac{\partial v_2}{\partial y} = \frac{-2xy}{2\pi(x^2 + y^2)^2} + \frac{2xy}{2\pi(x^2 + y^2)^2} = 0.$$

Thus, away from 0, we have an irrotational and source free flow, in spite of the fact that the vortex filament concentrated at 0 gives rise to a non-zero circulation along the border of the domain. Observe now that *the Pólya field of the vortex filament is just the defining function of the hyperfunction $\delta(x)$.*

From complex variable theory we know that a complex function $w = f(z)$ is holomorphic at z if and only if $\partial w/\partial \bar{z} = 0$. Therefore, we might state: *A flow in a domain $\mathcal{D}$ is irrotational and sourceless, if and only if the Pólya field of the flow is a holomorphic function in $\mathcal{D}$.* Let now I be a given interval, and $\mathcal{D}$ a full neighborhood of I with the upper and lower half-neighborhood $\mathcal{D}_+ = \mathcal{D} \cap \mathbb{C}_+$ and $\mathcal{D}_- = \mathcal{D} \cap \mathbb{C}_-$, respectively. Suppose that in $\mathcal{D}_+$ and $\mathcal{D}_-$ there is an irrotational and sourceless flow $\mathbf{V}_+$ and $\mathbf{V}_-$ specified by its Pólya field $F_+(z)$ and $F_-(z)$, respectively. In the interval I on the x-axis there may be a distribution of vortices and sources produced by the two flows. The Pólya field

$$w(z) := \begin{cases} F_+(z), & z \in \mathcal{D}_+ \\ F_-(z), & z \in \mathcal{D}_- \end{cases}$$

then determines a hyperfunction $f(x) = [w(z)] = [F_+(z), F_-(z)]$ which represents the distribution of vortices and sources on the x-axis determined by the given irrotational and sourceless flows in the upper and lower half-plane. In the special

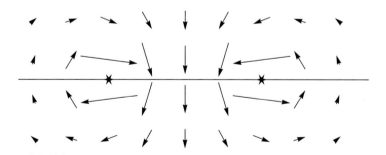

Figure A.1: Pólya field of the doublet $\delta(x + 1/2) - \delta(x - 1/2)$

case where $F_+(z)$ and $F_-(z)$ are analytic continuations of each other, the global analytic function $w = F(z)$ in $\mathcal{D}$ describes a flow $\mathbf{V}$ which is irrotational and sourceless in the upper and lower half-plane but may still have singularities (vortices and sources) on the x-axis. *The Pólya field of this flow is the defining function of the hyperfunction which describes the sources and vortices on the x-axis.*

Suppose that there is a vortex and source layer on the interval I described by the circulation density $\gamma(x)$ and the flux density $\phi(x)$ such that the circulation and the flux along I are given by

$$\Gamma(I) = \int_a^b y(x)\,dx, \quad \Phi(I) = \int_a^b \phi(x)\,dx.$$

The complex layer density $f(x) := \gamma(x) + i\phi(x)$ is now supposed to be an ordinary function represented as $f(x) = F_+(x+i0) - F_-(x-i0)$. We may write

$$\Gamma(I) + i\,\Phi(I) = \int_a^b f(x)\,dx = -\oint_C w(z)\,dz,$$

where the closed (positively directed) contour C is given by $C := -\gamma_{a,b}^+ + \gamma_{a,b}^-$. Under the above assumption we can write for the sum of all vortices and sources inside C,

$$\oint_C w(z)\,dz = \int_{\gamma_{a,b}^+} F_+(z)\,dz - \int_{\gamma_{a,b}^-} F_-(z)\,dz$$

$$= \int_a^b F_+(x+i0)\,dx - \int_a^b F_-(x+i0)\,dx = \int_a^b f(x)\,dx.$$

If we have vortices and sources concentrated at points, as in the above vortex filament, the layer density becomes a hyperfunction, and the integral of the layer density becomes the integral over the hyperfunction in accordance with its definition. Let us summarize:

An irrotational and sourceless flow in the upper and lower half-plane with a layer of vortices and sources on the real axis is a physical interpretation of a hyperfunction in the sense that the Pólya field of the flow is the defining function of the hyperfunction.

This offers a second possibility of visualizing a hyperfunction by plotting the Pólya field of the defining function of the hyperfunction. Figure A.1 shows the Pólya field of the defining function of the hyperfunction $\delta(x+1/2) - \delta(x-1/2)$.

A.2 Laplace Transforms in the Complex Plane

A.2.1 Functions of Exponential Type

Investigations of Laplace transforms of entire functions of exponential growth were inaugurated by Pólya, Macintyre and others. Especially, the Laplace transformation of an holomorphic function of exponential type in a sector of the complex plane was studied by [24, Macintyre]. A readable exposition of the theory can be found in [16, Henrici p. 305–321].

An entire function $F(z)$ of the complex variable z is holomorphic in the entire (finite) complex plane. Thus, the series

$$F(z) = \sum_{k=0}^{\infty} a_k z^k \quad (|z| < \infty)$$

converges and represents $F(z)$ for all finite z.

Let us now define a *closed sector with opening* $[\alpha, \beta]$ *and origin* $a \in \mathbb{C}$ by

$$S[a; \alpha, \beta] := \{z \in \mathbb{C} | \alpha \leq \arg(z - a) \leq \beta\}.$$

Remember that a function $F(z)$ defined in an arbitrary domain $\mathcal{D}$ is said to be holomorphic in $\mathcal{D}$, if it is holomorphic in some open domain that contains $\mathcal{D}$.

Definition A.1. A holomorphic function in the closed sector $S[a; \alpha, \beta]$ is said to be of *exponential type*, if there are constants $K > 0$ and σ such that with

$$\mu(\rho) := \max_{|z|=\rho} |F(z)|, \quad \alpha \leq \arg z \leq \beta$$

the inequality

$$\mu(\rho) < K e^{\sigma \rho} \tag{A.10}$$

holds.

The infimum γ of all numbers $\sigma > 0$ such that (A.10) is true will be called the *growth parameter of* $F(z)$. This means that for any $\epsilon > 0$ the estimate $\mu(\rho) < K e^{(\gamma+\epsilon)\rho}$ holds for all $\rho > 0$.

Examples. $\exp(z^2)$ is not of exponential type. $z^{10} \exp(-3z)$ is of exponential type in the entire plane $S[0; 0, 2\pi]$ with growth parameter $+3$. The same function is of exponential type with growth parameter -3 in any closed sector $S[0, \alpha, \beta]$ with $-\pi/2 < \alpha < \beta < \pi/2$. The function $\cosh \sqrt{z}$ and any polynomial are of exponential type in $S[0, 0, 2\pi]$ with growth parameter 0. The function $z^m \log(-z)$ is of exponential type in any closed sector $S[0, \alpha, \beta]$ with $0 < \alpha < \beta < 2\pi$ and the same is true for $z^m \log z$ in any closed sector $S[0, \alpha, \beta]$ with $-\pi < \alpha < \beta < \pi$, both with growth parameter zero.

Now let us consider the set

$$\mathcal{O}_{exp}(\mathbb{C} \setminus [0, \infty)) \tag{A.11}$$

consisting of all functions being holomorphic on $\mathbb{C} \setminus [0, \infty)$ and of exponential type in the sector $S[-\epsilon; \epsilon, 2\pi - \epsilon]$ for some arbitary small $\epsilon > 0$. Thus, a member of $\mathcal{O}_{exp}(\mathbb{C} \setminus [0, \infty))$ can only have singularities in $[0, \infty)$.

The growth parameter γ measures the overall growth of the function in the sector. If we want to measure the growth of $F(z)$ on an outward drawn ray starting at the vortex a of $S[a, \alpha, \beta]$, we use the *indicator function* which is defined and denoted by

$$\gamma_a(\phi) = \gamma(\phi) := \limsup_{t \to \infty} \frac{\log |F(a + t e^{i\phi})|}{t}, \tag{A.12}$$

where $\alpha \leq \phi \leq \beta$. This means that for any $\epsilon > 0$ we have for all sufficiently large t the estimate $|F(a + te^{i\phi})| \leq e^{(\gamma(\phi)+\epsilon)t}$. Note that $\gamma(\phi) \leq \gamma$ the growth parameter of $F(z)$.

Let us now define the Laplace transform of a function $F(z) \in \mathcal{O}_{exp}(\mathbb{C} \setminus [0, \infty))$ along an outward drawn ray, in a given direction.

Definition A.2. Let $F(z) \in \mathcal{O}_{exp}(\mathbb{C} \setminus [0, \infty))$. The *Laplace transform of* $F(z)$ *with origin* a *and direction* ϕ is defined and denoted by

$$\hat{F}_\phi(s) := \int_{\Gamma(\phi)} e^{-sz} F(z) \, dz. \tag{A.13}$$

Here the contour $\Gamma(\phi)$ is the line $z(t) := a + te^{i\phi}$, $t > 0$, $0 < \alpha \le \phi \le \beta < 2\pi$.

Explicitly, we have

$$\hat{F}_\phi(s) := e^{-as+i\phi} \int_0^\infty e^{-se^{i\phi} t} F(a + te^{i\phi}) \, dt$$

$$= e^{-as+i\phi} \mathcal{L}_+[F(a + te^{i\phi})](se^{i\phi}),$$

where $\mathcal{L}_+[\cdot](\cdot)$ denotes the ordinary right-sided Laplace transformation of a function of the real variable t. Note that $\hat{F}_\phi(s)$ is a holomorphic function for $\Re(se^{i\phi}) > \gamma_a(\phi)$, and since $\gamma_a(\phi) \le \gamma$ for all $\alpha \le \phi \le \beta$ we have that $\hat{F}_\phi(s)$ is a holomorphic function for $\Re(se^{i\phi}) > \gamma$.

In what follows, we shall be primarily interested in the case of the two following Laplace transforms with a real origin $a < 0$:

$$\hat{F}_\alpha(s) := \int_{\Gamma(\alpha)} e^{-sz} F(z) \, dz, \quad \hat{F}_{2\pi-\alpha}(s) := \int_{\Gamma(2\pi-\alpha)} e^{-sz} F(z) \, dz,$$

where α is a small positive angle (which, later on, we let tend to zero). The reason is that the defining function $F(z)$ of the hyperfunction $f(t)$ under consideration will be assumed to be a member of $\mathcal{O}_{exp}(\mathbb{C} \setminus [0, \infty))$ and the Laplace transform of this hyperfunction will then be defined as

$$\mathcal{L}[f(t)](s) := \lim_{\alpha \to 0+} \{\hat{F}_\alpha(s) - \hat{F}_{2\pi-\alpha}(s)\}. \tag{A.14}$$

Example A.2. We take

$$F(z) = -\frac{1}{2\pi i} \log(-z) \in \mathcal{O}_{exp}(\mathbb{C} \setminus [0, \infty)).$$

Here the small negative number a can be replaced by $a = 0$ which facilitates the computation due to the fact that log is integrable at 0. The function log denotes the principal branch of the logarithm, and the complex plane has a cut along the negative part of the real axis. We have for $t > 0$,

(i) $\log(-te^{i\phi}) = \log t + i(\phi - \pi)$ $(0 < \phi \le \pi)$,

(ii) $\log(-te^{i\phi}) = \log t + i(\phi + \pi)$ $(\pi \le \phi < 2\pi)$.

In the case (i) with $0 < \phi \le \pi$, we obtain with $p = se^{i\phi}$, and assuming $\Re[p] = \Re[se^{i\phi}] > 0$,

$$-2\pi i \, \hat{F}_\phi(s) := e^{i\phi} \{\mathcal{L}_+[\log t](p) + i(\phi - \pi)\frac{1}{p}\} = e^{i\phi} \{-\frac{1}{p}(\gamma + \log p) + i(\phi - \pi)\frac{1}{p}\}$$

$$= \frac{1}{s}\{-(\gamma + \log s + i\phi) + i(\phi - \pi)\} = -\frac{1}{s}(\gamma + \log s + i\pi),$$

and for the case (ii), where $\pi \le \phi < 2\pi$, we get for $\Re[se^{i\phi}] > 0$, $\Re[p] = \Re[se^{i\phi}] > 0$,

$$-2\pi i \, \hat{F}_\phi(s) = -\frac{1}{s}(\gamma + \log s - i\pi).$$

thus, the Laplace transform of the hyperfunction $f(t) = [-\log(-z)/(2\pi i)]$ gives

$$\mathcal{L}[f(t)](s) := \lim_{\alpha \to 0+} \{\hat{F}_\alpha(s) - \hat{F}_{2\pi-\alpha}(s)\} = \frac{1}{s} \quad (\Re s > 0).$$

The two expressions of $-2\pi i \, \hat{F}_\phi(s)$ for $0 < \phi \le \pi$, and for $\pi \le \phi < 2\pi$ only differs by $2\pi i$ and suggest that one is the analytic continuation of the other such that $\hat{F}_{0+}(s)$ and $\hat{F}_{2\pi-}(s)$ are the values on consecutive sheets of a multi-valued function.

We shall now show that this is the case, and return to the general setting with a function $F(z) \in \mathcal{O}_{exp}(\mathbb{C} \setminus [0, \infty))$ and a path of integration in the direction ϕ starting at a real number $a < 0$. Consider the following estimates, where $s = |s| \exp(i\psi)$, $\epsilon > 0$, and the growth parameter γ is used.

$$|\hat{F}_\phi(s)| \le \int_0^\infty e^{-\Re(s(a+t\exp(i\phi)))} \, |F(a + te^{i\phi})| \, dt$$

$$< K \, e^{-a\,\Re s} \int_0^\infty e^{-t\{|s|\cos(\phi+\psi))-\gamma-\epsilon\}} \, dt$$

$$= K \, \frac{e^{-a\,|s|\cos\psi}}{|s|\cos(\phi+\psi)) - \gamma - \epsilon}$$

provided that $|\phi + \psi| < \pi/2$ and $|s|$ is sufficiently large. It follows that

$$\frac{\log|\hat{F}_\phi(s)|}{|s|} < -a\cos\psi - \frac{\log\{|s|\cos(\phi+\psi)) - \gamma - \epsilon\}}{|s|},$$

therefore, under the restriction $|\phi + \psi| < \pi/2$, we have

$$\hat{\gamma}(\psi) := \limsup_{|s| \to \infty} \frac{\log|\hat{F}_\phi(|s|e^{i\psi})|}{|s|} \le -a\cos\psi. \tag{A.15}$$

Because the indicator function of the Laplace transform is bounded by $|a|$, we have established

Proposition A.1. *The Laplace transform of $F(z) \in \mathcal{O}_{exp}(\mathbb{C} \setminus [0, \infty))$ with origin $a < 0$ and direction ϕ is again a function of exponential growth and its indicator function $\hat{\gamma}(\psi)$ satisfies (A.15). Furthermore, $\hat{F}_\phi(s)$ is a holomorphic function in the half-plane $\Re(se^{i\phi}) > \gamma$ (independent of ϕ).*

Notice that $\Re(s\,e^{i\phi}) > \rho$ defines the half-plane $H_{\phi,\rho}$ in the s-plane with the normal vector $e^{-i\phi}$ pointing into the interior of the half-plane, whose boundary $\Re(s\,e^{i\phi}) = \rho$ has the distance $\rho > 0$ from the origin.

If in the above estimate we use the more accurate measure furnished by the indicator function $\gamma(\phi)$ instead of the overall growth parameter γ, we can write for any positive ϵ,

$$|\hat{F}_\phi(s)| \le e^{-a\,\Re s} \int_0^\infty e^{-t\{\Re(s\exp(i\phi))-\gamma(\phi)-\epsilon\}} \, |F(a + te^{i\phi})| \, dt$$

which shows that the integral converges absolutely provided that

$$\Re(s\,e^\phi) > \gamma(\phi) + \epsilon$$

holds, i.e., that $s \in H_{\phi,\gamma(\phi)}$. Thus, if we turn the direction of taking the Laplace transform by a positive angle ϕ, the normal of the corresponding half-plane of absolute convergence turns in the opposite sense by the angle $-\phi$.

Consider now two Laplace transforms $\hat{F}_{\phi'}(s)$ and $\hat{F}_{\phi''}(s)$. Figure A.2 depicts the situation in the z-plane, and Figure A.3 the corresponding situation in the s-plane.

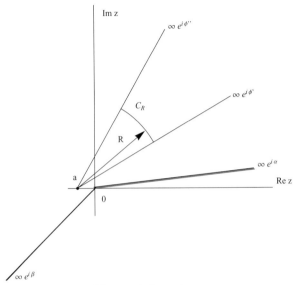

Figure A.2: z-plane

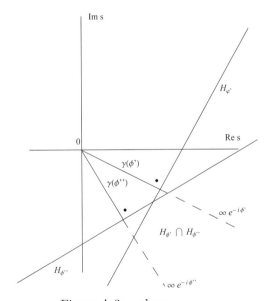

Figure A.3: s-plane

Consider the integral

$$\int_{C_R} e^{-sz} F(z)\, dz$$

along the circular arc C_R of radius R (Figure A.2). By using the growth index γ of $F(z)$, we have the estimates

$$\left| \int_{C_R} e^{-sz} F(z)\, dz \right| \leq K \int_{C_R} e^{-\Re(sz)} |F(z)|\, |dz|$$

$$\leq K \int_{C_R} e^{-\{|s||z|\cos(\arg s + \arg z) - |z|(\gamma - \epsilon)\}}\, |dz|$$

$$\leq K\,(\beta - \alpha) R\, e^{-R\{|s|\cos(\arg s + \arg z) - \gamma - \epsilon\}}$$

$$\leq K\,(\beta - \alpha) R\, e^{-R\{|s|\delta - \gamma - \epsilon\}},$$

where $\delta = \min_{\alpha \leq \arg z \leq \beta} \cos(\arg s + \arg z)$. This shows that this integral along C_R tends to zero as $R \to \infty$, provided that $\delta > 0$ and that $|s|$ is sufficiently large. However, we have $\phi' \leq \arg z \leq \phi''$, and $\delta > 0$ as long as $-\pi/2 - \arg z < \arg s < \pi/2 - \arg z$. The last inequality is satisfied if $-\pi/2 - \phi' < \arg s < \pi/2 - \phi''$ holds, and this is exactly the case if $s \in H_{\phi', \gamma(\phi')} \cap H_{\phi'', \gamma(\phi'')}$. This implies that the two Laplace transforms

$$\hat{F}_{\phi'}(s) := \int_{\Gamma(\phi')} e^{-sz} F(z)\, dz, \quad \hat{F}_{\phi''}(s) := \int_{\Gamma(\phi'')} e^{-sz} F(z)\, dz, \qquad (A.16)$$

coincide on the intersection $H_{\phi', \gamma(\phi')} \cap H_{\phi'', \gamma(\phi'')}$ provided that $|s|$ is sufficiently large. Because this identity between holomorphic functions holds in a non-empty set, it holds throughout. We have established

Lemma A.2. *For any $0 < \phi' < \phi'' < 2\pi$ the two Laplace transforms $\hat{F}_{\phi'}(s)$ and $\hat{F}_{\phi''}(s)$ are analytic continuations of each other.*

Consider now the union of all half-planes $H_\phi : \Re(se^{i\phi}) > \gamma(\phi)$, i.e.,

$$\Sigma := \bigcup_{0 < \phi < 2\pi} H_{\phi, \gamma(\phi)}. \qquad (A.17)$$

Then, gathering all preceding reasonings, we have proven

Proposition A.3. *Let $F(z) \in \mathcal{O}_{exp}(\mathbb{C} \setminus [0, \infty))$, and let $\gamma(\phi)$ be its indicator function. Then any Laplace transform $\hat{F}_\phi(s)$ with origin a and the direction ϕ can be extended to a holomorphic function throughout Σ. This function is called the Laplace transform of $F(z)$ and is denoted by $\hat{F}(s) = \mathcal{L}_a[F(z)](s)$.*

The complement of the set Σ where $\hat{F}(s)$ is holomorphic,

$$Z := \overline{\Sigma} = \bigcap_{\alpha < \phi < \beta} \overline{H_{\phi, \gamma(\phi)}}, \qquad (A.18)$$

is called the *convex hull of singularities* of $\hat{F}(s)$. This set being an intersection of closed half-planes $\overline{H_\phi} : \Re(se^{i\phi}) \leq \gamma(\phi)$, is a convex set. It can be proved (see [16, Henrici p. 318]) that the Laplace transform $\hat{F}(s) = \mathcal{L}[F(z)](s)$ cannot be

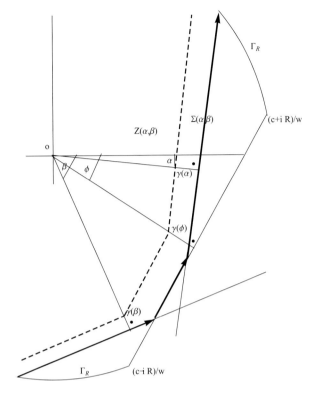

Figure A.4: Sets $\Sigma(\alpha, \beta)$, $Z(\alpha, \beta)$, contour Γ_R

continued across the boundary of Σ defined by the indicator function $\gamma(\phi)$, i.e., on the boundary there are always some singularities of $\hat{F}(s)$.

The sets Σ and Z corresponding to the situation of Figure A.2 are shown in Figure A.4.

Problem A.1. Show that the Laplace transform with origin a of z^n, $n \in \mathbb{N}_0$ in the direction of ϕ is independent of ϕ, and is given by

$$\hat{F}_\phi(s) = e^{-as} \sum_{k=0}^{n} \frac{n!}{(n-k)!} \frac{a^{n-k}}{s^{k+1}} =: \hat{F}(s), \quad |s| > 0.$$

In the above example $F(z) = z^n$ is entire and of exponential growth in the entire plane. It produces a Laplace transform with the property $\hat{F}_0(s) = \hat{F}_{2\pi}(s)$, i.e., a singled-valued function. It is clear that the same property holds for any entire function which is of exponential growth in the entire plane.

Next, we will establish the inversion formula. For any direction $0 < \phi < 2\pi$ we can write

$$\int_0^\infty e^{-p\,e^{i\phi}\,t} F(a + te^{i\phi})\,dt = e^{ap - i\phi}\hat{F}_\phi(p) \quad (\Re(pe^{i\phi}) > \gamma(\phi)).$$

Replacing p by $pe^{-i\phi}$ yields

$$G(p) := e^{ap\exp(-i\phi)-i\phi}\,\hat{F}_\phi(pe^{-i\phi}) = \int_0^\infty e^{-pt}\,F(a+te^{i\phi})\,dt$$

$$=: \int_0^\infty e^{-pt}g(t)\,dt.$$

The complex inversion formula for the one- sided Laplace transformation then yields

$$g(t) = F(a+te^{i\phi}) = \frac{1}{2\pi i}\lim_{R\to\infty}\int_{c-iR}^{c+iR} e^{pt}\,G(p)\,dp$$

$$= \frac{1}{2\pi i}\lim_{R\to\infty}\int_{c-iR}^{c+iR} e^{pt}\,e^{a\,p\exp(-i\phi)-i\phi}\,\hat{F}_\phi(pe^{-i\phi})\,dp$$

$$= \frac{1}{2\pi iw}\lim_{R\to\infty}\int_{c-iR}^{c+iR} e^{pt}\,e^{a\,p/w}\,\hat{F}_\phi(p/w)\,dp$$

with $w := \exp(i\phi)$ and $c > \gamma(\phi)$. A change of variables $s = p/w = se^{-i\phi}$ yields

$$F(a+wt) = \frac{1}{2\pi i}\lim_{R\to\infty}\int_{(c-iR)/w}^{(c+iR)/w} e^{swt}\,e^{as}\,\hat{F}_\phi(s)\,ds$$

$$= \frac{1}{2\pi i}\lim_{R\to\infty}\int_{(c-iR)/w}^{(c+iR)/w} e^{s(a+wt)}\,\hat{F}_\phi(s)\,ds.$$

The contour runs parallel to the boundary $\Re(se^{i\phi}) = \gamma(\phi)$ of the half-plane $H_{\phi,\gamma(\phi)}$ at distance $c > \gamma(\phi)$. Thus, we recover $F(a+e^{i\phi}t)$ by the above formula. Let now $\alpha \le \phi \le \beta$, and let us close the line segment from $(c-iR)/w$ to $(c+iR)/w$ by two circular arcs Γ_R with radius $\rho = \sqrt{R^2+c^2}$ and the two segments on the boundaries of the half-planes $H_{\alpha,\gamma(\alpha)}$ and $H_{\beta,\gamma(\beta)}$ (see Figure A.4). Inside the closed contour the Laplace transform $\hat{F}_\phi(s)$ is holomorphic. On the two circular arcs the contributions to the integral tends to zero due to Jordan's lemma, because on these arcs $\hat{F}_\phi(s)$ tends to zero uniformly with respect to $\arg s$ as $R \to \infty$. Thus, the line segment from $(c-iR)/w$ to $(c+iR)/w$, as $R \to \infty$, can be replaced by the corresponding infinite contour running parallel to the boundary of the convex hull. Therefore, for any $z = a + te^{i\phi}$ with $\alpha \le \phi \le \beta$ we obtain the inversion formula

$$F(z) = \mathcal{L}_a^{-1}[\hat{F}(s)](z) = \frac{1}{2\pi i}\lim_{R\to\infty}\int_{\gamma_R} e^{sz}\,\hat{F}(s)\,ds$$

$$=: \frac{1}{2\pi i}\,\text{pv}\int_{\gamma_\infty} e^{sz}\,\hat{F}(s)\,ds \tag{A.19}$$

where the infinite contour γ_∞ runs parallel to the boundary of the convex hull of singularities and inside Σ. The contour γ_R is indicated in Figure A.5 for the case $\alpha = 0$ and $\beta = 2\pi$. Notice that the lower vertical part of the contour is located in the s-half-plane with $-5\pi/2 < \arg s < -3\pi/2$, while the upper vertical part is in the half-plane with $-\pi/2 < \arg s < \pi/2$. If the Laplace transform $\hat{F}(s)$ is a multi-valued function, these two half-planes are located on different sheets of the Riemann surface of $\hat{F}(s)$.

Example A.3. Let $\hat{F}(s) = e^{-as}/s^{k+1}$. The convex hull of singularities is the set $\{0\}$. The given Laplace transform is single-valued which implies that the contour γ_∞ consists of two parts: a vertical straight line γ_1 from $c - i\infty$ to $c + i\infty$ for any $c > 0$, and a origin centered circle of radius c denoted by γ_0.

$$F(z) = \frac{1}{2\pi i} \, \mathrm{pv} \int_{\gamma_1} e^{s(z-a)} \frac{1}{s^{k+1}} \, ds + \frac{1}{2\pi i} \oint_{\gamma_0} e^{s(z-a)} \frac{1}{s^{k+1}} \, ds.$$

The upper half of the contour γ_1 can be rotated around c by $-\pi/2$ and the lower half by $+\pi/2$, becoming a contour which comprises the upper and the lower side of the real axis traveled in the opposite sense. If we assume temporarily $\Re z < \Re a$, then the first integral is zero. The second integral is evaluated by residues and yields

$$\frac{1}{2\pi i} \int_{\gamma_0} e^{s(z-a)} \frac{1}{s^{k+1}} \, ds = \frac{(z-a)^k}{k!}.$$

Thus,

$$\mathcal{L}_a^{-1}\left[\frac{e^{-as}}{s^{k+1}}\right](z) = \frac{(z-a)^k}{k!}.$$

By linearity,

$$\mathcal{L}_a^{-1}\left[\sum_{k=0}^n \frac{n!}{(n-k)!} \frac{e^{-as}}{(as)^{k+1}}\right](z) = \frac{1}{a} \sum_{k=0}^n \frac{n!}{(n-k)!} \frac{(z-a)^k}{a^k \, k!} = \frac{z^n}{a^{n+1}}$$

which is in agreement with the result of Problem A.1.

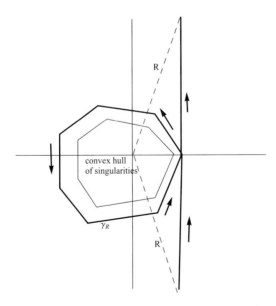

Figure A.5: Contour γ_R in the case of a sector $S(0, 2\pi)$

Notice that if the contour integral along the closed contour encircling the convex hull of singularities is a path of finite length, then it always yields a holomorphic function of z.

Also, if $F(z)$ is an entire function and of exponential growth in the whole complex plane with a growth parameter γ, its Laplace transform will always be a single-valued holomorphic function outside the circle $|s| = \gamma$ and the reasoning in the above example concerning the vanishing of the contour integral along the vertical part of the integration path holds. This gives

Proposition A.4 (Pincherle's Theorem). *Let $F(z)$ be an entire function of exponential growth in the whole complex plane with growth parameter $\gamma \geq 0$. Then $\hat{F}(s) = \mathcal{L}_0[F(z)](s)$ is holomorphic outside the circle $|s| = \gamma$, and $F(z)$ can be recovered for all z by*

$$F(z) = \frac{1}{2\pi i} \oint_{\gamma_0} e^{sz} \hat{F}(s)\,ds. \tag{A.20}$$

Here γ_0 is any simple closed curve encircling the set $|s| \leq \gamma$ in the positive sense.

Problem A.2. Show that the Laplace transform of $F(z) = ze^{-3z}$ with origin 0 and any direction ϕ is $1/(s+3)^2$. Recover $F(z)$ by using Pincherle's Theorem and the calculus of residues.

A.2.2   Laplace Hyperfunctions and their Transforms

Laplace hyperfunctions were introduced by [22, Komatsu]. His approach yields a concise theory of the Laplace transformation and its inverse for a certain subclass of hyperfunctions. It is based on the theory of Laplace transforms of functions from $\mathcal{O}_{exp}(\mathbb{C} \setminus [0, \infty))$. A function of this space can only have singularities in $[0, \infty]$. Let now $\mathcal{O}_{exp}(\mathbb{C})$ denote the subspace of all entire functions of exponential type in the complex plane, i.e., in $S[0; 0, 2\pi]$.

Definition A.3. The quotient space

$$\mathcal{B}_L(\mathbb{R}_+) := \mathcal{O}_{exp}(\mathbb{C} \setminus [0, \infty))/\mathcal{O}_{exp}(\mathbb{C}) \tag{A.21}$$

is called the *space of Laplace hyperfunctions with support in* $[0, \infty)$.

This abstract definition simply means that $\mathcal{B}_L(\mathbb{R}_+)$ consists of all equivalence classes of functions from $\mathcal{O}_{exp}(\mathbb{C}\setminus[0,\infty))$, where two such functions $F(z)$ and $G(z)$ are considered to be equivalent if $F(z) - G(z)$ is an entire function of exponential type in the whole complex plane, i.e., $F(z) \sim G(z) \Leftrightarrow F(z) - G(z) \in \mathcal{O}_{exp}(\mathbb{C})$. More concrete, a hyperfunction $f(t) = [F(z)]$ is a Laplace hyperfunction defined by a defining function $F(z) \in \mathcal{O}_{exp}(\mathbb{C} \setminus [0, \infty))$, and two defining functions define the same hyperfunction, if they differ at most by a function of $\mathcal{O}_{exp}(\mathbb{C})$. If $f(t) \in \mathcal{B}_L(\mathbb{R}_+)$ it vanishes on the negative part of the real axis. The zero Laplace hyperfunction is represented by any member of $\mathcal{O}_{exp}(\mathbb{C})$.

Definition A.4. We define the *Laplace transform of a* $f(t) \in \mathcal{B}_L(\mathbb{R}_+)$ by

$$\mathcal{L}[f(t)](s) := \lim_{\alpha \to 0+} \{\hat{F}_\alpha(s) - \hat{F}_{2\pi-\alpha}(s)\} =: \hat{F}_{0+}(s) - \hat{F}_{2\pi-}(s), \tag{A.22}$$

where $\hat{F}_\alpha(s)$ and $\hat{F}_{2\pi-\alpha}(s)$ are the Laplace transforms with origin $a < 0$ in the directions α and $2\pi - \alpha$ of the defining function $F(z) \in \mathcal{O}_{exp}(\mathbb{C} \setminus [0, \infty))$ of $f(t)$, respectively.

Example A.4.

$$f(t) = u(t)t^n = \left[-\frac{z^n}{2\pi i} \log(-z) \right], \quad n \in \mathbb{N}.$$

The function $\log z$ denotes here the principal value with a cut along the negative part of the real axis. On the upper and lower side of the real axis we have, for $\Re z > 0$,

$$\log(-z) = \log|z| + i \arg(-z) = \log|z| + i \arg(z) - i\pi,$$
$$\log(-z) = \log|z| + i \arg(-z) = \log|z| + i \arg(z) + i\pi,$$

respectively.

$$\hat{F}_{0+}(s) - \hat{F}_{2\pi-}(s) = -\frac{1}{2\pi i} \oint_{(-|a|,0)} e^{-sz} z^n \log(-z) \, dz$$
$$- \frac{1}{2\pi i} \int_0^\infty t^n e^{-st}(-2\pi i) \, dt = \int_0^\infty t^n e^{-st} \, dt = \frac{n!}{s^{n+1}}.$$

The first integral over the closed contour vanishes because the integrand is holomorphic inside the contour and continuous on its boundary.

Remember that $\hat{F}_\alpha(s)$ and $\hat{F}_{2\pi-\alpha}(s)$ are analytic continuation of each other of one global analytic function $\hat{F}(s) = \mathcal{L}_a[F(z)](s)$ holomorphic in the set Σ. Therefore, we may also state: The Laplace transform of the hyperfunction $f(t) = [F(z)] \in \mathcal{B}_L(\mathbb{R}_+)$ is the difference of $\hat{F}(s)$ on the two sheets $-\pi/2 < \arg s < \pi/2$ and $-5\pi/2 < \arg s < -3\pi/2$ of the Riemann surface of $\hat{F}(s)$. In formula this is expressed as

$$\mathcal{L}[f(t)](s) = \hat{F}(s)|_{-\pi/2 < \arg s < \pi/2} - \hat{F}(s)|_{-5\pi/2 < \arg s < -3\pi/2}. \tag{A.23}$$

If $\hat{F}(s)$ is single-valued, then $\mathcal{L}[f(t)](s) = 0$ which is right because in this case $F(z) \in \mathcal{O}_{exp}(\mathbb{C})$ which is the defining function of the null hyperfunction. Since the defining function $F(z)$ is a holomorphic function on the set $\mathbb{C} \setminus [0,\infty)$ the angle-contour formed by $z(t) = a + t \exp((2\pi - \alpha)i)$, $\infty > t \geq 0$ and $z(t) = a + t \exp(\alpha i)$, $0 \leq t < \infty$ can be deformed into an arbitrary loop starting at $\infty - i0$, running on the lower side of the real axis towards the origin, turning around 0 in the negative sense, and running back on the upper side of the real axis to $\infty + i0$. Therefore, we recover our definition of the Laplace transform of Chapter 3.

Let us now find the formula for the inverse Laplace transform. By (A.19) the defining function of $f(t)$ is restored in terms of the Laplace transform $\hat{F}(s)$ by means of

$$F(z) = \frac{1}{2\pi i} \, \text{pv} \int_{\gamma_\infty} e^{sz} \hat{F}(s) \, ds,$$

where the infinite contour γ_∞ runs inside Σ parallel to the boundary of the convex hull of singularities. We now choose a point s_0 on the positive real axis sufficiently far to the right such that it lies in the intersection of the half-planes of convergence of $\hat{F}_{0+}(s)$ and $\hat{F}_{2\pi-}(s)$. The contour γ_∞ is the sum of the contour γ_{1-} consisting of the lower vertical part, the closed contour γ_0 which encircles the convex hull of singularities, and the γ_{1+} being the upper vertical part, thus

$$\int_{\gamma_\infty} e^{sz}\,\hat{F}(s)\,ds = \int_{\gamma_{1-}} e^{sz}\,\hat{F}_{2\pi-}(s)\,ds + \int_{\gamma_0} e^{sz}\,\hat{F}(s)\,ds + \int_{\gamma_{1+}} e^{sz}\,\hat{F}_{0+}(s)\,ds$$

$$= -\int_{\gamma_{1-}} e^{sz}\,\mathcal{L}[f(t)](s)\,ds + \int_{\gamma_0} e^{sz}\,\hat{F}(s)\,ds$$

$$+ \int_{\gamma_{1+}} e^{sz}\,\hat{F}_{0+}(s)\,ds + \int_{\gamma_{1-}} e^{sz}\,\hat{F}_{0+}(s)$$

where the first integral was replaced by using (A.22). By assuming temporarily $\Re z < 0$, we can bend the contours γ_{1+} and γ_{1-} through 90 degrees to the right and obtain then for the two latter integrals

$$\int_{s_0}^{\infty} e^{sz}\,\hat{F}_{0+}(s)\,ds - \int_{s_0}^{\infty} e^{sz}\,\hat{F}_{0+}(s) = 0.$$

Finally, we get

$$F(z) = -\frac{1}{2\pi i}\int_{\gamma_{1-}} e^{sz}\,\mathcal{L}[f(t)](s)\,ds + \frac{1}{2\pi i}\int_{\gamma_0} e^{sz}\,\hat{F}(s)\,ds.$$

The second integral with a contour of finite length represents an entire function of exponential type and can be discarded, thereby yielding an equivalent defining function. The first integral is rewritten as

$$-\frac{1}{2\pi i}\int_{\gamma_{1-}} e^{sz}\,\mathcal{L}[f(t)](s)\,ds = \frac{1}{2\pi i}\int_{s_0}^{\infty} e^{sz}\,\mathcal{L}[f(t)](s)\,ds.$$

In order to force convergence of the last integral z must be assumed to be in a set located sufficiently far to the left, then, by evoking the identity principle, the obtained function is extended to $\mathbb{C}\setminus[0,\infty)$. We have proven

Proposition A.5. *Let $\mathcal{L}[f(t)](s)$ be the Laplace transform of the hyperfunction $f(t) = [F(z)] \in \mathcal{B}_L(\mathbb{R}_+)$, then the inverse Laplace transform is given by*

$$f(t) = \left[\frac{1}{2\pi i}\int_{s_0}^{\infty} e^{sz}\,\mathcal{L}[f(t)](s)\,ds\right], \tag{A.24}$$

where s_0 is an arbitrary point on the real axis sufficiently to the right to be located in the domain of holomorphy of $\mathcal{L}[f(t)](s)$, and the integration path runs to infinity such that $\Re s \to \infty$.

Example A.5. $\mathcal{L}[f(t)](s) = s^n$.

$$\frac{1}{2\pi i}\int_{s_0}^{\infty} e^{sz}\,s^n\,ds = \frac{1}{2\pi i}\int_{0}^{\infty} e^{sz}\,s^n\,ds - \frac{1}{2\pi i}\int_{0}^{s_0} e^{sz}\,s^n\,ds$$

$$= \frac{1}{2\pi i}\int_{0}^{\infty} e^{-s(-z)}\,s^n + \phi(z),$$

where $\phi(z) \in \mathcal{O}_{exp}(\mathbb{C})$. Assume now temporarily $\Re z < 0$, then the last integral becomes

$$\frac{1}{2\pi i}\frac{n!}{(-z)^{n+1}} = -\frac{1}{2\pi i}\frac{(-1)^n\,n!}{z^{n+1}}.$$

Finally,

$$\mathcal{L}^{-1}[s^n](t) = \left[-\frac{1}{2\pi i} \frac{(-1)^n \, n!}{z^{n+1}}\right] = \delta^{(n)}(t).$$

A.3 Some Basic Theorems of Function Theory

We mention here (without proofs) in the setting of Riemann integrals a few theorems about

1. the interchange of an infinite series with an improper integral,

2. reversing the order of integration of two improper integrals,

3. defining holomorphic functions by series and integrals.

A.3.1 Interchanging Infinite Series with Improper Integrals

It is well known that an infinite series and a proper Riemann integral, i.e., an integral with a finite integration interval $[a, b]$, can be interchanged, provided the series is uniformly convergent on $[a, b]$:

$$\sum_{k=1}^{\infty} \int_a^b f_k(x) \, dx = \int_a^b \sum_{k=1}^{\infty} f_k(x) \, dx.$$

Here all functions $f_k(x)$ are supposed to be continuous on $[a, b]$. This implies, due to the uniform convergence, that the sum

$$f(x) := \sum_{k=1}^{\infty} f_k(x)$$

is continuous on $[a, b]$ too. If we have an improper Riemann integral with an infinite integration domain the situation is more complicated. The following theorem (some sort of bounded convergence theorem in the setting of Riemann integrals) gives a sufficient condition for the validity of termwise integration. It covers the case of *a finite or infinite interval I. For definiteness we take $I = [0, \infty)$.*

Proposition A.6. *Let $f_1(x), f_2(x), \ldots$ be a sequence of complex-valued and continuous functions on I such that*

(i) *The series $\sum_{k=1}^{\infty} f_k(x)$ converges uniformly on every compact, i.e., finite and closed subinterval of I.*

(ii) *At least one of the following two conditions holds,*

$$\int_0^\infty \sum_{k=1}^{\infty} |f_k(x)| \, dx < \infty, \quad \sum_{k=1}^{\infty} \int_0^\infty |f_k(x)| \, dx < \infty.$$

Then, the series can be integrated termwise, and

$$\int_0^\infty \sum_{k=1}^{\infty} f_k(x) \, dx = \sum_{k=1}^{\infty} \int_0^\infty f_k(x) \, dx$$

holds.

Remark. Condition (i) is fullfilled if Weierstrass' M-Test can be applied, i.e., if there is a numerical sequence $c_1, c_2, \ldots$, such that

a) $|f_k(x)| \leq c_k$ for $k = 1, 2, \ldots$, and all x from any compact subinterval of I,

b) $\sum_{k=1}^{\infty} c_k < \infty$.

Condition (ii) is fullfilled, if there is a continuous function $g(x)$ on I such that

a) $\sum_{k=1}^{\infty} |f_k(x)| \leq g(x)$ for all $x \in I$,

b) $\int_0^{\infty} g(x)\, dx < \infty$;

or, if there is a numerical sequence $a_1, a_2, \ldots$ such that

a) $\int_0^{\infty} |f_k(x)|\, dx \leq a_k$ for $k = 1, 2, \ldots$,

b) $\sum_{k=1}^{\infty} a_k < \infty$.

You may check that $f_k(x) = (-1)^k\, e^{-kx}$ on $(0, \infty)$ yields an example where the series and the improper integral may be exchanged but where condition (ii) of the proposition is not fullfilled. This shows that the conditions of the proposition are only sufficient but not necessary. The sequence $f_k(x) = x\,\sin x/(x^2 + k^2)$ on $(0, \infty)$ is an example where

$$\sum_{k=1}^{\infty} \int_0^{\infty} f_k(x)\, dx$$

exists, but where

$$\int_0^{\infty} \sum_{k=1}^{\infty} f_k(x)\, dx$$

does not exist.

For sequences we also have some sort of bounded convergence theorem in the setting of the Riemann integral which gives a sufficient condition.

Proposition A.7. *Let $f_1(x), f_2(x), \ldots$ be a sequence of complex-valued and continuous functions on I, such that:*

(i) *The sequence $f_k(x)$ converges uniformly on every compact subinterval of I.*

(ii) *There is an $n_0 \in \mathbb{N}$ and a positive function $g(x)$ such that*

a) *$|f_k(x)| \leq g(x)$ for all $k \geq n_0$,*

b)

$$\int_0^{\infty} g(x)\, dx < \infty.$$

Then,

$$\lim_{k \to \infty} \int_0^{\infty} f_k(x)\, dx = \int_0^{\infty} \lim_{k \to \infty} f_k(x)\, dx.$$

A.3.2   Reversing the Order of Integration

Let us first give an example where the order of integration cannot be reversed. We have

$$\int_0^\infty \int_0^\infty \frac{x \sin x}{x^2 + y^2} \, dx \, dy = \frac{\pi}{2},$$

but if we change the order of integration, we find that

$$\int_0^\infty \int_0^\infty \frac{x \sin x}{x^2 + y^2} \, dy \, dx$$

does not exist.

Proposition A.8. *Let $f(x, y)$ be a continuous function on $(-\infty, \infty) \times (-\infty, \infty)$ and assume that the integrals*

$$\int_{-\infty}^\infty f(x, y) \, dx < \infty, \quad \int_{-\infty}^\infty f(x, y) \, dy < \infty$$

are uniformly convergent. The first one with respect to $y \in [c, d]$ and the second one with respect to $x \in [a, b]$, where $[c, d]$ and $[a, b]$ are any compact intervals. Then we have:
If at least one of the two following conditions holds

$$\int_{-\infty}^\infty \int_{-\infty}^\infty |f(x, y)| \, dx \, dy < \infty, \quad \int_{-\infty}^\infty \int_{-\infty}^\infty |f(x, y)| \, dy \, dx < \infty,$$

the order of integration can be reversed, i.e.,

$$\int_{-\infty}^\infty \int_{-\infty}^\infty f(x, y) \, dx \, dy = \int_{-\infty}^\infty \int_{-\infty}^\infty f(x, y) \, dy \, dx$$

holds.

The following simple example shows that reversing the order of integration may work without the conditions of the above proposition being satisfied,

$$\int_0^\infty \int_0^\infty \frac{e^{-x^2} \sin y}{y} \, dx \, dy = \int_0^\infty \int_0^\infty \frac{e^{-x^2} \sin y}{y} \, dy \, dx = \frac{\pi \sqrt{\pi}}{4}.$$

A.3.3   Defining Holomorphic Functions by Series and Integrals

In the following theorem the notion *convergent in the interior of a domain $\mathcal{D}$ or compact convergent in a domain $\mathcal{D}$* is used. This means that the convergence is uniform with respect to all z of any compact sub-domain of $\mathcal{D}$. Often, the term *uniformly convergent in $\mathcal{D}$, or inside $\mathcal{D}$* is also used.

First we cite a well-known theorem of complex variable theory (see for example [4, Behnke-Sommer], p. 87).

Proposition A.9. *A sequence of functions $f_n(z)$ is compact convergent in a domain $\mathcal{D}$, if for every point $z \in \mathcal{D}$, there is a neighborhood $\mathcal{N}(z)$ of z in which the sequence is uniformly convergent.*

Proposition A.10. *Let the sequence of holomorphic functions $f_k(z)$ be defined in a complex domain $\mathcal{D}$. If the series*

$$\sum_{k=0}^{\infty} f_k(z)$$

is compact convergent in $\mathcal{D}$, then

$$F(z) := \sum_{k=0}^{\infty} f_k(z)$$

is a holomorphic function in $\mathcal{D}$.

The uniform convergence can be checked by applying Weierstrass' M-Test: if there is a numerical sequence of positive constants $c_0, c_1, \ldots$, such that for all z from any compact subset of $\mathcal{D}$ it holds that

(i) $|f_k(z)| \leq c_k$ for all k,

(ii) $\sum_{k=1}^{\infty} c_k < \infty$,

then the series is compact convergent in $\mathcal{D}$.

Proposition A.11. *Let t be a real variable ranging over a finite or infinite interval (a, b) and z a complex variable ranging over a domain $\mathcal{D}$. If the function $f(z, t)$ satisfies*

(i) *$f(z, t)$ is continuous in both variables,*

(ii) *for each fixed value of t, $f(z, t)$ is a holomorphic function of z,*

(iii) *the integral*

$$F(z) := \int_a^b f(z, t)\, dt$$

converges uniformly at both limits in any compact subset of $\mathcal{D}$,

then, $F(z)$ is holomorphic in $\mathcal{D}$, and its derivatives of all orders may be computed by differentiating under the sign of integration.

Let, for example, a be finite, and $b = \infty$. Then the phrase "the integral converges uniformly at the upper limit in any compact subset $\mathcal{D}$" means that the sequence

$$F_n(z) := \int_a^{b_n} f(z, t)\, dt,$$

where $b_n \to \infty$, as $n \to \infty$, is compact convergent in $\mathcal{D}$. This is ensured, if for every $z \in \mathcal{D}$, there is a neighborhood $\mathcal{N}(z)$ of z such that the sequence $F_n(z)$ converges uniformly in this neighborhood.

In our text we often have to do with contour integrals depending on a parameter $z \in \mathcal{D} \subset \mathbb{C}$,

$$F(z) := \int_C f(z, \zeta)\, d\zeta,$$

where the contour C is of infinite length. In such a case think of the contour as having the parameter representation $C : \zeta = \zeta(\sigma)$, $a < \sigma < b$, where the parameter

σ is chosen to be the arc length of the curve with the parameter interval (a, b) being an infinite interval. Since

$$F(z) := \int_C f(z, \zeta) \, d\zeta = \int_a^b f(z, \zeta(\sigma)) \, \zeta'(\sigma) \, d\sigma,$$

the above proposition is then applied to the inegral of the right-hand side (where $|\zeta'(\sigma)|$ can be assumed to be bounded for all σ).

Appendix B

Tables

Convolution Properties of Hyperfunctions

$$(a_1 f_1 + a_2 f_2) \star (b_1 g_1 + b_2 g_2) \quad = a_1 b_1 (f_1 \star g_1) + a_1 b_2 (f_1 \star g_2)$$
$$+ a_2 b_1 (f_2 \star g_1) + a_2 b_2 (f_2 \star g_2)$$

$$D\{f(x) \star g(x)\} \quad = f'(x) \star g(x) = f(x) \star g'(x)$$

$$x\,(f(x) \star g(x)) \quad = (xf(x)) \star g(x) + f(x) \star (xg(x))$$

$$f(x + b) \star g(x + c) \quad = h(x + b + c), \quad h = f \star g$$

$$f(ax + b) \star g(ax + c) \quad = \tfrac{1}{|a|}\, h(ax + b + c)$$

$$\overline{f(x)} \star \overline{g(x)} \quad = \overline{f(x) \star g(x)}$$

$$\mathrm{fp}\,\tfrac{1}{(ax+b)^m} \star \mathrm{fp}\,\tfrac{1}{(ax+c)^n} \quad = -\tfrac{\pi^2}{|a|}\,\tfrac{(-1)^{m+n-1}}{(m-1)!\,(n-1)!}\,\delta^{(m+n-2)}(ax + b + c)$$

Operational Rules for the Laplace Transformation

$$f(t) \qquad \circ\!\!-\!\!\bullet \quad \hat{f}(s)$$

$$e^{ct} f(t) \qquad \circ\!\!-\!\!\bullet \quad \hat{f}(s - c)$$

$$t^n f(t) \qquad \circ\!\!-\!\!\bullet \quad (-1)^n \tfrac{d^n \hat{f}}{ds^n}$$

$$f(ct) \qquad \circ\!\!-\!\!\bullet \quad \tfrac{1}{|c|}\,\hat{f}(\tfrac{s}{c})$$

$$f^{(n)}(t) = D^n f(t) \qquad \circ\!\!-\!\!\bullet \quad s^n\,\hat{f}(s)$$

$$_a D^{-n} f(t) \qquad \circ\!\!-\!\!\bullet \quad \tfrac{1}{s^n}\,\hat{f}(s)$$

$$_{-\infty} D^{-1} f(t) \qquad \circ\!\!-\!\!\bullet \quad \tfrac{1}{s}\,\hat{f}(s)$$

$$_\infty D^{-1} f(t) \qquad \circ\!\!-\!\!\bullet \quad -\tfrac{1}{s}\,\hat{f}(s)$$

$$f(t - c) \qquad \circ\!\!-\!\!\bullet \quad e^{-cs}\,\hat{f}(s)$$

$$f(at + b) \qquad \circ\!\!-\!\!\bullet \quad \tfrac{1}{|a|}\,e^{(b/a)s}\,\hat{f}(\tfrac{s}{a})$$

$$f(t) \star g(t) \qquad \circ\!\!-\!\!\bullet \qquad \hat{f}(s)\,\hat{g}(s)$$

$$u(t-b)f(t-b) \qquad \circ\!\!-\!\!\bullet \qquad e^{-sb}\mathcal{L}_+[f(t)](s)$$

$$f(a\,t+b)\,u(-b/a-t) \qquad \circ\!\!-\!\!\bullet \qquad \frac{e^{s\,b/a}}{a}\,\mathcal{L}_+[f(t)](\tfrac{s}{a})$$

$$u(t)\,f(t+b) \qquad \circ\!\!-\!\!\bullet \qquad e^{bs}\,\{\hat{f}(s) + \int_{C_b} e^{-sz}\,F(z)\,dz\}$$

$$D^\alpha\left(\chi_{(0,\infty)}f(t)\right) \qquad \circ\!\!-\!\!\bullet \qquad s^\alpha\,\mathcal{L}_+[f(t)](s)$$

$$\chi_{(0,\infty)}Df(t) \qquad \circ\!\!-\!\!\bullet \qquad s\,\mathcal{L}[\chi_{(0,\infty)}f(t)](s) - f(0)$$

Some Laplace Transforms of Hyperfunctions

$f(t)$	$\hat{f}(s)$	
$\delta(x)$	1	
$\delta(x-a)$	e^{-as}	
$\delta^{(n)}(t-a)$	$e^{-as}\,s^n$	
$u(t)\dfrac{t^{m-1}}{(m-1)!}\,e^{at}$	$\dfrac{1}{(s-a)^m}$	$\Re s > \Re a$
$-u(-t)\dfrac{t^{m-1}}{(m-1)!}\,e^{at}$	$\dfrac{1}{(s-a)^m}$	$\Re s < \Re a$
$u(t)\exp(-a\,t^2)$	$\frac{1}{2}\sqrt{\frac{\pi}{a}}\,\exp(\frac{s^2}{4\,a})\,\mathrm{erfc}(\frac{s}{2\sqrt{a}})$	
$\exp(-a\,t^2)$	$\sqrt{\frac{\pi}{a}}\,\exp(\frac{s^2}{4\,a})$	
$u(t)\exp(-i\,b\,t^2)$	$\frac{1}{2}\sqrt{\frac{\pi}{2b}}(1-i)\,\exp(-i\,\frac{s^2}{4\,b})$ $\times\mathrm{erfc}(\sqrt{\frac{1}{2b}}\frac{1-i}{2}\,s)$	
$e^{-a|t|}$	$-\dfrac{2a}{s^2-a^2}$	$-a < \Re s < a$
$\sum_{k=-\infty}^{\infty} e^{-|t|}\,\delta(t-k)$	$\dfrac{1-e^{-2}}{(1-e^{-s-1})\,(1-e^{s-1})}$	$-1 < \Re s < 1$
$\sum_{k=0}^{\infty} \delta^{(n)}(t-kT)$	$\dfrac{s^n}{1-e^{-Ts}}$	$\Re s > 0$
$\sum_{k=0}^{\infty} \delta^{(n)}(t-(2k+1)T)$	$\dfrac{s^n}{2\sinh(Ts)}$	$\Re s > 0$
$\sum_{k=0}^{\infty}(-1)^k\,\delta^{(n)}(t-(2k+1)T)$	$\dfrac{s^n}{2\cosh(Ts)}$	$\Re s > 0$
$\delta^{(n)}(t)$ $+2\sum_{k=1}^{\infty}\delta^{(n)}(t-2kT)$	$s^n\coth(Ts)$	$\Re s > 0$
$\delta^{(n)}(t)$ $+2\sum_{k=1}^{\infty}(-1)^k\,\delta^{(n)}(t-2kT)$	$s^n\tanh(Ts)$	$\Re s > 0$
$u(t)\,t^m$	$\dfrac{m!}{s^{m+1}}$	$\Re s > 0$
$u(-t)\,t^m$	$-\dfrac{m!}{s^{m+1}}$	$\Re s < 0$
$\mathrm{fp}\dfrac{u(t)}{t^m}$	$(-1)^m\dfrac{s^{m-1}}{(m-1)!}\,\{\log s - \Psi(m)\}$	$\Re s > 0$
$\mathrm{fp}\dfrac{u(-t)}{t^m}$	$-(-1)^m\dfrac{s^{m-1}}{(m-1)!}$ $\times\{\log(-s)-\Psi(m)\}$	$\Re s < 0$
$(-1)^{m+1}\,m!\,\mathrm{fp}\dfrac{u(t)}{t^{m+1}}$ $+\Psi(m+1)\,\delta^{(m)}(t)$	$s^m\log s$	$\Re s > 0$

$u(t)\,t^{\alpha}$	$\frac{\Gamma(\alpha+1)}{s^{\alpha+1}}$	$\Re s > 0$
$u(-t)\,|t|^{\alpha}$	$-\frac{\Gamma(\alpha+1)}{s^{\alpha+1}}$	$\Re s < 0$
$u(t)\,\frac{\log t}{t}$	$\frac{1}{2}\left\{(\log s + \gamma)^2 - \frac{\pi^2}{6}\right\}$	$\Re s > 0$
$u(t)\,\frac{\log t}{t^m}$	$\frac{(-s)^{m-1}}{2(m-1)!}$ $\times\{(\log s - \Psi(m))^2 - \Psi'(m)\}$	$\Re s > 0$
$u(-t)\,\frac{\log|t|}{t^m}$	$-\frac{(-s)^{m-1}}{2(m-1)!}$ $\times\{(\log(-s) - \Psi(m))^2 - \Psi'(m)\}$	$\Re s < 0$
$2(-1)^m\,m!\,\{\log t - \Psi(m+1)\}\,\mathrm{fp}\frac{u(t)}{t^{m+1}}$ $+\{\Psi^2(m+1) + \Psi'(m+1)\}\,\delta^{(m)}(t)$	$s^m\,\log^2 s$	$\Re s > 0$
$u(t)t^{\alpha}\log t$	$\frac{\Gamma(\alpha+1)}{s^{\alpha+1}}\{\Psi(\alpha+1) - \log s\}$	$\Re s > 0$
$\frac{1}{\Gamma(-\alpha)}\,u(t)\frac{\Psi(-\alpha)-\log t}{t^{\alpha+1}}$	$s^{\alpha}\,\log s$	$\Re s > 0$
$u(t)t^n\log t$	$\frac{n!}{s^{n+1}}\left\{\sum_{k=1}^{n}\frac{1}{k} - \gamma - \log s\right\}$	$\Re s > 0$
$u(t)\,\frac{1}{t+b}$	$e^{b\,s}\,E_1(bs)$	$\Re s > 0$
$\frac{u(t-b)}{t}$	$E_1(bs)$	$\Re s > 0$
$\frac{u(t)}{(t+b)^n}$	$\frac{1}{b^{n-1}}\,e^{bs}\,E_n(bs)$	$\Re s > 0$
$\frac{u(t-b)}{t^n}$	$\frac{1}{b^{n-1}}\,E_n(bs)$	$\Re s > 0$
$u(t)\,\mathrm{Ein}(t)$	$\frac{1}{s}\,\log(1 + \frac{1}{s})$	$\Re s > 0$
$u(t)\,\frac{1}{t^2-a^2}$	$-\frac{1}{2x}\left\{e^{-as}\,\mathrm{Ei}(as) + e^{as}\,E_1(as)\right\}$	$\Re s > 0$
$u(t)\,E_n(t)$	$\frac{(-1)^{n-1}}{s^n}\{\log(1+s) + \sum_{k=1}^{n-1}\frac{(-1)^k\,s^k}{k}$	$\Re s > 0$
$u(t)\,\mathrm{Ei}(t)$	$-\frac{1}{s}\,\log(s-1)$	$\Re s > 1$
$(-1)^{n-1}\,u(t)\left\{E_n(t) - \sum_{k=0}^{n-2}\frac{(-1)^k\,t^k}{(n-1-k)\,k!}\right\}$	$\frac{\log(1+s)}{s^n}$	$\Re s > 0$
$u(t)\,\frac{\sinh\sqrt{at}}{t}$	$\pi\,\mathrm{erg}\sqrt{\frac{a}{4s}}$	$\Re s > 0$
$u(t)\,\frac{\cosh\sqrt{at}}{\sqrt{t}}$	$\sqrt{\frac{\pi}{s}}\,e^{a/(4s)}$	$\Re s > 0$
$\mathrm{fp}\frac{u(t/b)}{t-b}$	$-e^{-bs}\,\mathrm{Ei}(bs)$	
$\mathrm{fp}\frac{u(t/b)}{(t-b)^m}$	$(-1)^m\,e^{-bs}\{\frac{s^{m-1}}{(m-1)!}\,[\log(bs) - \Psi(m)]$ $+b^{1-m}\,\mathrm{Ein}^*(m,bs)\}$	$\Re s > 0$
$\mathrm{fp}\frac{u(t)}{t-b}$	$-e^{-bs}\,\{\mathrm{Ei}(bs) - \log b\}$	$\Re s > 0$
$\mathrm{fp}\frac{u(t)}{(t-b)^m}$	$(-1)^m\,e^{-bs}\{\frac{s^{m-1}}{(m-1)!}\,[\log(s) - \Psi(m)]$ $+b^{1-m}\,\mathrm{Ein}^*(m,bs)\}$	$\Re s > 0$
$\chi_{(0,b)}\mathrm{fp}\frac{1}{(t-b)^n}$	$(-1)^n\,e^{-bs}\,b^{1-n}\,\mathrm{Ein}^*(n,bs)$	$\Re s > 0$
$\chi_{(0,b)}\mathrm{fp}\frac{1}{t}$	$-\{E_1(bs) - \log(bs) + \gamma\}$	$\Re s > 0$
$\chi_{(0,b)}\mathrm{fp}\frac{1}{t^n}$	$-\frac{1}{b^{n-1}}\{E_n(bs) + \sum_{k=1}^{n-2}\frac{(-bs)^k}{(k-n+1)\,k!}\}$ $+\frac{(-s)^{n-1}}{(n-1)!}\{\Psi(n) - \log(bs)\}$	$\Re s > 0$
$\mathrm{fp}\frac{u(t)J_0(at)}{t}$	$\log(\frac{s+\sqrt{a^2+s^2}}{2}) - \gamma$	$\Re s > 0$

$\text{fp}\,\dfrac{u(t)J_0(at)}{t^2}$ $s\log\left(\dfrac{s+\sqrt{s^2+a^2}}{2}\right)+\gamma s-\sqrt{s^2+s^2}$ $\Re s>0$

$\text{fp}\,\dfrac{u(t)J_0(2\sqrt{at})}{t}$ $\log\left(\dfrac{s^2}{a}\right)-E_1\left(\dfrac{a}{s}\right)$ $\Re s>0$

$\text{fp}\,\dfrac{u(t)}{t^2\,(t-b)}$ $\dfrac{s}{b}+\left(\dfrac{1}{b^2}-\dfrac{s}{b}\right)(\log s+\gamma)-\dfrac{1}{b^2}\,e^{-bs}\,\text{Ei}(bs)$ $\Re s>0$

$\text{fp}\,\dfrac{e^{-a|t|}}{t}$ $\log\dfrac{a-s}{a+s}$ $-\Re a<\Re s<\Re a$

$\text{fp}\,\dfrac{e^{-a|t|}}{t^2}$ $s\log\dfrac{a+s}{a-s}+a\{\log((a^2-s^2)+2(\gamma-1)\}$ $-\Re a<\Re s<\Re a$

$\text{fp}\,\dfrac{u(t)\,\cosh(at)\,\cos(\omega t)}{t}$ $-\dfrac{1}{4}\log[(s^2-a^2)^2+2\omega^2(s^2+a^2)+\omega^4]-\gamma$ $\Re s>\Re a$

$\text{fp}\,\dfrac{u(t)\,\cosh(at)\,\cos(at)}{t}$ $-\dfrac{1}{4}\log[(s^4+4a^4]-\gamma$ $\Re s>\Re a$

$\text{fp}\,\dfrac{u(t)\,\sin^2(\omega t)}{t}$ $\dfrac{1}{4}\log(1+4\dfrac{\omega^2}{s^2})$ $\Re s>\omega>0$

$\text{fp}\,\dfrac{u(t)\,\cos^2(\omega t)}{t}$ $-\dfrac{1}{4}\log[s^2(s^2+4\omega^2)]-\gamma$ $\Re s>\omega>0$

$\text{fp}\,\dfrac{u(t)\,\cosh\sqrt{at}}{t^{3/2}}$ $\pi\sqrt{a}\,\text{erg}\sqrt{\dfrac{a}{4s}}-2\sqrt{\pi s}\,e^{a/(4s)}$ $\Re s>0$

Operational Rules for the Fourier Transformation

$f(x)$	$\lhd - \blacktriangleright$	$\hat{f}(\omega)$
$f(x-b)$	$\lhd - \blacktriangleright$	$e^{-bi\omega}\,\hat{f}(\omega)$
$f(ax)$	$\lhd - \blacktriangleright$	$\dfrac{1}{|a|}\hat{f}(\dfrac{\omega}{a})$
$f(ax+b)$	$\lhd - \blacktriangleright$	$\dfrac{1}{|a|}\,e^{i(b/a)\,\omega}\,\hat{f}(\dfrac{\omega}{a})$
$e^{ibx}\,f(x)$	$\lhd - \blacktriangleright$	$\hat{f}(\omega-b)$
$\dfrac{1}{|a|}\,e^{-i(b/a)\,x}\,f(\dfrac{x}{a})$	$\lhd - \blacktriangleright$	$\hat{f}(a\omega+b)$
$f(x)\cos bx$	$\lhd - \blacktriangleright$	$\dfrac{1}{2}\{\hat{f}(\omega-b)+\hat{f}(\omega+b)\}$
$f(x)\sin bx$	$\lhd - \blacktriangleright$	$\dfrac{1}{2i}\{\hat{f}(\omega-b)-\hat{f}(\omega+b)\}$
$\overline{f(x)}$	$\lhd - \blacktriangleright$	$\overline{\hat{f}(-\omega)}$
$\overline{f(-x)}$	$\lhd - \blacktriangleright$	$\overline{\hat{f}(\omega)}$
$D^n f(x)=f^{(n)}(x)$	$\lhd - \blacktriangleright$	$(i\,\omega)^n\,\hat{f}(\omega)$
$x^n\,f(x)$	$\lhd - \blacktriangleright$	$i^n\,\hat{f}^{(n)}(\omega)$
$\hat{f}(x)$	$\lhd - \blacktriangleright$	$2\pi\,f(-\omega)$

Some Fourier Transforms of Hyperfunctions

$\delta(x)$	1
$\delta^{(m)}(ax+b)$	$\dfrac{(i\omega)^m}{|a|\,a^m}\,e^{i(b/a)\,\omega}$
x^m	$2\pi\,i^m\delta^{(m)}(\omega)$

$u(x)$ $\qquad$ $\frac{1}{i}\,\mathrm{fp}\frac{1}{\omega} + \pi\,\delta(\omega)$

$u(x)\,x^n$ $\qquad$ $i^{n-1}\,(-1)^n n!\,\mathrm{fp}\frac{1}{\omega^{n+1}} + i^n\,\pi\,\delta^{(n)}(\omega)$

$|x|$ $\qquad$ $-2\,\mathrm{fp}\frac{1}{\omega^2}$

$\mathrm{sgn}(x)$ $\qquad$ $\frac{2}{i}\,\mathrm{fp}\frac{1}{\omega}$

$\mathrm{fp}\frac{1}{x^m}$ $\qquad$ $\frac{\pi}{i^m\,(m-1)!}\,\omega^{m-1}\,\mathrm{sgn}\,\omega$

$\frac{u(x)}{x^m}$ $\qquad$ $-\frac{(-i\omega)^{m-1}}{(m-1)!}\left\{\log|\omega| + i\frac{\pi}{2}\,\mathrm{sgn}\,\omega + \gamma - \sum_{k=1}^{m-1}\frac{1}{k}\right\}$

$\frac{u(-x)}{x^m}$ $\qquad$ $\frac{(-i\omega)^{m-1}}{(m-1)!}\left\{\log|\omega| - i\frac{\pi}{2}\,\mathrm{sgn}\,\omega + \gamma - \sum_{k=1}^{m-1}\frac{1}{k}\right\}$

$\frac{1}{(x-i\gamma)^m}$ $\qquad$ $\frac{2\pi i^m}{(m-1)!}\,u(-\omega)(-\omega)^{m-1}e^{\gamma\omega}$

$\frac{1}{(x+i\gamma)^m}$ $\qquad$ $\frac{2\pi i^m}{(m-1)!}\,u(\omega)(-\omega)^{m-1}e^{-\gamma\omega}$

$x^n\,\mathrm{sgn}(x)$ $\qquad$ $\frac{2n!}{i^{n+1}}\,\mathrm{fp}\frac{1}{\omega^{n+1}}$

$\frac{\mathrm{sgn}\,x}{x^m}$ $\qquad$ $-\frac{2\,\omega^{m-1}}{i^{m-1}\,(m-1)!}\left\{\log|\omega| - \Psi(m)\right\}$

$x^n\,e^{iax}$ $\qquad$ $2\pi\,i^n\delta^{(n)}(\omega - a)$

$x^n\,\sin ax$ $\qquad$ $\pi\,i^{n-1}\left\{\delta^{(n)}(\omega - a) - \delta^{(n)}(\omega + a)\right\}$

$x^n\,\cos ax$ $\qquad$ $\pi\,i^n\left\{\delta^{(n)}(\omega - a) + \delta^{(n)}(\omega + a)\right\}$

$u(x)\sin(\omega_0\,x)$ $\qquad$ $-i\frac{\pi}{2}\left\{\delta(\omega - \omega_0) - \delta(\omega + \omega_0)\right\} - \mathrm{fp}\frac{\omega_0}{\omega^2 - \omega_0^2}$

$u(x)\cos(\omega_0\,x)$ $\qquad$ $\frac{\pi}{2}\left\{\delta(\omega - \omega_0) + \delta(\omega + \omega_0)\right\} - i\,\mathrm{fp}\frac{\omega}{\omega^2 - \omega_0^2}$

$u(x)e^{i\omega_0\,x}$ $\qquad$ $\pi\delta(\omega - \omega_0) - i\,\mathrm{fp}\frac{1}{\omega - \omega_0}$

$u(x)x^n e^{-\gamma x}$ $\qquad$ $\frac{n!}{(i\omega+\gamma)^{n+1}}$

$u(-x)x^n e^{\gamma x}$ $\qquad$ $-\frac{n!}{(i\omega-\gamma)^{n+1}}$

$u(x)\,\frac{i^m}{(m-1)!}\,x^{m-1}\,e^{i\Omega x}$ $\qquad$ $\mathrm{fp}\frac{1}{(\omega-\Omega)^m}$

$-u(-x)\,\frac{i^m}{(m-1)!}\,x^{m-1}\,e^{i\Omega x}$ $\qquad$ $\mathrm{fp}\frac{1}{(\omega-\Omega)^m}$

$|x|^\alpha\,\mathrm{sgn}\,x$ $\qquad$ $-2i\,\Gamma(\alpha+1)\,\sin(\frac{\pi}{2}(\alpha+1))\,\frac{\mathrm{sgn}\,\omega}{|\omega|^{\alpha+1}}$

$|x|^\alpha$ $\qquad$ $2\,\Gamma(\alpha+1)\,\cos(\frac{\pi}{2}(\alpha+1))\,\frac{1}{|\omega|^{\alpha+1}}$

$\frac{\Psi(n+1)}{i^n}\,\delta^{(n)}(x) - i^n\,\frac{m!}{2}\,\frac{\mathrm{sgn}\,x}{x^{n+1}}$ $\qquad$ $\omega^n\,\log|\omega|$

$|x|^\alpha u(x)$ $\qquad$ $\Gamma(\alpha+1)\,e^{i\frac{\pi}{2}(\alpha+1)\,\mathrm{sgn}\,\omega}\,\frac{1}{|\omega|^{\alpha+1}}$

$|x|^\alpha u(-x)$ $\qquad$ $\Gamma(\alpha+1)\,e^{-i\frac{\pi}{2}(\alpha+1)\,\mathrm{sgn}\,\omega}\,\frac{1}{|\omega|^{\alpha+1}}$

$u(x)x^n\,\log|x|$ $\qquad$ $\pi\,i^n\,\Psi(n+1)\,\delta^{(n)}(\omega) + \frac{n!}{i^{n+1}}\,\Psi(n+1)\,\mathrm{fp}\frac{1}{\omega^{n+1}}$
$\qquad\qquad\qquad$ $+ n!\left[0, \frac{\log(i\zeta)}{(i\zeta)^{n+1}}\right]$

$u(-x)x^n\,\log|x|$ $\qquad$ $\pi\,i^n\,\Psi(n+1)\,\delta^{(n)}(\omega) - \frac{n!}{i^{n+1}}\,\Psi(n+1)\,\mathrm{fp}\frac{1}{\omega^{n+1}}$
$\qquad\qquad\qquad$ $+ n!\left[\frac{\log(-i\zeta)}{(i\zeta)^{n+1}}, 0\right]$

$x^n\,\log|x|$ $\qquad$ $2\pi\,i^n\,\Psi(n+1)\,\delta^{(n)}(\omega) - \frac{\pi n!}{i^n}\,\frac{\mathrm{sgn}\,\omega}{\omega^{n+1}}$

$x^n\,\mathrm{sgn}\,x\,\log|x|$ $\qquad$ $\frac{2n!}{i^{n+1}}\left\{\Psi(n+1)\,\mathrm{fp}\frac{1}{\omega^{n+1}} - \frac{\log|\omega|}{\omega^{n+1}}\right\}$

$\operatorname{sgn} x \, \log\|x\|$	$\frac{2}{i}\left\{-\gamma\operatorname{fp}\frac{1}{\omega}-\frac{\log\|\omega\|}{\omega}\right\}$
$\log\|x\|$	$-2\pi\gamma\,\delta(\omega)-\pi\frac{1}{\|\omega\|}$
$\frac{\log\|x\|}{x^m}$	$\pi\,\frac{1}{i^m\,(m-1)!}\,\omega^{m-1}\operatorname{sgn}\omega\{\Psi(m)-\log\|\omega\|\}$
$\frac{\log\|x\|}{x}$	$\pi\,i\,\operatorname{sgn}\omega\{\gamma+\log\|\omega\|\}$
$\frac{\operatorname{sgn} x \, \log\|x\|}{x^m}$	$\frac{1}{i^{m-1}\,(m-1)!}\,\omega^{m-1}\left\{(\Psi(m)-\log\|\omega\|)^2+\Psi'(m)-\frac{\pi^2}{4}\right\}$
$u(x)\,\|x\|^\alpha\,\log\|x\|$	$\frac{\Gamma(\alpha+1)}{\|\omega\|^{\alpha+1}}\,e^{-i\pi(\alpha+1)\operatorname{sgn}(\omega)/2}$ $\times\left\{\Psi(\alpha+1)-i\frac{\pi}{2}\operatorname{sgn}(\omega)-\log\|\omega\|\right\}$
$\|x\|^\alpha\,\log\|x\|$	$\frac{2\Gamma(\alpha+1)}{\|\omega\|^{\alpha+1}}\left\{[\Psi(\alpha+1)-\log\|\omega\|]\cos\frac{\pi}{2}(\alpha+1)\right.$ $\left.-\frac{\pi}{2}\sin\frac{\pi}{2}(\alpha+1)\right\}$
$\|x\|^\alpha\,\operatorname{sgn} x\,\log\|x\|$	$-\frac{2i\,\Gamma(\alpha+1)\operatorname{sgn}(\omega)}{\|\omega\|^{\alpha+1}}\times\left\{[\Psi(\alpha+1)-\log\|\omega\|]\sin(\frac{\pi}{2}(\alpha+1))\right.$ $\left.+\frac{\pi}{2}\cos(\frac{\pi}{2}(\alpha+1))\right\}$
$\operatorname{fp}\frac{1}{\|x\|^m}$	$-\frac{2\,\omega^{m-1}}{i^{m-1}\,(m-1)!}\left\{\log\|\omega\|+\gamma-\sum_{k=1}^{m-1}\frac{1}{k}\right\}$

Operational Rules for the Hilbert Transformation

$f(t)$	$\circ\!\!-\!\!\overset{\mathcal{H}}{\bullet}$	$\mathcal{H}[f](t)$
$f(t)$	$\circ\!\!-\!\!\overset{\mathcal{H}}{\bullet}$	$-\frac{1}{\pi}\operatorname{fp}\frac{1}{t}\otimes f(t)$
$f(t)=[F(z)]$	$\circ\!\!-\!\!\overset{\mathcal{H}}{\bullet}$	$2i\left[\tilde{F}(z)\,\mathbf{1}(z)\right]$
$\mathcal{H}[f](t)$	$\circ\!\!-\!\!\overset{\mathcal{H}}{\bullet}$	$-f(t)$
$f(t)=[\tilde{F}(z)]$	$\circ\!\!-\!\!\overset{\mathcal{H}}{\bullet}$	$i\{\tilde{F}_+(t+i0)+\tilde{F}_-(t-i0)\}$
$(t-a)^n\,f(t)$	$\circ\!\!-\!\!\overset{\mathcal{H}}{\bullet}$	$(t-a)^n\,\mathcal{H}[f(t)](t)$ $+\frac{1}{\pi}\sum_{k=1}^n m_{k-1}(a)\,(t-a)^{n-k}$
$t\,f(t)$	$\circ\!\!-\!\!\overset{\mathcal{H}}{\bullet}$	$t\,\mathcal{H}[f(t)](t)+\frac{1}{\pi}\int_{-\infty}^\infty f(t)\,dt$
$c_1\,f_1(t)+c_2\,f_2(t)$	$\circ\!\!-\!\!\overset{\mathcal{H}}{\bullet}$	$c_1\,\mathcal{H}[f_1(t)](t)+c_2\,\mathcal{H}[f_2(t)](t)$
$f'(t)$	$\circ\!\!-\!\!\overset{\mathcal{H}}{\bullet}$	$\frac{d}{dt}\mathcal{H}[f](t)$
$f(at+b)$	$\circ\!\!-\!\!\overset{\mathcal{H}}{\bullet}$	$\operatorname{sgn} a\,g(at+b)$
$f(-t)$	$\circ\!\!-\!\!\overset{\mathcal{H}}{\bullet}$	$-\mathcal{H}[f](-t)$
$f(t-b)$	$\circ\!\!-\!\!\overset{\mathcal{H}}{\bullet}$	$\mathcal{H}[f](t-b)$
$f(t)\star g(t)$	$\circ\!\!-\!\!\overset{\mathcal{H}}{\bullet}$	$\mathcal{H}[f](t)\star g(t)$
$f(t)$	$\circ\!\!-\!\!\overset{\mathcal{H}}{\bullet}$	$i\,\mathcal{F}^{-1}[\operatorname{sgn}(\omega)\,\hat{f}(\omega)](t)$
$\operatorname{pv}\sum_{k=-\infty}^\infty c_k\,e^{ik\omega t}$	$\circ\!\!-\!\!\overset{\mathcal{H}}{\bullet}$	$\operatorname{pv}\sum_{k=-\infty,k\neq0}^\infty i\,\operatorname{sgn}(k\omega)\,c_k\,e^{ik\omega t}$

$$\frac{A_0}{2} + \sum_{k=1}^{\infty}(A_k \cos k\omega t + B_k \sin k\omega t) \quad \circ\!\!\overset{\mathcal{H}}{-\!\!\!-}\!\!\bullet \quad \sum_{k=1}^{\infty}(B_k \cos k\omega t - A_k \sin k\omega t)$$

$$f(t) = f(t+T) \qquad\qquad\qquad \circ\!\!\overset{\mathcal{H}}{-\!\!\!-}\!\!\bullet \quad -\frac{1}{T} \operatorname{pv} \int_a^{a+T} \cot\left(\frac{\pi(t-x)}{T}\right) f(x)\, dx$$

Some Hilbert Transforms of Hyperfunctions

$f(t)$	$\mathcal{H}[f](t)$
$\delta^{(n)}(t-a)$	$\frac{(-1)^{n+1}\, n!}{\pi} \operatorname{fp} \frac{1}{(t-a)^{n+1}}$
$\operatorname{fp}\frac{1}{(t-a)^n}$	$\frac{(-1)^{n-1}\pi}{(n-1)!}\, \delta^{(n-1)}(t-a)$
e^{-t^2}	$-\frac{2}{\sqrt{\pi}} \operatorname{Daw}(t)$
$\chi_{(-1,1)} P_n(t)$	$\frac{2}{\pi} \sum_{k=0}^{n-1} \frac{(n+k)!}{(n-k)!\,(k!)^2\, 2^k} \{\Psi(n+1) - \Psi(k+1)\}(t-1)^k$
$\frac{1}{(t-c)^m}$	$-i\operatorname{sgn}(\Im c) \frac{1}{(t-c)^m},\quad (\Im c \neq 0, m \in \mathbb{N})$
$\exp(i\Omega t)$	$i\operatorname{sgn}(\Omega)\exp(i\Omega t)$
$\cos t$	$-\sin t$
$\sin t$	$\cos t$
$\operatorname{pv}\sum_{k=-\infty}^{\infty}\delta(t-kT)$	$-\operatorname{fp}\frac{1}{T}\cot(\frac{\pi t}{T})$
$\operatorname{fp}\cot(\frac{\pi t}{T})$	$T\operatorname{pv}\sum_{k=-\infty}^{\infty}\delta(t-kT) - 1$
1	0
$\chi_{a,b}(x)$	$\frac{1}{\pi}\log\left|\frac{b-t}{a-t}\right|$
$\frac{t}{t^2+a^2}$	$\frac{a}{t^2+a^2}$
$\frac{t^2-a^2}{(t^2+a^2)^2}$	$\frac{2at}{(t^2+a^2)^2}$
$\frac{t^3-3a^2t}{(t^2+a^2)^3}$	$\frac{3at^2-a^3}{(t^2+a^2)^3}$
$\frac{1}{t+i0}$	$\frac{i}{t+i0}$
$\frac{1}{t+ia}$	$\frac{i}{t+ia}$
$\chi_{(-1,1)}\frac{1}{\sqrt{1-t^2}}$	$-\frac{\operatorname{sgn}(t;-1,1)}{\sqrt{t^2-1}}$
$\chi_{(a,b)}\frac{1}{\sqrt{(t-a)(b-t)}}$	$-\frac{\operatorname{sgn}(t;a,b)}{\sqrt{(t-a)(t-b)}}$
$\chi_{(a,b)}\frac{1}{(t-a)^n\sqrt{(t-a)(b-t)}}$	$-\frac{\operatorname{sgn}(t;a,b)}{(t-a)^n\sqrt{(t-a)(t-b)}}$
$\chi_{(a,b)}\sqrt{(t-a)(b-t)}$	$\operatorname{sgn}(t;a,b)\sqrt{(t-a)(t-b)} + t - \frac{a+b}{2}$
$\chi_{(-1,1)}\sqrt{1-t^2}$	$\operatorname{sgn}(t;-1,1)\sqrt{t^2-1} + t$
$\frac{\sin t}{t}$	$\frac{\cos t - 1}{t}$
$\frac{1-\cos t}{t}$	$\frac{\sin t}{t}$
$\frac{\cos t}{t}$	$\pi\,\delta(t) - \frac{\sin t}{t}$
$\chi_{(a,b)} t^n$	$\frac{t^n}{\pi}\log\left|\frac{b-t}{a-t}\right| + \frac{1}{\pi}\sum_{k=1}^{n}\frac{b^k-a^k}{k}\, t^{n-k}$

$$\chi_{(a,b)}\frac{(t-a)^n}{\sqrt{(t-a)(b-t)}} \qquad -\frac{\operatorname{sgn}(t;a,b)\,(t-a)^n}{\sqrt{(t-a)(t-b)}}$$
$$+\frac{1}{\sqrt{\pi}}\sum_{k=1}^{n}\frac{(b-a)^{k-1}\,\Gamma(k-1/2)}{(k-1)!}\,(t-a)^{n-k}$$

$$\chi_{(a,b)}\frac{(t-b)^n}{\sqrt{(t-a)(b-t)}} \qquad -\frac{\operatorname{sgn}(t;a,b)\,(t-b)^n}{\sqrt{(t-a)(t-b)}}$$
$$+\frac{1}{\sqrt{\pi}}\sum_{k=1}^{n}\frac{(a-b)^{k-1}\,\Gamma(k-1/2)}{(k-1)!}\,(t-a)^{n-k}$$

$$\chi_{(-1,1)}\frac{t^n}{\sqrt{1-t^2}} \qquad -\frac{\operatorname{sgn}(t;-1,1)\,t^n}{\sqrt{t^2-1}}$$
$$+\frac{1}{\sqrt{\pi}}\sum_{k=1}^{[(n+1)/2]}\frac{\Gamma(k-1/2)}{(k-1)!}\,t^{n+1-2k}$$

$$\chi_{(a,b)}\frac{t}{\sqrt{(t-a)(b-t)}} \qquad -\frac{t\,\operatorname{sgn}(t;a,b)}{\sqrt{(t-a)(t-b)}}+1$$

$$\Delta_a(t) \qquad \frac{1}{\pi a}\left\{(a+t)\log\left|\frac{t}{a+t}\right|-(a-t)\log\left|\frac{t}{a-t}\right|\right\}$$

$$\frac{\cos t}{t^2} \qquad \pi\,\delta'(t)+\frac{\sin t-t}{t^2}$$

$$|t|^\alpha\,\operatorname{sgn}t \qquad -\cot(\tfrac{\pi}{2}\alpha)\,|t|^\alpha$$

$$\frac{u(t)}{t^m} \qquad -\frac{1}{\pi}\frac{\log|t|}{t^m}+\frac{(-1)^{m-1}\pi}{2(m-1)!}\,\delta^{(m-1)}(t)$$

$$\frac{u(-t)}{t^m} \qquad \frac{1}{\pi}\frac{\log|t|}{t^m}+\frac{(-1)^{m-1}\pi}{2(m-1)!}\,\delta^{(m-1)}(t)$$

$$e^{-a|t|} \qquad -u(t)\{\tfrac{1}{\pi}\{e^{-at}\operatorname{Ei}(at)+e^{at}E_1(at)\}\}$$
$$+u(-t)\{\tfrac{1}{\pi}\{e^{at}\operatorname{Ei}(-at)+e^{-at}E_1(-at)\}\}$$

$$|t|^\alpha \qquad \tan(\tfrac{\pi}{2}\alpha)\,|t|^\alpha\,\operatorname{sgn}t$$

$$\frac{\log|t|}{t^m} \qquad \frac{\pi}{2}\frac{\operatorname{sgn}t}{t^m}$$

$$\frac{\operatorname{sgn}t}{t^m} \qquad -\frac{2}{\pi}\frac{\log|t|}{t^m}$$

$$u(t)\,t^\alpha \qquad -u(t)\cot(\pi\alpha)\,t^\alpha-u(-t)\frac{1}{\sin(\pi\alpha)}\,|t|^\alpha$$

$$u(-t)\,|t|^\alpha \qquad u(-t)\cot(\pi\alpha)\,|t|^\alpha+u(t)\frac{1}{\sin(\pi\alpha)}\,t^\alpha$$

$$\frac{u(t)}{t^{n+1/2}} \qquad u(-t)\,(-1)^{n+1}\frac{1}{|t|^{n+1/2}}$$

$$\frac{u(-t)}{|t|^{n+1/2}} \qquad u(t)\,(-1)^n\frac{1}{t^{n+1/2}}$$

$$\operatorname{sgn}(\cos\omega t) \qquad -\frac{2}{\pi}\log|\tan(\tfrac{\omega t}{2}+\tfrac{\pi}{4})|$$

$$\operatorname{sgn}(\sin\omega t) \qquad -\frac{2}{\pi}\log|\tan(\tfrac{\omega t}{2})|$$

Operational Rules for the Mellin Transformation

$$f(x) \qquad \overset{\mathcal{M}}{\circ\!-\!\bullet}\;\tilde{f}(s) \qquad\qquad \alpha<\Re s<\beta$$

$$f(x) \qquad \overset{\mathcal{M}}{\circ\!-\!\bullet}\;\mathcal{L}[f(e^{-t})](s)$$

$$\sum_{k=1}^{n}c_k\,f_k(x) \qquad \overset{\mathcal{M}}{\circ\!-\!\bullet}\;\sum_{k=1}^{n}c_k\,\mathcal{M}[f_k(x)](s) \quad \max\{\alpha_1,\dots,\alpha_n\}<\Re s$$
$$<\min\{\beta_1,\dots,\beta_n\}$$

$$f(ax) \qquad \overset{\mathcal{M}}{\circ\!-\!\bullet}\;a^{-s}\,\tilde{f}(s)$$

$$f(x^a) \qquad \overset{\mathcal{M}}{\circ\!-\!\bullet}\;\frac{1}{|a|}\,\tilde{f}(\tfrac{s}{a}) \qquad\qquad a\,\alpha<\Re s<a\,\beta$$
$$a\,\beta<\Re s<a\,\alpha\;\;(a<0)$$

$$(\log x)^n f(x) \qquad \circ\!\!-\!\!\overset{\mathcal{M}}{-}\!\!\bullet \ \tilde{f}^{(n)}(s)$$

$$x^\mu f(x) \qquad \circ\!\!-\!\!\overset{\mathcal{M}}{-}\!\!\bullet \ \tilde{f}(s+\mu) \qquad\qquad \alpha - \Re\mu < \Re s < \beta - \Re\mu$$

$$\tfrac{1}{x} f(\tfrac{1}{x}) \qquad \circ\!\!-\!\!\overset{\mathcal{M}}{-}\!\!\bullet \ \tilde{f}(1-s) \qquad\qquad 1 - \beta < \Re s < 1 - \alpha$$

$$f(\tfrac{1}{x}) \qquad \circ\!\!-\!\!\overset{\mathcal{M}}{-}\!\!\bullet \ \tilde{f}(-s) \qquad\qquad -\beta < \Re s < -\alpha$$

$$f'(x) \qquad \circ\!\!-\!\!\overset{\mathcal{M}}{-}\!\!\bullet \ -(s-1)\,\tilde{f}(s-1) \qquad \alpha + 1 < \Re s < \beta + 1$$

$$D^n f(x) \qquad \circ\!\!-\!\!\overset{\mathcal{M}}{-}\!\!\bullet \ (1-s)_n\,\tilde{f}(s-n) \qquad \alpha + n < \Re s < \beta + n$$

$$x^2\,D^2 f(x) + xDf(x) \qquad \circ\!\!-\!\!\overset{\mathcal{M}}{-}\!\!\bullet \ s^2\,\tilde{f}(s)$$

$$D^2 f(x) + \tfrac{1}{x}\,Df(x) \qquad \circ\!\!-\!\!\overset{\mathcal{M}}{-}\!\!\bullet \ (s-2)^2\,\tilde{f}(s-2) \qquad \alpha - 2 < \Re s < \beta - 2$$

$$_0 D^{-1} f(x) \qquad \circ\!\!-\!\!\overset{\mathcal{M}}{-}\!\!\bullet \ -\tfrac{1}{s}\,\tilde{f}(s+1) \qquad \alpha - 1 < \Re s < \min[\beta - 1, -1]$$

$$_\infty D^{-1} f(x) \qquad \circ\!\!-\!\!\overset{\mathcal{M}}{-}\!\!\bullet \ -\tfrac{1}{s}\,\tilde{f}(s+1) \qquad \alpha - 1 < \Re s < \beta - 1$$

$$\left[\tfrac{1}{2\pi i} \int_{c-i\infty}^{c} z^{-s}\,\tilde{f}(s)\,ds, \right.$$
$$\left. -\tfrac{1}{2\pi i} \int_{c}^{c+i\infty} z^{-s}\,\tilde{f}(s)\,ds \right] \ \circ\!\!-\!\!\overset{\mathcal{M}}{-}\!\!\bullet \ \tilde{f}(s) \qquad\qquad \alpha < c < \beta$$

$$f(x) * g(x) \qquad \circ\!\!-\!\!\overset{\mathcal{M}}{-}\!\!\bullet \ \tilde{f}(s)\tilde{g}(s)$$

$$f_1(x) * x^\mu f_2(x) \qquad \circ\!\!-\!\!\overset{\mathcal{M}}{-}\!\!\bullet \ \tilde{f}_1(s)\tilde{f}_2(s+\mu) \qquad \begin{array}{l} \max[\alpha_1 - \Re\mu, \alpha_2] < \Re s \\ < \min[\beta_1 - \Re\mu, \beta_2] \end{array}$$

$$\int_0^\infty f_1(\tau x)\,f_2(\tau)\,d\tau \qquad \circ\!\!-\!\!\overset{\mathcal{M}}{-}\!\!\bullet \ \tilde{f}_1(s)\,\tilde{f}_2(1-s) \qquad \begin{array}{l} \max[\alpha_1, 1-\beta_2] < \Re s \\ < \min[\beta_1, 1-\alpha_2] \end{array}$$

$$f(x) \circ g(x) \qquad \circ\!\!-\!\!\overset{\mathcal{M}}{-}\!\!\bullet \ \tilde{f}(s)\,\tilde{g}(1-s) \qquad \begin{array}{l} \max[\alpha_1, 1-\beta_2] < \Re s \\ < \min[\beta_1, 1-\alpha_2] \end{array}$$

$$f_1(x)f_2(x) \qquad \circ\!\!-\!\!\overset{\mathcal{M}}{-}\!\!\bullet \ \begin{array}{l} \tfrac{1}{2\pi i} \int_{c-i\infty}^{c+i\infty} \tilde{f}_2(p) \\ \times \tilde{f}_1(s-p)\,dp \end{array} \quad \alpha_1 + \alpha_2 < \Re s < \beta_1 + \beta_2$$

Some Mellin Transforms of Hyperfunctions

$$e^{-ax} \qquad\qquad a^{-s}\,\Gamma(s) \qquad\qquad\qquad \Re s > 0$$

$$e^{ix} \qquad\qquad e^{i\frac{\pi}{2} s}\,\Gamma(s) \qquad\qquad\qquad 0 < \Re s < 1$$

$$\cos x \qquad\qquad \cos(\tfrac{\pi}{2} s) \qquad\qquad\qquad 0 < \Re s < 1$$

$$\sin x \qquad\qquad \sin(\tfrac{\pi}{2} s) \qquad\qquad\qquad -1 < \Re s < 1$$

$$\tfrac{1}{1+x} \qquad\qquad \tfrac{\pi}{\sin(\pi s)} \qquad\qquad\qquad 0 < \Re s < 1$$

$$\tfrac{1}{(1+x)^n} \qquad\qquad \tfrac{\Gamma(s)\,\Gamma(n-s)}{\Gamma(n)} \qquad\qquad\qquad 0 < \Re s < n$$

$$\tfrac{1 + x\cos\alpha}{x^2 + 2x\cos\alpha + 1} \qquad \tfrac{\pi\,\cos(\alpha s)}{\sin(\pi s)} \qquad\qquad\qquad 0 < \Re s < 1$$

$$\tfrac{x\sin\alpha}{x^2 + 2x\cos\alpha + 1} \qquad \tfrac{\pi\,\sin(\alpha s)}{\sin(\pi s)} \qquad\qquad\qquad 0 < \Re s < 1$$

$$u(x - a)x^{-\mu} \qquad -\tfrac{a^{s-\mu}}{s-\mu} \qquad\qquad\qquad \Re s < \Re\mu$$

$u(a-x)x^{-\mu}$ $\qquad \dfrac{a^{s-\mu}}{s-\mu}$ $\qquad\qquad \Re s > \Re\mu$

$J_\nu(x)$ $\qquad 2^{s-1}\dfrac{\Gamma(\frac{\nu+s}{2})}{\Gamma(\frac{\nu-s+2}{2})}$ $\qquad -\Re\nu < \Re s < \Re\nu + 2$

$\delta(x)$ $\qquad 0$

$\delta(x-a)$ $\qquad a^{s-1}$ $\qquad\qquad a > 0$

$\sum_{k=-\infty}^{\infty}\delta(x-kT)$ $\quad T^{s-1}\,\zeta(1-s)$ $\qquad\qquad \Re s < 0$

$\delta'(x-a)$ $\qquad (1-s)\,a^{s-2}$ $\qquad\qquad a > 0$

$x^n u(x-a)$ $\qquad -\dfrac{a^{n+s}}{n+s}$ $\qquad\qquad \Re s < -n$

$\chi_{(a,b)}x^n$ $\qquad \dfrac{1}{n+s}\{b^{n+s}-a^{n+s}\}$ $\qquad \Re s < -n$

$\chi_{(a,b)}$ $\qquad \dfrac{1}{s}\{b^s - a^s\}$ $\qquad\qquad \Re s < 0$

$\chi_{(a,b)}x^n$ $\qquad \dfrac{1}{n+s}\{b^{n+s}-a^{n+s}\}$ $\qquad \Re s < -n$

$x^n u(a-x)$ $\qquad \dfrac{a^{n+s}}{n+s}$ $\qquad\qquad \Re s > -n$

e^{-ax^b} $\qquad \dfrac{a^{-s/b}}{|b|}\Gamma(\frac{s}{b})$ $\qquad\qquad \Re s > 0$

$(\log x)^n\,\delta(x-a)$ $\quad (\log a)^n\,a^{s-1}$ $\qquad\qquad a > 0$

$(\log x)^n\,e^{-x}$ $\qquad \dfrac{d^n}{ds^n}\Gamma(s)$ $\qquad\qquad 0 < \Re s$

$x^\mu\,\delta(x-a)$ $\qquad a^{s+\mu-1}$ $\qquad\qquad a > 0$

$x^\mu\,e^{-x}$ $\qquad \Gamma(s+\mu)$ $\qquad\qquad -\Re\mu < \Re s$

$\dfrac{e^{-a/x}}{x}$ $\qquad a^{s-1}\,\Gamma(1-s)$ $\qquad\qquad \Re s < 1$

$\chi_{(a,b)}\dfrac{1}{x^n}$ $\qquad \dfrac{b^{n-s}-a^{n-s}}{n-s}$ $\qquad\qquad n < \Re s$

$\delta^{(n)}(x-a)$ $\qquad (-1)^n(s-1)(s-2)\ldots(s-n)\,a^{s-n-1}$ $\quad a > 0$

$_0D^{-1}\delta(x-a)$ $\qquad -\dfrac{a^s}{s}$ $\qquad\qquad \Re s < -1$

$\mathrm{erfc}(x)$ $\qquad \dfrac{1}{\sqrt{\pi}\,s}\Gamma(\frac{s+1}{2})$ $\qquad\qquad 0 < \Re s$

Operational Rules for the Hankel Transformation

$f(x)$ $\qquad \overset{H_\nu}{\circ\!-\!\bullet}\quad F_\nu(y)$

$F_\nu(x)$ $\qquad \overset{H_\nu}{\circ\!-\!\bullet}\quad f(y)$

$f(ax)$ $\qquad \overset{H_\nu}{\circ\!-\!\bullet}\quad \dfrac{1}{a^2}F_\nu(\frac{y}{a})$

$\dfrac{f(x)}{x}$ $\qquad \overset{H_\nu}{\circ\!-\!\bullet}\quad \dfrac{y}{2\nu}\{F_{\nu-1}(y)+F_{\nu+1}(y)\}$

$\dfrac{f(x)}{x^2}$ $\qquad \overset{H_\nu}{\circ\!-\!\bullet}\quad \dfrac{y^2}{4\nu}\{\dfrac{1}{\nu-1}F_{\nu-2}(y)+\dfrac{2\nu}{\nu^2-1}F_\nu(y)+\dfrac{1}{\nu+1}F_{\nu+2}(y)\}$

$f'(x)$ $\qquad \overset{H_\nu}{\circ\!-\!\bullet}\quad y\{(\frac{\nu-1}{2\nu})F_{\nu+1}(y)-(\frac{\nu+1}{2\nu})F_{\nu-1}(y)\}$

$\dfrac{1}{x}f'(x)$ $\qquad \overset{H_\nu}{\circ\!-\!\bullet}\quad -\dfrac{y^2}{4}\{\dfrac{1}{\nu-1}F_{\nu-2}(y)+\dfrac{2}{\nu^2-1}F_\nu(y)-\dfrac{1}{\nu+1}F_{\nu+2}(y)\}$

$$D^2 f(x) \qquad \overset{H_\nu}{\circ\!-\!\bullet} \quad \frac{y^2}{4} \left\{ \frac{\nu+1}{\nu-1} F_{\nu-2}(y) - \frac{2(\nu^2-3)}{\nu^2-1} F_\nu(y) + \frac{\nu-1}{\nu+1} F_{\nu+2}(y) \right\}$$

$$(D^2 + \tfrac{1}{x} D - \nu^2) f(x) \quad \overset{H_\nu}{\circ\!-\!\bullet} \quad -y^2 F_\nu(y)$$

Some Hankel Transforms of order ν of Hyperfunctions

$\dfrac{e^{-ax}}{x}$	$\dfrac{(\sqrt{a^2+y^2}-a)^\nu}{y^\nu \sqrt{a^2+y^2}}$	
$\dfrac{1}{\sqrt{a^2+x^2}}$	$\dfrac{e^{-ay}}{y}$	$\nu = 0$
e^{-ax}	$\dfrac{a}{(a^2+y^2)^{3/2}}$	$\nu = 0$
$x^\nu e^{-ax}$	$\dfrac{2^{\nu+1}\,\Gamma(\nu+3/2)\,a\,y^\nu}{\sqrt{\pi}\,(a^2+y^2)^{\nu+3/2}}$	$a > 0$
$\dfrac{x^\nu}{(a^2+x^2)^{\nu+3/2}}$	$\dfrac{\sqrt{\pi}}{2^{\nu+1}\,\Gamma(\nu+3/2)\,a}\, y^\nu e^{-ay}$	
$x^{\nu+2n}\, e^{-ax^2}$	$n!\,\left(\frac{y}{2}\right)^\nu \dfrac{\exp(-y^2/(4a))}{2a^{\nu+n+1}} L_n^\nu\left(\frac{y^2}{4a}\right)$ Laguerre's polynomial	
e^{-ax^2}	$\dfrac{\exp(-y^2/(4a))}{2a}$	$\nu = 0$
$x\, e^{-ax^2}$	$\dfrac{y}{4}\, \dfrac{\exp(-y^2/(4a))}{a^2}$	$\nu = 1$
$e^{-\kappa x^2}\cos(\omega x^2)$	$2\Omega\, e^{-\kappa\,\Omega\,y^2}\{\kappa\,\cos(\omega\Omega y^2) + \omega\,\sin(\omega\Omega y^2)\}$	$\nu = 0$
$e^{-\kappa x^2}\sin(\omega x^2)$	$2\Omega\, e^{-\kappa\,\Omega\,y^2}\{-\omega\,\cos(\omega\Omega y^2) + \kappa\,\sin(\omega\Omega y^2)\}$	$\nu = 0$
$e^{-ax^2} J_\nu(bx)$	$\dfrac{1}{2a}\, e^{-(y^2+b^2)/(4a)}\, I_\nu\left(\frac{by}{2a}\right)$	
x^μ	$2^{\mu+1}\dfrac{\Gamma(\frac{\nu+\mu+2}{2})}{\Gamma(\frac{\nu-\mu}{2})}\, y^{-\mu-2},\ \ (-\Re\nu - 2 < \Re\mu < -1/2)$	
$\dfrac{1}{x}$	$\dfrac{1}{y}$	$\nu = 0$
$x^\nu (a^2 - x^2)^\mu\, u(a-x)$	$2^\mu\, a^{\nu+\mu+1}\dfrac{\Gamma(\mu+1)}{y^{\mu+1}}\, J_{\nu+\mu+1}(ay)$	
$\delta(x-a)$	$a\, J_\nu(ay)$	$a > 0$
$\delta^{(m)}(x-a)$	$(-1)^m\, y^{m-1}\{a\, y\, J_\nu^{(m)}(ay) + m\, J_\nu^{(m-1)}(ay)\}$	$a > 0$
$J_\nu(ax)$	$\dfrac{1}{a}\delta(y-a)$	$a > 0$
x^n	$0,\ n\ \text{odd}$	$\nu = 0$
x^n	$(-1)^\ell\, \dfrac{(1\cdot3\cdot5\cdots(2\ell-1))^2}{y^{2\ell+1}},\ n = 2\ell\ \text{even}$	$\nu = 0$
$x^\alpha\, \delta(x-a)$	$a^{\alpha+1}\, J_\nu(ay)$	$a > 0$
$x^\alpha\, \delta^{(m)}(x-a)$	$(-1)^m \sum_{k=0}^m \binom{\alpha}{k} \dfrac{m!}{(m-k)!}\, a^{\alpha-k}\, y^{m-k-1}$	
	$\times \{ay\, J_\nu^{(m-k)}(ay) + (m-k)\, J_n u^{(m-k-1)}(ay)\}$	$a > 0$
$\delta(ax-b)$	$\dfrac{b}{a^2}\, J_\nu\left(\frac{b}{a}y\right)$	$a > 0$
$\delta^{(m)}(ax-b)$	$\dfrac{(-1)^m\, y^{m-1}}{a^2}\{\frac{b}{a}\, y\, J_\nu^{(m)}\left(\frac{b}{a}y\right) + m\, J_\nu^{(m-1)}\left(\frac{b}{a}y\right)\}$	$a > 0$

Bibliography

[1] M. Abramowitz, I.A. Stegun (eds.). *Pocketbook of Mathematical Functions.* Harri Deutsch. Thun-Frankfurt/Main 1984.

[2] Lars V. Ahlfors. *Complex Analysis.* McGraw-Hill Book Company. New York 1966.

[3] P. Antosik, J. Mikusinski, R. Sikorski. *Theory of Distributions, The Sequential Approach.* Elsevier Scientific Publishing Company. Amsterdam 1973.

[4] H. Behnke, F. Sommer. *Theorie der analytischen Funktionen einer komplexen Veränderlichen.* Springer-Verlag. Berlin 1965.

[5] K. Chandrasekharan. *Classical Fourier Transforms.* Springer-Verlag. New York 1980.

[6] B. Davies. *Integral Transforms and Their Applications.* Springer-Verlag. New York 1978.

[7] L. Debnath and D. Bhatta. *Integral Transforms and Their Applications.* Chapman & Hall. New York 2007.

[8] G. Doetsch, W. Nader (Translater). *Introduction to the Theory and Application of the Laplace Transformation.* Springer-Verlag. New York 1974.

[9] G. Doetsch. *Einführung in Theorie und Anwendung der Laplace-Transformation.* Birkhäuser Verlag. Basel 1976.

[10] R. Estrada, R.P. Kanwal. *A Distributional Approach to Asymptotics.* Birkhäuser Verlag. Boston 2002.

[11] I.M. Gel'fand, G.E. Shilov. *Generalized Functions Vol. 2.* Academic Press New York 1968.

[12] F.D. Gakhov. *Boundary Value Problems.* Dover Publications Inc. New York 1990.

[13] C.Gasquet, P. Witomski. *Fourier Analysis and Applications.* Springer-Verlag. New York 1998.

[14] Urs Graf. *Applied Laplace Transforms und z-Transforms for Scientists and Engineers.* Birkhäuser Verlag. Basel-Boston-Berlin 2004.

[15] K.F. Graff *Wave Motion in Elastic Solids.* Dover Publications Inc. New York 1991.

[16] Peter Henrici *Applied and Computational Complex Analysis, Vol. II.* John Wiley & Sons. New York 1977.

[17] Peter Henrici. *Applied and Computational Complex Analysis, Vol. III.* John Wiley & Sons. New York 1986.

[18] I.I. Hirschmann, D.V. Widder. *The Convolution Transform.* Dover Publications Inc. New York 2005.

[19] Isac Imai. *Applied Hyperfunction Theory.* Kluwer Academic Publ. New York 1992.

[20] A. Kaneko. *Introductions to Hyperfunctions.* Kluwer Academic Publishers. Boston 1988.

[21] R.P. Kanwal. *Generalized Functions.* Birkhäuser Verlag. Boston 2004.

[22] H. Komatsu. *Laplace Transforms of Hyperfunctions.* J. Fac. Sci. Univ. Tokyo, Sect. IA. Math., 34 (1987), 805-820. Tokyo 1987.

[23] M. J. Lighthill. *Fourier Analysis and Generalized Functions.* Cambridge University Press. New York 1958.

[24] A. J. Macintyre. *Laplace's Transformation and Integral Functions.* Proc. London Math. Soc. 45 (1938). London 1938.

[25] Mitsuo Morimoto. *An Introduction to Sato's Hyperfunctions.* American Mathematical Society Volume 129. Providence, Rhode Island 1993.

[26] T. Needham. *Visual Complex Analysis.* Clarendon Press, Oxford. New York 2000.

[27] F. Oberhettinger, L. Badii *Table of Laplace Transforms.* Springer-Verlag. Heidelberg 1973.

[28] F. Oberhettinger. *Table of Fourier Transforms and Fourier Transforms of Distributions.* Springer-Verlag. Heidelberg 1990.

[29] F.W.J. Olver. *Asymptotics and Special Functions.* Academic Press. New York 1974.

[30] A.C. Pipkin *A Course on Integral Equations.* Springer-Verlag. Berlin 1991.

[31] M. Sato. *Theory of hyperfunctions I.* J. Fac. Sci. Univ. Tokyo Sect. I 8, 133 - 193. Tokyo 1959.

[32] Laurent Schwartz. *Méthodes mathématiques pour les sciences physiques.* Hermann. Paris 1965.

[33] Laurent Schwartz. *Théorie des distributions.* Hermann. Paris 1966.

[34] Ian N. Sneddon. *Fourier Transforms.* Dover Publications Inc. New York 1995.

[35] B. Stankovic *Laplace Transform of Laplace Hyperfunctions and its Applications.* Novi Sad J. Math. Vol. 31, No 1, 2001, 9 -17 Novi Sad 2001.

[36] V.S. Vladimirov *Equations of Mathematical Physics.* Mir Publishers Moscow 1984.

[37] V.S. Vladimirov *Methods of the Theory of Generalized Functions.* Taylor & Francis London 2002.

[38] E.T.Whittaker, G.N. Watson. *A Course of Modern Analysis.* Cambridge University Press. London 1973.

[39] A.H. Zemanian. *Distribution Theory and Transform Analysis.* Dover Publications Inc. New York 1987.

[40] A.H. Zemanian. *Generalized Integral Transformations.* Dover Publications Inc. New York 1987.

List of Symbols

I	real interval
$\mathbb{C}$	complex plane
$\mathbb{C}_+$	upper complex half-plane
$\mathbb{C}_-$	lower complex half-plane
$\mathcal{D}(I)$	complex neighborhood of the interval I
$\mathcal{D}_+,\ \mathcal{D}_+(I)$	upper half-neighborhood of the interval I
$\mathcal{D}_-,\ \mathcal{D}_-(I)$	lower half-neighborhood of the interval I
$\mathcal{O}(\mathcal{D})$	ring of all holomorphic functions in the domain $\mathcal{D}$
$\mathcal{A}(I)$	ring of all real analytic functions on the interval I
$\mathcal{F}(I)$	$= \mathcal{O}(\mathcal{D}(I) \setminus I)$
$\mathcal{O}_0(\mathbb{C}_\pm)$	upper and lower half-plane functions (Def. 2.4)
$[F(z)]$	hyperfunction defined by the defining function $F(z)$
$F_+(z)$	upper component of the defining function $F(z)$
$F_-(z)$	lower component of the defining function $F(z)$
$[F_+(z), F_-(z)]$	hyperfunction defined by the upper and lower components of the defining function $F(z)$
$F(z) \sim G(z)$	equivalent defining functions
$\mathcal{B}(I)$	space of all hyperfunctions defined on I
$\mathcal{B}_+(I)$	space of all upper hyperfunctions defined on I
$\mathcal{B}_-(I)$	space of all lower hyperfunctions defined on I
$\mathcal{B}_{\mathcal{O}}(I)$	space of all holomorphic hyperfunctions defined on I
$\mathcal{B}_0(\mathbb{R})$	space of all hyperfunctions with upper and lower component of their defining functions being upper and lower half-plane functions.
$\mathcal{A}(\mathbb{R}_+, \star)$	convolution algebra on $\mathbb{R}_+$
$\mathcal{B}_L(\mathbb{R}_+)$	space of all Laplace hyperfunctions with support in $[0, \infty)$
$1_+(z)$	function equals 1 for $\Im z > 0$, and 0 for $\Im z < 0$
$1_-(z)$	function equals 0 for $\Im z > 0$, and -1 for $\Im z < 0$
$1(z)$	function equals 1/2 for $\Im z > 0$, and -1/2 for $\Im z < 0$
$u(x)$	unit-step hyperfunction
$Y(x)$	unit-step ordinary function or Heaviside function
$\mathrm{sgn}(x)$	$= u(x) - u(-x)$ sign-hyperfunction
$\phi(x), \psi(x), \dots$	real analytic functions on $\mathbb{R}$
$\delta(x - a)$	Dirac impulse hyperfunction at $x = a$
$\delta_{(n)}(x - a)$	generalized delta-hyperfunction of order n at $x = a$
$\oint_{(a,b)}$	contour integral over the closed contour from a to b below the real axis, and back from b to a above the real axis

$\gamma_{a,b}^{+}$	contour from $a \in \mathbb{R}$ to $b \in \mathbb{R}$ above the real axis
$\gamma_{a,b}^{-}$	contour from $a \in \mathbb{R}$ to $b \in \mathbb{R}$ below the real axis
$Df(x) = f'(x)$	first derivative in the sense of hyperfunction
$D^n f(x) = f^{(n)}(x)$	derivative of order n in the sense of hyperfunction
$\frac{d^n F}{dz^n}$	ordinary derivative of an ordinary function
$\Re z, \Im z$	real part, imaginary part of the complex number z
$\overline{z}$	complex conjugate of z
$\mathrm{fp} f(x)$	finite part hyperfunction
FP	Hadamard's Finite Part of an integral
$\chi_{a,b}(x)$	characteristic (hyper-)function on the interval (a, b)
$\chi_{(a,b)} f(x)$	projection of the hyperfunction $f(x)$ on (a, b)
Res	Residue
Σ_0	largest open subset of $\mathbb{R}$ where the hyperfunction $f(x)$ vanishes
$\mathrm{supp} f(x)$	$\mathbb{R} \setminus \Sigma_0$ support of the hyperfunction $f(x)$
Σ_1	largest open subset of $\mathbb{R}$ where the hyperfunction $f(x)$ is holomorphic
Σ_2	largest open subset of $\mathbb{R}$ where the hyperfunction $f(x)$ is micro-analytic
$\mathrm{sing\ supp} f(x)$	$\mathbb{R} \setminus \Sigma_1$ singular support of the hyperfunction $f(x)$
$\mathrm{sing\ spec} f(x)$	$\mathbb{R} \setminus \Sigma_2$ singular spectrum of the hyperfunction $f(x)$
$f(x) \cdot g(x)$	product of the hyperfunctions $f(x)$ and $g(x)$
$\mathcal{H}[f], \mathcal{H}[f(x)](t)$	Hilbert transform of f
$\mathcal{L}[f], \mathcal{L}[f(t)](s)$	Laplace transform of f
$\mathcal{F}[f], \mathcal{F}[f(x)](\omega)$	Fourier transform of f
$f(t) \circ\!\!-\!\!\bullet \hat{f}(s)$	correspondence between the original and its Laplace-image
$f(x) \triangleleft\!\!-\!\!\blacktriangleright \hat{f}(\omega)$	correspondence between $f(x)$ and its Fourier transform $\hat{f}(\omega)$
$f(x) \circ\!\!\xrightarrow{\mathcal{H}}\!\!\bullet \hat{f}(y)$	correspondence between $f(x)$ and its $\mathcal{H}$-transform $\hat{f}(y)$
pv	principal value of a sum or integral
$O(\mathbb{R}_{+})$	set of right-sided originals
$O(\mathbb{R}_{-})$	set of left-sided originals
$f(x) \star g(x)$	convolution of $f(x)$ and $g(x)$, also denoted by $(f \star g)(x)$
$f(x) \otimes g(x)$	pv-convolution of $f(x)$ and $g(x)$, also denoted by $(f \otimes g)(x)$
$\mathrm{sgn}(x, a, b)$	$= -1$ for $x < a$; $= 0$ for $a < x < b$; $= 1$ for $b < x$
$P_n(t)$	Legendre polynomial of the first kind of degree n
$\Delta_a(t)$	triangle function

Index

Applied Laplace Transforms and z-Transforms for Scientists and Engineers

A Computational Approach Using a Mathematica Package

Graf, U., Hochschule für Technik und Architektur, Biel, Switzerland

2004. X, 500 p. With CD-ROM.
Hardcover
ISBN 978-3-7643-2427-8

Laplace and z-transformations are important parts of the mathematical background required for engineers, physicists and mathematicians, particularly for solving classes of differential equations and difference equations. This book presents theory and applications of Laplace and z-transforms together with a Mathematica package developed by the author. The package substantially enhances the built-in facilities of Mathematica and includes algorithms for the numerical inversion of Laplace transforms.

The emphasis lies on the computational and applied sides, particularly in the fields of control engineering, electrical engineering, and mechanics (heat conduction, diffusion, vibrations). Many worked out examples from applied sciences and engineering illustrate the applicability of the theory and the usage of the package.

Students, instructors, practical engineers and researchers working in the field of control, electricity or mechanics will find this textbook a most valuable resource and will profit from the package and further examples and Mathematica notebooks on the included CD-ROM.

From the Contents:
Preface.- Laplace Transformation.- z-Transformation.- Laplace Transforms with the Package.- z-Transforms with the Package.- Applications to Automatic Control.- Laplace Transformation: Further Topics.- z-Transformation: Further Topics.- Examples from Electricity.- Examples from Control Engineering.- Heat Conduction and Vibration Problems.- Further Techniques.- Numerical Inversion of the Laplace Transform.- Appendix: Package Commands by Subjects.- Bibliography.

www.birkhauser.ch